The (p,n) Reaction
and the
Nucleon-Nucleon Force

The (p,n) Reaction and the Nucleon–Nucleon Force

Edited by
Charles D. Goodman
Oak Ridge National Laboratory
Oak Ridge, Tennessee

Sam M. Austin
Michigan State University
East Lansing, Michigan

Stewart D. Bloom
Lawrence Livermore Laboratory
Livermore, California

J. Rapaport
Ohio University
Athens, Ohio

and
G. R. Satchler
Oak Ridge National Laboratory
Oak Ridge, Tennessee

Plenum Press • New York and London

Library of Congress Cataloging in Publication Data

Conference on the (p,n) Reaction and the Nucleon—Nucleon Force, Telluride, Colo.,
    1979.
    The (p,n) reaction and the nucleon—nucleon force.

    Includes bibliographies and index.
    1. Neutron-proton interactions – Congresses. 2. Nucleon-nucleon interactions –
Congresses. I. Goodman, Charles D. II. Title.
QC794.8.N5C66 1979   539.7'54   79-27785
ISBN 978-1-4684-8862-3   ISBN 978-1-4684-8860-9 (eBook)
DOI 10.1007/978-1-4684-8860-9

Proceedings of the Conference on The (p, n) Reaction and the
Nucleon—Nucleon Force, held in Telluride, Colorado, March 29–31, 1979.

© 1980 Plenum Press, New York
Softcover reprint of the hardcover 1st edition 1980
A Division of Plenum Publishing Corporation
227 West 17th Street, New York, N.Y. 10011

PREFACE

This volume contains the proceedings of the "Conference on the (p,n) Reaction and the Nucleon-Nucleon Force" held in Telluride, Colorado, March 29-31, 1979.

The idea to hold this conference grew out of a program at the Indiana University Cyclotron Facility to study the (p,n) reaction in the 50-200 MeV energy range. The first new Indiana data, in contrast to low energy data, showed features suggestive of a dominant one pion exchange interaction. It seemed desirable to review what was known about the free and the effective nucleon-nucleon force and the connection between the low and high energy (p,n) data. Thus the conference was born.

The following people served as the organizing committee:

> S. M. Austin, Michigan State University
> W. Bertozzi, Massachusetts Institute of Technology
> S. D. Bloom, Lawrence Livermore Laboratory
> C. C. Foster, Indiana University
> C. D. Goodman, Oak Ridge National Laboratory
>   (Conference Chairman)
> D. A. Lind, University of Colorado
> J. Rapaport, Ohio University
> G. R. Satchler, Oak Ridge National Laboratory
> G. E. Walker, Indiana University
> R. L. Walter, Duke University and TUNL

The sponsoring organizations were:

> Indiana University, Bloomington, Indiana
> University of Colorado, Boulder, Colorado
> Oak Ridge National Laboratory, Oak Ridge, Tennessee
> Triangle Universities Nuclear Laboratory, Durham,
>   North Carolina

Of course, the major credit for the success of the conference must go to the speakers who diligently prepared their talks that are reproduced in this volume. Next, credit must go to the conference delegates whose participation in discussions

gave spirit to the conference. For practicality, only fragments
of the discussions have been retained for these proceedings.

Credit for almost flawless conference arrangements is due to
many people. Judy Lankford of the Indiana University Conference
Bureau oversaw the arrangements before, during and after the conference, and with help from Virginia Bowers of Lawrence Livermore
Laboratory and Joan Sonnenberg of the University of Colorado
managed an on-site conference office.

Christine Wallace from Oak Ridge National Laboratory handled
all the conference mailings, correction of manuscripts and much
of the work of putting the proceedings into their final form.

I must also thank many people of Telluride who were involved
in a variety of ways in aiding the conference. With apologies to
anyone I have accidentally left out I thank (without significance
to the order) Bill and Stella Pence, Walter McClennan, Pat
Schuler, Ross and Ursula Willis, Terry Selby, Craig and Barb
Chapot, Salli McLaughlin, Sue Gray, D. D. Richmond, and Hope Marno.

I also thank the CPT Corporation for providing at no cost a
CPT-8000 word processor for use at the conference. The con-
ference mailings and some of the manuscript work was done on a
CPT-8000 system.

The scheduling of technical sessions was modeled to some
degree after the Gordon Conferences with morning and evening
sessions and free afternoons. Sessions took place in the
Sheridan Opera House, with delegates housed in the adjoining
hotel and in a nearby inn providing compact siting to enhance the
opportunities for communication among the delegates.

                                          Charles D. Goodman

CONTENTS

(The name in capitals indicates the author who presented the talk)

CONTENTS

WHAT WE THINK WE KNOW ABOUT THE FREE NUCLEON-NUCLEON INTERACTION

Peter Signell

Physics Department
Michigan State University
East Lansing, MI   48824

TABLE OF CONTENTS

INTRODUCTION

Outline of the Talk

The purpose of this talk is to show what we currently believe about the free NN system, and to compare it with what is actually being used in nuclear calculations. I will touch on four aspects: (i) present theory; (ii) the potentials produced by that theory as well as by NN experiment; (iii) the relevant NN amplitudes (phase shifts) and (iv) how to choose NN potentials and amplitudes for nuclear calculations.

Theoretical Regions in V, δ

The NN potentials V and phase shifts δ break naturally into three regions (see Fig. 1) representing past, present and future theoretical knowledge. We have long known how to calculate the long

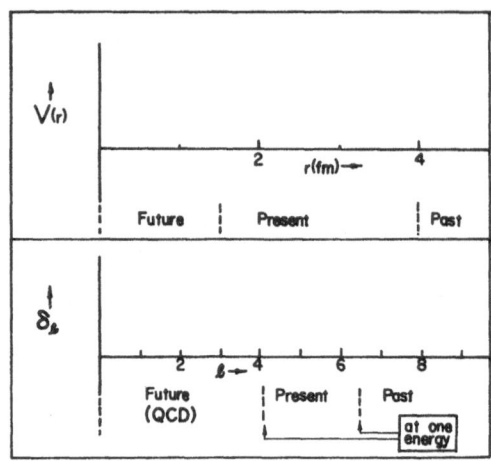

Fig. 1. An indication of theoretical knowledge of NN potentials and phases.

range 1π—exchange part of the potential, we now believe we can calculate the intermediate range part with some precision, and there is some evidence that Quantum Chromodynamics or its successor may enable us to calculate the short range part a few years from now. Until then, we take the short range part from NN experiment. Now in principle, of course, experiment could be used to fix V at all ranges. In fact, however, I know of no serious proposal to carry out such a program at any energy. The cost in dollars and effort would be vastly too high.

Field Theory Principles + Data = Answers

crossing unitarity analyticity	+	πN ππ eN	=	V(r>1.5 fm) δ (ℓ >2)

We believe that present theory can give quite precise answers for the intermediate and long range parts of the NN potential as well as for the higher angular momentum phase shifts. The theory is in solid standing in high energy physics, it is not likely to be replaced in the near future and it has no adjustable parameters. It uses only general principles of quantum field theory (crossing, unitarity and analyticity) along with experimental data from other processes (πN, ππ, eN scattering). Most importantly, it is fairly unique in passing a set of tests in nearby channels, tests which we think any serious candidate for the NN interaction must pass.

PRESENT THEORY

Outline

Present theory[1] generates potentials and phase shifts from the nucleon-nucleon scattering amplitude, <NN:NN>. This amplitude is, in

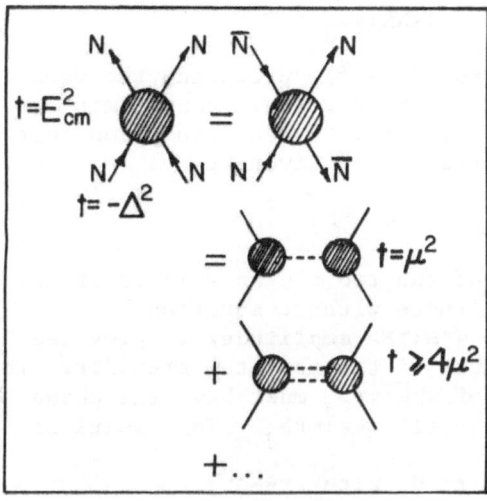

Fig. 2.   Crossing and unitary expansion for the NN amplitude (see text).

turn, generated by "crossing" from the $N\bar{N}$ scattering amplitude, indicated by $\langle N\bar{N}:N\bar{N}\rangle$. The $N\bar{N}$ amplitude is used because the insertion of a unitarity expansion in the middle of it produces terms representing 1-, 2-,...-π exchange:

$$\langle N\bar{N}:N\bar{N}\rangle = \langle N\bar{N}:\pi\rangle\langle\pi:N\bar{N}\rangle + \langle N\bar{N}:\pi\pi\rangle\langle\pi\pi:N\bar{N}\rangle + \ldots.$$

Thus the theoretical program calls for calculating 1-, 2-,...-π exchange, summing those terms to get $\langle N\bar{N}:N\bar{N}\rangle$, then crossing the latter to get $\langle NN:NN\rangle$ along with its potentials and phase shifts. This process is indicated schematically in Fig. 2.

Exchanged Particles, t-Value Ranges

The relationship between exchanged particle mass, momentum transfer, and range of interaction is easily seen by rotating Fig. 2 90 degrees so that the pion lines stretch away from you. Look at the first term in the expansion, the 1π term: $N\bar{N}\to\pi\to N\bar{N}$. The pion is at rest in this process's CM frame, where the amplitude is evaluated. In this frame the pion's kinetic energy is zero and the square of its total energy can only have the single value $t=\mu^2$, where μ is the pion mass. However, in the 2π intermediate state the pions can have nonzero relative momenta. Thus for the 2π contribution there is a continuum of allowed energies beginning at two pion masses. The allowed range of squared energy is then: $t>4\mu^2$.

Spectral Functions: Relation to V(r)

The theoretical calculation of the potential can be indicated by an integral over a spectrum of Yukawa potentials[7]:

$$V(r) = \sum_J \int_0^\infty \rho_J(t) \frac{\exp(-r\sqrt{t})}{r}\, dt.$$

The sum is over the angular momenta of the intermediate states and each ρ_J is a sum of ρ's over numbers of intermediate state particles. The 1π-exchange spectral functions,

$$\rho_J(1\pi) \sim \langle N\bar{N}:\pi\rangle_J \langle \pi:N\bar{N}\rangle_J = |\langle N\bar{N}:\pi\rangle_J|^2$$

contain single t value delta functions, $\delta(t-\mu^2)$, producing the usual
1π-exchange Yukawa potential. For 2π-exchange, the lower limit of
the integral is effectively $4\mu^2$ since their ρ_J's are zero below that
point. The 2π-exchange spectral functions are given by

$$\rho_J(2\pi) \sim |\langle N\bar{N}:\pi\pi\rangle_J|^2$$

where J is now the angular momentum of the two π's as well as of the
$N\bar{N}$ pair. If we add the $\langle N\bar{N}:\pi\pi\rangle$ amplitudes without squaring,
"crossing" says that we now have the $\langle \pi N:\pi N\rangle$ amplitude... provided
we reinterpret t as the negative square of the momentum transfer. It
is also easy to show that the phase of $\langle N\bar{N}:\pi\pi\rangle_J$ must have the phase
of the corresponding ππ scattering state:[2] $\langle \pi\pi:\pi\pi\rangle_J$. To summarize:

(i) $\langle N\bar{N}:\pi\pi\rangle$ is $\langle \pi N:\pi N\rangle$ for t< 0 with $t=-\Delta^2$;

(ii) $\delta\langle N\bar{N}:\pi\pi\rangle$ is $\delta\langle \pi\pi:\pi\pi\rangle$ for t>4μ² with $t=E_{CM}^2$.

Thus the observed ππ phases and observed πN scattering amplitudes can

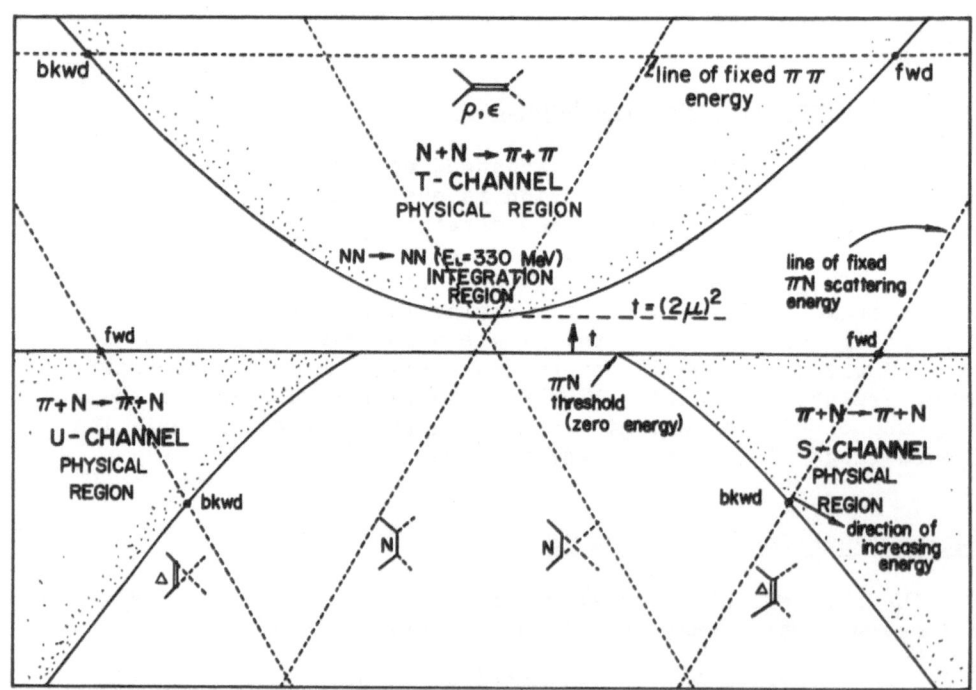

Fig. 3. Locations of nearby contributions to the NN spectral functions. Each small diagram is half an NN diagram
and is shown only to help identify regions of the space.

be used as requirements which proper $\langle N\bar{N}:\pi\pi\rangle_J$'s must satisfy[9].
Alternatively, we can use these $\pi\pi$ and πN scattering data as input in
order to construct $\langle N\bar{N}:\pi\pi\rangle$'s which automatically satisfy the above
two requirements. The latter is the path taken by present theory[10].

The Major ("Nearby") Contributors

For intermediate and large impact parameters, the exponential function in the potential integral cuts off all but the small-t region of the ρ's[7]. Thus it is very important to us that our input $\pi\pi$ phases and πN amplitudes be accurate for low energies. This can be seen graphically in Fig. 3 where the integration boundary in the $\pi\pi$ channel is shown as a dotted area. It rises up from $t=4\mu^2$. Note how close the physical πN region is to the integration region, making it a big contributor. Note also, however, that as one proceeds toward high mass baryon states (to the far right or left) or to high mass mesons (up), the contributions fade away due to distance from our integration region.

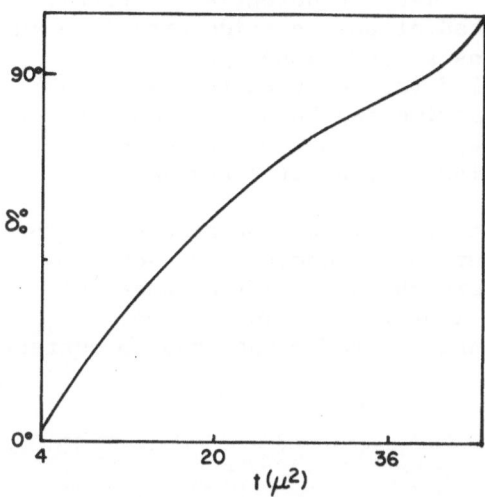

Fig. 4. The $\pi\pi$ S-wave phase shift used in calculating the NN spectral function ρ_0. The data have been smoothed via a dispersion relation.

$\pi\pi$ and Form Factor Input

Fig. 4 shows the observed $\pi\pi$ s-wave phase shifts used as input to the calculation of the s-wave spectral function, ρ_0[11]. Note that the spectral function uses the amplitude's magnitude squared, not the phase shift. However, the $\pi\pi$ phase shift helps determine the magnitude through a dispersion relation which also includes πN contributions. The nucleon electromagnetic form factors, obtained from eN scattering, help determine the p-wave spectral function, ρ_1[11].

CURRENT THEORY vs V(r)'S

V(r) Choices for Comparison

We will show the theoretical NN potentials from two different groups and compare them to two widely used phenomenological potentials. The theoretical potentials are from the MSU[4] and PARIS[5] groups. These two groups' calculations differ slightly in detail but their basic approaches are the same. The MSU potential is purely theoretical and hence should be looked at only outside about 1.5 fm. However, the PARIS potential contains an additional purely phenomenological part inside about 1.5 fm, designed to give it a state-of-the-art fit to a set of experimental NN data. Thus the PARIS potential is an all-radius one, ready for use in nuclear calculations. One slight complication is that it contains p^2-dependent terms.

The phenomenological potentials we plot for comparison are the soft core one of Reid[12] and the older one of Hamada and Johnston[13]. These two, particularly the first, are the basis for a number of calculations being reported in this conference. That is why we picked them, rather than other potentials, for comparison to current theory.

Comparison: MSU, PARIS, Reid SC, HJ

Fig. 5 compares the four potentials, labeled by the NN system's spin 0 and 1 "singlet" and "triplet" central, tensor and spin-orbit components. The parity of each component is also labeled. The $^1V_C^+$ potentials in Fig. 5a show the theoretical and phenomenological potentials all coming into nice agreement outside about 1.5 fm. Fig. 5b displays the same set but with an expanded vertical scale, showing the 2-3 fm region more clearly. Note that the Reid potential has about half the non-1π strength shown by theory and HJ at 3 fm. Fig.'s 5c and 5d show our current knowledge to be in strong disagreement with the phenomenological potentials, especially the Reid one. In the next section we will see that this problem arose from phenomenological phase shifts of an earlier day which have recently been shown to be erroneous. Of course guidance from theoretical potentials could also have prevented the problem from arising, had they been available at the time the phenomenological potentials were constructed.

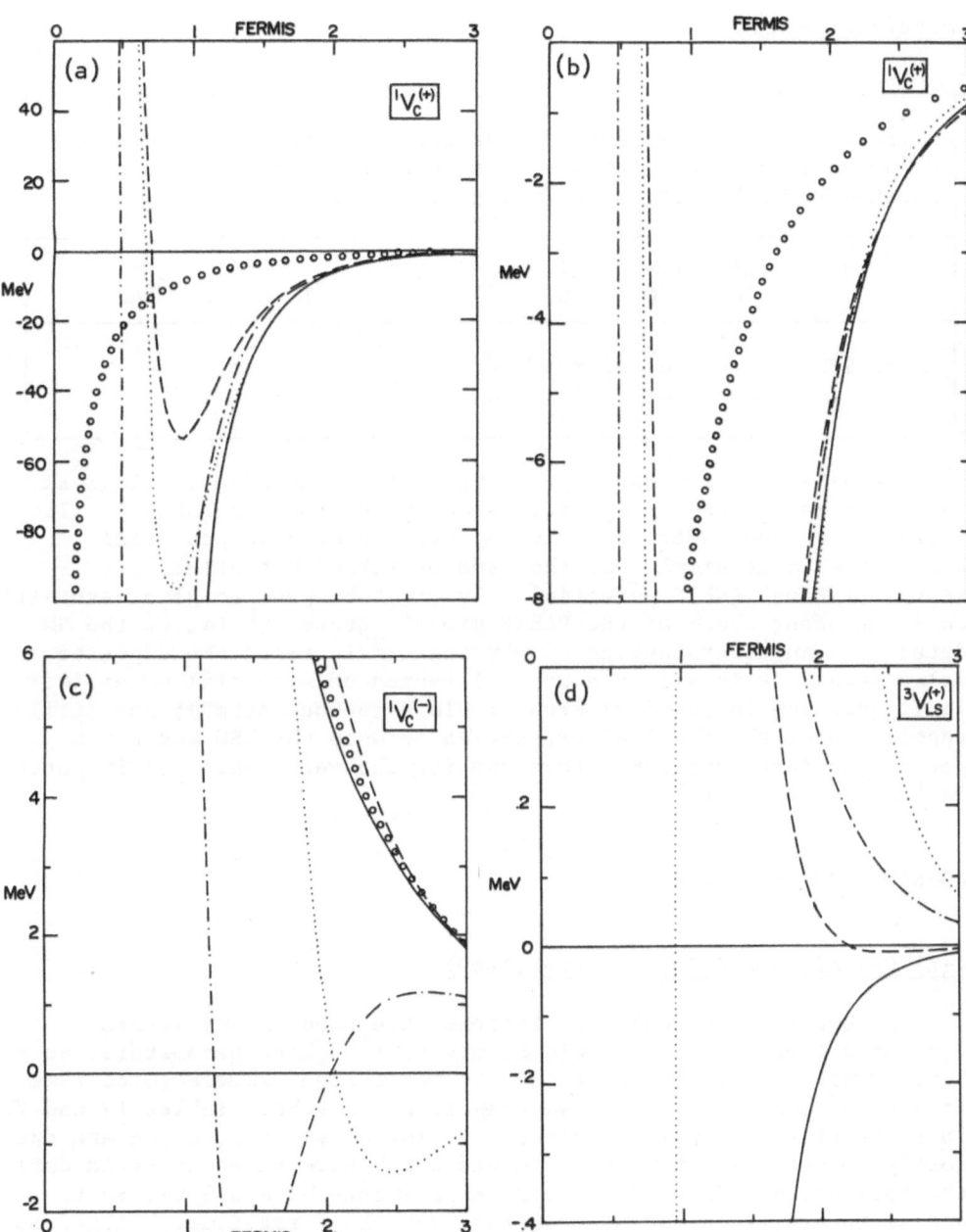

Fig. 5. Comparison of potentials due to 1π exchange (ooo), PARIS (---), MSU (——), Reid (•••), HJ (-•-), For other states see Appendix A.

Comparison Summary

	$^1V_C^+$	$^1V_C^-$	$^3V_C^+$	$^3V_C^-$	$^3V_T^+$	$^3V_T^-$	$^3V_{LS}^+$	$^3V_{LS}^-$
Reid SC	✓	ɑ	ɑ	✓	✓	✓	ɑ	✓
HJ	✓	ɑ	✓	ɑ	✓	✓	x	✓

Table I.   An impression of the extent to which the Reid and HJ potentials agree with our current theoretical beliefs.   A check mark is "good", x is "not good" and ɑ is "bad".

Detailed comparisons of the theoretical and phenomenological potentials are shown in Appendix A and summarized in Table I.   It would seem a good idea to either update the Reid SC potential parameters or to simply use the "sum of Yukawa potentials representation" PARIS potential.   It might be good to also have: (i) an independent check of the PARIS group's quoted χ^2 fit to the NN data; (ii) an understanding of why that χ^2 is twice the expected value (e.g. is it all in a crucial region of a crucial potential? is it just due to the inclusion of old erroneous data?); and (iii) a resolution of the small discrepancies between the MSU and PARIS potentials (note the opposite signs for the very small non-1π parts in the $^1V_C^-$ potential).

CURRENT THEORY vs δ's

δ(E) Comparison: Updating Liv-X(1969)

A number of nuclear calculations have used NN amplitudes generated from the 10 year old Livermore-X[14] phase parameters, so we here compare those "Liv-X" values to our current knowledge of them from experiment.   In Fig. 6 we compare Liv-X (their Tables IV and VI) to those of recent phase analyses[15].   The differences shown are due mostly to some very fine experiments which have been performed during the past decade.   Sometimes those experiments have allowed us to define the relevant phase parameters for the first time.   Also, we now know that erroneous early data caused Liv-X's $^1P_1$ to have an anomalous behavior at low energy (Fig. 6c)[16].   This, in turn, must have led to the anomalous shape in the corresponding Reid potential[23].   In Fig. 6a we see that the $^3S_1$ hard core now appears to be not as hard as implied by Liv-X, and the same may be true for the $^1S_0$ core.   The $^3S_1-^3D_1$ mixing parameter, ε_1, has changed its shape considerably.   The Liv-X negative ε_1 values at low energy would

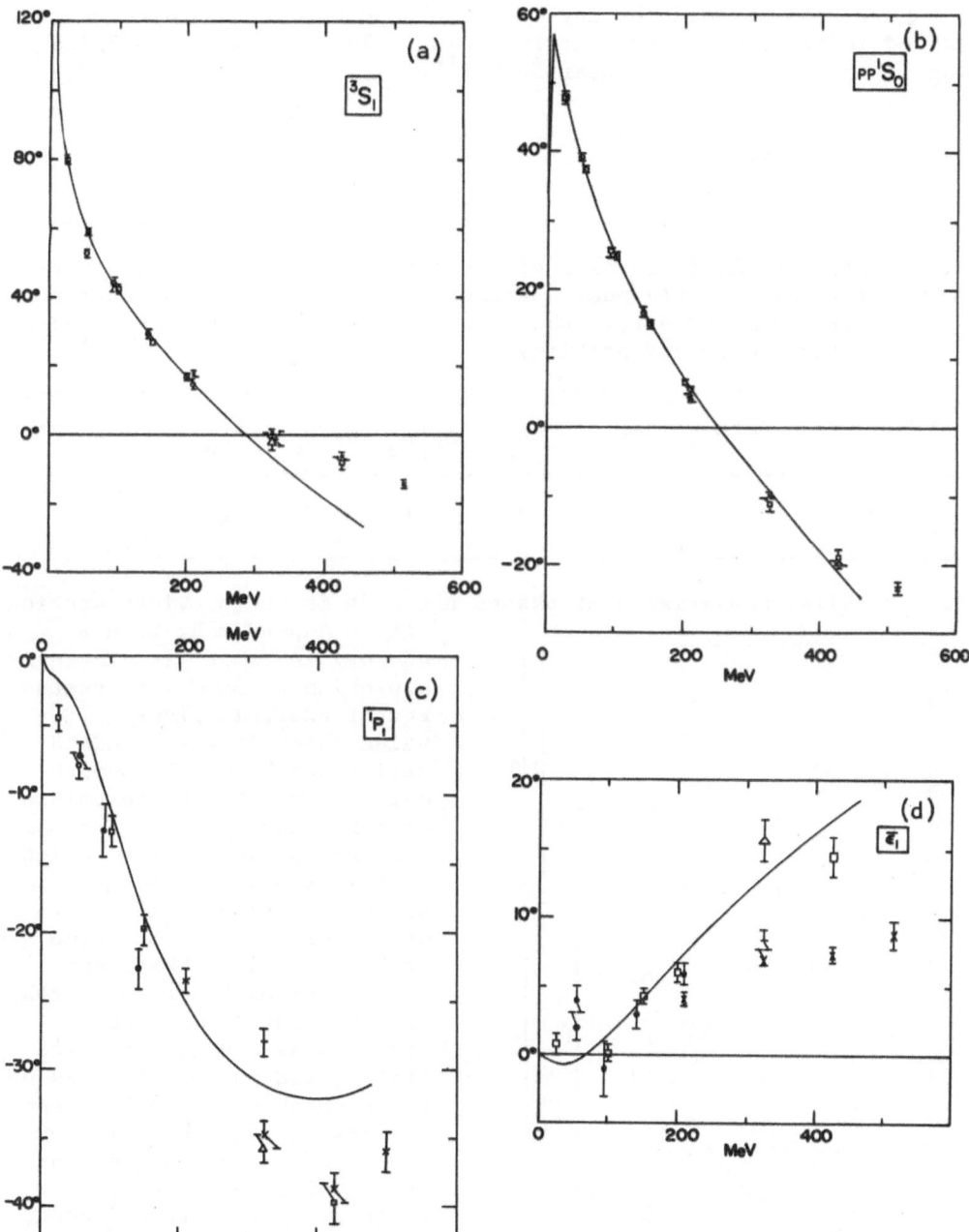

Fig. 6. Comparison of Liv-X(1969) phases (smooth curve) with those from recent analyses: MSU-LASL-Davis[16](o), MSU[17] (•), AHR-II[18](□), AHR-III[19](Δ), Bugg[20](×), TxA&M[21](+). For other states, see Appendix B.

not have been matched by any reasonable potential and may have been
caused indirectly by the sharply rising values at higher energies
and/or some tainted data near 50 MeV[16].

Comparison Summary

Table II. An impression of the extent to which the 1969 Liv-X (their Tables IV, VI) phase parameters agree with recent analyses of the accumulated data. Checks are "good", x's "not", *'s "good except for the 50 MeV problem."										
$^3S_1$	ε_1	$^1P_1$	$^3D_1$	$^3D_2$	np^1S_0	pp^1S_0	$^3P_0$	$^3P_1$	$^3P_2$	
*	xx	xx	✓	✓	*	✓	x	✓	✓	

Detailed comparisons of phases are made in the previous section

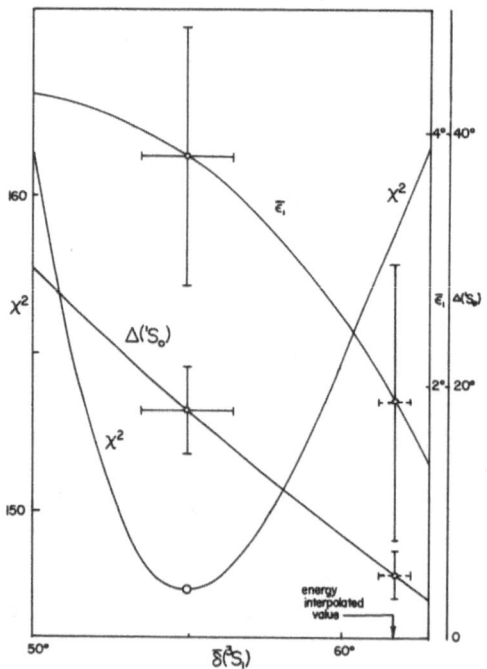

and in Appendix B, with a summary in Table II. There is a problem at 50 MeV in that a recent analysis gives a $^3S_1$ value which is six standard deviations below the smooth curve drawn though the points at other energies[16]. The same analysis produces an ε_1 which is a little high and an $np\ ^1S_0$ which is outlandishly high. If we interpolate a $^3S_1$ value and uncertainty at 50 MeV, using the smooth multi-energy curve, and then put it into the analysis as a datum, we find that ε_1 and the $np\ ^1S_0$ wind up at reasonable values (the open circles in the figures). Of course χ^2 is much higher for this "solution" (Fig.7)[16]. Further experiments are needed in order to straighten out this unsatisfactory situation.

Fig. 7. Problems at 50 MeV.
Here $\Delta(^1S_0) = np^1S_0 - pp^1S_0$. The
energy-interpolated $^3S_1$ is $\simeq 62°$.

THEORETICAL PROBLEMS

Comparisons of Theoretical Groups' δ's

The present method used to calculate the intermediate range interaction is believed to be quite good, but it does not follow that all calculations using it are in agreement. One might think that the longer range predictions would be in good agreement, but Table III shows alarming discrepancies for G waves at 330 MeV.

	$^3G_4$	$^3G_5$
Table III. Three groups' non-1π predictions for two NN G phases at 330 MeV. The three groups appear to be in agreement on basic theoretical requirements.		
MSU[24]	-0.06	0.52
STONY BROOK[8]	0.08	0.92
PARIS[6]	1.02	1.34

Possible Origins of the Discrepancies

The differences in theoretical predictions for high partial waves may have their origins in: (i) different input data; (ii) different πN amplitude extrapolation techniques; (iii) insufficient computational accuracy;and (iv) errors of formula or computation. The main input is via the $N\bar{N} \rightarrow \pi\pi$ S-wave spectral function, ρ_0. The apparent ρ_0's used by the PARIS and MSU groups are compared in Fig. 8. The low energy parts, which are the determiners of the high partial waves, are seen to be virtually identical for the two groups. Thus the predictions from them should agree. It would be a good idea to similarly compare the P-wave and higher-J spectral functions. Of course the theoretical discrepancies may also be due to any of the other causes listed above. We feel that the task of pinpointing the origins of the differences is an important one.

A New $\pi N \rightarrow \pi N / \pi\pi \rightarrow N\bar{N}$ Continuation

Bohannon has recently proposed a new technique for computing ρ_0 from the πN and $\pi\pi$ data[25]. His method appears to get around some of the uncertainty arising from long analytic continuations through

unphysical regions. His resulting ρ_0, shown as a dashed line in
Fig. 8, is significantly different. It will be interesting to see
the effect of this ρ_0 on the intermediate range $V(r)$'s and δ's shown
in Fig.'s 5 and 6. So far, no critical commentary has appeared on
Bohannon's ρ_0.

N̄N DATA ANALYSIS PROBLEMS

There are some interesting discrepancies among the phase
parameter analyses published in the past few years. For example,
analyses of the 50 MeV data have produced very different $^3S_1$, ε_1,
$^1P_1$ and np $^1S_0$ values. Another case is the $^1F_3$, shown in Fig. 9.
At some energies this phase has fluctuated wildly with time. Are
these fluctuations due to the rapid replacement of grossly incorrect
data by newly measured data? Are they due to the existence of
multiple solutions into which analyses randomly wind up? Are they
due to very different choices for which phases are made free to fit
the data and which are fixed at some theoretical values? Are there
computational errors which become corrected in successive
publications? Are they due to differences in inelastic assumptions?

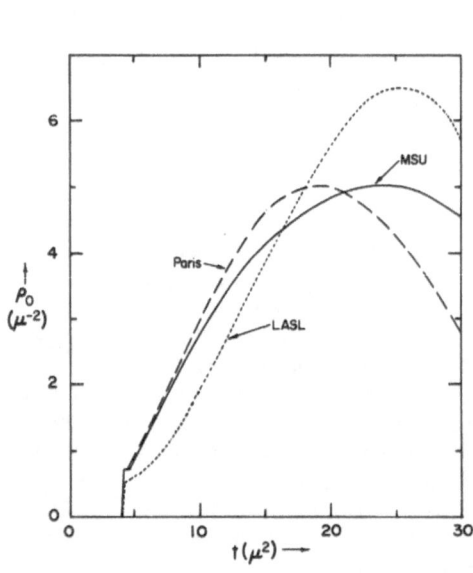

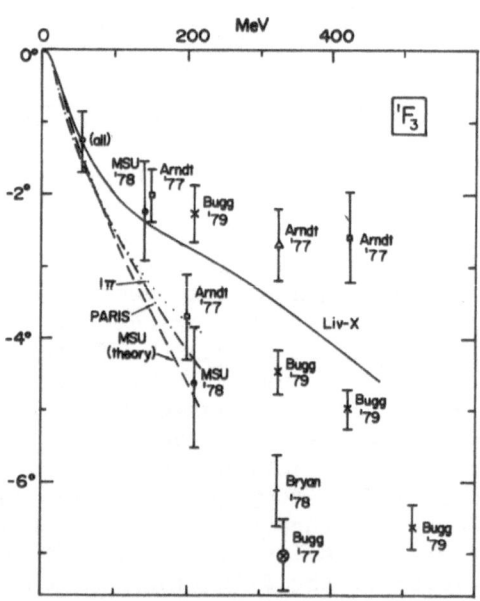

Fig. 8. The S-wave-exchange
spectral functions used by the
PARIS and MSU groups, along
with Bohannon's improved one
("LASL").

Fig. 9. $^1F_3$ values from these
data analyses: MSU'78($\cdot$);
Arndt'77[18]($\square$); Arndt'77[19](Δ);
Bugg'79[20]($\times$); Bryan'78[21](+);
and Bugg'77[22]($\otimes$).

In an effort to track down the origins of these discrepancies, several of the analysis groups have agreed to try communicating via an "electronic mail" conference. The merit of this approach is that one can quickly accumulate printed comments on posted items (very short papers), along with new responses to previously-printed responses. We hope that this approach will also result in convergence to a new standard data set.

CHOOSING AN NN POTENTIAL FOR NUCLEAR CALCULATIONS

If I needed an NN potential for a nuclear calculation, I think I would want a precise one, not the most interesting one or one based perhaps on some appealing but crude theoretical idea. In evaluating a candidate for my "realistic" NN potential, I would carefully examine its printed literature to see how precisely it matched: (i) the low- to moderate-energy πN phase shifts; (ii) the low- to moderate-energy $\pi\pi$ phase shifts; (iii) the low- to moderate-t nucleon isovector electromagnetic form factors; and (iv) the NN data. Finally, I would try to see to what extent the calculation stays out of unphysical regions, especially those where there are poorly known singularities. Such guidelines may be only one person's view but I hope they can serve as a basis for discussion and refinement.

ACKNOWLEDGEMENTS

I would like to thank Dr. R. Vinh Mau for kindly sending listings of the PARIS potential in advance of publication. M. Brandl, G. Bohannon, R. Signell, and C. Washburn helped with the figures. T. Burt, R. Tinker, G. Edwards and F. Zerilli helped with the design of the electronic scientific text processor used on this manuscript. Charles Goodman suggested the presentation's topic. The National Science Foundation provided valuable support.

REFERENCES

1. See Ref.'s 2-6.
2. D. Amati, E. Leader and B. Vitale, Nuovo Cimento 17:68 (1960).
3. G. Epstein and B. McKellar, Lettere al Nuovo Cimento 5:807 (1972).
4. G. Bohannon and P. Signell, Phys. Rev. D10:815 (1974).
5. M. Lacombe, B. Loiseau, J. Richard, R. Vinh Mau, J. Cote, P. Pires, and R. de Toureil, Orsay Preprint IPNO/TH 78-46, Nov. 1978 and private communication from R. Vinh Mau.
6. G. Bohannon, Phys. Lett. 53B:11 (1974).
7. P. Signell in: "Advances in Nuclear Physics," M. Baranger and E. Vogt, Ed., Plenum Press, New York (1969).
8. A. Jackson, D. Riska and B. Ver West, Nuclear Physics, A249:397 (1975), quoted in Ref. 2.

9. P. Signell in: "Proc. Amsterdam International Conference on
    Elementary Particles," A. Tenner and M. Veltman, Ed.,
    North–Holland, Amsterdam (1972).

10. See Ref.'s 4–6.

11. See Ref.'s 4 and 5 and Ref.'s therein.

12. R. Reid, Annal. Phys. ,New York, 50:41 (1968).

13. T. Hamada and I. Johnston, Nucl. Phys. 34:382 (1962).

14. M. MacGregor, R. Arndt and R. Wright, Physical Review
    182:1714 (1969).

15. See Ref.'s 16–22.

16. P. Signell, T. Burt, N. King and D. Fitzgerald (to be
    published).

17. G. Bohannon, T. Burt and P. Signell, Physical Review
    C13:1816 (1976) and subsequent analyses (to be published).

18. R. Arndt, R. Hackman and L. Roper, Phys. Rev. C15:1002 (1977).

19. R. Arndt, R. Hackman and L. Roper, Phys. Rev. C15:1021 (1977).

20. D. Bugg, J. Edgington, W. Gibson, N. Wright, N. Stewart,
    A. Clough, D. Axen, G. Ludgate, C. Oram, L. Robertson,
    J. Richardson and C. Amsler, Rutherford Laboratory preprint
    RL–79–023, to be published.

21. R. Bryan, R. Clark and B. VerWest, Phys. Lett. 74B:321 (1978).

22. C. Amsler, D. Axen, J. Beveridge, D. Bugg, A. Clough,
    J. Edgington, S. Jaccard, G. Ludgate, C. Oram,
    J. Richardson, L. Robertson, N. Stewart and J. VA'VRA,
    Phys. Lett. 68B:419 (1977).

23. N. King, J. Reber, J. Romero, D. Fitzerald, J. Ullman,
    T. Subramanian and F. Brady, Crocker Nuclear Lab
    preprint (to be published).

24. See Ref.'s 4,6.

25. G. Bohannon, Phys. Rev. 14:126 (1976).

APPENDIX A.

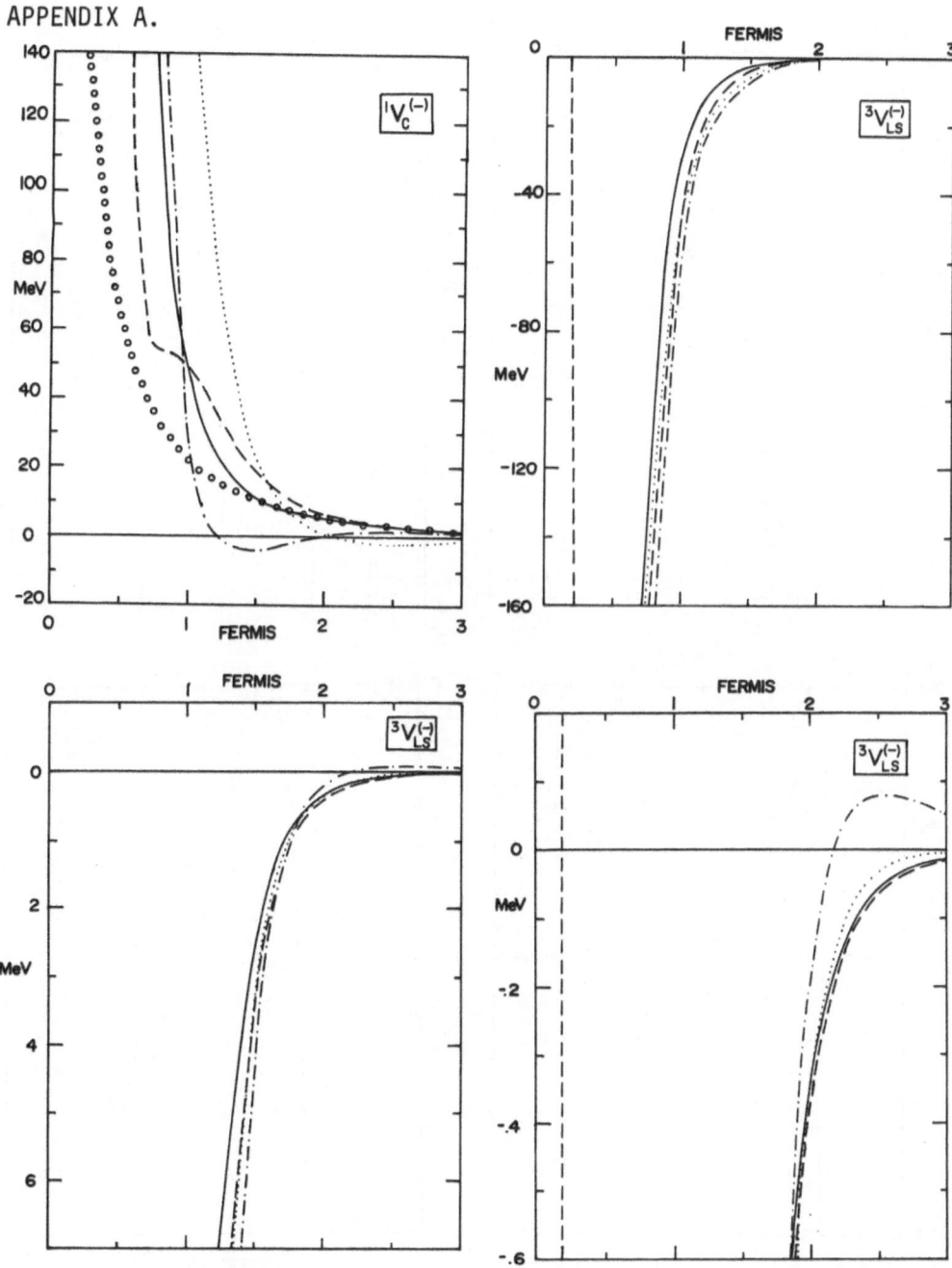

Fig. 10.   Comparisons of potentials due to 1π exchange(ooo),
PARIS (---), MSU (——), Reid (•••), HJ (—•—).   For other
states see Fig. 5.

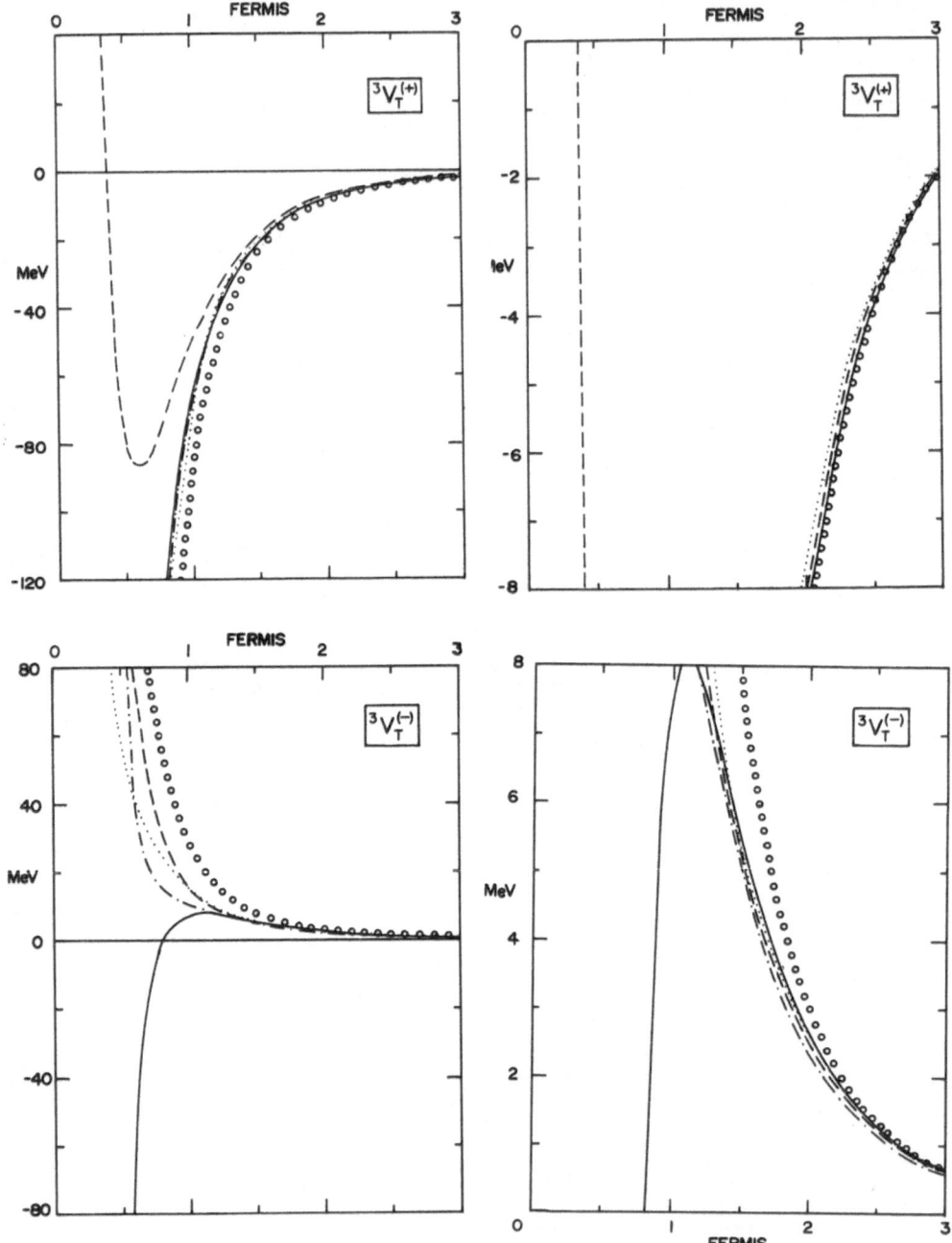

Fig. 10.   (continued)

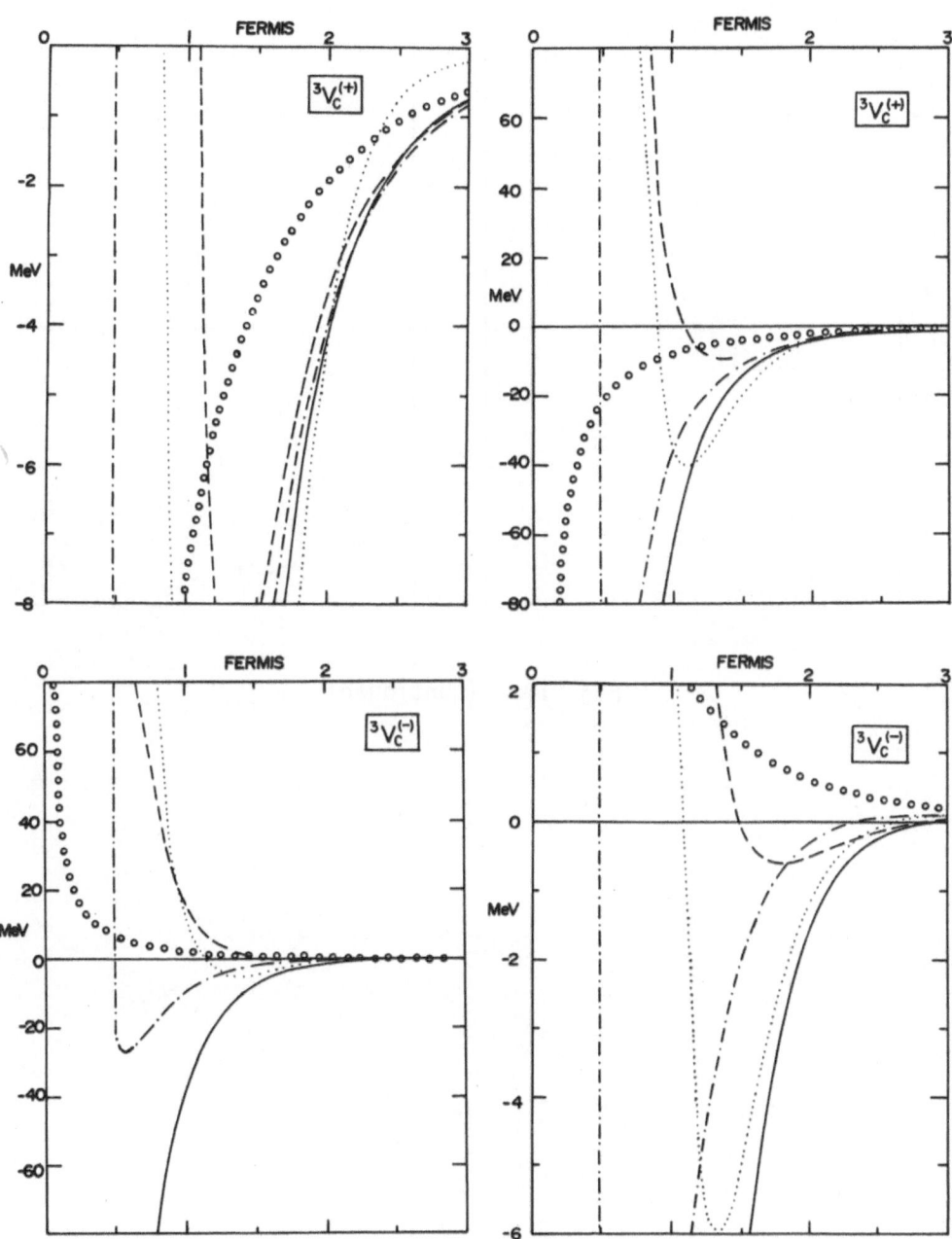

Fig. 10.   (continued)

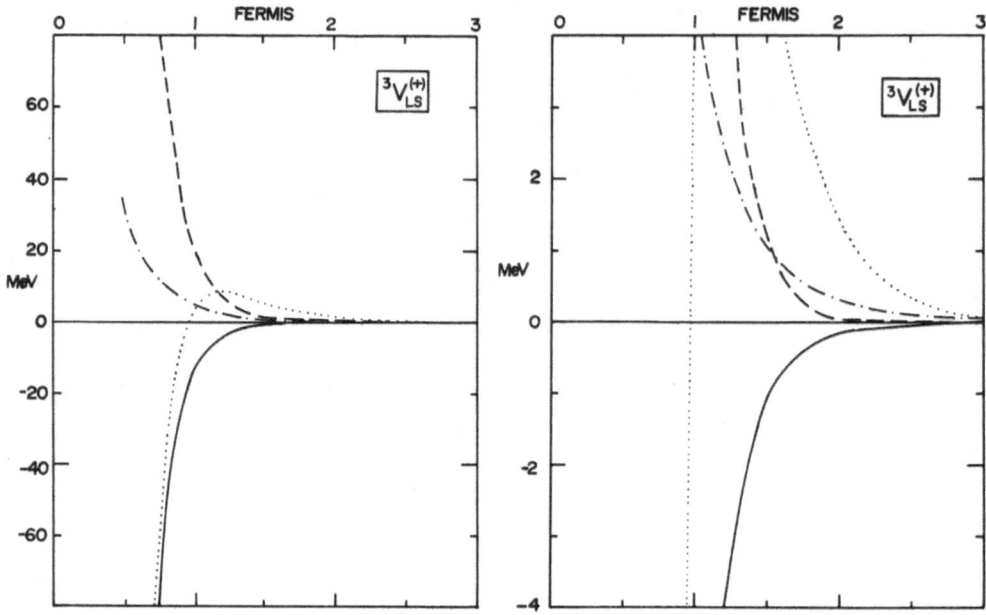

Fig. 10.   (continued)

APPENDIX B.

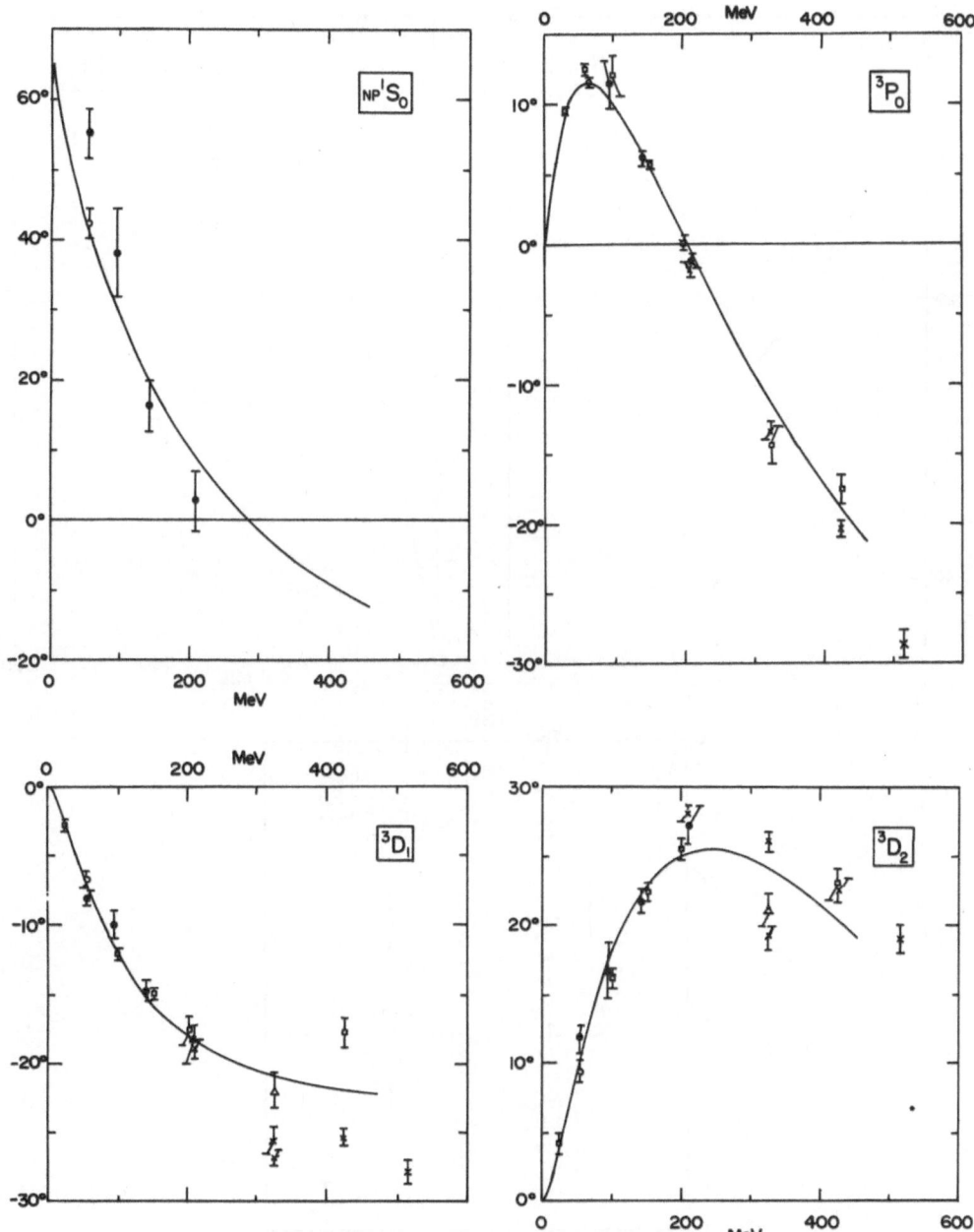

Fig. 11. Comparison of Liv-X(1969) phases (smooth curve) with those from recent analyses: MSU-LASL-Davis[16](o), MSU[17] (•), AHR-II[18](□), AHR-III[19](▲), Bugg[20](×), TxA&M[21](+). For other states see Fig. 6.

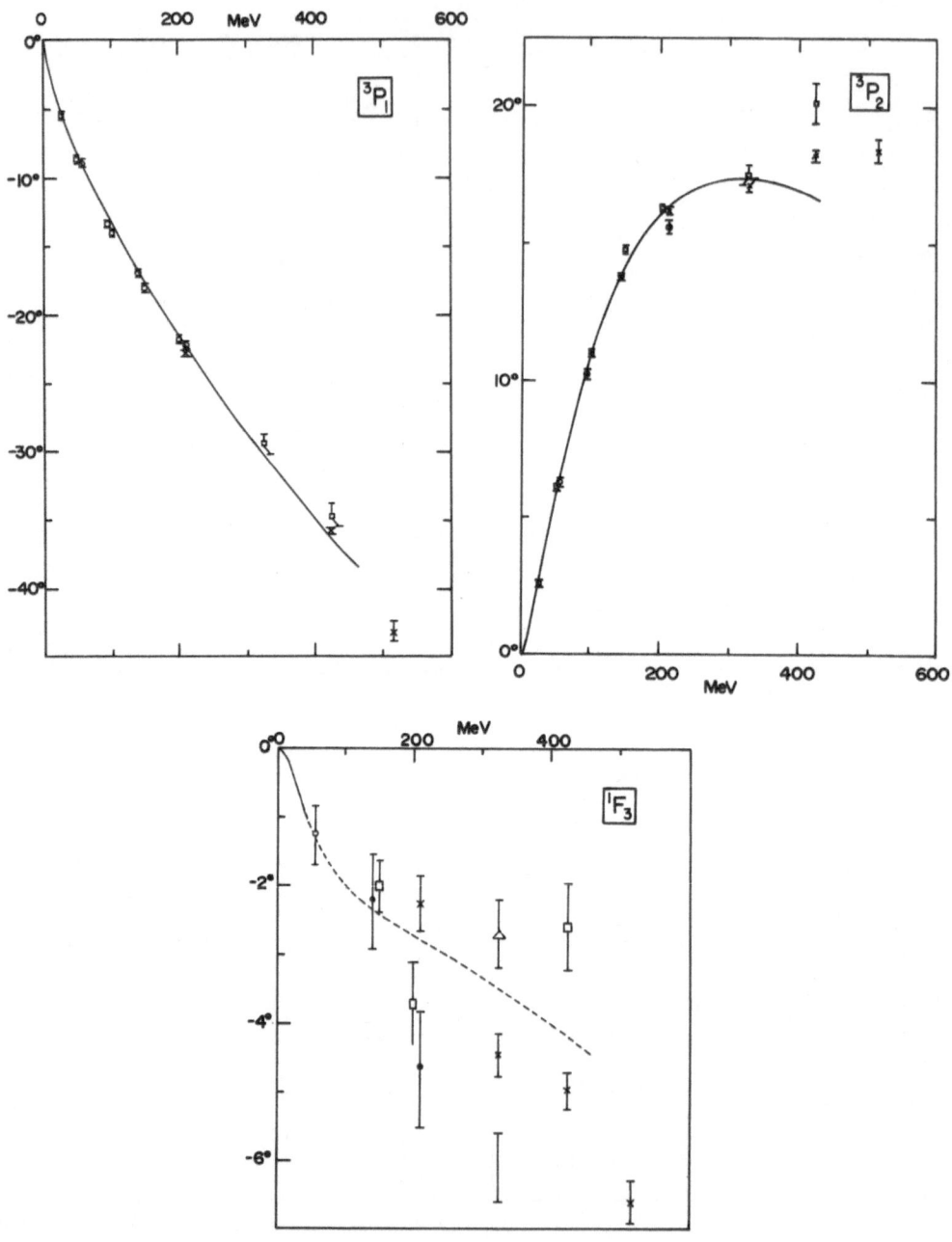

Fig. 11. (continued)

DISCUSSION

   Brady: What is the nature of the anomaly at 50 MeV to which you referred?

   Signell: If we do a straightforward phase analysis of the data near 50 MeV, the $^3S_1$ phase is six standard deviations away from a smooth curve drawn through the points at other energies. The $^1S_0$ np/pp splitting is also way off but the $^1P_1$ and ε_1 are not strongly affected. If we force the $^3S_1$ to be at the inter-polated value, χ^2 goes up but the increase is spread out over a lot of different data; there is no single culprit. This certainly calls for new experimental work.

   Barrett: Nuclear-structure calculations are now sophisti-cated enough to distinguish among different models for the nucleon-nucleon (N-N) force and indicate the need for a new generation of N-N forces. It is very important to understand the true nature of the short-range repulsion and of the tensor com-ponent of the N-N force, since these two properties of the N-N force greatly complicate nuclear-structure calculations, making them long and expensive. Therefore, we need to better determine the N-N force at short distances, say at roughly less than 0.7 fermi. It would be nice, if the tensor component of the N-N force were found to be weaker than in the present forces.

PROPERTIES AND APPLICATIONS OF EFFECTIVE INTERACTIONS DERIVED FROM

FREE NUCLEON-NUCLEON FORCES

W. G. Love

Department of Physics and Astronomy
University of Georgia
Athens, Georgia 30602

ABSTRACT

Techniques are outlined for deriving effective internucleon interactions for nucleon-nucleus and nucleus-nucleus scattering at bombarding energies (E_p) between 10 and 200 MeV per nucleon. Below E_p=65 MeV, the effective interaction is based on G matrices constructed from realistic internucleon potentials. For $E_p \gtrsim 100$ MeV a complex effective interaction is derived directly from the free nucleon-nucleon t-matrix. These interactions are applied to a variety of nucleon-nucleus scattering processes and charge exchange reactions which sample some of the most important characteristics of the interactions and illustrate the roles played by the central, spin-orbit and tensor parts of the force.

1. INTRODUCTION

An integral part of understanding nucleon-nucleus scattering (including (p,n) reactions) is that of relating the reaction mechanism to the underlying nucleon-nucleon (N-N) interaction. From a purely practical viewpoint, it is essential to understand the characteristics of the N-N force when it is used as a probe of the nucleus. While nuclear structure information can be (and has been) obtained[1] without direct reference to the free N-N interaction, a more satisfactory and unified understanding of nucleon-nucleus scattering should make this connection. Although the details for carrying out this program differ from group to group, the end product is an effective N-N interaction which may be used (usually in perturbation theory) within some restricted portion of the full Hilbert space. The necessity for introducing an effective N-N interaction can be

23

attributed to the strong, short range repulsion present in the bare
N-N potential which renders it unsuitable for perturbative calculations. Unfortunately, the effective N-N interaction which results
from a "straightforward" transformation of the bare N-N interaction
acting in the full Hilbert space is still prohibitively complicated
to be used routinely in calculations of nucleon-nucleus scattering.
For example, such an effective interaction is invariably non-local
and depends on energy, relative angular momentum, density, density
gradients, etc. The dependence of v^{eff} on bombarding energy and
density is not necessarily objectionable if it is smooth and pre-
dictable.

For the calculation and interpretation of nucleon-nucleus scattering it is desirable to have an effective interaction which is
simple and yet realistic enough to represent the important and
interesting characteristics of the underlying N-N interaction in a
variety of nuclear transitions. It should be clear that the desired
properties, "simple" and "realistic" are not only relative and sub-
jective but may also be to some extent incompatible. Nevertheless,
approaches along these lines have been encouraging and are discussed
below.

2. DESIRABLE PROPERTIES OF AN EFFECTIVE INTERACTION

Before describing in detail some of the techniques which have
been used to determine effective interactions, let me first enumerate
some of the properties which I find desirable for such an interaction.
As one may note, many of these properties (or constraints) are
dictated primarily by the calculational schemes currently used
(available) to treat nucleon-nucleus scattering. In particular, in
either the distorted wave or coupled channel approaches, two-body
matrix elements are required in which one of the particles is in the
continuum. This situation has precluded any widespread use of a
transformation from individual to relative and CM coordinates so
useful in nuclear matter and nuclear structure calculations. It is
particularly unfortunate that we have been unable to exploit this
transformation in nucleon-nucleus scattering calculations since
information about the free N-N interaction is most transparent in
relative coordinates. Consequently, it is highly desirable to try
and represent the effective N-N interaction in each N-N channel
(triplet-even, triplet-odd etc.) by a local operator of the form

$$v^{eff} = V_C(r) + V_{LS}(r)\vec{L}\cdot\vec{S} + V_T(r)S_{12} \quad , \tag{1}$$

where V_C denotes the central interactions and $\vec{L}\cdot\vec{S}$ and S_{12} are the
usual spin-orbit and tensor operators. Even this form for v^{eff}
becomes non-local when the space exchange terms[2-3] arising from the
Pauli principle are included, but non-local terms of this particular

type have been routinely calculated[2-5] for some years and simple
local estimates[6-7] of them are also available. Since in the calculational schemes currently available the multipoles of v^{eff} are
required, it is desirable for V_C, V_{LS}, and V_T to have a simple analytic form such as a Yukawa, a Gaussian or superposition of these. This
restriction to analytic forms is most important when the space exchange
terms are to be calculated exactly since a large number of multipoles
are required in that case. The restriction to Yukawa forms exclusively (with the exception of V_T where r^2x a Yukawa is more convenient) is especially convenient when the space exchange terms are
to be calculated exactly since each Yukawa term gives rise to a semi-
separable form of multipoles $[f(r_<) \cdot g(r_>)]$.

Although reasonably complete N-N scattering data is available[8]
for $E_p \lesssim 450$ MeV, most of the highest quality nucleon-nucleus scattering
data had until very recently been available primarily for bombarding
energies below ~65 MeV. Historically, this led to an emphasis on
deriving effective interactions based on what may be termed low-
energy techniques.[3,6] Recent data taken at IUCF and LAMPF has,
however, rejuvenated high-energy techniques[9] for obtaining effective
interactions. While there are many features common to both techniques, I will discuss each energy regime separately.

3. EFFECTIVE N-N INTERACTIONS: $E_p \lesssim 65$ MeV

In this energy regime the derivation of v^{eff} for scattering
processes has been strongly influenced by the work of Brown[10] and
others on the analogous problem for bound states. In this approach
v^{eff} is chosen to represent as well as possible the G-matrix generated
by some N-N potential which describes N-N scattering. By definition
G satisfies the equation

$$G\phi = V\psi \tag{2}$$

where V is the bare N-N potential and $\phi(\psi)$ is the uncorrelated (correlated) relative N-N wave function. Although a proper treatment of
the boundary conditions for the case when one particle is in the
continuum yields a complex G-matrix, the assumption is often made[3,6]
that the real part of the interaction (G-matrix) between a projectile
nucleon and a target nucleon should be quite similar to that for two
bound nucleons provided the energy of the projectile is not too much
larger than the Fermi energy. This assumption precludes any determination of the imaginary part of v^{eff} since the G-matrix for two
interacting bound nucleons is real. A rough estimate of the relative
importance of Rev^{eff} compared to Imv^{eff} may be obtained by examining
phenomenological optical potentials which, at relatively low bombarding energies, are predominately real. It is worth noting that
those calculations[11,12] in which both the real and imaginary parts

of the optical potential are calculated from a complex v^{eff} are more successful at predicting the real part than the imaginary part.

With the above limitations in mind the path from N-N scattering data to an effective interaction may be written schematically as:

$$\text{N-N data} \rightarrow \text{free V} \rightarrow \text{G-matrix} \rightarrow v^{eff} . \tag{3}$$

The above procedure has been carried out by a number of groups at various levels of sophistication. The most comprehensive effective interaction is probably that of Brieva, Geramb and Rook[12] who obtain a complex v^{eff} from a G-matrix appropriate for one particle in the continuum and which depends (as it should) on both bombarding energy and density. Here I will describe a less comprehensive but somewhat simpler technique[13] for obtaining v^{eff}.

Since G is a very complicated operator the last step of the program implied by eq. (3) is carried out by first assuming some local form for v^{eff} and then adjusting its parameters until some selected subset of relative oscillator or momentum space matrix elements of v^{eff} are matched to those of the exact operator G. With the exception of the strong even-state part of the tensor force, scattering below E_p=65 MeV is dominated by the interaction in relative S and P states so that the even (odd) -state part of v^{eff} was determined from the relative S(P)-state matrix elements of G. The $\vec{L}\cdot\vec{S}$ force in even states was determined by the D-state G-matrix. To insure the simplicity of v^{eff} and to avoid a proliferation of parameters the following ansatz was adapted for v^{eff}.

$$V_C = \sum_{i=1}^{3} v_i^C Y(r/R_1) , \tag{4a}$$

$$V_{LS} = \sum_{i=1}^{2} v_i^{LS} Y(r/R_i) \; \vec{L}\cdot\vec{S} \tag{4b}$$

$$V_T = \sum_{i=2,4} v_i^T r^2 Y(r/R_i) \; S_{12} \tag{4c}$$

or

$$V_T = v_3^T [Z(r/R_3) - (R_3/R_2)^3 Z(r/R_2)] S_{12}, \tag{4d}$$

where

$$Y(x) = e^{-x}/x$$

and

$$Z(x) = (1 + 3/x + 3/x^2) \ Y(x) \ .$$

In addition to providing reasonably good fits to the G-matrix elements, we imposed the additional constraint that V_C have the correct OPEP tail. This constraint fixes the strengths V_3^C and the range R_3 (1.414 fm). A number of earlier effective interactions based on G-matrices did not impose this constraint. While to date there is little evidence that this is critical for nucleon-nucleus scattering, it has been shown[14] to be extremely important when v^{eff} is used to construct heavy-ion potentials. The regularized OPEP form[15,16] of the tensor force (4d) also has R_3 fixed at the OPEP value but V_3^T was allowed to vary. For the form of V_T in eq.(4c) the longest range part should not be (and was not) fixed at R_3, the OPEP value, due to the extra power of r^2 which is introduced to yield simple multipoles for the tensor force. Instead, the range R_4 (0.7 fm) was introduced in this form to simulate the OPEP term. The range R_2 (0.4 fm) was chosen to crudely simulate multiple-pion exchange processes; R_1 (0.25 fm) was chosen for computational convenience. With these restrictions there remain two strengths in V_C, V_{LS} and V_T to be determined in each of the nucleon-nucleon channels in which they act. The form (4d) for the tensor force has only one adjustable strength.

A number of sources of G-matrices were considered for determining the interaction strengths described above. For the particular interaction we discuss here, the singlet-even and triplet-even G-matrix elements were generated by the Reid soft-core potential in a harmonic oscillator basis. For the odd-state (and the even-state $\vec{L}\cdot\vec{S}$) interaction the Sussex oscillator matrix elements[17] were used. The resulting effective interaction strengths[13] are given in Table 1 for each of the four N-N channels: singlet-even (SE), triplet-even (TE), singlet-odd (SO) and triplet-odd (TO). This v^{eff} will be referred to as M3Y.

With the exception of the triplet-odd states, reasonable fits to the oscillator matrix elements were obtained[13] for each of the N-N channels. For nucleon-nucleus scattering below 60 MeV the TO part of the force is rather unimportant, but in an effort to accommodate heavy-ion scattering we simply assume the OPEP interaction in this channel.

A persistent (and expected) characteristic of local effective interactions such as the one presented here is the presence of a short-range repulsion and a longer-range attraction in even states. This characteristic manifests itself as a change in sign of the interaction as a function of momentum-transfer near $q=2$ fm^{-1}. Calculations for transitions sensitive to momentum transfers near this

Table 1. G-matrix Effective Interaction Strengths (M3Y)

Nucleon-Nucleon State	V_i (MeV)		
	$R_1=0.25$ fm	$R_2=0.40$ fm	$R_3=1.414$ fm
Central			
SE	12454.	-3835.	-10.463
TE	21227.	-6622.	-10.463
SO	26941.	-2777.	31.389
TO	0.0	0.0	3.488
Spin-Orbit			
TE	0.0	-813.	0.0
TO	-2672.	-620.	0.0
Tensor			R_{max} [c]
TE[a]	0.0	-1259.6	-28.41
TO[a]	0.0	283.0	13.62
TE[b]	0.0	[b]	-12.16
TO[b]	0.0	[b]	3.255

[a] r^2xYukawa form (4c) of V_T with V_i in MeV-fm^{-2}.
[b] Regularized OPEP form (4d) of V_T with V_i in MeV.
[c] $R_{max}=0.7$ fm for form (4c) and 1.414 fm for form (4d).

value (usually those requiring large angular momentum transfers)
should therefore be made and assessed with caution.

To better illustrate the roles of the various parts of v^{eff} in
nuclear excitations it is convenient to rewrite it in a way that more
directly displays the strengths for definite spin and isospin transfers. In particular,

$$v^{eff} = V_0(r) + V_\sigma(r)\vec{\sigma}_1\cdot\vec{\sigma}_2 + V_\tau(r)\vec{\tau}_1\cdot\vec{\tau}_2 + V_{\sigma\tau}(r)\vec{\sigma}_1\cdot\vec{\sigma}_2\vec{\tau}_1\cdot\vec{\tau}_2$$

$$+ V_{LS}(r)\vec{L}\cdot\vec{S} + V_{LS\tau}(r)\vec{L}\cdot\vec{S}\vec{\tau}_1\cdot\vec{\tau}_2$$

$$+ V_T(r)S_{12} + V_{T\tau}(r)S_{12}\vec{\tau}_1\cdot\vec{\tau}_2 \quad , \tag{5}$$

where,

$$V_o = \frac{1}{16}[3v^{SE} + 3v^{TE} + v^{SO} + 9v^{TO}]$$

$$V_\sigma = \frac{1}{16}[-3v^{SE} + v^{TE} - v^{SO} + 3v^{TO}]$$

$$V_\tau = \frac{1}{16}[v^{SE} - 3v^{TE} - v^{SO} + 3v^{TO}]$$

$$V_{\sigma\tau} = \frac{1}{16}[-v^{SE} - v^{TE} + v^{SO} + v^{TO}]$$

$$V_{LS} = \frac{1}{4}[v^{LSE} + 3v^{LSO}] \quad , \quad V_T = \frac{1}{4}[v^{TNE} + 3v^{TNO}]$$

$$V_{LS\tau} = \frac{1}{4}[-v^{LSE} + v^{LSO}] \quad , \quad V_{T\tau} = \frac{1}{4}[-v^{TNE} + v^{TNO}]. \tag{6}$$

Here v^{TNE}, for example, denotes the even-state part of the tensor component of v^{eff}. In the absense of knock-on exchange terms V_τ, for example, would give the entire central force contribution for transferring 1 unit of isospin and 0 units of spin. The knock-on exchange terms arising from antisymmetrization of the projectile with the target nucleus are, however, known to be essential for inter-actions which include odd-state forces[13,14] and these were calculated exactly using the code DWBA-70[18] in the results shown here. It is nevertheless, often helpful to have some relatively simple estimate of the overall strengths of the various types of terms (S and T transfers) appearing in eq. (6). To get such an estimate we use the fact that the strengths of the various components of the interaction operative[3] in the knock-on exchange terms ($\hat{V}$) are obtained by simply changing the sign of the odd-state parts of the interaction in eq.(6). An approximate effective interaction may then be obtained by making the replacement

$$V(r) \rightarrow V(r) + \hat{J}(E)\delta(\vec{r}) \equiv u^{eff} \quad , \quad E = \frac{\hbar^2 k^2}{2\mu} \tag{7}$$

for the central parts of V in eq.(6) where to lowest order[6,7] $\hat{J}(E)$ is the Fourier transform of $\hat{V}(r)$ at momentum k, the asymptotic momentum of the projectile. In this lower energy regime, we have calibrated $\hat{J}(E)$ by comparison with a few exact calculations. The results are given in Table 2 in the form of volume integrals (J) and mean-square radii ($<r^2>$) of the spin transfer and isospin transfer components of u^{eff}. These quantities determine the characteristics of u^{eff} at small momentum transfer which are usually its best deter-mined properties. For larger momentum transfers the spin-orbit and tensor parts of the interaction are known[16,19] to be important, especially for isoscalar and isovector transitions respectively. The information on u^{eff} contained in Table 2 is intended to help facilitate comparison of it with effective forces determined

Table 2. Volume Integrals and Mean-Square Radii of
          an Effective Interaction Derived from a
          G-matrix (M3Y)

σ τ[a)]	E_p=10 MeV J[b)]	E_p=10 MeV $<r^2>$[b)]	E_p=40 MeV J	E_p=40 MeV $<r^2>$	E_p=61 MeV J	E_p=61 MeV $<r^2>$
0 0	-410.	2.58	-364.	2.91	-299.	3.54
1 0	54.	3.17	43.	4.03	42.	4.12
0 1	204.	2.68	181.	3.03	155.	3.55
1 1	253.	7.21	247.	7.40	211.	8.64

[a)] σ and τ label spin and isospin transfer components
respectively.
[b)] Volume integrals (J) are in MeV-fm^3; Mean-square
radii ($<r^2>$) are in fm^2. J and $<r^2>$ refer to the
approximate form of the interaction in eq. 7 in
which the exchange terms are included. In particular, $J=J_V + \hat{J}(E)$.

empirically from calculations in which no exchange terms are explicitly included. The energy dependence of the M3Y interaction represented in Table 2 arises entirely from the energy dependence of the
knock-on exchange amplitudes as contained in the $\hat{J}(E)$ coefficient
and does not include any intrinsic dependence of the G-matrix upon
bombarding energy.

One of the most striking features of u^{eff}, the effective
interaction indicated in Table 2, is the relative smallness of u_σ^{eff}.
Recalling that the cross section for inelastic scattering goes as
$|J|^2$, we can understand why empirical determinations[1] of u_σ^{eff} have
been rather unsuccessful. This characteristic also suggests that
the non-central parts of v^{eff} (and higher-order processes) are likely
to be relatively more important when a state requiring $\Delta S=1$ and $\Delta T=0$
is excited.   In particular, it helps explain why microscopic descriptions of the excitation of the 1^+ T=0 level in ^{12}C (12.7 MeV)
and the 2^- T=0 level in ^{16}O (8.88 MeV) have encountered considerable
difficulty.[1,20]

It should also be noted that the values of J_{oo} (and hence the
energy dependence) are consistent with those determined from phenomenological analyses.[21] On the other hand the values of $<r^2>_{oo}$ are
somewhat smaller than those found empirically[21] and this may partially reflect our omission of any explicit density dependence of
v^{eff}.

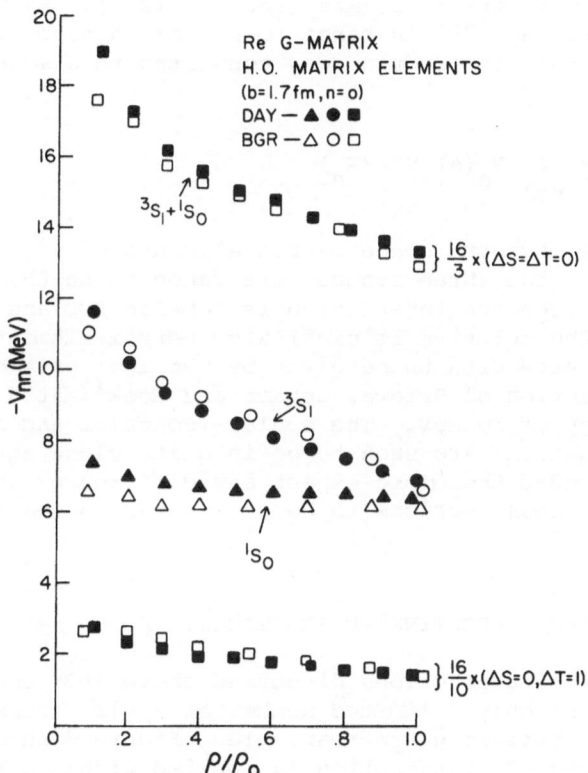

Fig. 1.   Relative S-state harmonic oscillator matrix elements as a
function of density.

Before going to some specific applications, I would like to introduce another v^{eff} based on a density-dependent G-matrix constructed by Day and coworkers[22] for two nucleons interacting in nuclear matter. In coordinate space a trivially local equivalent G-matrix is, for a particular Fermi momentum k_F, given by

$$G(r,k_F) = \frac{V(r)\psi(r,k_F)}{\phi(r)} \qquad (8)$$

averaged over the <u>relative</u> momenta (see eq. (2)). By relating k_F to ρ in the usual way v^{eff} is taken to be this average $G(r,\rho)$. The central part of this interaction was converted to a more convenient form by taking

$$v^{eff}(r,\rho) = \sum_{n=1}^{3} V_n(\rho) Y(r/R_n) \qquad (9)$$

and matching the momentum-space matrix elements of v^{eff} at each ρ to those of $G(r,\rho)$. The three ranges were taken to be the same as in eq. (4a). This effective interaction is labeled DDD and is real. Figure 1 shows the relative 1s oscillator matrix elements of this interaction compared with those given by the real part of the density-dependent interaction of Brieva, Geramb and Rook[12] at a proton bombarding energy of 10 MeV. The scalar-isoscalar and scalar-isovector combinations are seen to be in quite close agreement. Near this energy we would therefore expect any differences in predicted nucleon-nucleus cross sections to be attributed to the imaginary coupling.

4. APPLICATIONS OF EFFECTIVE INTERACTIONS: $E_p < 65$ MeV

The effective interactions discussed above (M3Y and DDD) are relatively new and only a limited number of applications of them to nucleon-nucleus scattering are available. Figure 2 shows the results[23] when the M3Y interaction is applied within a folding-model context to the elastic scattering of neutrons and protons from ^{60}Ni between 5 and 15 MeV. The imaginary and spin-orbit terms were taken from the global potential of Becchetti and Greenlees.[24] Although the fits are not excellent by optical model standards, they are quite reasonable with only a 5% reduction in the strength of the interaction (N=1 would correspond to no adjustment of the real folded potential). As indicated earlier the mean-square radius of the calculated potential is somewhat too small, an effect that is amplified for somewhat higher energy collisions which are more sensitive to the shape of the potential in the surface. While use of the explicitly density-dependent interaction (DDD) gives almost identical scattering, it has been shown[25] that a density-dependent modification of the M3Y interaction consistent with nuclear saturation leads to improved results.

Figure 3 shows results for excitation of the lowest 3^- state in ^{208}Pb by both neutrons[26] and protons[27]. The proton data is actually for the L=3 doublet in ^{207}Pb which has a cross section virtually identical to that of the 3_1^- state in ^{208}Pb. Like the elastic scattering result this transition is primarily sensitive to the scalar-isoscalar part of u^{eff} and was studied since its transition density (taken from the work of Ring and Speth[28]) is known to be in quite reasonable agreement with (e,e') scattering and B(E3) values. For the neutron results the DDD interaction yields results quite similar to those of the M3Y force in reasonable agreement with the data. At E_n=25.7 MeV there is some indication that a density dependence somewhat stronger than that of the DDD interaction might be preferred. For the proton case only the M3Y result is shown.

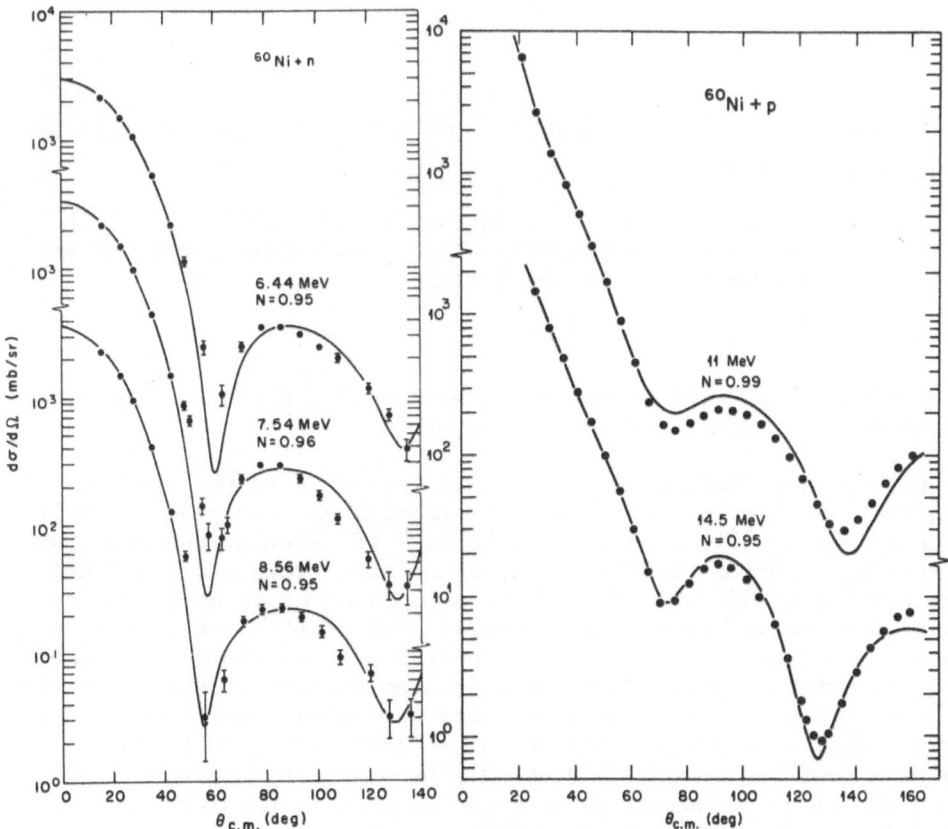

Fig. 2.   Comparison of experimental and calculated elastic scattering cross sections. The real part of the central potential was calculated using the M3Y interaction described in the text.

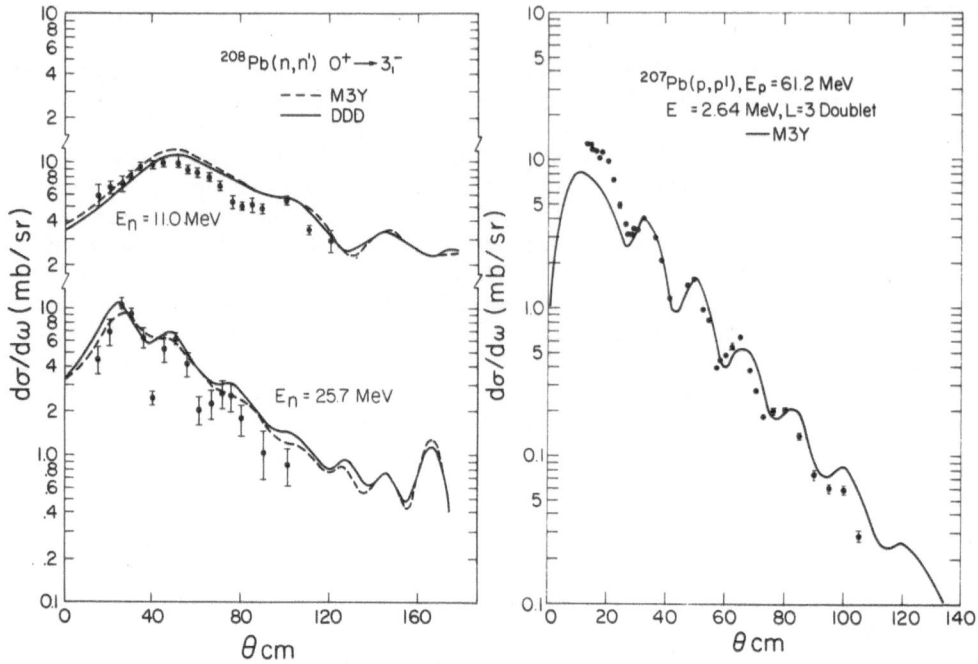

Fig. 3. Microscopic calculations of (n,n') and (p,p') cross sections
         compared with experimental data. The interactions M3Y and
         DDD are described in the text.

There is good agreement except at forward angles where the imaginary
part of v^{eff} is known[29] to be most important.

      Only a very limited number of calculations have been made using
the effective interactions above for transitions sensitive to their
isovector parts. Figure 4 shows results[30] of calculations made for
the Sn(p,n) reaction near E_p=25 MeV where both cross section[31] and
asymmetry[32] data are available. The calculated cross sections are
sensitive primarily to the real central part of u^{eff}, while the
asymmetry is sensitive to both the central and non-central parts of
u^{eff}. Unfortunately, the asymmetry is also discomfortingly sensitive,
not only to the choice of optical model parameters, which were taken
from the work of Becchetti and Greenlees (BG), but also to the
imaginary coupling for which we simply used the isovector part of
the BG potential. The M3Y interaction is seen to provide a reasonable description of both cross section and asymmetry data when the
imaginary coupling is included. Inclusion of the imaginary coupling
only slightly alters the magnitude of the calculated cross section.
The isotopic shape change of the cross section is not reproduced at
large scattering angles; this may reflect an oversimplified shell-
model description of the neutron excess. Due to their volume-like

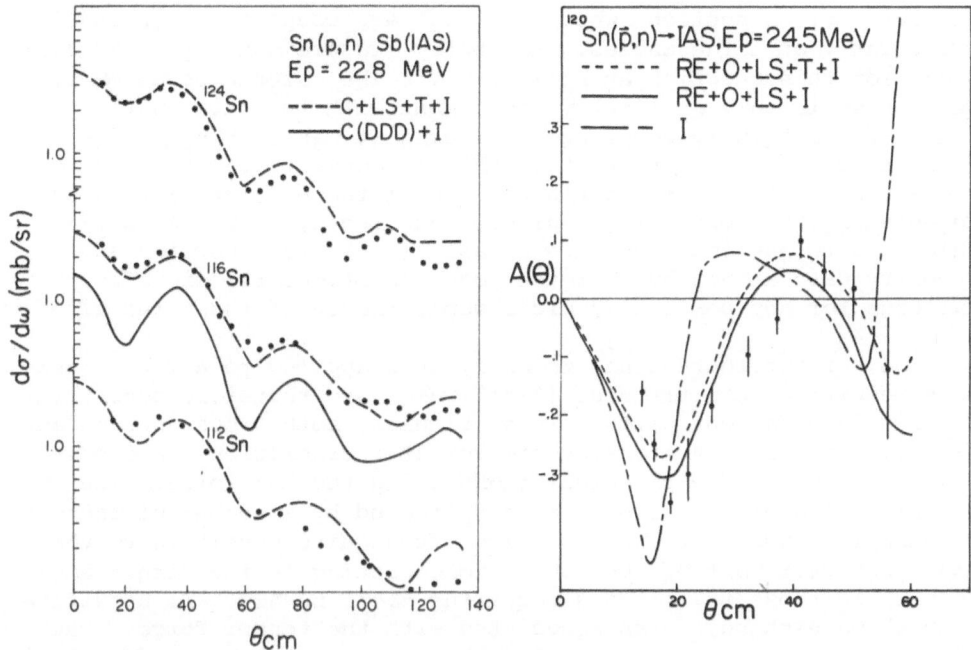

Fig. 4.   Calculated and measured cross sections and asymmetry for the
Sn(p,n) reaction. With the exception of the solid cross
section curve the M3Y interaction in Table 1 was used. C,
LS, T and I denote central, spin-orbit, tensor and imaginary
contributions respectively. RE+O indicates that the even-
state part of the N-N interaction is derived from the Reid
potential and that odd-state forces were included.

nature, (p,n) reactions should be especially sensitive to any density
dependence in v^{eff}, especially since the difference between TE and
SE components enters (see eq. (6)). Results using the DDD interaction suggest it is too weak in the interior.

Similar calculations using the M3Y interaction have been made
for the ^{64}Ni (p,n) reaction at 22.8 MeV. A comparison of the experimental[33] and calculated cross sections are shown in Fig. 5. Two
different configurations for the neutron excess are shown in order
to exhibit the sensitivity of the cross section (or lack of it) to
the microscopic constituency of the neutron excess. In the upper
part of Fig. 5 no imaginary coupling was included and the calculated
cross sections are too small by ~30-40%. Inclusion of the macroscopic imaginary coupling suggested by Carlson et al.[33] in Table 4 of
ref. 33 yields the calculated cross sections in the lower part of
Fig. 5 which agree reasonably well with experiment without renormalization. There is, however, somewhat too much structure in the

calculated cross sections when the imaginary coupling is included,
particularly at large angles. Asymmetry measurements are also available[32] for this reaction at E_p=22.8 MeV. The calculated asymmetries
are at best in only qualitative agreement with the data, with or
without the imaginary coupling (including it makes things worse).
There is some sensitivity to the orbitals constituting the neutron
excess. It would be interesting to see if the M3Y interaction sup-
plemented by the imaginary coupling suggested by Schery et al.[34]
would lead to any improvement. Clearly (p,n) cross section and
asymmetry data at nearby energies pose a particularly significant
challenge for any completely microscopic theory of these reactions.

The M3Y interaction has recently been applied to a few transitions requiring spin-transfer (ΔS=1) which are therefore sensitive
to its spin-dependent parts. In particular, both cross section and
spin-flip probability measurements for ΔS=1 transitions have been
made[20] and compared with calculations using the M3Y force. The
interpretation of the results is complicated by a number of factors.
For example, $\Delta S=\Delta T$=1 transitions are often quite sensitive to the
large isovector part of the tensor force acting in the direct amplitudes while those transitions requiring ΔS=1 and ΔT=0 are sensitive
to knock-on exchange terms associated with the tensor force. The
latter type of transition can also be sensitive to the predominantly

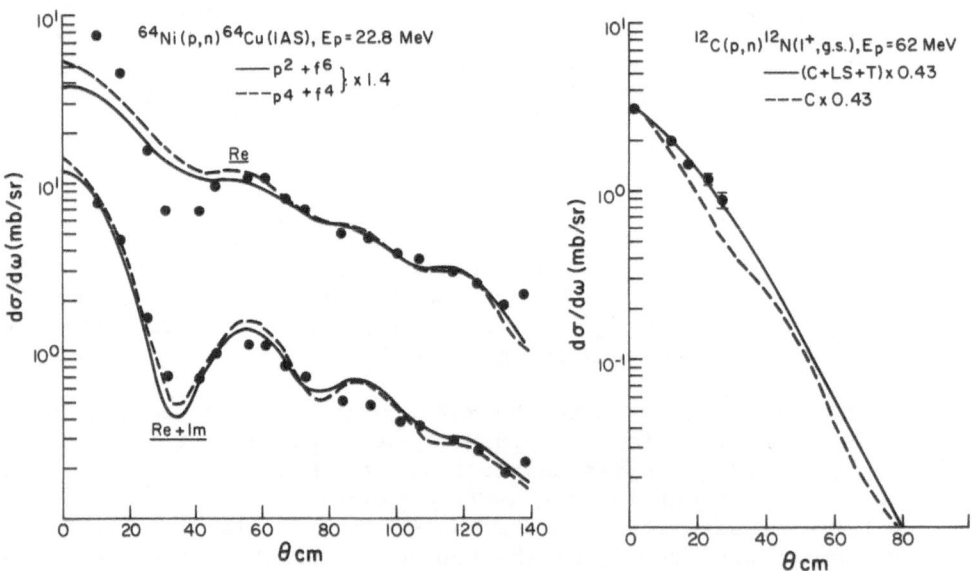

Fig. 5. Calculated and measured cross sections for the ^{64}Ni (p,n)
^{64}Cu and ^{12}C (p,n) ^{12}N reactions. The state in ^{64}Cu is the
IAS of the ^{64}Cu ground state; the state in ^{12}N is the
isobaric analogue of the 15.1 MeV state in ^{12}C.

isoscalar $\vec{L} \cdot \vec{S}$ force.

Our understanding of $\Delta S=1$, $\Delta T=0$ transitions seems especially cloudy. One reason for this is the relative smallness of u^{eff}; another is the fact that (e,e') data for such transitions is largely unavailable due to the predominately isovector nature of the electron-nucleon magnetic coupling. The absence of (e,e') data for $\Delta S=1$, $\Delta T=0$ transitions precludes our calibrating the transition density, leaving a considerable uncertainty in the wavefunctions used. Moss et al.[20] have recently discussed some of these difficulties in an article reporting spin-flip probability measurements for the inelastic excitation of the 2^- T=0 state in ^{16}O at E_x=8.88 MeV. Their results suggest that the M3Y interaction may be slightly too strong. More cases will have to be considered in order to obtain a completely definitive result.

Of those $\Delta S=1$ transitions requiring $\Delta T=1$ the strong M1-type transitions are likely to provide the most definitive test of the central part of $u_{\sigma\tau}^{eff}$. Not only can the wave functions be checked against electron scattering, but the importance of the tensor force decreases with lower momentum transfer and hence lower multipolarity. Excitation of the 1^+, T=1 level in ^{12}C (or its analog in ^{12}N by the (p,n) reaction) should be an excellent transition to test the $\Delta S=\Delta T=1$ part of the force at low momentum transfers since the wave functions of Cohen and Kurath[35] are known to describe most of the characteristics of this transition quite well. Although (p,p') data is available for this transition, Fig. 5 shows a comparison of the analogous (p,n) transition[36] at E_p=62 MeV with calculations using the M3Y interaction. The relative importance of the central and tensor forces appears to be given correctly by this interaction; however, the calculated cross section is too large by roughly a factor of 2. A number of similar M1-type (p,n) cross sections at this bombarding energy are also overestimated by roughly this same factor when the M3Y force is used. There will be discussed in more detail in another talk at this meeting by F. Petrovich.

The overestimate of the cross section at E_p=62 MeV is not understood; it is particularly puzzling since Howell et al.[37] find reasonable agreement with the ^{12}C (p,p') cross section for this transition at E_p=25 MeV with essentially no renormalization. The shape of the cross section at this lower energy is not, however, well reproduced. It is amusing to note that omission of the (poorly known) odd-state parts of the force reduces $u_{\sigma\tau}^{eff}$ by ~30%. It should be added that these are all very recent results and no study of the effects of varying the optical model parameters etc. has been made.

Based on the very limited number of cases considered here we may tentatively conclude that for nucleon-nucleus collisions below 65 MeV the u_0 and u_τ parts of the M3Y interaction seem to work reasonably well. The $u_{\sigma\tau}$ part is ~40% too large and u_σ has not been definitively tested. The non-central parts of this interaction have not been studied adequately to reach any meaningful conclusions. Excitation of high-spin states of both natural and unnatural parity, together with measurements of spin-flip probability should be most sensitive to these non-central components.

6.   EFFECTIVE INTERACTIONS: $E_p \gtrsim 100$ MeV

One of the primary motivations for extending (p,p') and (p,n) measurements to higher energies is to avoid many of the complicated reaction mechanisms believed to be important at lower energies. In particular, it has been suggested that above ~100 MeV bombarding energy the impulse approximation might be appropriate for under-standing nucleon-nucleus scattering. Since in the simplest version of the impulse approximation v^{eff} is taken to be the free N-N t-matrix, the appeal of this technique is evident; we can regard the interaction as known and focus our attention on extracting nuclear structure information. By impulse approximation I am referring to the distorted-wave impulse approximation (DWIA) rather than the plane-wave version (PWIA) since the effects of distortion are known to be essential for quantitative comparisons of theory with experi-ment even at 1 GeV. The validity of the DWIA at energies between 100 and 200 MeV is not well established (see ref. 38) giving rise to a number of ambiguities in relating N-N information to the scat-tering of nucleons by nuclei. For example, the Fermi motion of nucleons in the target may be non-negligible in this range of bom-barding energies. In addition, in the DWIA, the projectile moves in a complex well suggesting the need for a free t-matrix evaluated at complex energies. Here we simply adopt the philosophy that the IA provides a reasonable starting point; we must be prepared to make corrections when they are required.

Although it is not essential to make reference to a bare N-N potential when using the impulse approximation (another of its advantages) it is instructive to do so for comparison with the low-energy techniques for containing an effective interaction. The N-N t-matrix (operator) satisfies:

$$t\phi = V_{bare}\psi \tag{2'}$$

just as is eq. (2) except that here ψ is the correlated wave function of relative motion in the absence of other nucleons. In the DWIA reported here ψ and ϕ are to be evaluated at an energy corresponding to the struck target nucleon being at rest so that, in principle, a new effective interaction will have to be constructed for each bom-barding energy. (Of course if this is necessary for changes in the energy less than the optical-model well depth, the approximation is highly suspect anyway.) On the N-N energy shell, t is related to the scattering amplitude[9] (M) via:

$$t(E,\theta) = \eta M(E,\theta) \quad , \quad \eta = \frac{-(\hbar c)^2}{2\pi^2 E_{cm}} \quad , \quad E_{cm}^2 = m^2 c^4 + (\hbar c k_{cm})^2 \qquad (10)$$

where k_{cm} is the nucleon momentum in the N-N center-of-mass system, m is the nucleon mass and

$$M(E,\theta) = A + B\vec{\sigma}_1 \cdot \hat{n}\vec{\sigma}_2 \cdot \hat{n} + C(\vec{\sigma}_1 + \vec{\sigma}_2) \cdot \hat{n}$$

$$+ E\vec{\sigma}_1 \cdot \hat{q}\vec{\sigma}_2 \cdot \hat{q} + F\vec{\sigma}_1 \cdot \hat{Q}\vec{\sigma}_2 \cdot \hat{Q} \quad . \qquad (11)$$

A, B, C, E and F are functions of bombarding energy, scattering angle and isospin; θ is the scattering angle in the N-N center-of-mass system; the unit vectors $[\hat{q}, \hat{Q}, \hat{n}]$ form a right-handed coordinate system with

$$\vec{q} = \vec{k} - \vec{k}' \quad , \quad \vec{Q} = \vec{k} + \vec{k}' \quad , \quad q = 2k\sin\frac{\theta}{2} \quad , \quad Q = 2k\cos\frac{\theta}{2} \quad , \qquad (12)$$

and $\vec{k}(\vec{k}')$ is the initial (final) momentum in the center-of-mass system. At IUCF energies only a few percent error is made by taking $E_{cm} = mc^2$ in eq. (10). For purposes of identifying terms of various spin and spatial ranks (scalar, vector and tensor) appearing in t it is convenient to use the identity

$$\vec{\sigma}_1 \cdot \hat{u}\vec{\sigma}_2 \cdot \hat{u} = \frac{1}{3} [S_{12}(\hat{u}) + \vec{\sigma}_1 \cdot \vec{\sigma}_2] \quad , \quad \hat{u} = \hat{q} \text{ or } \hat{Q} \qquad (13)$$

together with the completeness of $[\hat{q}, \hat{Q}, \hat{n}]$ giving

$$M(E,\theta) = A'P_S + B'P_T + C(\vec{\sigma}_1 + \vec{\sigma}_2) \cdot \hat{n}$$

$$+ E'S_{12}(\hat{q}) + F'S_{12}(\hat{Q}) \qquad (14)$$

where S_{12} is the usual tensor operator, $P_S (P_T)$ is the singlet (triplet) spin-projection operator, and

$$A' = A - B - E - F, \quad B' = A + \frac{B+E+F}{3}, \quad E' = \frac{E-B}{3} \quad , \quad F' = \frac{F-B}{3} \quad . \qquad (15)$$

The amplitudes A' and B' are to be identified with the central parts of the interaction; the C-term which is linear in the Pauli-spin matrices can be regarded as a spin-orbit-like term, and E' and F' are to be associated with the tensor parts of the effective interaction.

Since most codes for calculating nucleon-nucleus scattering are written for coordinate-space interactions we ask, just as we did at lower energies, can we find a reasonably accurate local representation of the free N-N t-matrix? In particular, can we find a local $t(\vec{r})$ in the form of eq.(1) such that its antisymmetrized momentum-space matrix elements describe the on-shell t-matrix deduced, say, from a phase-shift analysis? V_C, V_{LS} and V_T will now be complex. The effective interaction in coordinate space is obtained by equating:

$$\eta M(E,\theta) = \frac{1}{(2\pi)^3} \int e^{-i\vec{k}'\cdot\vec{r}} t(\vec{r}) [1 + (-)^{1+S+T} P^x] e^{i\vec{k}\cdot\vec{r}} \quad , \tag{16}$$

where P^x is the space exchange operator which converts $\vec{r} \rightarrow -\vec{r}$ on the right and S(T) labels the two-body spin (isospin) state. Inserting $t(\vec{r})$ in the form of eq.(1) into the right-hand side of eq.(16) gives:

$$(2\pi)^3 \langle\vec{k}'|t|\vec{k}\rangle = [\tilde{v}_C^S(q) + (-)^\ell \tilde{v}_C^S(Q)] P_S + [\tilde{v}_C^T(q) + (-)^\ell \tilde{v}_C^T(Q)] P_T$$

$$+ \frac{i}{4} [Q\tilde{v}_{LS}(q) - (-)^\ell q\tilde{v}_{LS}(Q)] (\vec{\sigma}_1 + \vec{\sigma}_2) \cdot \hat{n}$$

$$- [\tilde{v}_T(q) S_{12}(\hat{q}) + (-)^\ell \tilde{v}_T(Q) S_{12}(\hat{Q})] \tag{17}$$

where ℓ is even(odd) for even(odd) states of relative motion. The Fourier transforms ($\tilde{v}$) are defined by:

$$\tilde{v}_C(p) = 4\pi \int_0^\infty r^2 dr j_0(pr) v_C(r) \quad , \tag{18a}$$

$$\tilde{v}_{LS}(p) = 4\pi \int_0^\infty r^3 dr j_1(pr) v_{LS}(r) \quad , \tag{18b}$$

$$\tilde{v}_T(p) = 4\pi \int_0^\infty r^2 dr j_2(pr) v_T(r) \quad , \quad p=q \text{ or } Q \tag{18c}$$

Comparing eq.(14) and eq.(17) and letting $\eta'=(2\pi)^3\eta$

$$\tilde{v}_C^S(q) + (-)^\ell \tilde{v}_C^S(Q) = \eta' A' \tag{19a}$$

$$\tilde{v}_C^T(q) + (-)^\ell \tilde{v}_C^T(Q) = \eta' B' \tag{19b}$$

$$Q\tilde{v}_{LS}(q) - (-)^\ell q\tilde{v}_{LS}(Q) = -4i\eta' C \tag{19c}$$

$$\tilde{v}_T(q) = -\eta' E' \quad , \quad \tilde{v}_T(Q) = (-)^{\ell+1} \eta' F' \quad . \tag{19d}$$

On the energy shell (see eq.(12)) q and Q are not independent so that eq.(19d) implies

$$E'(\pi-\theta) = (-)^\ell F'(\theta) \tag{20}$$

which is just the appropriate interdependence of these amplitudes[38] for the scattering of identical particles.

In anticipation of the necessity for including knock-on exchange amplitudes in the range of bombarding energies available at the IUCF, an analytic coordinate-space representation for V_C, V_{LS} and V_T was sought in the form of a sum of Yukawa terms as described by eq.(4). Table 3 lists complex effective interaction strengths at a bombarding energy of 140 MeV derived from $M(E,\theta)$ constructed directly from the N-N phase shifts of ref. 8. As at lower bombarding energies, each N-N channel was constrained to have the appropriate OPEP form at large distances. We will refer to this interaction as G3Y. The imaginary part of the tensor amplitudes are not well described by a 3-Yukawa ansatz and are much smaller than the real part over most of the angular range anyway (but see ref. 40). It is interesting to note that the tensor force derived from the Sussex oscillator matrix elements also provides a reasonable representation of the real parts of the effective tensor force at this energy and is also included in Table 3.

Table 3. Complex Effective Interaction Strengths for 140 MeV Protons Derived from Nucleon-Nucleon Phase Shifts from tables 4 and 6 of ref. 8.

Nucleon-Nucleon State		Real V_i (MeV) [a]			Imaginary V_i (MeV) [a]	
		R_1	R_2	R_{max} [b]	R_1	R_2
Central	SE	9128.	−2810.	−10.463	1208.	−425.
	TE	6883.	−2249.	−10.463	11028.	−3504.
	SO	36300.	−4913.	31.389	−143.	−512.
	TO	6065.	−1325.	3.488	−399.	−312.
Spin-Orbit	TO	−2320.	−537.	0	−831.	111.
	TE	11064.	−1509.	0	−5866.	1282.
Tensor	TE	29546.	−3263.	5.15	5466.	−580.3
	TO	1199.	90.1	16.08	−3018.	200.4
	TE[c]	0	−171.7	−78.03	0	0
	TO[c]	0	283.0	13.62	0	0

[a] For the tensor force V_i are in MeV-fm^{-2} (see eq.(4c)).
[b] R_{max} = 1.414(0.7) fm for the central (tensor) force.
[c] Tensor force fit to Sussex matrix elements.

Table 4. Volume Integrals and Mean-square Radii of Effective
Interactions Derived from N-N Phase Shifts (t-matrix)

σ	τ [a]	Re J	$<r^2>_{Re}$	ImJ	$<r^2>_{Im}$	$|J|$	$|\eta'M(E_p,0°)|$ [b]	E_p(MeV)
0	0	−303.	3.69	−229.	2.81	380.	381.	100
1	0	6.3	23.4	−19.9	6.47	21.	22.	100
0	1	37.5	5.39	86.3	4.08	94.	98.	100
1	1	151.	10.46	34.8	3.02	155.	152.	100
0	0	−254.	3.56	−210.	2.75	330.	325.	140
1	0	−2.9	−57.	−26.3	3.76	26.	28.	140
0	1	17.7	5.58	70.2	4.48	72.	71.	140
1	1	155.	10.0	13.4	6.78	156.	149.	140
0	0	−165.	4.24	−214.	2.29	270.	266.	200
1	0	−9.1	−19.	−24.	2.61	26.	25.	200
0	1	−17.7	−0.67	52.4	4.89	55.	60.	200
1	1	155.	9.90	−9.1	−7.95	155.	142.	200

[a] The notation is the same as used in Table 2 except that at these
energies v^{eff} is explicitly complex and the exchange terms were
evaluated in the asymptotic energy approximation as described in
refs. 6,7.

[b] $|\eta'M|$ is the magnitude of the exact (not-fitted) N-N scattering
amplitude at 0° normalized (see eq. (19)) for comparison with
$|J|$ which is derived from the fitted interaction. For a perfect
fit, $\eta'M = J$.

Just as at lower bombarding energies it is useful to have a
rough estimate of the importance of the terms in v^{eff} which mediate
a definite spin and isospin transfer. Following the earlier discussion we have constructed a complex u^{eff} (see eq.(7)) for the
central part of the effective interaction at this energy (E_p=140 MeV).
It was verified that no calibration of $\hat{J}(E)$ in eq. (7) appears
necessary at this higher energy; it may be calculated directly from
the Fourier transform of $v^{eff}(r)$. Table 4 shows the resulting real
and imaginary volume integrals and mean square radii of that part of
u^{eff} operative for each spin and isospin transfer. Analogous effective interactions have been constructed at 100 and 200 MeV bombarding
energies and results for these are also given in Table 4 to display
the energy dependence of v^{eff} predicted by this simple model. As at
lower energies u^{eff}_σ is predicted to be much smaller than the other
parts of the interaction. The real part of J_{00} given by the IA is
in rough agreement with values found[41] empirically; the imaginary
part of J_{00} is, however, considerably overestimated.

The relative importance of u_τ^{eff} and $u_{\sigma\tau}^{eff}$ as a function of bombarding energy is of special significance for the (p,n) reaction. In particular, the value of $|J_\tau|^2$ is seen to drop by a factor of ~13 between 20 and 200 MeV, while $|J_{\sigma\tau}|^2$ drops by only a factor of 3. The decrease in $|J_{\sigma\tau}|$ is likely an overestimate since the G-matrix effective interaction overestimates the cross sections for $\Delta S=\Delta T=1$ transitions at lower energies and the impulse approximation appears to correctly predict their magnitudes for bombarding energies above ~100 MeV. Also, above 100 MeV the IA yields a J_τ which is predominately imaginary and a $J_{\sigma\tau}$ which is predominantly real and nearly constant. The stability of $J_{\sigma\tau}$ reflects the presence of the OPEP which is both real and of long range. If the IA estimate of the relative strengths of the isovector terms is correct, this could explain the prominence of 1^+ states[36] seen in the (p,n) spectrum at IUCF energies. The predicted suppression of $\Delta S=0$ states in the (p,n) reaction may also permit the study of unnatural parity states of higher multipolarity. Also, any isovector S=1 strength in the region of the giant dipole resonance might be exposed which could be important for extracting the "true" ($\Delta S=0$, $\Delta T=1$) giant dipole strength at bombarding energies when the $\Delta S=0$, $\Delta T=1$ terms are much less suppressed.

At any particular bombarding energy, the relative importance of the different terms in v^{eff} as a function of momentum transfer is more reliably given by the magnitudes of the momentum components of the t-matrix (the right hand side of eq.(17)). The results for both natural and unnatural parity transitions using the 140 MeV t-matrix from Table 3 at a bombarding energy of 135 MeV are shown in Fig. 6. The tensor force in Fig. 6 is that derived from the Sussex matrix elements and only its direct terms are shown. The OPEP result is also shown in Fig. 6 for comparison; for small q the OPEP clearly provides most of the vector-isovector strength.

Pickelsimer and Walker[40] have also derived a local effective interaction based on the N-N t-matrix which is very similar in form to that discussed above. Their v^{eff} is somewhat more global than the one given here since it is fitted to N-N amplitudes at bombarding energies between 50 and 400 MeV and has the desirable property of being energy dependent only through the exchange terms. This energy independence is achieved through the superposition of a much larger number of Yukawas than for the interaction discussed above. Preliminary calculations[42] suggest that there may be some important differences between the two interactions and these differences need further study. Although the energy range over which the "fixed energy" t-matrix effective interaction may be used has not been established, preliminary calculations using an analogous effective interaction derived from the Hamada-Johnston phase shifts at 155 MeV gives results very similar to the G3Y interaction derived from 140 MeV phase shifts.

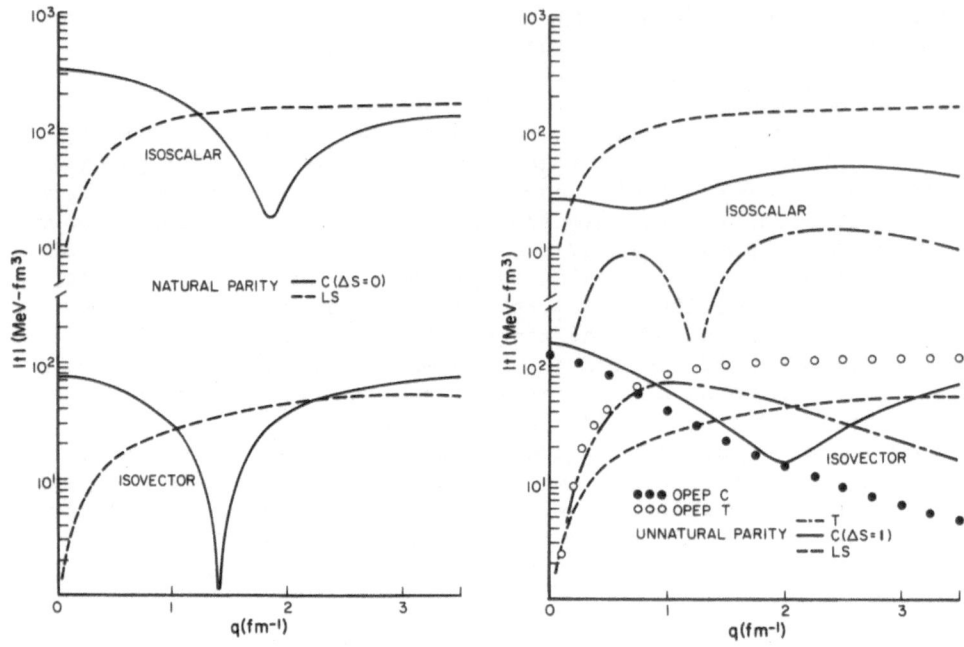

Fig. 6. Momentum-transfer dependence of the magnitude of the G3Y
         interaction near E_p=135 MeV. The knock-on exchange terms
         for the central and spin-orbit (LS) parts of the force
         have been included in the asymptotic energy approximation.
         The tensor (T) interaction includes only the direct terms.

7. APPLICATIONS OF t-MATRIX EFFECTIVE INTERACTIONS: E_p>100 MeV

Due to their relative newness, neither of the t-matrix effective interactions described above has been thoroughly tested. A
number of preliminary results are, however, available for elastic
scattering, inelastic scattering and charge exchange reactions.

Figure 7 shows the elastic scattering cross section for protons
on ^{208}Pb at 182 MeV compared with calculations of Schwandt et al.[42]
who construct the entire optical potential (central and LS parts)
by folding the complex G3Y interaction in Table 3 with the density
distribution for ^{208}Pb. The scattering is then calculated by solving
the Schrodinger equation with this calculated optical potential in
which the exchange terms are included in the asymptotic energy
approximation.[6,7] Although the volume integral of the absorptive
potential is considerably larger than that found phenomenologically,[41]
very reasonable agreement with experiment is obtained for a calculation without any free parameters. The fit is not perfect and as
shown in Fig. 7 it can be improved by a relatively small adjustment
of the imaginary strength together with a more substantial increase

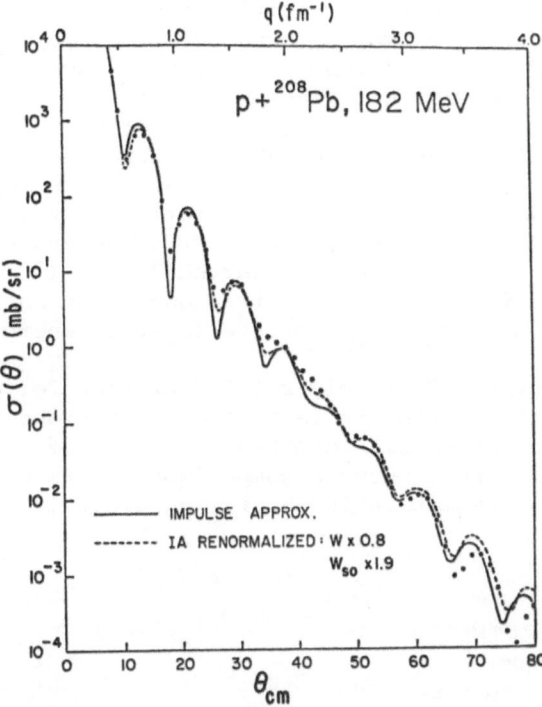

Fig. 7.   Measured elastic scattering cross sections compared with
          calculations using the G3Y interaction at E_p=182 MeV.

in the imaginary spin-orbit interaction. Results in which the
non-local Schrodinger equation is solved exactly are reported[42] to
be forthcoming. Similar results have also been obtained[42] for
p + ^{40}Ca and p + ^{90}Zr. Just as with the M3Y and DDD interactions
at lower energies, the mean-square radius of the calculated optical
potential is somewhat too small and the scattering is sensitive
primarily to the scalar-isoscalar part of u^{eff}.

The G3Y interaction has also been used[31] to calculate the cross
section for excitation of the collective L=3 doublet at E_x=2.64 MeV in
^{207}Pb by 135 MeV protons. It is well established that the unresolved
cross section for this doublet is essentially identical to that for
excitation of the 3^-_1 state in ^{208}Pb at 2.61 MeV. Consequently we
used the $3\hbar\omega$ transition density of Ring and Speth[28] calculated for
the 3^-_1 state in ^{208}Pb. A comparison of the DWIA results compared
with the experimental data of Scott et al.[43] is shown in Fig. 8.
The magnitude of the calculated cross section is in reasonable agreement with the data. The shape is not reproduced in detail, however,
and this will likely be more serious when a comparison of the theoretical and experimental asymmetries can be made. It should be noted

that simply taking the neutron transition density to be N/Z times
that for the protons (which may be obtained from e,e' measurements)
leads to calculated cross sections too large by ~50%. An important
point to be noted from Fig. 8 is the important role of the N-N spin-
orbit interaction for the cross section. At 61 MeV the effects of
v_{LS}^{eff} are much less important. The enhnaced importance of the N-N
spin-orbit interaction with increasing bombarding energy and increasing multipolarity has been discussed earlier;[19] very simply, v_{LS}^{eff}
is proportional to $\vec{r} \times \vec{p}$ and $\vec{p}$ increases with increasing energy.
A similar effect has recently been observed[44] near 1 GeV where,
however, the effect is less dramatic than one might anticipate due
to the relatively small v_{LS}^{eff} appropriate for that energy.

Another important and closely related problem for microscopic
theories of inelastic scattering is that of understanding the collective model deformed-spin-orbit potential in terms of the underlying
N-N interaction. Figure 8 shows a comparison of the partial cross
section for the L=3 transition in Pb arising exclusively from the

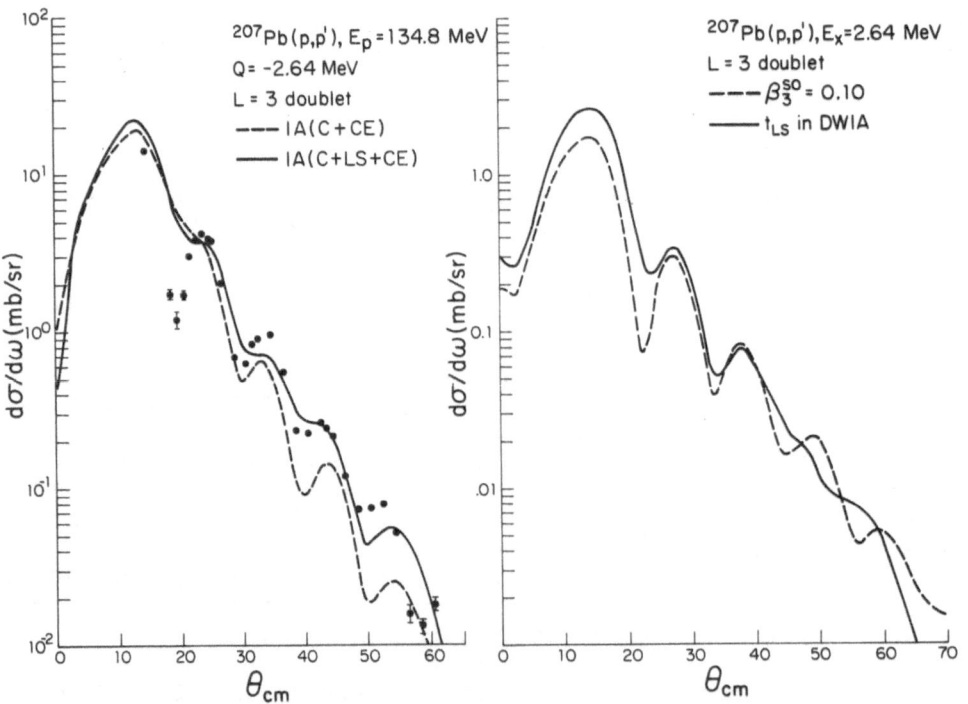

Fig. 8. Comparison of measured and DWIA cross sections for excitation of the octupole doublet in ^{207}Pb at E_x=2.64 MeV by
134.8 MeV protons. On the right is a comparison of the
cross section predicted by the DSO potential with that
predicted by the LS part of the N-N interaction (G3Y).

DSO potential compared with that due to the N-N spin-orbit part of v^{eff}. In addition to yielding quite similar partial cross sections, the microscopic and macroscopic amplitudes are almost completely in phase, establishing a semi-quantitative microscopic basis for the DSO potential, at least for this multipolarity.

The most definitive measure of the $\Delta T=1$ part of v^{eff} (i.e. u_{τ}^{eff}) could likely be obtained from good (p,n) data for a number of ground state analogue transitions. Unfortunately, such data is not yet available at IUCF energies. Future experiments using polarized protons in $(\vec{p},n)$ reactions should prove especially helpful in understanding the effective isovector part (u_{LS}^{eff}) of the spin-orbit component of the interaction.

The inelastic excitation of a number of states in ^{12}C provides an interesting test of the spin and isospin components of the N-N interaction. In particular, DWIA calculations were made at E_p=122 MeV using the G3Y interaction and the Cohen-Kurath[35] (C-K) wavefunctions for excitation of: the 2^+, T=0 state at E_x=4.43 MeV, the 1^+, T=0 state at E_x=12.71 MeV, the 1^+, T=1 state at 15.11 MeV and the 2^+, T=1 state at E_x=16.11 MeV. The calculated cross sections are compared with the experimental results of Comfort et al.[45] in Fig. 9. The 1^+, T=1 state at 15.11 MeV is sensitive to $\Delta S=\Delta T=1$ part of u^{eff} and the calculated cross section is in good agreement with the experimental data without any adjustment. With the C-K wavefunctions the shoulder in the angular distribution is almost entirely attributable to the tensor part of the force. Very recent[46] (e,e') results indicate a deficiency in the C-K transition density corresponding to momentum transfers in the region of this shoulder, suggesting that part of it should be attributed to the central part of v^{eff}; the small momentum transfer part of the (e,e') cross section is well described by the C-K wavefunctions. As at lower energies the transition to the 1^+, T=0 state, which is sensitive to the very weak u_{σ}^{eff}, is very poorly described by the DWIA. As discussed earlier, we have no cross check on this $\Delta T=0$ transition density from (e,e'); moreover, multistep processes may be relatively more important for this more weakly excited state.

Excitation of the 2^+, T=0 state at 4.43 MeV is sensitive, not only to the $\Delta S=\Delta T=0$ part of u^{eff} but also to u_{LS}^{eff}, especially beyond the peak cross section. Inclusion of the spin-orbit part of v^{eff} increases the integrated cross section by nearly 60%. The cross section for this transition is underestimated in the DWIA by a factor of 1.4 which is to be compared with an enhancement factor of 2 required for these wavefunctions to reproduce the correct B(E2) and electron scattering cross section. To see if this is a genuine inconsistency, further study of the sensitivity of these results to the choice of oscillator constant and optical model parameters is necessary.

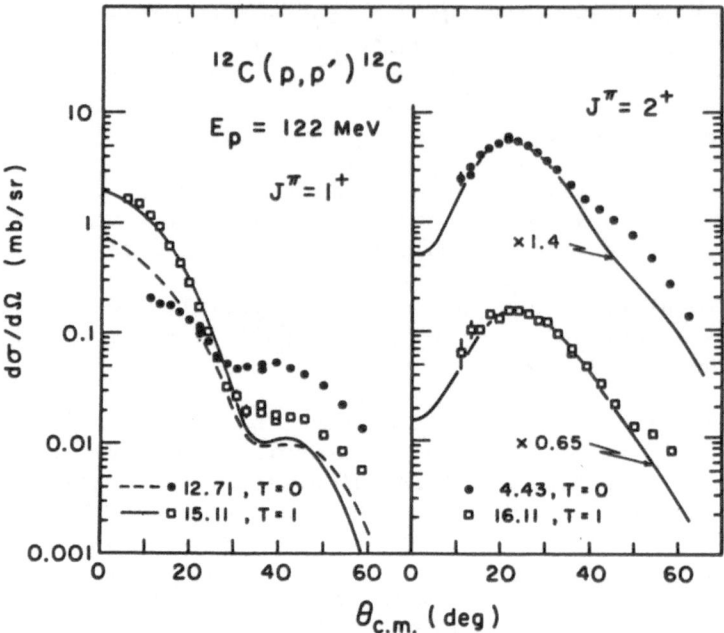

Fig. 9. Measured and calculated cross sections for the inelastic
         excitation of four states in ^{12}C by 122 MeV protons. The
         G3Y interaction was used.

Excitation of the 2^+, T=1 state at 16.11 MeV can proceed via
u_τ^{eff}, $u_{\sigma\tau}^{eff}$ or via the strong isovector part of the tensor force;
the isovector part of the spin-orbit force is quite small. Due to
the smallness of u_τ^{eff}, the reaction is mediated primarily by $u_{\sigma\tau}^{eff}$
and u_T^{eff}. The shape of the calculated cross section is in excellent
agreement with experiment but its magnitude is ~35% too large.
Inelastic electron scattering also suggests that the C-K transition
density may be too large for this transition by ~10-20%.

The inelastic excitation of the 0^+, T=1 state at 2.31 MeV in
^{14}N(p,p') is known[1] to be quite sensitive to the tensor part of
v^{eff}. This results from the extremely small transition density (in
momentum space) for small momentum transfers together with the relatively greater importance of the tensor force at larger momentum
transfers (see Fig. 6). This is an essential feature of any

transition density which is consistent with the strongly inhibited Gammov-Teller matrix element for the analogous β-decay of ^{14}C. This is illustrated in Fig. 10 where the quantity

$$M_{LJ}(q) \equiv \langle J_f || \sum_i j_L(qr_i) \, T_{L1J}(\Omega_i, \vec{\sigma}_i) \tau_{iz} || J_i \rangle \tag{21}$$

is plotted with J=1, L=0,2 using the C-K p-shell wavefunctions as well as those deduced recently[47] from electron scattering. The tensor T_{L1J} is constructed from a spherical harmonic of rank L and the Pauli spin operator;[3] M_{LJ} is proportional to that part of the transverse-magnetic form factor used in electron scattering which is associated with the spin density. The amplitude for each orbital (L) and total (J) angular momentum transfer to the target in (p,p') is proportional to M_{LJ}. In Fig. 10 the (p,p') cross sections obtained by Comfort et al.[45] at E_p=122 MeV are compared with DWIA calculations using the G3Y interaction together with each set of wavefunctions. The cross section associated with the tensor force is roughly 5 times as large as that due to the central force near θ_{cm}=30° (q≈1.2 fm^{-1}). Although the results are not completely satisfactory, the magnitude of the second peak is only given correctly when the wavefunctions deduced from (e,e') are used. From Fig. 10

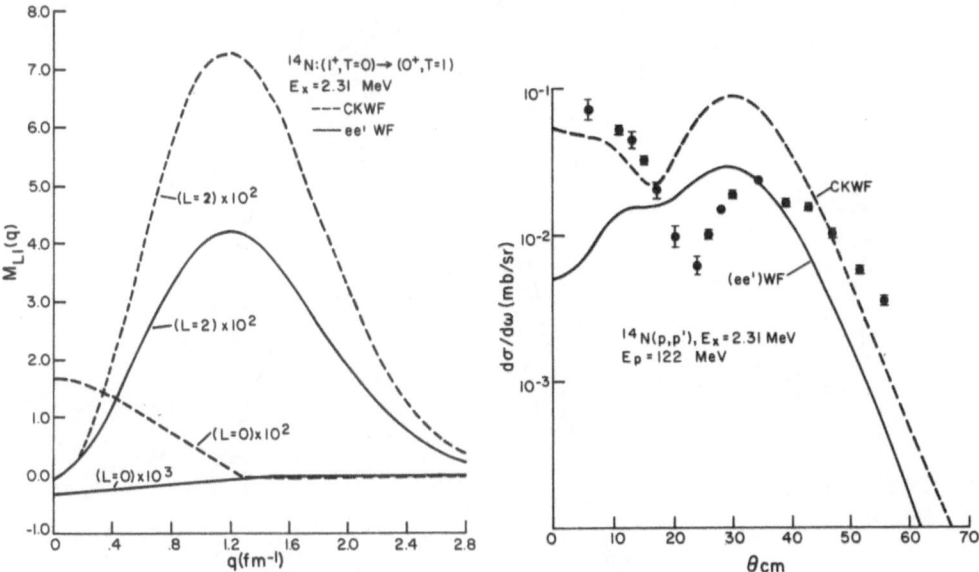

Fig. 10. Fourier transforms of the L=0 and L=2 transition densities for the ^{14}N(g.s.)→$^{14}N(E_x$=2.31 MeV) transition density. CKWF (ee'WF) denotes the wavefunctions of ref. 35 (ref. 47). The corresponding measured and calculated cross sections for the ^{14}N(p,p') reaction at 122 MeV is shown on the right.

we see that this is a consequence of the relative sizes of $M_{21}(q)$ for the two sets of wavefunctions. Of course the C-K wavefunctions also overestimate the form factor deduced from electron scattering. The (p,p') cross section at forward angles is better described by the C-K wavefunctions; however, unlike the peak near 30°, the (p,p') cross section at forward angles has been shown to be quite sensitive to distortion effects. This follows from the inhibition of small momentum transfers indicated in Fig. 10 and suggests that in using this reaction to test effective forces considerably greater weight should be given to describing the peak near $q=1.2$ fm^{-1}. The relative stability of this peak as a function of bombarding energy has been noted by Austin.[48]

For another application quite sensitive to the tensor force we consider the recent results of Bacher et al.[49] at $E_p=135$ MeV for excitation of a state in ^{208}Pb at $E_x=6.72$ MeV which, on the basis of energetics and (e,e') results has been attributed to the $j_{15/2}$ particle-$i_{13/2}$ hole neutron configuration with $J^{\pi}=14^-$. Figure 11 shows the experimental cross section for this state compared with

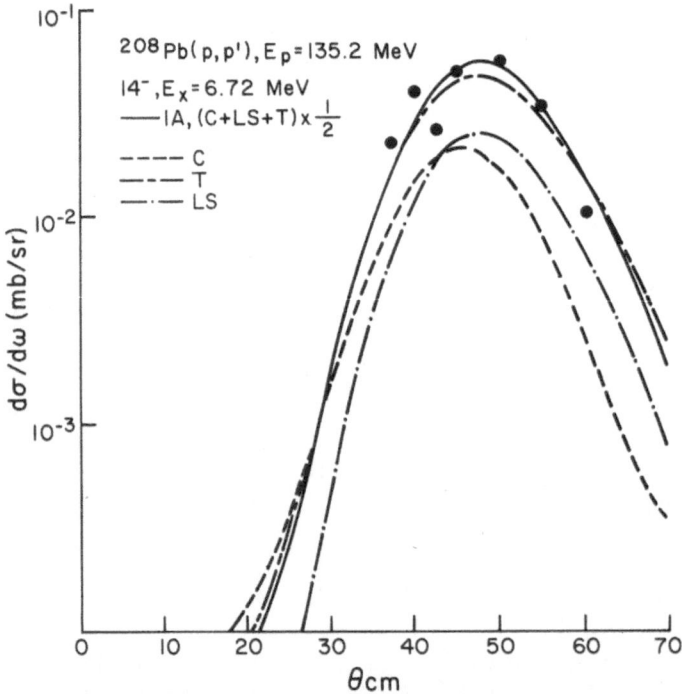

Fig. 11. Experimental and theoretical (DWIA) cross sections for excitation of the 14⁻ state in ^{208}Pb at $E_x=6.72$ MeV by 135 MeV protons.

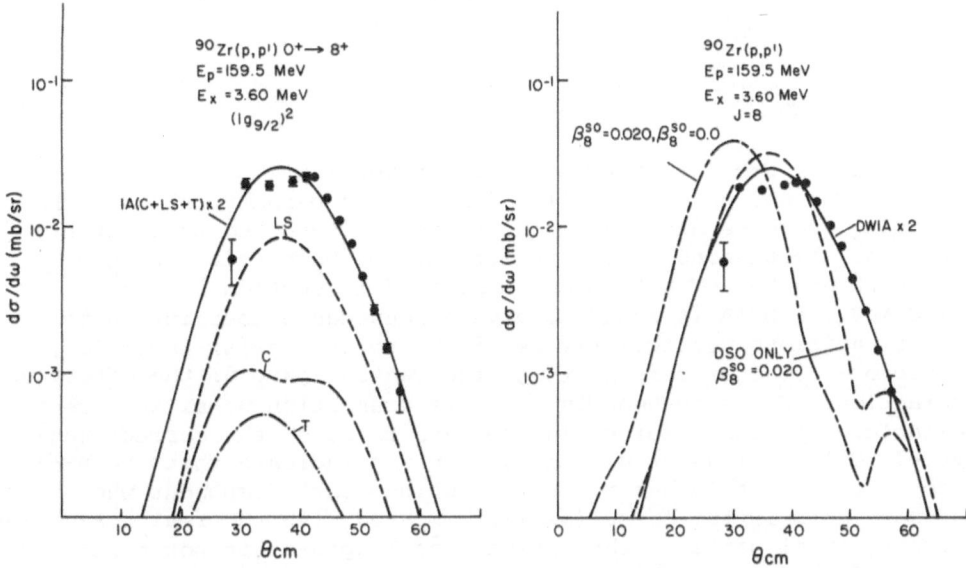

Fig. 12.   Experimental and calculated cross sections for excitation of the 8$^+$ state in ^{90}Zr at E_x=3.6 MeV by 159.5 MeV protons.

DWIA calculations using this simple wave function. This is a particular example of a stretched configuration,[50] i.e., of the form $J=j_p+j_h$ with $j_p=\ell_p+\frac{1}{2}$ and $j_h=\ell_h+\frac{1}{2}$. For this case only form factors of the type M_{J-1J} in eq.(20) enter in the direct terms and since the reaction is dominated by momentum transfers near 2 fm^{-1}, the tensor force term dominates. Although the cross section is over-predicted by a factor of 2, this is consistent with the calculated (e,e') cross section which has to be multiplied by ~0.6 to match experiment.

For an application extremely sensitive[19] to the N-N spin-orbit interaction we consider the recent data of Scott et al.[51] for the inelastic excitation of the 8_1^+ state in ^{90}Zr at E_x=3.6 MeV using 159.5 MeV protons. In the simplest shell-model picture the 8$^+$ state is formed by simply recoupling two $1g_{9/2}$ protons. This configuration excludes[19] the participation of the tensor force in the direct terms; moreover, the short range repulsive part of the interaction is quite effective in cancelling (see Fig. 6) the long range attraction for the observed momentum transfers (q≃1.9 fm^{-1}) leaving an unusually large N-N spin-orbit contribution as shown in Fig. 12. The underestimate of the cross section may be due to the omission of core polarization effects. In collective model calculations for this transition (see Fig. 12) Scott et al.[51] had previously found that the optimum fit to the shape of the cross section is obtained when the DSO contribution greatly exceeds that of the deformed central potential $(\beta_{SO} \gg \beta_N)$.

8.   CONCLUSIONS

Effective N-N interactions based on realistic N-N potentials
at low energy and on phase shifts (and mixing parameters) at higher
bombarding energies have been constructed and applied to a variety
of nuclear transitions.   Results for nucleon bombarding energies
between 10 and 65 MeV have been summarized in section 5.   For
$E_p \gtrsim 100$ MeV the DWIA is found to give a reasonable description of
those transitions particularly sensitive to the scalar-isoscalar,
and vector-isovector components of the central part of the effective
interaction.   Where reasonably complete transition densities are
available, the tensor force also appears to be of the correct magnitude.   Calculations of elastic scattering indicate that the v^{eff}
based on the t-matrix has a central (spin-orbit) component whose
imaginary part is too large (small) compared with the real term.   The
calculated real parts of the optical model spin-orbit potential and
of the deformed spin-orbit potential are found to be in good agreement with those found empirically.   The scalar-isovector part of the
interaction has not been tested but is predicted to be much smaller
than at lower energies.   As at lower energies that part of v^{eff}
important for $\Delta S=1$, $\Delta T=0$ transitions is predicted to be extremely
small.

Although the impulse approximation does not work perfectly near
$E_p=140$ MeV (and this will likely be more conspicuous when asymmetry
data are confronted) it is relatively simple to apply and appears
to provide at least a semi-quantitative scheme for the calculation
of nucleon-nucleus scattering processes at intermediate energies.

ACKNOWLEDGMENTS

I have had the pleasure and benefit of numerous stimulating
discussions with F. Petrovich and G. R. Satchler regarding much of
this material.   I would also like to thank numerous experimentalists
for the use of their data, some of it prior to publication.   This
work was supported in part by the National Science Foundation.

REFERENCES

1.   S. M. Austin, in: "The Two-Body Force in Nuclei," ed. S. M.
         Austin and G. M. Crawley, Plenum, New York (1972).
2.   J. Atkinson, and V. A. Madsen, Phys. Rev. C1:1377 (1970).
3.   W. G. Love and G. R. Satchler, Nucl. Phys. A159:1 (1970).
4.   R. Schaeffer, Nucl. Phys. A132:186 (1969).
5.   H. V. Geramb and K. Amos, Nucl. Phys. A163:337 (1971).

6. F. Petrovich, H. McManus, V. A. Madsen and J. Atkinson, Phys. Rev. Lett. 22:895 (1969).

7. W. G. Love, Part. and Nuclei 3:318 (1972); Nucl. Phys. A312: 160 (1978).

8. M. H. MacGregor, R. A. Arndt and R. M. Wright, Phys. Rev. 182: 1714 (1969).

9. A. K. Kerman, H. McManus and R. M. Thaler, Ann. Phys. 8:551 (1959).

10. G. E. Brown,"Unified Theory of Nuclear Models and Forces," John Wiley & Sons, New York (1967); B. L. Scott and S. A. Moszkowski, Ann. of Phys. 14:107 (1961).

11. J. P. Jeukenne, A. Lejeune and C. Mahaux, Phys. Rev. C16:80 (1977).

12. F. A. Brieva and J. R. Rook, Nucl. Phys. A297:206 (1978); F. A. Brieva, H. V. Geramb and J. R. Rook, to be published.

13. G. Bertsch, J. Borysowicz, H. McManus and W. G. Love, Nucl. Phys. A284:399 (1977).

14. G. R. Satchler and W. G. Love, Phys. Lett. 65B:415 (1976).

15. C. Wong, J. D. Anderson, V. A. Madsen, F. A. Schmittroth and M. J. Stomp, Phys. Rev. C3:1904 (1971).

16. W. G. Love and L. J. Parish, Nucl. Phys. A157:625 (1970).

17. J. P. Elliot, A. D. Jackson, H. A. Mavromatis, E. A. Sanderson and B. Singh, Nucl. Phys. A121:241 (1968).

18. R. Schaeffer and J. Raynal, unpublished.

19. W. G. Love, Phys. Lett 35B:371 (1971); Nucl. Phys. A192:49 (1972).

20. J. M. Moss, W. D. Cornelius and D. R. Brown, Phys. Rev. Lett. 41:930 (1978).

21. W. T. H. Van Oers and H. Haw, Phys. Lett. 45B:227 (1973).

22. Y. Eisen and B. Day, Phys. Lett 63B:253 (1976).

23. G. R. Satchler and W. G. Love, to be published.

24. F. D. Becchetti and G. W. Greenlees, Phys. Rev. 182:1190 (1969).

25. F. Petrovich in: "Microscopic Optical Potentials," vol. 89, Springer-Verlag, New York (1979).

26. D. E. Bainum, R. W. Finlay, J. Rapaport, J. D. Carlson and W. G. Love, Phys. Rev. C16:1377 (1977).

27. Alan Scott, M. Owais and W. G. Love, Nucl. Phys. A289:123 (1977).

28. P. Ring and J. Speth, Nucl. Phys. A235:315 (1974).

29. E. C. Halbert and G. R. Satchler, Nucl. Phys. A233:265 (1974).

30. W. G. Love, Phys. Rev. C15:1261 (1977).

31. S. D. Schery, D. A. Lind and Howard Weiman, Phys. Rev. C14:1800 (1976).

32. J. M. Moss, C. Brassard, R. Vyse and J. Gosset, Phys. Rev. C6:168 (1972); J. Gosset, B. Mayer and J. L. Escudie, Phys. Rev. C14:878 (1976).

33. J. D. Carlson, C. D. Zafiratos and D. A. Lind, Nucl. Phys. A249:29 (1975).

34. S. D. Schery, S. M. Austin, A. Galonsky, L. E. Young and U. E. P. Berg, Phys. Lett. 79B:30 (1978).

35. S. Cohen and D. Kurath, Nucl. Phys. 73:11 (1965).

36. B. D. Anderson, D. E. Bainum, A. Baldwin, C. C. Foster, C. D. Goodman, C. A. Goulding, M. B. Greenfield, J. N. Knudson, R. Madey, J. Rapaport and T. R. Whitten, IUCF Techn. and Scient. Report, p. 117 (1978) and private communication.

37. R. H. Howell, F. S. Dietrich, D. W. Heikkinen and F. Petrovich, to be published.

38. L. S. Rodberg and R. M. Thaler, "The Quantum Theory of Scattering," Academic Press, New York (1967); M. L. Goldberger and K. M. Watson, "Collision Theory," John Wiley & Sons, New York (1964).

39. W. G. Love, Alan Scott, F. Todd Baker, W. P. Jones and J. D. Wiggins, Jr., Phys. Lett. 73B:277 (1978).

40. A. Picklesimer and G. Walker, Phys. Rev. C17:237 (1978).

41. P. Schwandt, A. Nadasen, P. P. Singh, M. D. Kaitchuck, W. W. Jacobs, J. Meek, A. D. Bacher and P. T. Debevec, IUCF Techn. and Scient. Report, p. 79 (1978).

42. P. Schwandt, F. Petrovich, W. G. Love and A. Picklesimer, to be published.

43. Alan Scott, F. Todd Baker, W. G. Love, W. P. Jones and J. D. Wiggins, Jr., to be published.

44. R. P. Liljestrand, G. S. Blanpied, W. R. Coker, C. Harvey, G. W. Hoffmann, L. Ray, C. Glashausser, G. S. Adams, T. S. Bauer, G. Igo, G. Pauletta, C. A. Whitten, Jr., M. A. Oothoudt, B. E. Wood and H. Mann, Phys. Rev. Lett. 42:363 (1979).

45. J. R. Comfort, S. Austin, P. Debevec, G. Moake, R. Finlay and W. G. Love, IUCF Techn. and Scient. Report, p. 91 (1978); J. R. Comfort in: "Proc. International Symposium on Nuclear Direct Reaction Mechanisms," Fukuoka, Japan, Oct. 25-28 (1978), to be published.

46. "Proc. International Symposium on Nuclear Direct Reaction Mechanisms," Fukuoka, Japan, Oct. 25-28 (1978), to be published.

47. N. Ensslin, W. Bertozzi, S. Kowalski, C. P. Sargent, W. Turchinetz, C. F. Williamson, S. P. Fivozinsky, J. W. Lightbody, Jr. and S. Penner, Phys. Rev. C9:1705 (1974).

48. S. M. Austin, private communication.

49. A. D. Bacher, G. T. Emery, W. P. Jones, D. W. Miller, G. S. Adams, F. Petrovich and W. G. Love, Bull. Am. Phys. Soc. 23:945 (1978) and to be published.

50. R. A. Lindgren, W. J. Gerace, A. D. Bacher, W. G. Love and F. Petrovich, to be published.

51. Alan Scott, F. Todd Baker, R. C. Styles, W. P. Jones and J. D. Wiggins, Jr., Bull. Am. Phys. Soc. 23:13 (1978) and private communication.

DISCUSSION

Rapaport: Two groups in Europe (Mahaux and Lejeune in Belgium and Brieva and Rook in England) have done microscopic calculations to obtain the real and imaginary terms of the effective interaction. How do their real terms compare with your real effective interactions?

Love: For the interaction of Brieva and Rook see fig. 1; I have made no similar comparison for the Mahaux-Lejeunne interaction. The M3Y interaction is density independent and roughly matches the even state DDD interaction near $\rho/\rho_0 = \frac{1}{3}$.

DETERMINATION OF THE EFFECTIVE INTERACTION

FOR BOUND STATE CALCULATIONS IN NUCLEI

Bruce R. Barrett

Department of Physics
University of Arizona
Tucson, Arizona   85721

INTRODUCTION

Although this is a conference principally on the interaction between a free nucleon and a nucleon bound in a nucleus, it is perhaps appropriate at the beginning to survey the present status of calculating microscopically the effective interaction among nucleons bound inside a nucleus. The basic theory, which relates the free nucleon-nucleon interaction to the effective interaction appropriate for shell-model calculations, is now well-established and will not be repeated here. Readers interested in the details of this theory are referred to several excellent review articles on this subject.[1-4] Length restrictions do not permit a complete list of references in the present paper.

The effective interaction $\mathcal{V}$ for <u>any</u> <u>number</u> of valence nucleons outside of the closed-shell core is given by

$$\mathcal{V} = \left\{ V + V \frac{Q}{E_o^V - H_o^V} \mathcal{V} \right\}_{\substack{\text{totally linked valence} \\ \text{terms only}}} \tag{1}$$

In Eq. (1) V denotes the free nucleon-nucleon interaction which is the principal part of the interaction Hamiltonian, Q is the projection operator which allows only intermediate states outside the shell-model space, and H_o^V is the unperturbed many-body Hamiltonian for the intermediate states. The subscript in Eq. (1) indicates that only totally linked valence terms are to appear in the expansion for $\mathcal{V}$. Consequently, the energy denominator contains the unperturbed valence energies E_o^v, but the matrix elements of $\mathcal{V}$ form a non-hermitian matrix, because of the folded diagrams which appear in the totally linked expansion.

The series expansion for $\mathcal{V}$ can be regrouped in terms of the reaction matrix G,

$$\mathcal{V} = \left\{ G + G \frac{Q'}{E_o^v - H_o^v} \mathcal{V} \right\}_{\substack{\text{totally linked valence} \\ \text{terms only}}} \tag{2}$$

in order to treat the strong, short-range correlations produced by V. The reaction matrix is defined by[5]

$$G(\omega) = V + V \frac{Q_{2p}}{\omega - H_o'} G(\omega) \tag{3}$$

where the dependence of G on the starting energy ω has been explicitly indicated and H_o' is the unperturbed Hamiltonian for the two-particle intermediate states contributing to G. The operator Q_{2p} represents the two-particle projection operator which defines the physically allowed intermediate states for G. Consequently, in Eq. (2) the prime on Q indicates that ladders of G matrices are forbidden.

The problem of calculating the G matrix from V has been extensively studied and discussed in the literature,[5] and elaborate and accurate methods for computing the two-body matrix elements of G now exist,[6,7] but will not be reviewed here.

Although Eqs. (1) and (2) are completely general and hold for any number of valence particles, large-scale calculations have been performed for only two valence particles. More extensive calculations of $\mathcal{V}$ for three or more particles would be of interest and worth doing, since they have the possibility of telling us something about the effective three- and more-body forces in nuclei.

CALCULATING THE EFFECTIVE TWO-PARTICLE INTERACTION

Historically, the principal method that has been employed for calculating the effective two-particle interaction $\mathcal{V}_{2p}$ (note the subscript, since $\mathcal{V}$ in its general form is a many-body operator) is a perturbation-series expansion in orders of G. The successes and failures of the early perturbation-theory calculations are well-known.[1-4] The problems still facing perturbation-theory calculations will now be considered, followed by a brief discussion of alternate approaches for computing $\mathcal{V}_{2p}$.

Convergence Properties of the Effective Two-Particle Interaction

When the calculation of $\mathcal{V}_{2p}$ is attempted for a given nucleus using the perturbation-theory approach, two convergence problems immediately appear. First, does the perturbation-theory expansion for $\mathcal{V}_{2p}$ converge order by order in G? Secondly, does the intermediate-state summation for each individual term in the perturbation expansion converge? Let us look at these problems in detail one at a time.

Convergence Properties Order by Order in G. The general convergence properties of an operator, similar to Eq. (1), were investigated by Schucan and Weidenmüller,[8] who demonstrated that the perturbation expansion for $\mathcal{V}$ will, in general, diverge due to the overlap in energy of states in the shell-model space with states outside the shell-model space. The assumption of no overlap among states in these two spaces was an important part of the derivation of Eq. (1).[1-4] States outside the shell-model space are defined by Q. States in the Q space which have energies in the same energy range as shell-model states are referred to as "intruder states." Since the intruder-state phenomenon appears to exist in the spectra of all nuclei, we must expect the perturbation expansion for $\mathcal{V}$ to diverge, although the expansion may be asymptotically convergent.

Hofmann et al.[9] have made a systematic study of how to handle the intruder-state divergence problem, mainly by using Padé approximants or by explicitly including the intruder states in the model space. In particular, Hofmann[10] has shown that the Padé approximants will generally converge to that branch of the effective interaction which reproduces the model space states. Unfortunately, Padé approximants of a rather high order are required to obtain an accurate result, and computing $\mathcal{V}_{2p}$ beyond third order appears impossible at this time.

Much formal work has also been carried out to find theoretical techniques for predicting how badly the perturbation series for $\mathcal{V}_{2p}$ diverges and to find upper and lower bounds on the eigenenergies for any given order of perturbation theory. The major work here has been done by Vincent and coworkers,[11] and their results so far are encouraging.

Convergence Properties of the Intermediate-State Summation. Vary et al.[12] were the first to demonstrate in detail that the intermediate-state summation in the second-order (in G) core-polarization term[1-4] does not converge rapidly because of the tensor component of the nucleon-nucleon force. However, convergence by excitations of 10 $\hbar\omega$ (where $\hbar\omega$ is the energy of the harmonic-oscillator level spacing) occurs for the second-order core-polarization term, because of a natural cut-off on momentum for this term. This so-

called Vary-Sauer-Wong effect for second order has been further
investigated and substantiated in later work.[13]

Sandel et al.[14] have shown that the intermediate-state summation also converges by 6 or 8 $\hbar\omega$ for most third order (in G) terms,
due to a natural momentum cut-off and to energy-denominator damping.
The problem is that not all third-order terms have a momentum cut-off
restriction, and hence their intermediate-state summations do not
seem to converge until very high excitation energies.[15] This appears
to be a hard-core effect rather than a tensor-force effect. For
these particular terms some kind of infinite-partial-summation technique will probably be required in order to obtain the converged
result.

Other Problems in Computing $\mathscr{V}_{2p}$. There are other difficulties
in calculating $\mathscr{V}_{2p}$ besides the convergence problems, such as choosing the best single-particle basis for the calculation, the spurious
center-of-mass motion problem, and the importance of three- and
higher-body forces. It appears that using a Hartree-Fock basis will
improve the convergence properties of the perturbation expansion[16]
for $\mathscr{V}_{2p}$ but will not make the expansion convergent.[17]

Kirson and Starkand[18] have shown that the spurious center-of-
mass effects in computing the terms contributing to $\mathscr{V}_{2p}$ can be
large, of the order of hundreds of keV. This is a problem, which
must be treated accurately in future calculations of $\mathscr{V}_{2p}$.

Finally, Coon et al.[19] have calculated the contribution of a
true three-nucleon force to $\mathscr{V}_{2p}$ and have found its contribution to
be rather small. There is also the separate problem of effective
three-body (and higher-body) forces when three or more valence particles are present. In these cases investigations by Barrett et
al.[20] have shown that these higher-body effective forces must be included in $\mathscr{V}$, if one wants to obtain accurate agreement with experimental data.

Summary Concerning Convergence Properties of $\mathscr{V}_{2p}$.

The perturbation expansion for $\mathscr{V}_{2p}$ diverges because of the intruder-state problem. However, the expansion does appear to be
asymptotically convergent, so that one can estimate the "converged"
result, which is probably well approximated by the sum of the first
plus second plus third order terms in $\mathscr{V}_{2p}$. Using these perturbation-
theory results, one can also employ Padé approximants to predict the
converged results, although the [2,1] Padé approximant given by
third-order calculations of $\mathscr{V}_{2p}$ is not of high enough order to yield
a reasonably accurate result.

In any case, one wants to compute the first, second and third
order terms in $\mathscr{V}_{2p}$ as accurately as possible. There is the inter-

mediate-state summation problem, but it can probably be handled through third order, but not in higher orders. There are other problems, such as the correct choice of single-particle basis, the spurious center-of-mass motion problem and the importance of true three- and higher-body forces. These problems must be accurately treated in future calculations of $\mathscr{V}_{2p}$.

Other Techniques for Computing the Effective Two-Particle Interaction

Several other techniques have been used to solve for $\mathscr{V}_{2p}$ besides perturbation theory. There is the Q box method of Kuo and collaborators,[2] which first calculates the non-folded diagrams using a perturbation-series expansion. This sum is called the Q box. The folded diagrams are then calculated by taking derivatives of the Q box. Obviously, this technique has convergence problems similar to those for the ordinary perturbation-theory expansion.

Attempts have been made to determine the major processes contributing to $\mathscr{V}_{2p}$ by carrying out infinite partial summations[21] of selected terms in $\mathscr{V}_{2p}$ (similar to the two-particle ladder summation which makes up the G matrix). On the other hand, it is not obvious that the contribution of terms not included in these infinite partial summations is unimportant.

The idea of solving for $\mathscr{V}_{2p}$ by iteration has been intriguing for some time. Lee and Suzuki[22] have recently suggested possible iteration schemes for computing $\mathscr{V}_{2p}$, but these ideas need to be studied and developed in more detail.

Finally, Chang[23] has developed equations for $\mathscr{V}_{2p}$ using spectral distribution methods, but no numerical calculations of $\mathscr{V}_{2p}$ using this technique have been reported so far.

The main problem with all these other techniques is that they do not indicate specifically from where $\mathscr{V}_{2p}$ comes. They simply provide a final result for the matrix elements of $\mathscr{V}_{2p}$ (as the least-square-fit method does). Hence, one has no real feeling for the physics behind $\mathscr{V}_{2p}$. Thus, we still need to use perturbation theory to tell us the physics behind the origin of $\mathscr{V}_{2p}$. Once we understand the physics, then we can use the more powerful mathematical techniques of Padé approximants or of iteration methods to obtain an accurate value for $\mathscr{V}_{2p}$.

THE CALCULATION OF NUCLEAR PROPERTIES

Dalton et al.[23] have used effective two-particle interactions computed in perturbation theory to determine the first two moments of the total nuclear Hamiltonian. They first use these moments to determine the shell-model level densities and then employ these

level densities to compute the binding energies of nuclei throughout
the entire sd shell with rather impressive success. Their technique
is referred to as the moment method.

A very successful method for computing the energy spectra of
finite nuclei is the e^S method of Coester and Kümmel as applied by
Zabolitzky.[25] This technique does not calculate individual terms in
ψ but computes the perturbed wavefunction for the fully interacting
system to fourth order in the number of correlated particles. This
wavefunction is then used to evaluate the expectation value of the
nuclear Hamiltonian and thereby to obtain the energy spectrum. At
the present time this technique appears to be the most accurate one
available for computing the spectra of finite nuclei. Unfortunately,
the results obtained do not agree very well with the experimental
data, even when the effects of a true three-nucleon force are included. Since this technique appears to be correct and accurate,
we must conclude that the fault lies with the free nucleon-nucleon
interaction and that we need better theories of this force, than now
exist.

REFERENCES

1. B. R. Barrett and M. W. Kirson, in "Advances in Nuclear Physics,"
    Vol. 6, M. Baranger and E. Vogt, ed., (Plenum, New York,
    1973) p. 219.
2. T. T. S. Kuo, Ann. Rev. Nucl. Sci. 24, 101 (1974).
3. B. R. Barrett, "Proceedings of the International Topical Con-
    ference on Effective Interactions and Operators in Nuclei,"
    Lecture Notes in Physics, Vol. 40, (Springer, Berlin, 1975).
4. P. J. Ellis and E. Osnes, Rev. Mod. Phys. 49, 777 (1977).
5. R. L. Becker, in "Lecture Notes in Physics," Vol. 40, B. R.
    Barrett, ed., (Springer, Berlin, 1975) p. 96.
6. B. R. Barrett, R. G. L. Hewitt and R. J. McCarthy, Phys. Rev.
    C3, 1137 (1971).
7. E. M. Krenciglowa, C. L. Kung, T. T. S. Kuo and E. Osnes, Ann.
    Phys. (N.Y.) 101, 154 (1976).
8. T. H. Schucan and H. A. Weidenmüller, Ann. Phys. (N.Y.) 73, 108
    (1972); 76, 483 (1973).
9. H. M. Hofmann, S. Y. Lee, J. Richert, H. A. Weidenmüller and
    T. H. Schucan, Phys. Lett. 45B, 421 (1973 and Ann. Phys.
    (N.Y.) 85, 410 (1974).
10. H. M. Hofmann, Ann. Phys. (N.Y.) 113, 400 (1978).
11. S. Pittel, C. M. Vincent and J. D. Vergados, Phys. Rev. C13,
    412 (1976).
    C. M. Vincent, Phys. Rev. C14, 660 (1976).
    F. Darema-Rogers and C. M. Vincent, Phys. Rev. C17, 1461 (1978).
12. J. P. Vary, P. U. Sauer and C. W. Wong, Phys. Rev. C7, 1776
    (1973).

13. M. S. Sandel, R. J. McCarthy and B. R. Barrett, Z. Phys. A280,
    259 (1977).
    J. P. Vary and S. N. Yang, Phys. Rev. C15, 55 (1977).
    C. L. Kung, T. T. S. Kuo and K. F. Ratcliff, Phys. Rev. C19,
    (1979).
14. M. S. Sandel, R. J. McCarthy, B. R. Barrett and J. P. Vary,
    Phys. Rev. C17, 777 (1978).
15. P. R. Goode and B. R. Barrett, Phys. Rev. C17, 1848 (1978).
    P. R. Goode, B. R. Barrett and O. Portilho, Phys. Rev. C19, 542
    (1979).
16. J. M. Leinaas and T. T. S. Kuo, Phys. Lett. 62B, 275 (1976).
17. P. R. Goode and B. R. Barrett, Phys. Lett. 68B, 77 (1977).
18. M. W. Kirson and Y. Starkand, Centre-of-Mass Motion and Micro-
    scopic Nuclear Effective Interactions, in Weizmann Institute
    of Science preprint WIS-78/15 (1978).
19. S. A. Coon, R. J. McCarthy and C. P. Malta, J. Phys. G: Nucl.
    Phys. 4, 183 (1978).
20. B. R. Barrett, E. C. Halbert and J. C. McGrory, Ann. Phys.
    (N.Y.) 90, 321 (1975).
21. D. W. L. Sprung, in "Lecture Notes in Physics," Vol. 40, B. R.
    Barrett, ed., (Springer, Berlin, 1975) p. 207.
22. S. Y. Lee and K. Suzuki, The Effective Interaction of Two
    Nucleons in the sd Shell, in SUNY at Stony Brook preprint
    (1979).
23. B. D. Chang, Nucl. Phys. A304, 127 (1978).
24. B. J. Dalton, J. P. Vary and W. J. Baldridge, Phys. Rev. Lett.
    38, 1348 (1977).
    B. J. Dalton, W. J. Baldridge and J. P. Vary, Realistic
    Effective Interactions for the sd Shell: Global and Dis-
    crete Nuclear Properties, in Ames Laboratory preprint
    (1979).
25. H. Kümmel, K. H. Lührmann and J. G. Zabolitzky, Phys. Rep.
    38C, 1 (1978).
    J. G. Zabolitzky and E. Ey, Theory of the Effective Interaction
    in Nuclei, in Ruhr-Universität Bochum preprint (1978).

LINEAR LEAST-SQUARES ANALYSIS AND EFFECTIVE INTERACTIONS

S. D. Bloom*

Lawrence Livermore Laboratory and the Department of
Applied Science
University of California (Davis/Livermore)
Livermore, California 94550

R. F. Hausman, Jr.

Los Alamos Scientific Laboratory
Los Alamos, New Mexico

Jon Larsen*

Lawrence Livermore Laboratory
Livermore, California 94550

In the following we exhibit a few results from the application
of the method of singular value analysis (SVA), a particular form
of linear least-squares analysis,[1] to two sets of experimental
two-body matrix elements, the 63 Chung-Wildenthal (CW) matrix
elements[2] for the 1d5/2, 2s1/2, 1d3/2 shells (A $\approx$ 20-40) and the
best 84 matrix elements from the work of Schiffer and True
(ST).[3] The latter cover the atomic weight range of A $\approx$ 40 to 200
and nine shells from 1d3/2 to 1i13/2. Our purpose is the
extraction of a first-order set of strength parameters for an
effective interaction of minimum rank and simple form for each set
or both sets of data. The ranges, being non-linear in their
effects, are investigated separately. In this work we consider
mainly local potential forms. (This is not a limitation of the
approach.) The SVA method is well-suited to this problem since it

---

*Work performed under the auspices of the U.S. Department of
Energy by the Lawrence Livermore Laboratory under contract No.
W-7405-ENG-48.

was developed for and extensively applied to the analysis of satellite trajectories and the like,[1] where similar over-determined problems arise.

Several of the powerful features of SVA may be shown via the following over-determined problem,

$$A_{mn} x_n \approx b_m \; ; \; m > n \qquad\qquad (1)$$

where A is an m by n matrix, x is a vector of n unknowns, and b is a vector of m data points. In our case A is the array of theoretical two-body matrix elements, of which there are m for each term of the potential form being studied, and b consists of the m two-body matrix elements, CW or ST. The usual solution to this classic problem is the minimization of the Euclidean norm of r, the m-dimensional residual vector, defined by

$$r \approx Ax - b \qquad\qquad (2)$$

which yields the least-squares result. The SVA, while ultimately giving the standard result, also yields an ordered but non-unique sequence of solutions consisting of linear combinations of the model force components which contribute independently to the squared r-norm $||r||^2$. The judicious choice of the character and cut-off of this sequence will in general yield a more realistic and far different solution than the full least-squares result, as we shall show. This cut-off is essentially the same as the effective rank of the model matrix A, which is determined by the quasi-diagonal singular-value matrix.[1] The effective rank or cut-off number (c) is the best approximation for the number of independent potential parameters.

Using SVA one may determine the minimum set of decoupled potential parameters necessary to produce the "best" result, i.e. the result at cut-off. Practically, this is done by examining how $||r||^2$ approaches $||r_n||^2$, where n is the maximum possible number of independent potential parameters and $||r_n||^2$ is the minimum or least-squares result. In the present work we have set cut-off at the 1% level, i.e.,

$$\rho_c^2 - \rho_n^2 \approx 1\%; \; c \leq n \qquad\qquad (3a)$$

$$\rho^2 \equiv ||r||^2 / ||b||^2 \; (\text{i.e. } \rho^2 \equiv 1 \text{ for no interaction}) \qquad (3b)$$

where c denotes the solution-sequence number at cut-off. The SVA method takes advantage of the non-uniqueness of the path by which ρ_n^2 is approached by arranging the initial weighting so that the coupling between potential terms is minimized or even eliminated. This has the double advantage of simplifying the form of the best result as well as making it possible to compare different models and data sets in a meaningful way. (This will be illustrated shortly for the CW and ST matrix element sets.) These features are essential to making a judgment as to the general tenability or "universality" of an effective interaction.

In Figs. 1 and 2 we show the results of the SVA for ρ^2 in the case of the 63 CW and the (selected) 84 ST matrix elements for several different force models. (The analysis of the latter did not include the tensor force.) Each figure displays the fractional effect of the individual major components on ρ^2. The definition for the symbols are as follows: $(S,T) \equiv$ (singlet, triplet); $(O,E) \equiv$ (odd, even); $TT \equiv$ tensor; $(Y,G) \equiv$ (Yukawa, gaussian); $SD \equiv$ surface-delta interaction. The symbols CY(4), CY(8), SD(2) mean 4-term central potential (Yukawa), 8-term central-potential (Yukawa), surface delta interaction for $T = 0$ and $T = 1$, etc.

Although our analysis is not exhaustive of all force forms certain constant features of Fig. 1 and 2 are evident and clearly independent of the potential forms employed here. For the CW analyses we note the following: First, $\Delta\rho^2$, the change in ρ^2 associated with each central term, is quite constant and ordered as follows: $\Delta\rho^2 = 70\%, 15\%, 5\%, 3.5\%$ for the TE, SE, TO, SO terms. Thus the central terms account for about 95% of the reduction in ρ^2. Second, the improvement of an 8-parameter or two-range central force over a one-range or 4-parameter central force, even with the addition of the SDI, is well below the 1% level. Third the effect of the tensor force is quite noticeable and, we believe, denotes a real effect, because of its large fractional effect on the residual ρ^2. For the ST matrix elements, where our analysis is not yet complete, only the first two terms, TE and SE, replicate the behavior observed in the CW analysis. They account for 90% of the squared r-norm. Beyond this point much more complicated force forms are required, as shown in Fig. 2 for the CY(8) plot and discussed below.

In Tables 1 and 2 we exhibit the parameter values given by the SVA. These bear out the implications of Figs. 1 and 2 as well as our earlier remarks. We discuss first the CW results of Table 1. In this case the central force parameters at cut-off are substantially the same in the SVA of the CY(2) and CY(4) models as well as the CY(4) model plus either the SDI or tensor-force additions. The same consistency pertains to the CY(8) parameters as well as the tensor parameters TTE and TTO (these are characterized by gaussian form factors). The CY(8) models require only 4 central potential terms, as can be seen from Figs. 1 and 2. Nonetheless the two-range formulation clearly shows the strongly repulsive character of the long-range TO component, as noted by Schiffer and True.[3] For the CW matrix elements, at least, no significant values for an SDI interaction could be

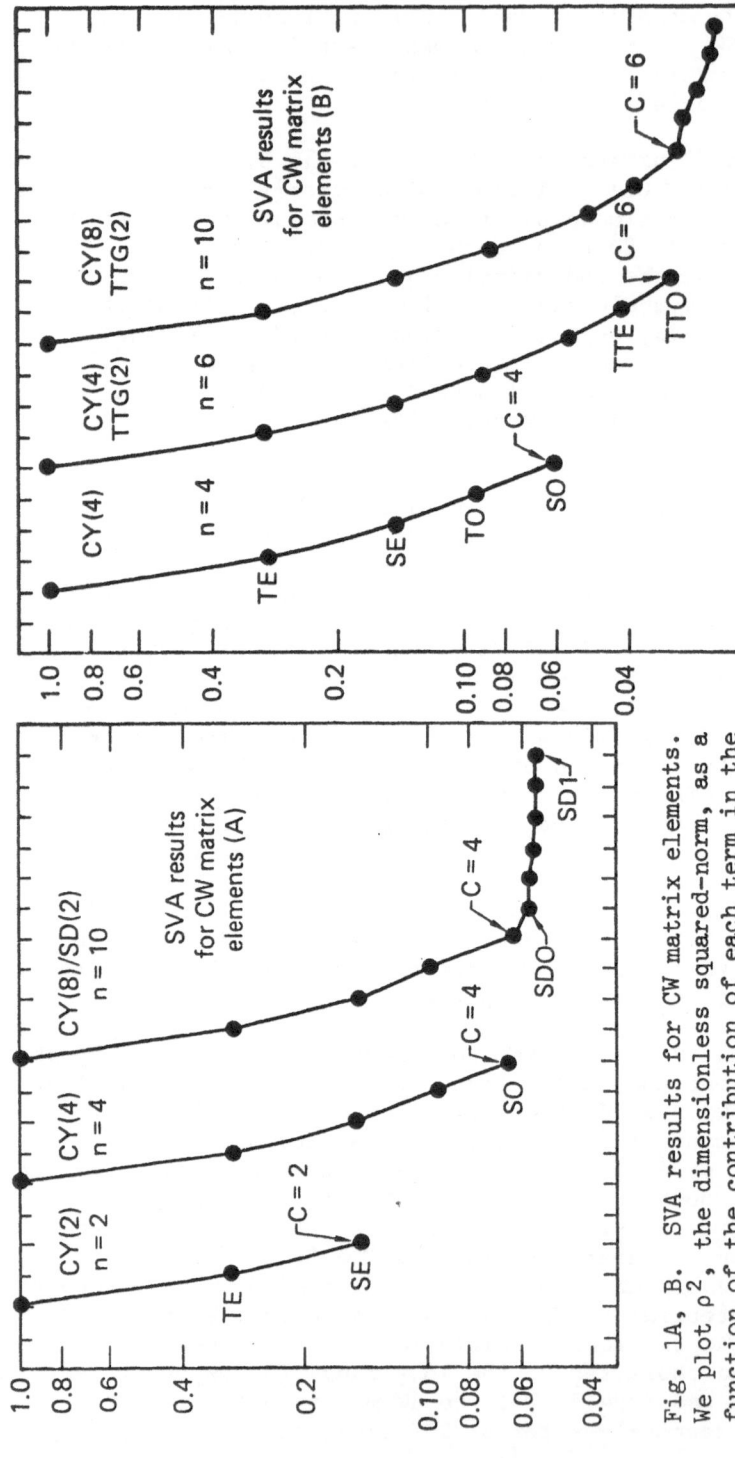

Fig. 1A, B. SVA results for CW matrix elements.
We plot ρ^2, the dimensionless squared-norm, as a
function of the contribution of each term in the
potential form for the CW matrix elements. The
letter n denotes the full-order of the potential model and the last point on each plot is the
least-squares result. The cut-off point (c) is indicated and is the effective rank of the problem. The contributions of each term in the potential are labeled with symbols defined in the text
and are given in decreasing order of importance except where $\Delta\rho^2$ is < 1%. The labels at the top
of each plot describe the force model. The similarity of the curves arises from the decoupled
character of the cut-off solutions and is reflected in the corresponding strengths of the parameters (Table 1). The plot for CY(4) is shown on both (a) and (b) for comparison.

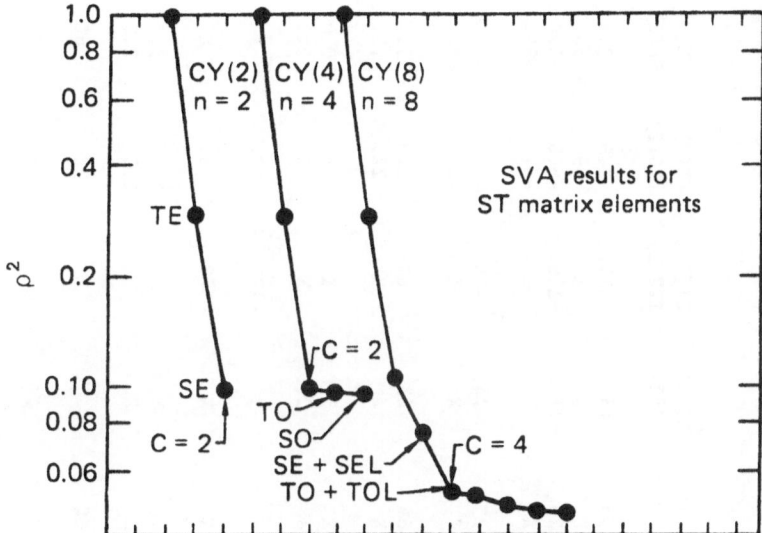

Fig. 2   Plot of ρ^2 in the case of the SVA results for the 84 (selected) ST matrix elements. See the caption to Fig. 1. Note that the third and fourth points on the CY(8) plots refer to highly coupled solutions. The symbols SEL, TOL refer to the long-range components $\lambda_2 = 2.0$ fm. See the text and Table 2.

Table 1. SVA best results ($\rho_c^2 - \rho_n^2 \lesssim 1\%$; see text) for the CW matrix elements. The symbols are defined in text. All potential entries are in MeV and are given to two significant figures. The two-range entries (e.g., SO1, SO2) correspond to the ranges λ_1, λ_2. The tensor-gaussian form factor had a range parameter of 1.0 fm. The symbol c/n refers to the SVA cut-off and rank of the potential form, as described in text. The oscillator parameter was 11.6 MeV, the same as used in reference 3.

	SD(2)	TTG(2)	CY(2)	CY(4)	CY(4), SD(2)	CY(4), TTG(2)	CY(8), TTG(2)	CY(8), SD(2)	CY(8)	CY(8)[c]
SO1	X	X	X	150.	150.	137.	2.6	2.5	2.5	390.
SO2	X	X	X	X	X	X	49.	54.	54.	190.
TE1	X	X	-120.	-120.	-117.	-116.	-6.8	-5.0	-15.	-110.
TE2	X	X	X	X	X	X	-51.	-52.	-47.	16.
SE1	X	X	-75.	-75.	-74.	-78.	-3.1	-3.5	-3.8	24.
SE2	X	X	X	X	X	X	-36.	-34.	-34.	-43.
TO1	X	X	X	64.	63.	60.	1.5	1.1	1.1	-73
TO2	X	X	X	X	X	X	22.	23.	23.	49.
SD0	.85	X	X	X	0.0	X	X	0.0	X	X
SD1	.67	X	X	X	0.0	X	X	0.0	X	X
TTE	X	-230.	X	X	X	-181.	-183.	X	X	X
TTO	X	22.	X	X	X	79.	78.	X	X	X
λ_1, fm	X	X	1.0[a]	1.0[a]	1.0	1.0	1.0[b]	1.0	1.0	1.0
λ_2, fm	X	X	X	X	2.0	X	2.0	2.0	2.0	2.0
ρ^2	72%	96.4%	9%	6%	6%	3%	3%	6%	6%	5.8%
c/n	2/2	2/2	2/2	4/4	4/6	6/6	6/10	4/10	4/8	8/8

a $\lambda = 1.0$ fm is optimal for both CY(2) and CY(4).

b Very insensitive to λ ; $\lambda_\sigma = 1.50$ reduces ρ^2 by 0.3% and $\lambda_\tau = 1.50$ by 0.1%.

c This column shows the full least-squares result for CY(8). Note large-scale changes from cols. 8-10.

Table 2. **SVA best results for the 84 selected ST matrix elements.** The oscillator parameters are the same as in reference 3. See caption to Table 1 for an explanation of the symbols. The tensor force was not included in this analysis.

	SD(2)	CY(2)	CY(2)	CY(4)	CY(8)[a]
SO1	x	x	x	− 31	0
SO2	x	x	x	x	0
TE1	x	− 110	x	− 104	0
TE2	x	x	− 49	x	− 49
SE1	x	− 69	x	− 76	− 280
SE2	x	x	− 33	x	+ 110
TO1	x	x	x	20	− 660
TO2	x	x	x	x	240
SD0	.58	x	x	x	x
SD1	.46	x	x	x	x
λ_1, fm	x	1.0	x	1.0	1.0
λ_2, fm	x	x	2.0	x	2.0
ρ^2	42%	10%	10.7%	9.4%	5.2%
c/n	2/2	2/2	2/2	4/4	4/4

[a]The SE components in this column result from the combination of two linearly-independent components. The first component is shown in column 3 under CY(2).

deduced since the central components completely subsume the SDI
role.

The last column of Table 1 exhibits what happens (in the case
of CY(8)) if we pursue the solution to the 8-parameter or
least-squares limit. The result for the parameters is not easily
recognizable. However at the cut-off limit the TE and SE
components are not only essentially the same in all comparable
models but, in the case of the single range models, equal to the
PMM parameters,[4] within a few percent. This is significant
because these parameters give the local-potential equivalent of
the Kallio-Koltveit potential,[5] a realistic interaction based on
the Scott-Moszkowski prescription[6] for treating the short-range
repulsive part of the free interaction.

The results for the ST analysis (Table 2) are in good
agreement with the CW results insofar as the first two terms are
concerned, i.e. the TE and SE components. (The CY(2) results with
range 2.0 fm are included to exhibit the insensitivity of ρ^2 to
range.) The results for the odd components as well as for the SE
component in the CY(4) and CY(8) models show that something more
complex is required to reach the 5% level or lower. This will be
our next undertaking.

In sum, the initial results of the application of SVA to the
CW and ST data sets strongly support a single effective
interaction which accounts for 90% to 95% of the squared-norm of
the residual vector (r, Eq. (2)). The form of this potential is
not unique but, at the 95% level, its rank is no more than 6 nor
less than 4. Furthermore 85% to 90% is accounted for by the two
central-even parameters which are produced by the Scott-Moszkowski
range cut-off procedure.[6] It is notable that this procedure
makes no direct reference to the nuclear environment except for the
postulated effect of the Pauli principle. Finally, in the case of
the CW analysis, the SVA results support the utility of both
tensor-force components and a long-range triplet-odd component
which is strongly repulsive, the latter in conformity with the
findings of Schiffer and True.[3] It is presumably these
components which would raise the rank of the effective interaction
from 4 to 6 or possibly 7. Although it may seem from the
foregoing that no more complexity is required, the burden of this
conclusion is carried by the last 5-10% of the squared-norm of r.
It may well be that this is due to the limitation of the
experimental data to the two-body energy spectra. Other types of
nuclear data (three body spectra, electromagnetic transitions,
etc.) may not be so limited and will be considered in future
studies.

We acknowledge fruitful discussions with F. N. Fritsch and
W. W. True.

REFERENCES

1. C. L. Lawson and R. T. Hanson, "Solving Least Squares Problems," Prentice-Hall, Inc., 1974.
2. W. Chung and B. H. Wildenthal, unpublished.
3. J. P. Schiffer and W. W. True, Revs. Mod. Phys. 48:191 (1976).
4. F. Petrovich, H. McManus, V. A. Madsen, and J. Atkinson, Phys. Rev. Letters 22:895 (1969).
5. A. Kallio and K. Koltveit, Nucl. Phys. 53:87 (1964).
6. B. L. Scott and S. A. Moszkowski, Nucl. Phys. 29:665 (1962).

DISCUSSION

Philpott: 1. Is it possible to relate the interactions which you have found to those discussed earlier by Love? 2. What can you say about this relationship? 3. Does your work imply that Love's interactions could be considerably simplified?

Bloom: Questions 1 and 2. It is highly likely that there is a fairly direct relationship between the M3Y interaction (which is the Love-MSU interaction) and the parameters produced by our SVA approach. In fact one of our first projects will be the exploration of this relationship. This can be done most simply (and directly) by using the M3Y form and extracting the strength parameters exactly as I have described. The $\Delta\rho^2$ plot of the type shown in Figs. 1 and 2 will then reveal the differences and likenesses among any set of interactions. Needless to say the analysis need not be confined to bound state spectroscopic data as done here. Question 3. As to the simplification of the M3Y interaction, I can give my expectation that it will simplify considerably at very low energies and become more complex as we proceed to the continuum.

Love: The interaction of Dr. Bloom has been tailored for use in the s-d shell and as such includes core polarization effects of all kinds. The M3Y interaction includes a more limited class of extra-shell excitations and may be expected to differ from an interaction restricted to the s-d shell. The fact that the Chung-Wildenthal matrix elements are different for the upper and lower halves of the shell places a limit on the closeness of the M3Y and Bloom interactions. Although the M3Y interaction may be simplified for use within a single major shell, an accurate representation of the N-N information requires at least the level of complexity used in its construction.

EFFECTIVE INTERACTIONS: DERIVATION OF MOMENT CRITERIA

B. J. Dalton

Ames Laboratory-USDOE
Department of Physics, Iowa State University
Ames, Iowa 50011

INTRODUCTION

Meaningful spectral calculations for a many particle system usually require severe truncation of the vector space and renormalization of the interaction to account for effects of truncation[1]. Traditional methods for imposing spectral criteria to find the effective Hamiltonian (indicated here by $\hat{h}$) have centered around the use of projection operators and linked-folded perturbation theory as, for example, in the Block-Horowitz-Brandow[2] formulation. In most cases, the difficulties of calculating the significant third and higher order diagram contributions have made these approaches impractical.[3]

Moment expansions have been suggested as a means to overcome this convergence difficulty[4-6] as recent developments make practical the calculation of traces (moments) in very large spaces. Chang[5] has recently described a moment expansion within the Block-Horowitz equation, and more recently Chang and Draayer[6] have made an initial study of the convergence of this expansion using a simple Lipkin model.

In the present work we describe the derivation of moment criteria from the spectral criteria for the effective interaction and briefly discuss one possible method for imposing the moment criteria to find $\hat{h}$. The motivation for this significantly different approach is the numerically demonstrated fact that the average shell model spectrum (level density) is very near to a Gaussian in shape.[7,8] A typical shell model level density is shown in Figure 1 in comparison with a two moment (Gaussian)

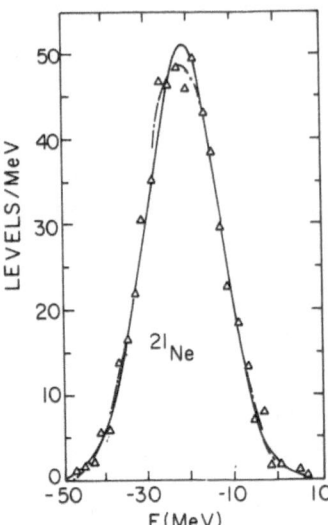

Fig. 1.   Comparison of a shell model (Δ) level density with that
         calculated with a two moment (———) and an eight moment
         (·——·) expansion.   The space is five nucleons ^{21}Ne in
         the sd shell with an ^{16}O core.   (From Ref. 8)

expansion and an eight moment expansion.[8] From this figure it is
clear that the first two moments essentially determine the shape
of the average shell model spectrum and that the higher moments have
much less effect on the latter. This fact suggests that the many-
particle diagrams contained in lowest moments have the largest
average effect on the effective interaction. An algorithm for
finding $\hat{h}$ which contains the lowest moments as the leading terms
should have rapid convergence. In addition, the effects on the
average shell model spectrum of truncations via moments can be
easily assessed.

DERIVATION

         In this section we start with the usual spectral criteria for
an effective interaction and derive corresponding moment criteria.
We are considering effects of truncation from a large but <u>finite</u>
basis space B to a subspace P. The large basis space is indicated
by

$$B = \{ \ |\alpha> \ \left| \ \ \alpha = 1\text{--}N, \ \ <\alpha|\beta> = \delta_{\alpha\beta} \ \right\}$$

We partition B as follows

$$B = P \cup Q, \quad P \cap Q = \emptyset, \quad \emptyset \equiv \text{null set}$$

$$P = \{ \ |p > \ \Big| \ p = 1 \text{ -- } M \ \}, \ Q = \{ \ |q > \ \Big| \ q = M + 1 \text{ -- } N \ \}.$$

Throughout we will use $|p >$ for an element of P and $|q >$ for an element of Q. Sums on the index p will be from 1 to M and sums on the index q will be from M + 1 to N. All other sums (i.e. on α and i) will be from 1 to N. We use the conventional projection operators P and Q defined by

$$P = \Sigma \ |p >< p| \quad , \quad Q = \Sigma \ |q >< q| \ . \tag{1}$$

These operators have the useful properties

$$P^2 = P \ , \quad P^\dagger = P \ , \ Q^2 = Q \ , \ Q^\dagger = Q \tag{2}$$

$$P + Q = 1 \ , \ PQ = 0 \tag{3}$$

We consider a finite set S of eigenstates of the Hamiltonian H.

$$S = \{ \ |i > \ \Big| \ i = 1 \text{ -- } N, \quad < i|j > \ = \delta_{ij} \ \} \tag{4}$$

$$H|i > \ = E_i |i > \tag{5}$$

We will assume that the states of S can be related to the states of B by a similarity transformation.

$$|i > \ = \Sigma \ a_{i\alpha}| \alpha > \quad , \ \Sigma \ a_{i\alpha}^* a_{i\beta} = \delta_{\alpha\beta} \tag{6}$$

We have the relation

$$|i > \ = P|i > + Q|i > \ . \tag{7}$$

From (7) and (5) one can easily obtain the following relation

$$P[H + HQ(E_i - QHQ)^{-1}QH]P \ P|i> \ = E_i P|i> \tag{8}$$

We will use the following notation

$$h(i) \equiv H + HQ(E_i - QHQ)^{-1} QH \tag{9}$$

$$\hat{h}(i) \equiv P \ h(i)P \ , \ \hat{H} = PHP \tag{10}$$

where we explicitly indicate that $\hat{h}(i)$ depends on the index i.
With (9) and (10) we have from (8) the relation

$$\hat{h}(i)P|i> = E_i P|i> .\qquad(11)$$

With the notation

$$\hat{S} \equiv \{P|i> \;\Big|\; i = 1 -- N\} ,$$

we see from (11) that $\hat{h}$ acting in $\hat{S}$ has the same eigenvalue
spectrum as H acting in S.

We now consider trace relations between $\hat{h}$ and H. We follow
recent convention of denoting the trace of K by $<< K >>$

$$<< K >>_S \equiv \Sigma <i|K|i> \;,\;\; << K >>_{\hat{S}} \equiv \Sigma <i|PKP|i>$$

$$<< K >>_B \equiv \Sigma <\alpha|K|\alpha> \;,\;\; << K >>_P \equiv \Sigma <p|K|p>$$

$$<< K >>_Q \equiv \Sigma <q|K|q> \;.\qquad(12)$$

Notice that the trace over S for an operator of form PKP is equal
to the trace of K over $\hat{S}$.

With $1 = \Sigma |i><i| = \Sigma |\alpha><\alpha|$ we have the usual cyclic
relation

$$<< K \cdot\cdot\cdot LM >>_S = << MK \cdot\cdot\cdot L >>_S\qquad(13)$$

for operators which do not have an explicit dependence on the
index i. We also have relations like (13) for traces defined
over B. For an operator K which does not have an explicit
dependence on the index i (recall $\hat{h}(i)$), we have from (6) the
relationship

$$<< K >>_S = << K >>_B .\qquad(14)$$

Consider an operator L(i) (like $\hat{h}(i)$) which depends upon the
index i only via the eigenvalue E_i, that is, $L(i) = L(E_i)$ and
for which a suitable expansion can be found in terms of powers
of E_i, that is,

$$L(E_i) = \Sigma L_\ell E_i^\ell .\qquad(15)$$

For this type of operator we have

$$<< L >>_S \equiv \Sigma < i|L(i)|i >$$

$$= << \Sigma L_\ell H^\ell >>_S = << \Sigma L_\ell H^\ell >>_B . \tag{16}$$

Returning to (11) one can easily show the relationship

$$\hat{h}^n(i)P|i > = PH^n|i > . \tag{17}$$

For any operator $\hat{K} = PHP$ which is not state dependent we have from (17) the following trace relation

$$<< \hat{K}\hat{h}^n P >>_S = << \hat{K}PH^n >>_S$$

$$= << \hat{K}PH^n >>_B = << \hat{K}H^n >>_B$$

$$= << \hat{K}H^n >>_P = << H^n\hat{K} >>_P \tag{18}$$

where we have used the relations in (3) with (13) and (14). For the left-hand side of (18) we have

$$<< \hat{K}\hat{h}^n P >>_S = << \hat{K}\hat{h}^n >>_S = << \hat{K}\hat{h}^n >>_{\hat{S}} \tag{19}$$

With (19) and (18) we have

$$<< \hat{K}\hat{h}^n >>_{\hat{S}} = << \hat{K}H^n >>_P \tag{20}$$

For the operator $\hat{K}\hat{h}^n(i)$ (18)-(20) mean that we have a relationship between the trace over the eigenstates of the effective interaction $\hat{h}(i)$ and the trace over the states of P. These relationships are direct consequences of the definition of $\hat{h}(i)$ as given in (8), (9) and (10), and as such represent <u>necessary</u> restrictions or <u>trace criteria</u>, for the effective interaction.

Given a formal expansion of the resolvent operator in powers of E_i, the operator $\hat{K}\hat{h}^n(i)$ may be written in the form (15).

$$\hat{K}\hat{h}^n(i) = \Sigma \hat{K}\hat{h}_\ell(n)E_i^\ell . \tag{21}$$

With $E_i^n|i > = H^n|i >$ we have

$$\hat{K}\hat{h}^n(i)|i > = \Sigma\hat{K}\hat{h}_\ell(n)H^\ell|i >$$

and using (16) we obtain the relation

$$<< \hat{K}\hat{h}^n >>_S = << \hat{K}\Sigma h_\ell(n)H^\ell >>_S$$

$$= << \hat{K}\Sigma h_\ell(n)H^\ell >>_B$$

$$= << \hat{K}\Sigma h_\ell(n)H^\ell >>_P . \tag{22}$$

From (22) and (18) we have

$$<< \hat{K}\Sigma h_\ell(n)H^\ell >>_P = << \hat{K}H^n >>_P . \tag{23}$$

The content of the above trace criteria for $\hat{h}(i)$ can be made more clear by considering the lowest order criteria. We have

$$<< \hat{h} >>_S = << H >>_P , \tag{24}$$

and using the following abbreviation

$$\hat{\delta}(i) = PHQ(E_i - QHQ)^{-1}QHP, \tag{25}$$

we have

$$<< \hat{H} >>_S + << \hat{\delta} >>_S = << H >>_P . \tag{26}$$

Since

$$<< PHP >>_S = << PHP >>_B = << H >>_P \tag{27}$$

we must have

$$<< \hat{\delta} >>_{\hat{S}} = 0 . \tag{28}$$

This expression states that the $\hat{S}$ space trace of $\hat{h}$ is contained in $\hat{H}$. It also means that the $\hat{S}$ space trace of $\delta(i)$ must be <u>zero</u>, a rather interesting result. If $\delta(i)$ is given by some conventional truncated expansion, the closeness of $<< \delta >>_{\hat{S}}$ to zero represents a measure of convergence (or non-convergence) obtained by the given truncation.

Now consider the criteria for $\hat{h}^2(i)$

$$<< \hat{h}^2 >>_S = << \hat{H}^2 >>_S + << \hat{H}\hat{\delta} >>_S + << \hat{\delta}\hat{H} >>_S + << \hat{\delta}^2 >>_S$$

$$= << \hat{H}^2 >>_P + \Sigma < p|H|q >< q|H|p > \tag{29}$$

Since

$$<< \hat{H}^2 >>_S \ = \ << \hat{H}^2 >>_B \ = \ << \hat{H}^2 >>_P$$

we have

$$<< \hat{H}\hat{\delta} >>_{\hat{S}} \ + \ << \hat{\delta}\hat{H} >>_{\hat{S}} \ + \ << \hat{\delta}^2 >>_{\hat{S}} \ = \ \Sigma \ < p|H|q><q|H|p > \quad (30)$$

For the case where P represents valence space orbits and B includes the core orbits, the right-hand side of (30) includes contributions up through 2 particle − 2 hole diagrams.

FINDING $\hat{h}$

To find an approximation to $\hat{h}$ using these trace or moment criteria, a model or expansion for $\hat{h}$ is needed. One model which shows promise is that used in the semirealistic[11] approach in which the following form for $\hat{h}$ is assumed

$$\hat{h} \ = \ \hat{H} \ + \ \Sigma A_\mu \hat{O}_\mu$$

Here, the $\hat{O}_\mu = P O_\mu P$ are selected operators. The moment criteria can be used to help determine the coefficients A_μ, especially if the set $\{\hat{O}_\mu\}$ is restricted, as in a number of past applications[11], to only a few operators such as the pairing interaction plus some low multiple terms. For instance the centroid criteria $<< \hat{\delta} >>_{\hat{S}} = 0$ mentioned above leads to

$$\Sigma \ A_\mu \ << O_\mu >>_{\hat{S}} \ = \ 0 \ .$$

The above development represents some initial work in an effort to use moment criteria to help renormalize operators to account for truncation.

ACKNOWLEDGEMENT

This work was supported by the U. S. Department of Energy, Contract No. W-7405-Eng-82, Office of Basic Sciences, Nuclear Physics Division.

REFERENCES

1. P. O. Lowdin, J. Math. Phys. $\underline{3}$ 969 (1962). H. Feshbach, Ann. of Phys. $\underline{19}$ 287 (1962).
2. C. Block and J. Horowitz, Nucl. Phys. $\underline{8}$. 91 (1958). B. H. Brandow, Rev. Mod. Phys. $\underline{39}$, 771 (1967).

3. M. S. Sandel, R. J. McCarthy, B. R. Barrett and J. P. Vary,
    Phys. Rev. C 17, 777 (1978).
4. B. J. Dalton, BAPS 22, 1012 (1977), B. D. Chang, BAPS 22, 1012
    (1977), S. S. M. Wong, Nucl. Phys. A295, 275 (1978).
5. B. D. Chang, Nucl. Phys. A304, 127 (1978).
6. B. D. Chang and J. P. Draayer, to be published.
7. F. S. Chang, J. B. French, and T. H. Thio, Ann. of Phys. N.Y.,
    66, 137 (1971). A review article on this subject is
    presently being prepared by French and co-workers.
8. S. M. Grimes, S. D. Bloom, R. F. Hausman, Jr. and B. J. Dalton,
    Phys. Rev. C 19, 2389 (1979).
9. J. P. Draayer, J. B. French, and S. S. M. Wong, Ann. Phys.
    N.Y.) 196, 472 (1978).
10. S. M. Grimes, C. H. Poppe, C. Wong and B. J. Dalton, Phys. Rev.
    C 18, 1100 (1977).
11. W. J. Baldridge and B. J. Dalton, Phys. Rev. C 18, 539 (1978),
    W. J. Baldridge and J. P. Vary, Phys. Rev. C 14, 2246 (1976),
    N. Freed and J. Gibbons, Nucl. Phys. A136, (1969), N. Freed
    and W. Rhodes, Nucl. Phys. A126, 481 (1969), F. Tabakin,
    Ann. Phys. (N.Y.) 30, 51 (1964).

DISCUSSION

     Dalton: I would like to strongly urge research which will
determine more accurately the free nucleon-nucleon interaction
and of a G-matrix derived from it. Recent developments of
spectral distribution theory make practical the calculations
of nuclear level densities $\rho(E,J)$, one-body densities $\rho(\vec{R})$,
two-body correlations $\rho(\vec{R},\vec{R}')$ and various strength distributions in very large spaces (say 10^{20} Slater determinants).
With large space, no-core calculations of these quantities,
the G-matrix can be used directly, thereby avoiding the uncertainties which occur in renormalizing the interaction for
truncated spaces. The development of codes to calculate $\rho(\vec{R})$
and $\rho(\vec{R},\vec{R}')$ from the G-matrix is presently a major program of
mine at the Ames Laboratory. With these new theoretical
methods the measured propertites of the free nucleon-nucleon
interaction will be directly reflected in the calculated
values of the above mentioned important quantities. I emphasize however, that the accuracy of the calculated values can
be no better than the accuracy of the free nucleon-nucleon
interaction used.

FIELD-THEORETIC ORIGIN OF THE ISOVECTOR CENTRAL (LANE)

AND SPIN-ORBIT POTENTIALS FOR QUASIELASTIC (P,N) REACTIONS

J.V. Noble[*]

Department of Physics, University of Washington
Seattle, Washington 98195

As several authors have pointed out,[1-8] a relativistic description of nuclear one-body motion improves on the usual Schroedinger theory in several ways: First, the spin-orbit potential arises naturally from the velocity dependence of the forces,[1,7,8] as does the "full-Thomas" form of the spin-orbit potential in deformed nuclei.[9] Second, the observed energy dependence of the real part of the nucleon-nucleus optical potential can be understood as mainly a relativistic kinematic effect.[1,2,5,10] And third, the relativistic approach is more fundamental, in the sense that it relates the nuclear one-body potentials to the average meson fields present in nuclei.[1-4,6] This note is a study of the consequences of the nuclear ρ-meson field for charge-exchange reactions, and particularly (p,n) reactions.

Consider a nucleon scattering from a spinless nucleus. Although the vector mesons are described by 4-vector fields

$$\omega^\mu \equiv (\omega^0, \vec{\omega}) \quad , \quad \underline{\rho}^\mu \equiv (\underline{\rho}^0, \underline{\vec{\rho}}) \tag{1}$$

only the timelike component can be generated by a spinless nucleus, for reasons of symmetry. (That is, the spacelike components will exist only as quantum fluctuations associated with virtual transitions from the nuclear ground state to excited states. We ignore these for elastic scattering, to lowest order.) The one-body equation of motion for an external nucleon is then

$$[-i \, \vec{\alpha} \cdot \nabla + \beta M + \beta U(r) + V(r) + G_\rho^V \, \underline{\tau} \cdot \underline{\rho}^0(r) +$$

* Permanent address: Department of Physics, University of Virginia
Charlottesville, Virginia 22901

$$+ (G_\rho^T /2M)\ i\beta\ \vec{\alpha}\cdot\nabla(\underline{\tau}\cdot\underline{\rho}^0(r))]\ \psi(\vec{r}) = E\ \psi(\vec{r}) \qquad (2)$$

where $\vec{\alpha}$ and β are the usual 4×4 Dirac matrices and $\psi(\vec{r})$ is a 4-component Dirac-spinor wave function. The potentials $U(r)$ and $V(r)$ we identify with the average neutral scalar (σ) and neutral vector (ω) meson fields of the nucleus (times their respective coupling constants). The tensor coupling of the ω is omitted because it is known to be small[11] (and would be expected to be small by analogy with the isoscalar anomalous magnetic moment, in the vector-meson-dominance (VMD) picture), whereas that of the ρ must be included because it is known to be big[11] (and corresponds to the large isovector anomalous magnetic moment in VMD). The magnitude of the vector coupling constant, G_ρ^V, is not well known empirically.[11]

It is easy to see that the average ρ-field will lead to an isovector interaction, including quasielastic charge exchange (although this is an inelastic process, the near-degeneracy of IAS leads us to lump it with elastic scattering — put another way, the quantum fluctuations between IAS will be of low frequency, hence quasistatic). In order to make contact with the usual phenomenology, it is useful to re-express Eq. (2) in terms of (2-component) Pauli spinors. Writing

$$\psi(\vec{r}) = \begin{pmatrix} \psi_L(\vec{r}) \\ \psi_S(\vec{r}) \end{pmatrix}$$

as usual, we find upon keeping terms through first order in small quantities

$$[- \nabla^2/2M + U + VE/M + V_\rho E/M + (U^2 - V^2)/2M - V V_\rho/M - V_\rho^2/2M$$

$$+ \vec{\sigma}\cdot\vec{\ell}\ (4M^2 r)^{-1}\ \{(V' - U' + V_\rho')/(1 + (U-V)/2M)$$

$$- 2\kappa\ V_\rho'/(1 - V/M)^2\}]\ \psi_L$$

$$= (E - M)\psi_L \qquad (3)$$

where

$$\kappa = G_\rho^T/G_\rho^V \qquad (4)$$

$$V_\rho = G_\rho^V\ \underline{\tau}\cdot\underline{\rho}^0 \qquad (5)$$

and the extra factor $(1 - V/M)^{-2}$ comes from the underlying structure of the ρNN coupling, using VMD with the analogy to the iso-

vector anomalous magnetic moment.[12] From Eq. (3) it is easy to
identify the phenomenological nonrelativistic (isoscalar) central
potential as

$$v_C^{(0)} = U + VE/M + (U^2 - V^2)/2M \qquad (6)$$

and the corresponding spin-orbit potential as

$$v_{s.o.}^{(0)} = (4M^2r)^{-1} \vec{\sigma}\cdot\vec{\ell} \ (V' - U')/(1+ \qquad (7)$$

$$(U - V)/2M).$$

By comparing Eq.(6) and Eq. (7) with the usual fits to nucleon-
nucleus scattering, we can determine the empirical scalar and vector potential strengths U(0) and V(0), assuming both have the same
shape as the nucleus itself (i.e., taking the range of the heavy-
meson propagators to be zero). Elsewhere, it has been shown[2,13]
that these numbers are consistent with what would have been predicted from first principles using the empirical meson masses and
coupling constants. The isovector central potential deduced from
Eq. (3) is

$$v_C^{(1)} \ f(r) \ \underline{\tau}\cdot\underline{T}/A$$

where f(r) is the nuclear shape function and

$$v_C^{(1)} \simeq (3/4\pi)(G_\rho^V)^2 \ [\ E - V(0) + m_\rho \ (G_\rho^V)^2/8\pi]/($$

$$M \ m_\rho^2 r_0^3) \ . \qquad (8)$$

Taking for the ρNN vector coupling constant the value[11]

$$(G_\rho^V)^2/4\pi = 0.8,$$

we obtain $v_C^{(1)} = 18 \pm 2$ MeV. This is the number which should be
compared with charge-exchange experiments.[14]

Before turning to the isovector spin-orbit interaction, we
should examine briefly the question of the symmetry term in the
semi-empirical mass formula, about 28 $(N - Z)^2/A$ MeV. As is well
known, this term derives partly from the exclusion effect on kinetic energies[15] which gives

$$0.3 \ \varepsilon_F \ (N - Z)^2/A \simeq 10 \ (N - Z)^2/A \quad MeV,$$

partly from the exclusion effect on potential energies (this comes
from the fact that a neutron can get close to Z + ½N nucleons, a
proton to N + ½Z nucleons, so in a nucleus with N ≠ Z there is

an induced symmetry term having nothing to do with charge exchange) which gives

$$- v_c^{(0)} \ (N - Z)^2/6A \simeq 9 \ (N - Z)^2/A \quad \text{MeV},$$

and finally, from true isovector meson-exchange, which gives

$$v_c^{(1)} \ (N - Z)^2/2A = 9 \ (N - Z)^2/A \quad \text{MeV}.$$

As we see, adding up all three terms gives the empirical symmetry energy of the mass formula with good accuracy. The corresponding single-neutron well-depth is

$$- 54 + 36 \ (N - Z)/A \ \text{MeV},$$

in reasonable agreement with nuclear spectra.[16]

Let us now examine the isovector spin-orbit interaction. It derives from the V'_ρ terms in Eq. (3); additionally there is an exclusion-induced symmetry contribution which does not involve charge exchange. Since V'_ρ is surface-peaked, we evaluate the associated functions at the nuclear surface, to find

$$v_{s.o.}^{(1)} = - \ (3/4\pi)(G_\rho^V)^2 \ [\ \kappa/(1 - V(0)/2M)^2 - 0.5/$$

$$(1 + (U(0) - V(0))/4M)]/($$

$$M^2 m_\rho^2 \ r_0^5) \qquad (9)$$

giving

$$v_{s.o.}^{(1)} = \begin{cases} - \ 2.8 \ \text{MeV} & \text{(VMD)} \\ - \ 4.8 \ \text{MeV} & \text{(exp't)}, \end{cases}$$

where (VMD) refers to $\kappa = 3.7$, and (exp't) refers to[11] $\kappa \simeq 6$. The exclusion-induced symmetry term is

$$- \tau_3 [(Z - N)/3A] \ v_{s.o.}^{(0)} \quad .$$

The empirical spin-orbit interaction for bound neutrons is[16]

$$22 - 14 \ (N - Z)/A \quad \text{MeV} \ ;$$

the symmetry term of the above is not in serious disagreement with the sum of Eq. (9) and the exclusion-induced value, which together give $- 10.5 \ (N - Z)/A$ MeV. The charge exchange term, $- 4.8$ MeV, agrees well with a recent experimental determination of -4.3 MeV.[17] It is worth noting that the charge exchange term is derived mainly from the anomalous (i.e., tensor) ρNN coupling, and would be both considerably smaller and of the opposite sign, in the absence of this coupling.

We may conclude that the Lane potential and the charge exchange spin-orbit potential are well described by the low-frequency components of the nuclear ρ-meson field. We may therefore feel some confidence in the prediction of charge exchange excitation of natural-parity collective excitations mediated by ρ-exchange, as we presently do in the corresponding prediction of unnatural-parity transitions mediated by π-exchange.

I am grateful to the Department of Physics of the University of Washington for its kind hospitality during the course of this work.

References

1. H.P. Duerr,Relativistic Effects in Nuclear Forces, Phys. Rev. 103:469 (1956).
2. L.D. Miller and A.E.S. Green, Relativistic Self-Consistent Meson Field Theory of Spherical Nuclei, Phys. Rev. C5:241 (1972).
3. T.D. Lee and G.C. Wick, Vacuum Stability and Vacuum Excitation in a Spin-0 Field Theory, Phys. Rev. D9:2291 (1974).
4. J.D. Walecka, A Theory of Highly Condensed Matter, Ann. of Phys. 83:491 (1974).
5. R.L. Mercer, L.G. Arnold and B.C. Clark, Phenomenological Optical Model for p-^{4}He Elastic Scattering, Phys. Lett. 73B: 9 (1978).
6. J. Boguta and J. Rafelski, Thomas-Fermi Model of Finite Nuclei, Phys. Lett. 71B:22 (1977).
7. D.R. Inglis, Spin-Orbit Coupling in Nuclei, Phys. Rev. 50:783 (1936).
8. W.H. Furry, On the Introduction of Non-Electric Forces into Dirac's Equations, Phys. Rev. 50:784 (1936).
9. H. Sherif and J.S. Blair, Spin-Dependent Effects in Inelastic Scattering of High Energy Protons, Nucl. Phys. A140:33 (1970).
10. R. Humphries, The 1/γ Velocity Dependence of Nucleon-Nucleus Optical Potentials, Nucl. Phys. A182:580 (1972).
11. M.M.Nagels, et al., Compilation of Coupling Constants and Low Energy Parameters, Nucl. Phys. B109:1 (1976).
12. J.V. Noble, Axial and Magnetic Tests for Nuclear Dirac Wave Functions, Phys. Lett. B (to be published).
13. J.V. Noble, Consistency of Nuclear Dirac Phenomenology with Meson-Nucleon Interactions, Nucl. Phys. A (to be published).
14. D.M. Patterson, R.R. Doering and A. Galonsky, An Energy-Dependent Lane Model Nucleon- Nucleus Optical Potential, Nucl. Phys. A263:261 (1976).
15. A. deShalit and H. Feshbach, "Theoretical Nuclear Physics," John Wiley and Sons, Inc., New York (1974), p. 127.
16. A. Bohr and B.R. Mottelson, "Nuclear Structure, v. I,"

W.A. Benjamin, Inc., New York (1969), p.239.
17. J. Gosset , B. Mayer and J.L. Escudié, Quasielastic (p,n)
    Reactions Induced by Polarized Protons, Phys. Rev. C14:
    878 (1976).

MICROSCOPIC STRUCTURE OF $\Delta J^{\pi} = 1^{+}$ EXCITATIONS IN sd-SHELL NUCLEI

B.H. Wildenthal and W. Chung

Cyclotron Laboratory
Michigan State University
East Lansing, MI 48824

INTRODUCTION

The subject of this paper is the theoretical structure of
even-parity dipole excitations in light nuclei. The defining characteristics of "light" nuclei in the present context are that they
have approximately equal numbers of neutrons and protons and that
the "active" nucleon orbits needed to describe their structure
in the conventional shell-model representation are those of a
single, multiply-degenerate, level of the three-dimensional harmonic oscillator. Physical phenomena by which this class of excitations can be experimentally observed include magnetic dipole
(M1) electromagnetic excitations, Gamow-Teller beta decay, radiative meson capture, and hadronically-induced charge-exchange processes such as the (p,n) reaction.

Insight into the relationships of these phenomena to each
other and to the underlying microscopic structure of the nuclear
states involved can be obtained from shell-model calculations as
discussed in the following sections. The specific calculations
to be discussed are of the matrix elements of the one-body operators
which connect nuclear states of $J^{\pi} = 0^{+}$ with those of $J^{\pi} = 1^{+}$.
Results are presented for examples corresponding to transitions
commencing from the $J^{\pi} = 0^{+}$ ground states of the double-even nuclei
$18 \leq A \leq 38$. These calculations are the common foundation for
predictions about all of the various physical processes mentioned
above. The present discussion will concentrate upon M1 phenomena.
The presentation will treat in turn the model from which the nuclear
wave functions used in the present study are derived, the special
characteristics of these wave functions, the techniques by which

these wave functions are used in the calculation of the even-parity
dipole strengths, some general aspects of the microscopic structure
of these excitations,and the relationship of the theoretical structures to experimental observations.

MODEL WAVE FUNCTIONS FOR $18 \leq A \leq 38$ NUCLEI

The present discussion is concerned specifically with transi-
tions from states with $J^{\pi} = 0^+$ to those with $J^{\pi} = 1^+$ in the region
$18 \leq A \leq 38$. The model wave functions which are utilized in the
calculations for these transitions were obtained in the course
of a more general project[1] directed at producing wave functions
for the positive-parity states of all spins (J) and isospins (T)
in this same region. The model space assumed for the calculation
of these wave functions was comprised by the single-nucleon orbits
with quantum numbers $0d_{5/2}$, $1s_{1/2}$ and $0d_{3/2}$, i.e., the conventional
sd-shell. The complete set of basis vectors allowed in this space
by the Pauli Principle was utilized in all cases, a fact which
is particularly relevant in the context of calculations of matrix
elements of the orbital and spin angular momentum operators, operators which play a dominant role in the physical processes which
will be of interest in the present discussion. The calculations
were carried out in a j—j coupling, J—T representation with the
Oak Ridge-Rochester shell model codes[2] as modified by Chung[1] to
incorporate the Lanczos matrix diagonalization technique of Whitehead and Watt.

The model Hamiltonian for the present calculations was assumed
to be comprised of one-body plus two-body terms. For nuclei with
$A \leq 28$, the values of the one-body terms (single-particle energies)
were taken from the experimental data on ^{17}O and the values of the
two-body terms were obtained by adjusting the values of Kuo[4] so
as to produce a least-squares fit to 200 experimental level energies in the $18 \leq A \leq 24$ region. For nuclei with $A \geq 28$, the values
of the one-body terms were taken from the experimental data on
^{39}K and the values of the two-body terms were obtained by adjusting
values calculated by Kuo for the $A = 40$ region to produce a least-
squares fit to 140 experimental level energies in the $32 \leq A \leq 38$
region. The two Hamiltonians yield rather similar wave functions
for $A = 28$.

The wave functions for nuclear states in the $18 \leq A \leq 38$ mass
region obtained in the shell-model calculations just described
have been used to calculate a variety of nuclear observables.
Comparison of these predictions to experimental results indicates
that the preponderance of qualitative structural features experimentally observed in this region are accounted for by using these
wave functions together with the conventional forms of the operators

presumed to correspond to the experimental phenomena measured.[5,6] Quantitative agreement between theory and experiment for matrix elements which have magnitudes of the order of a significant fraction of a single particle unit is typically 25% or better. Phenomena which have been studied so far include single-nucleon transfer,[5] electric quadrupole moments and transitions,[6] magnetic dipole moments,[7] and Gamow-Teller beta decay.[8] It seems reasonable to conclude from these examinations of the present shell-model wave functions that the conventional sd-shell space suffices to incorporate the degrees of freedom which are most important in a description of the static and dynamic features of the large majority of bound, positive-parity states in the $20 \leq A \leq 36$ region and that the Chung-Wildenthal Hamiltonians give a reasonable guide to the proper configuration mixing within this space.

MATRIX ELEMENTS OF ONE-BODY-TRANSITION OPERATORS

The prediction from a "shell-model calculation" for the value of a particular nuclear observable incorporates two quite distinct components, of which only one depends upon the detailed structures of the nuclear states involved. It is this "nuclear structure" component of such predictions that is the topic of this section. For a given initial state $\psi^{NJ'T'}$ and final state ψ^{NJT} this component is the set of matrix elements of the one-body-transition operators connecting the two states which have the angular momentum and isospin rank of the observable in question. These matrix elements are expressed as

$$
D_{\Delta J, \Delta T}^{N, JT, J'T'}(j,j') = \frac{\langle \psi^{NJT} ||| (a_j^+ \otimes a_{j'})_{\Delta J, \Delta T} ||| \psi^{NJ'T'} \rangle}{[(2\Delta J+1)(2\Delta T+1)]^{1/2}}
\tag{1}
$$

where N is the number of active model nucleons, J,T and J',T' are the angular momentum and isospin quantum numbers of the final and initial N-body states, respectively, ΔJ and ΔT indicate the angular momentum and isospin rank of the observable of interest, and a_j^+ and $a_{j'}$ are, respectively, the creation operator for a nucleon in an orbit "j" of the model basis and an annihilation operator for a nucleon in orbit "j'".

The values of the matrix elements of expression (1) for a given pair of nuclear states constitute a complete microscopic foundation for any process characterized by the appropriate rank $\Delta J, \Delta T$ which effects a transition connecting these states. All of the information which is encoded into the wave function amplitudes of ψ^{NJT} and $\psi^{NJ'T'}$ via choice of the model space and diagonalization of the chosen model Hamiltonian which is relevant to a physical process of rank $\Delta J, \Delta T$ is distilled into these quantities. Conversely, they contain only the information relevant to

a process of this rank, since an undetermined amount of information about the structure of these states does not explicitly emerge
from these overlaps. Each matrix element (squared) is proportional
to the probability that the state ψ^{NJT} is formed by annihilating
from the state $\psi^{NJ'T'}$ a nucleon occupying shell model orbit j'
and creating a nucleon in orbit j as its replacement.

In a given model space (given set of single-particle orbits)
the number of "D" elements is obviously fixed. In the
$d_{5/2} - s_{1/2} - d_{3/2}$ space the $\Delta J = 1$, $\Delta T = 1$ operators number nine,
including the $d_{5/2} - s_{1/2}$ elements which are identically zero
from the angular momentum constraint. A transition between nuclear
states which are represented in the model space by single-particle
wave functions (e.g., for the present space, transitions such
as $^{17}O(5/2^+) \rightarrow {}^{17}O(3/2^+)$) are represented by D values of unity
for the relevant pair (3/2 – 5/2 in the example just noted) and
zero for the others. A transition between (in the model space)
many-particle states which, to a good approximation, differ only
in the quantum state of a single nucleon, would have D values
qualitatively similar to those of the "single-particle" nuclei.
If the dominant components of state ψ^{NJT} cannot be formed by action
of the one-body operators upon the dominant components of state
$\psi^{NJ'T'}$ (e.g., if ψ^{NJT} is to a good approximation a 2 particle-2
hole excitation built upon $\psi^{NJ'T'}$) then none of the j–j' matrix
elements will have magnitudes larger than a small fraction of
unity. Finally, if the pair of nuclear states in question is
connected by a transition corresponding to a coherent superposition
of strong one-body transitions, then several of the values of
the corresponding set of D elements will have magnitudes which
are a significant fraction of unity.

Transitions from $J^\pi = 0^+$ ground states to $J^\pi = 1^+$ excited
states in nuclei of the region $18 \leq A \leq 38$, transitions which
necessarily are of rank $\Delta J = 1$, are the subject of the present
investigation. As will be elaborated upon in the following sections, the nature of presently utilized probes of positive-parity
dipole nuclear excitations strongly emphasizes transitions characterized by $\Delta T = 1$. This is true obviously for transitions
induced by charge-exchange processes such as beta decay and (p,n)-
type reactions and holds approximately for electromagnetically
induced (magnetic dipole, to be precise) transitions. Hence,
the "D" matrix elements of $\Delta J = 1$, $\Delta T = 1$ between the shell-model
wave functions of the $J^\pi = 0^+$ ground states of the $18 \leq A \leq 38$
nuclei (which have T = 1 and 0, alternatively) and the excited
states of $J^\pi = 1^+$, T = 1, should constitute the dominant theoretical factor in the microscopic interpretation of all such phenomena.
These quantities are presented in Table 1.

Table 1.   One-body-transition-density matrix elements for
single nucleon orbits j and j' (5/2, 1/2 and 3/2)
corresponding to $\Delta J=1$, $\Delta T=1$ transitions from the
J=0 ground states of $18<A<38$ nuclei to the #)th
J=1, T=1 states.

#)	5/2←5/2	5/2←3/2	3/2←5/2	1/2←1/2	3/2←3/2	1/2←3/2	3/2←1/2
			(A=18; J=0,T=1 → J=1,T=1)[a]				
1)	0.0000	-0.1969	-0.4924	0.0000	0.0000	-0.0342	-0.0708
2)	0.0000	0.0342	0.0857	0.0000	0.0000	-0.1969	-0.4070
			(A=20; J=0,T=0 → J=1,T=1)[a]				
1)	0.3012	0.2026	0.1164	-0.0589	0.0570	-0.0250	-0.0612
2)	0.1310	-0.1219	-0.0453	-0.3280	0.0010	-0.0214	-0.0527
3)	-0.0100	-0.0217	0.0838	-0.1141	0.0040	0.1951	0.1990
4)	0.0067	0.0215	-0.0435	0.0700	0.0158	0.1274	-0.0213
5)	-0.0969	0.0447	0.2613	-0.0400	0.0018	0.0041	0.0869
6)	-0.1750	0.0165	0.2582	-0.0451	-0.0827	-0.0003	-0.2312
			(A=22; J=0,T=1 → J=1,T=1)[a]				
1)	-0.1839	-0.1602	-0.0291	0.0079	-0.0194	-0.0393	0.0025
2)	0.2531	-0.0420	-0.0370	-0.0417	0.0285	-0.1544	-0.0418
3)	-0.0168	-0.0485	-0.0759	-0.0480	-0.0373	0.0325	-0.0036
4)	0.1291	0.1260	0.4188	0.0239	-0.0811	0.0206	0.0592
5)	0.0201	0.0430	0.1211	0.0026	-0.0173	-0.0454	-0.2037
6)	-0.0570	-0.0714	-0.2214	0.0483	-0.0005	0.0596	0.0896
7)	0.0675	-0.0985	-0.3754	-0.0576	-0.0972	-0.0081	-0.0511
8)	0.0359	-0.0134	-0.1376	0.0322	-0.0380	-0.1094	-0.1381
			(A=24; J=0,T=0 → J=1,T=1)[a]				
1)	-0.1964	-0.1862	-0.0137	-0.0576	-0.0147	-0.0612	-0.0419
2)	-0.1960	-0.1135	-0.2273	-0.0090	-0.0107	-0.1437	-0.0030
3)	0.1302	-0.0410	-0.3310	-0.0138	0.0785	-0.1204	-0.0419
4)	-0.0822	0.1049	0.1876	0.1733	0.0328	-0.0262	0.0215
5)	0.1586	0.0372	0.1219	-0.2418	-0.0038	0.0117	0.0323
6)	0.0717	0.0436	-0.0273	-0.0058	0.0534	-0.1180	-0.0350
7)	-0.0266	0.0154	0.0347	0.0717	0.0459	0.1222	0.0693
8)	-0.0514	-0.0293	-0.0674	0.0285	-0.0481	-0.0831	-0.2081
9)	-0.0157	-0.0037	-0.0842	0.0362	-0.0450	0.0735	-0.0779
10)	0.0946	0.0547	-0.0261	-0.0088	-0.0683	-0.0087	-0.0094
11)	0.0736	0.0545	0.1874	-0.0011	-0.0562	-0.0444	-0.2027
12)	0.0277	-0.0095	-0.0847	0.0471	0.0812	-0.0121	0.0384

Table 1. Continued

#)	5/2←5/2	5/2←3/2	3/2←5/2	1/2←1/2	3/2←3/2	1/2←3/2	3/2←1/2
			$(A=26; \ J=0,T=1 \ \to \ J=1,T=1)^{a}$				
1)	-0.0384	-0.0665	-0.0557	0.0025	0.0102	-0.1220	-0.1035
2)	0.0307	-0.0041	-0.0095	-0.0256	-0.0254	0.1090	0.1823
3)	-0.0041	0.0181	0.0601	-0.0064	0.0015	-0.1348	-0.1975
4)	-0.1053	-0.1008	-0.1413	0.0262	-0.0379	0.0986	0.0916
5)	-0.0419	-0.0024	-0.0185	0.0125	0.0155	0.0286	-0.0171
6)	0.1194	0.0596	0.1358	-0.0225	-0.0047	-0.0461	0.0262
7)	0.0572	0.0328	0.0803	-0.0002	0.0244	-0.0018	0.0466
8)	0.0062	-0.0741	-0.1902	-0.0595	-0.0011	-0.0716	-0.0976
			$(A=28; \ J=0,T=0 \ \to \ J=1,T=1)^{a}$				
1)	0.2058	0.0072	0.0400	-0.2392	0.0314	-0.1270	-0.1079
2)	0.0222	0.1084	0.2759	0.1570	0.1027	-0.0879	-0.1190
3)	-0.1260	-0.1687	-0.4039	-0.0198	0.0795	0.0569	-0.2305
4)	-0.0393	-0.0385	-0.1506	0.0215	0.0103	-0.0614	-0.1574
			$(A=30; \ J=0,T=1 \ \to \ J=1,T=1)^{b}$				
1)	0.0347	0.0406	0.0908	-0.0249	-0.0918	0.5064	0.2371
2)	0.0437	-0.0938	-0.0578	-0.0192	-0.0614	-0.3082	-0.1900
3)	0.0358	-0.0458	-0.0544	-0.0177	0.0902	0.3218	0.2154
4)	-0.0083	-0.0994	-0.0728	0.0501	0.0442	0.1173	0.0412
5)	-0.0744	-0.0452	-0.0442	0.1366	0.0238	-0.0440	-0.0562
6)	-0.0774	0.5210	0.1369	-0.1576	-0.0659	0.0019	-0.0752
7)	0.0036	0.2911	0.1004	-0.0265	0.0376	0.0146	0.0583
8)	0.0240	-0.0331	0.0082	-0.0094	0.0693	-0.0723	-0.0759
			$(A=32; \ J=0,T=0 \ \to \ J=1,T=1)^{b}$				
1)	-0.0140	-0.0293	-0.0877	0.0674	0.1076	-0.5205	-0.1755
2)	-0.0423	-0.1337	-0.1328	0.2212	0.0226	0.0785	0.0758
3)	0.0145	-0.1813	-0.0382	-0.0612	-0.1048	-0.1796	0.0905
4)	-0.0364	0.1613	0.0009	-0.0080	-0.2149	-0.1354	0.0528
5)	-0.0273	0.1388	0.0125	0.0227	0.0670	0.2557	0.1397
6)	-0.0360	0.3726	0.1201	-0.0281	0.0513	0.0109	0.0653
7)	0.0436	0.2050	0.0373	-0.0382	-0.0658	-0.0403	-0.0508
8)	0.0245	0.1350	0.0810	-0.0112	0.0553	-0.0423	-0.0292
9)	-0.0479	0.2537	0.0480	0.1019	0.0044	-0.0069	-0.0518
10)	-0.0283	-0.0644	-0.0254	-0.0553	-0.0335	0.0482	-0.0412
11)	-0.0065	0.0004	-0.0126	-0.0144	-0.1102	-0.1622	-0.0173
12)	-0.0566	-0.1181	-0.0356	0.0484	-0.0453	0.0568	-0.0433

Table 1.　Continued

#)	5/2←5/2	5/2←3/2	3/2←5/2	1/2←1/2	3/2←3/2	1/2←3/2	3/2←1/2
			$(A=34; \ J=0,T=1 \to J=1,T=1)^b$				
1)	0.0165	-0.0213	0.0467	-0.0125	-0.1310	0.6561	0.2018
2)	0.0068	-0.0314	0.0159	-0.0040	0.0039	-0.3285	-0.2131
3)	0.0012	-0.0377	0.0220	-0.0073	0.0128	0.0458	-0.0025
4)	-0.0274	0.1079	0.0320	-0.0034	0.1005	0.1523	-0.0049
5)	0.0261	-0.5790	-0.1784	0.1021	0.0498	0.0460	0.0151
6)	-0.0110	-0.0043	-0.0055	-0.0359	-0.0250	0.0400	0.0191
7)	-0.0255	0.3285	0.0527	0.0483	0.1157	0.1952	0.0636
8)	0.0413	0.1232	0.0509	-0.0253	0.0408	-0.0335	0.0159
			$(A=36; \ J=0,T=0 \to J=1,T=1)^b$				
1)	0.0195	-0.0088	0.0204	0.0199	0.1917	-0.3562	-0.2297
2)	0.0153	0.2088	0.1193	-0.0742	0.0719	0.1577	-0.1108
3)	0.0248	0.0466	0.0633	0.0337	0.2718	0.3241	0.0832
4)	-0.0349	0.1601	0.1159	-0.1810	-0.0970	-0.1009	0.0437
5)	-0.0143	-0.0466	-0.0033	-0.0524	-0.0406	-0.0089	0.0371
6)	-0.0233	0.3852	0.0446	0.1314	-0.0341	-0.0978	0.0479
7)	-0.0131	0.0334	0.0413	-0.0279	0.0678	-0.0017	0.0443
8)	0.0428	-0.1632	-0.0797	-0.0145	0.0821	0.0047	-0.0027
9)	0.0395	0.0539	0.0135	-0.0112	0.0103	0.0058	0.0005
10)	0.0107	0.0778	0.0019	0.0035	-0.0475	0.0670	-0.0171
11)	-0.0427	0.0374	0.0044	0.0132	-0.0775	0.0645	-0.0157
12)	-0.0551	-0.0187	0.0194	0.0028	-0.0625	0.0313	-0.0285
			$(A=38; \ J=0,T=1 \to J=1,T=1)^b$				
1)	0.0000	-0.0096	-0.0023	0.0000	0.0000	0.6711	0.1483
2)	0.0000	0.6711	0.1602	0.0000	0.0000	0.0096	0.0021

^{a}As calculated for particle notation (N=A-16).

^{b}As calculated for hole notation (N=40-A).

B. H. WILDENTHAL AND W. CHUNG

Table 2.   One-body-transition-density matrix elements for
            single nucleon orbits j and j' (5/2, 1/2 and 3/2)
            corresponding to ΔJ=1, ΔT=1 transitions from the
            J=0 ground states of A=18, 22, 26, 30, 34 and 38
            to the #)th J=1,T=0 states.

#)	5/2←5/2	5/2←3/2	3/2←5/2	1/2←1/2	3/2←3/2	1/2←3/2	3/2←1/2
			(A=18; J=0,T=1 → J=1,T=0)[a]				
1)	-0.3019	0.0924	-0.2310	-0.1813	0.0103	0.0159	-0.0330
2)	0.2479	-0.0167	0.0418	-0.3509	-0.0211	0.0121	-0.0250
3)	-0.2232	-0.0700	0.1751	-0.1151	0.0306	-0.0952	0.1969
			(A=22; J=0,T=1 → J=1,T=0)[a]				
1)	-0.2837	0.0521	-0.1286	0.0355	-0.0237	-0.0483	0.0537
2)	-0.0661	0.0858	-0.2776	-0.0429	0.0692	0.0730	-0.0671
3)	0.1919	0.1345	0.0539	0.1395	0.0007	0.0229	-0.0219
4)	-0.2280	-0.0342	0.0258	0.1668	0.0547	0.0067	0.0285
5)	-0.1518	-0.0427	0.1002	0.0556	-0.0368	0.0618	-0.0299
6)	-0.0739	-0.0353	0.1195	-0.0634	0.0472	0.0117	0.0482
			(A=26; J=0,T=1 → J=1,T=0)[a]				
1)	-0.4328	-0.0790	-0.0266	-0.0532	0.0472	-0.0591	0.0707
2)	0.1010	0.0418	-0.1623	-0.1392	0.0331	0.0011	-0.0117
3)	0.0388	-0.0198	0.1623	0.0760	0.0482	-0.0585	0.0816
4)	-0.0292	-0.0869	0.1743	-0.1094	0.0054	0.0178	-0.0373
5)	-0.0849	-0.0678	-0.0549	-0.1842	-0.0380	0.0570	-0.0510
6)	0.0469	-0.0169	0.2085	-0.1203	-0.0498	-0.0495	0.0822
7)	0.1565	0.0311	-0.1135	-0.1315	0.0202	-0.0507	0.0388
8)	0.0481	0.0106	0.0378	-0.0448	-0.0923	-0.0540	0.0933
			(A=30; J=0,T=1 → J=1,T=0)[b]				
1)	-0.0515	-0.2002	0.1671	-0.4023	0.1223	-0.1149	0.0865
2)	-0.0272	-0.0132	0.1200	-0.2677	-0.0594	0.3801	-0.2194
3)	0.0897	-0.1623	0.0182	0.1484	-0.2103	0.0756	0.0048
4)	0.1189	0.2006	-0.0190	-0.2766	-0.2571	-0.0867	0.1035
5)	0.0301	0.0421	-0.0055	0.0324	0.1266	0.2359	-0.0595
6)	0.1132	-0.2391	-0.0300	0.0138	-0.0223	-0.0100	0.0911
7)	-0.1003	0.0173	0.0071	0.0760	-0.0581	-0.0060	0.0542
8)	0.0024	-0.0641	0.0170	-0.0854	-0.0524	-0.0922	-0.0052
			(A=34; J=0,T=1 → J=1,T=0)[b]				
1)	-0.0095	0.2096	-0.0538	-0.0153	-0.4272	0.2494	-0.0192
2)	0.0335	-0.0420	0.1136	-0.1374	0.2119	0.3371	-0.1269
3)	-0.0329	0.1427	-0.0129	-0.1899	-0.1472	-0.2159	0.1041
4)	-0.0674	0.2586	-0.0127	-0.0052	0.3610	0.0170	-0.0655
5)	-0.0194	0.0806	-0.0129	0.0308	0.0039	0.1381	-0.0055
6)	0.0001	-0.1403	0.0064	-0.0395	0.0365	-0.0948	-0.0433

Table 2.   Continued

#)	5/2←5/2	5/2←3/2	3/2←5/2	1/2←1/2	3/2←3/2	1/2←3/2	3/2←1/2
			(A=38; J=0,T=1 → J=1,T=0)^b				
1)	-0.0105	0.2554	-0.0610	-0.0408	-0.2350	0.3382	-0.0747
2)	0.0324	-0.2011	0.0480	0.0286	-0.5783	-0.0048	0.0011
3)	0.0178	-0.1997	0.0477	0.0724	0.2237	0.3227	-0.0713

^{a,b}Particle and hole notations as in Table 1.

Table 3.   One-body-transition-density matrix elements for single-nucleon orbits j and j' (5/2, 1/2 and 3/2) corresponding to ΔJ=1, ΔT=1 transitions from the J=0 ground states of A=22, 26 and 34 to the #)th J=1,T=2 states.

#)	5/2←5/2	5/2←3/2	3/2←5/2	1/2←1/2	3/2←3/2	1/2←3/2	3/2←1/2
			(A=22; J=0,T=1 → J=1,T=2)^a				
1)	-0.2165	0.0102	-0.0268	0.3319	-0.0217	-0.0384	0.0098
2)	-0.0236	0.1838	0.2277	0.1790	0.0240	-0.0086	0.0893
3)	0.1568	-0.0202	-0.0235	-0.2009	0.0093	-0.1016	0.0556
4)	-0.1197	0.0482	0.3568	-0.0955	-0.0241	0.0349	0.1788
5)	-0.0986	0.0165	0.2770	-0.0821	-0.0297	-0.0444	0.0136
6)	0.0901	-0.0154	-0.1198	0.0068	0.0473	0.1106	0.3819
			(A=26; J=0,T=1 → J=1,T=2)^a				
1)	-0.1546	-0.1520	-0.3826	0.0311	0.0010	0.1036	-0.0180
2)	0.2190	-0.0560	-0.2041	-0.2490	0.0373	0.0089	-0.0134
3)	-0.0399	0.1403	0.3616	0.0771	0.0261	0.0890	0.1112
4)	0.0004	0.0103	0.1991	-0.1591	-0.1269	-0.0329	-0.0666
			(A=34; J=0,T=1 → J=1,T=2)^b				
1)	0.0074	0.1786	0.1450	-0.0650	0.1037	0.6110	0.1628
2)	0.0316	-0.2388	-0.1574	0.2356	0.2056	0.2055	0.0649
3)	0.0473	0.0329	-0.0070	0.0683	0.0819	0.0490	-0.0930
4)	-0.0455	0.5801	0.0952	0.1726	-0.0941	-0.1088	0.0711
5)	0.0421	0.0573	0.0548	-0.0446	0.1520	-0.0178	0.0566
6)	0.0830	0.0003	-0.0072	-0.0405	-0.0293	0.0089	-0.0190

^{a,b}Particle and hole notation as in Table 1.

Table 4.   One-body-transition-density matrix elements for
              single nucleon orbits j and j' (5/2, 1/2 and 3/2)
              corresponding to ΔJ=1, ΔT=0 transitions from the
              J=0 ground states of A=20, 24, 28, 32 and 36 to
              the #)th J=1,T=0 states.

#)	5/2←5/2	5/2←3/2	3/2←5/2	1/2←1/2	3/2←3/2	1/2←3/2	3/2←1/2
			(A=20; J=0,T=0 → J=1,T=0)[a]				
1)	-0.0302	0.2356	0.2538	0.2061	-0.0087	0.0034	0.0169
2)	-0.0386	-0.0438	-0.2926	0.2499	-0.0067	-0.0692	-0.1832
3)	0.0194	0.0320	0.0768	-0.0674	-0.0150	-0.2348	-0.1793
4)	0.0729	0.0385	0.0653	-0.1311	-0.0950	-0.0196	-0.2774
			(A=24; J=0,T=0 → J=1,T=0)[a]				
1)	-0.0122	0.0744	-0.0264	0.0319	0.0128	0.0891	-0.0122
2)	0.0447	0.1819	0.4000	0.0649	-0.1042	0.0056	-0.0173
3)	0.0356	-0.0672	-0.1123	-0.0935	-0.0310	-0.0203	-0.0897
4)	-0.0205	-0.0097	-0.0248	0.0603	0.0192	-0.0846	-0.0630
			(A=28; J=0,T=0 → J=1,T=0)[b]				
1)	0.0607	0.0204	0.0473	-0.1023	-0.0812	0.1003	0.1961
2)	-0.0291	-0.2468	-0.6420	-0.0810	0.0800	0.0364	0.0039
3)	0.0868	-0.0192	-0.1425	-0.0859	-0.1353	-0.1585	-0.2093
4)	0.0276	-0.0193	-0.2858	0.1061	-0.0851	0.0308	0.1470
			(A=32; J=0,T=0 → J=1,T=0)[b]				
1)	-0.0356	0.0168	-0.0354	0.0132	0.0623	-0.5196	-0.2640
2)	-0.0086	0.5613	0.2827	-0.0435	0.0298	0.0051	-0.0062
3)	-0.0020	0.3184	0.1238	0.0063	0.0017	-0.0233	0.0028
4)	0.0499	-0.1236	-0.0126	-0.0160	-0.0883	-0.2884	-0.1622
			(A=36; J=0,T=0 → J=1,T=0)				
1)	0.0063	-0.1047	-0.0751	0.0139	-0.0162	0.4571	0.2754
2)	0.0067	-0.4350	-0.2394	0.0556	-0.0302	-0.1290	-0.0587
3)	0.0222	-0.1048	-0.0493	0.0044	-0.0430	-0.0325	0.0678
4)	-0.0113	-0.1323	-0.0098	-0.0559	0.0388	0.1167	0.0134

[a,b]Particle and hole notations as in Table 1.

Table 5. One-body-transition-density matrix elements for single nucleon orbits j and j' (5/2, 1/2 and 3/2) corresponding to $\Delta J=1$, $\Delta T=0$ transitions from the J=0 ground states of A=18, 22, 26, 30, 34 and 38 to the #)th J=1, T=1 states.

#)	5/2←5/2	5/2←3/2	3/2←5/2	1/2←1/2	3/2←3/2	1/2←3/2	3/2←1/2
			(A=18; J=0,T=1 → J=1,T=1)[a]				
1)	0.0000	-0.2411	-0.6031	0.0000	0.0000	-0.0420	-0.0867
2)	0.0000	0.0420	0.1049	0.0000	0.0000	-0.2411	-0.4985
			(A=22; J=0,T=1 → J=1,T=1)[a]				
1)	0.0014	-0.1209	-0.0413	-0.0526	0.0139	-0.1042	-0.0308
2)	0.0032	-0.0569	-0.0541	0.0398	-0.0185	-0.1270	0.0025
3)	0.0231	0.2663	0.6586	0.1037	-0.0760	0.0172	0.0294
4)	0.0748	0.0266	0.2404	-0.0434	-0.1261	0.0258	-0.0646
5)	-0.0129	-0.0111	0.1484	-0.0724	0.0471	0.1421	0.3567
6)	0.0071	-0.0870	-0.2365	-0.0269	-0.0048	0.0002	-0.0465
7)	0.0567	-0.1377	-0.4289	-0.0104	-0.1028	0.0348	0.0234
8)	0.0284	-0.0309	-0.0462	-0.0663	-0.1745	-0.1745	-0.1089
			(A=26; J=0,T=1 → J=1,T=1)[a]				
1)	0.0079	0.1140	0.1560	0.0247	-0.0227	0.1602	0.1560
2)	0.0304	0.0031	-0.0008	-0.0141	-0.0523	0.0276	0.0946
3)	0.0074	0.0410	0.0592	-0.0131	-0.0098	-0.1554	-0.2274
4)	-0.0254	0.1112	0.1650	0.0506	0.0316	-0.1283	-0.1828
5)	0.0291	0.2846	0.9002	0.0622	-0.0741	0.0198	0.0582
6)	-0.0406	0.0655	0.1843	0.0416	0.0629	0.0491	0.0310
7)	0.0220	0.0916	0.3538	-0.0080	-0.0386	-0.1453	-0.1957
8)	0.0010	-0.1123	-0.2938	-0.0371	0.0099	-0.0441	-0.0024
			(A=30; J=0,T=1 → J=1,T=1)[b]				
1)	-0.0695	0.0532	-0.0753	0.0380	0.1181	-0.5480	-0.2686
2)	0.0014	0.7399	0.2567	-0.1466	0.0437	-0.0708	0.0025
3)	0.1017	0.2762	0.1521	-0.1900	-0.1302	-0.3068	-0.1856
4)	-0.0656	0.5703	0.2413	-0.0012	0.1233	0.3696	0.2055
5)	-0.0251	0.2890	0.1328	0.1279	0.0066	-0.3467	-0.1893
6)	-0.0196	0.2485	0.0609	0.0113	0.0331	-0.0143	-0.0546
7)	-0.0206	0.6074	0.1458	0.0613	0.0191	0.0124	-0.0277
8)	-0.0548	-0.1637	-0.0755	0.0356	0.0914	-0.2802	-0.1260

Table 5. Continued

#)	5/2←5/2	5/2←3/2	3/2←5/2	1/2←1/2	3/2←3/2	1/2←3/2	3/2←1/2
			(A=34; J=0,T=1 → J=1,T=1)				
1)	0.0513	0.0127	0.0887	-0.0190	-0.0900	0.7144	0.1756
2)	0.0286	-0.1818	-0.0840	0.0107	-0.0568	0.5733	0.3206
3)	0.0254	-0.9479	-0.3959	0.0558	-0.0652	-0.2204	-0.1178
4)	-0.0292	-0.2218	-0.1124	0.0046	0.0531	0.4124	0.2115
5)	-0.0067	-0.3017	-0.0830	0.0522	-0.0039	0.0743	0.0402
6)	-0.0133	-0.0205	0.0186	-0.0384	0.0371	0.0933	0.0190
7)	-0.0607	0.3333	0.0419	0.0542	0.0964	0.1842	0.0441
8)	0.0539	0.0327	0.0481	-0.0567	-0.0823	-0.0755	0.0969
			(A=38; J=0,T=1 → J=1,T=1)				
1)	0.0000	-0.0118	-0.0028	0.0000	0.0000	0.8220	0.1816
2)	0.0000	0.8220	0.1962	0.0000	0.0000	0.0118	0.0026

^{a,b}Particle and hole notations as in Table 1.

The entries of Table 1 are the full shell-model contribution to that class of transitions in which the value of either T or T_z is changed. Such transitions for the T = 1 initial states of A = 18, 22, 26, 30, 34, and 38 are those corresponding to beta decay or (p,n)-type processes, which lead from (T,T_z) = (1,−1) to (1,0). For the T = 0 targets of A = 20, 24, 28, 32, and 36 such transitions can proceed either via the charge-exchange processes or via M1 excitation, leading from (T,T_z) = (0,0) to (1,±1) or (1,0), respectively. For the class of transitions in which both T and T_z remain unchanged, that is, the (T,T_z) = (1,−1) to (1,−1) transitions for the A = 18, 22, 26, 30, 34, and 38 nuclei induced by the M1 field, a ΔT = 0 contribution must be added (coherently) to the ΔT = 1 contribution based on the entries of Table 1. (In this text, the convention of isospin is followed in which the neutron has T_z = -1/2. However, as noted, entries for A ≥ 30 in Tables 1-5 have been calculated in "hole" rather than "particle" formalism. Hence, in using the numbers as listed for A ≥ 30, the above convention for T_z must be reversed; e.g. ^{34}S, the neutron-rich T = 1 nucleus in A = 34, must be taken to have T_z = +1, corresponding to its excess of "proton holes".)

There are additional classes of ΔJ = 1, ΔT = 1 transitions which can originate from the J^π = 0^+ ground states considered in Table 1. One of these classes is comprised by the charge-exchange transitions which connect the J^π, T = 0^+,1 ground states of the T_z = −1 nuclei to the J^π,T = 1^+,0 states of the corresponding T_z = 0 nuclei. The matrix elements of the one-body operators representing these transitions are presented in Table 2. These same physical processes of course will simultaneously populate the J^{π}, $T = 1^+$,1 states of the same residual $T_z = 0$ nuclei via the one-body transitions represented by the entries of Table 1, so that the observed spectra should exhibit a combination of $T = 0$ and $T = 1$ final states, both reached by $\Delta T = 1$ transitions. The other class of $\Delta T = 1$ transitions lead from the J^{π}, $T = 0^+$,1 initial states to $T = 2$ final states. These can have $T_z = 0, -1$ or -2, depending upon whether the physical probe is of the (p,n), M1, or (n,p) type. The matrix elements of the corresponding one-body operators are given in Table 3. Little is experimentally known at present about this class of transitions. In the $T_z = 0$ and $T_z = -1$ systems, formed respectively by the (p,n)-type and M1 processes, the $T = 2$ states occur several MeV higher in excitation than the lower-T states, but even so there could be some ambiguity in making their identification in the background of other $\Delta J = 1$ excitations. On the other hand, the $T = 2$ states comprise the entire low-lying spectrum of the $T_z = -2$ nuclei formed by processes of the (n,p) type. Comparison of such data to the spectra of the $T_z = -1$ and 0 systems should help in the sorting out of the different isospin states in these more complex situations.

There are two additional sets of $\Delta J = 1$ one-body-operator matrix elements which are necessary to a complete analysis of these dipole phenomena in the sd-shell. In Table 4 are presented the matrix elements for the $\Delta J = 1$, $\Delta T = 0$ operators which connect the J^{π}, $T = 0^+$,0 ground states of A = 20, 24, 28, 32, and 36 to the J^{π}, $T = 1^+$,0 states of these same systems. Identification of dipole states is made on the basis of angular momentum transfer, of course, and $T = 0$ and $T = 1$ states are not necessarily distinguishable unless the correspondences with the $T_z = -1$ analogs are precise and complete. The entries of Table 4 allow the complete theoretical spectra of positive-parity dipole excitations to be drawn for these nuclei, which should help in analyzing the experimental data.

The isoscalar ($\Delta J = 1$, $\Delta T = 0$) contributions to the transitions from the J^{π}, $T = 0^+$,1 ground states of A = 18, 22, 26, 30, 34, and 38 to the J^{π}, $T = 1^+$,1 final states of the same T_z can be obtained from the matrix elements of the corresponding one-body operators presented in Table 5. While it is expected that these isoscalar contributions are small relative to the isovector ($\Delta T = 1$) contributions, it seems nonetheless advisable to have explicit theoretical confirmation of this supposition. In comparing the results of M1 excitation ($T_z = -1$ to $T_z = -1$) to those of beta decay or (p,n)-type processes ($T_z = -1$ to $T_z = 0$), one of the various possible sources of difference is that the isoscalar matrix elements contribute to the M1 but not to the charge-exchange transitions.

In and of themselves, little in the way of recognizable pat-
terns can be immediately discerned in the magnitudes of the one-
body-transition densities of Tables 1—5. General aspects are
as follows. First, the different categories of transitions, as
classified by the groupings in Tables 1—5, do not appear to have
distinguishing characteristics, so the remaining comments apply
equally well to each table. Second, the transitions are characterized in almost all instances by several $a_j a_j'$ routes of comparable
importance. For transitions with "significant" (≥ 0.2) one-body-
transition matrix elements, typically two to four different matrix
elements are important. For transitions not exhibiting significant
one-body connections, most or all of the seven elements are of
the same magnitude. Thus, most transitions, strong or weak, are
characterized by a significant degree of interference, whether
constructive or destructive. However, this superposition of different one-body transition routes falls short of the sort of maxi-
mally possible coherence in which all of the one-body-transition
routes are significantly large.

The influence upon the entries of Tables 1—5 of the energy
spectrum of the single-nucleon orbits, which determines that,
as the sd shell is filled, the $d_{5/2}$ orbit tends to fill first,
followed by the $s_{1/2}$, with the $d_{3/2}$ orbit being last filled to
close out the shell at ^{40}Ca, seems to be small. The clearest
evidence of this underlying energy ordering appears to be that
$d_{3/2} - d_{3/2}$ transitions are unimportant below A = 32 and $d_{5/2} - d_{5/2}$
transitions unimportant above A = 28. Otherwise, the significant
participation of all other routes in all the cases A = 20—36 is
a striking illustration of the important role configuration mixing
plays in the present shell model calculations and strongly implies
that truncations of the full $d_{5/2} - s_{1/2} - d_{3/2}$ model space, with
their attendant impacts upon the details of the configuration
mixing, might significantly alter the predicted features of this
family of transitions.

SINGLE-PARTICLE MATRIX ELEMENTS

As was noted, the matrix elements of the one-body transition
operators $(a_j^+ \otimes a_j')$ presented and discussed in the preceding section yield the probability that the initial model wave function
can be transformed into the final via the various routes of annihilating a "j'" nucleon and creating back a "j" nucleon to take
its place. These matrix elements above, however, do not suffice
for the calculation of observable transition probabilities. To
obtain such predictions it is necessary to combine the shell-model
probabilities that the transitions $(a_j^+ \otimes a_j')$ occur between the
initial and final states with the probabilities that the physical
operator effecting the observed transition can convert a nucleon

from single-particle state "j'" to single-particle state "j".
For processes that occur at fixed (in practical examples, at zero)
momentum transfer these latter probabilities are called the "single-
particle matrix elements", (SPME), and are expressed as

$$\text{SPME } [\vec{O}_p(\Delta J, \Delta T); \; j, j'] = <j|||\vec{O}_p(\Delta J, \Delta T)|||j'> \tag{2}$$

where j and j' represent single-particle wave functions identified
by quantum numbers $n\ell j \equiv$ "j" and $\vec{O}_p(\Delta J, \Delta T)$ represents an operator,
of rank ΔJ and ΔT in angular momentum and isospin, respectively,
on the single nucleon state.

The single-particle matrix elements of the intrinsic spin
and orbital angular momentum operators, $\vec{s}$ and $\vec{\ell}$, are of central
importance to the physical processes by which $\Delta J^\pi = 1^+$ excitations
can be studied. Under the assumption that the effective operator
for finite nuclear systems is the same as that for free nucleons,
the magnetic dipole (M1) operator can be written as a combination
of $\vec{s}$ and $\vec{\ell}$ operators, with the coefficients taking values obtained
from the electric and magnetic properties of the free neutron and
proton as follows:

$$\vec{M1}(1, \Delta T) = (\frac{3}{16\pi})^{1/2} \{(g_p^\ell + (-1)^{\Delta T} g_n^\ell)\vec{\ell} + (g_p^s + (-1)^{\Delta T} g_n^s)\vec{s}\} \tag{3}$$

where $g_p^\ell = 1.0$, $g_n^\ell = 0.0$, $g_p^s = +5.586$, and $g_n^s = -3.826$ in units
of nuclear magnetons. The operator corresponding to Gamow-Teller
beta decay is, in the same "free nucleon" assumption, simply proportional to $\vec{s}$. The values of the matrix elements of $\vec{s}$, $\vec{\ell}$, and
M1 for the single-particle wave functions of the sd-shell orbits
are listed in Table 6.

The characteristics of the SPME values of Table 6 are fundamental to understanding how the microscopic structures of the many-
nucleon shell-model wave functions, as expressed by the condensa-
tion into the one-body transition densities of Tables 1–5, affect
the prediction of the strength of a transition initiated by one
or another type of operator. Essential points are that for $\vec{s}$,
the matrix elements 5/2 − 5/2, 3/2 − 5/2, and 1/2 − 1/2 are positive
and of approximately the same magnitude while the 5/2 − 3/2 and
3/2 − 3/2 elements are negative, the 3/2 − 3/2 being only about
half the magnitude of the others. For the $\vec{\ell}$ operator the 5/2 − 5/2,
5/2 − 3/2, and 3/2 − 3/2 are positive, with the 5/2 − 3/2 element
being significantly smaller in magnitude than the diagonal values,
and the 1/2 − 1/2 element is zero. Since neither s nor ℓ can connect $\ell = 2$ and $\ell = 0$ states, the 3/2 − 1/2 and 1/2 − 3/2 matrix
elements are identically zero in all the instances of Table 6.

Table 6. Matrix elements of the operators $\vec{s}$, $\vec{\ell}$, and $\vec{M1}$ evaluated
with the single-particle wave functions $0d_{5/2}$ (j = 5/2),
$1s_{1/2}$ (j = 1/2), and $0d_{3/2}$ (j = 3/2).

operator	ΔT	j-j'				
		5/2-5/2	5/2-3/2	3/2-5/2	1/2-1/2	3/2-3/2
$(\frac{3}{16\pi})^{\frac{1}{2}}\langle j|||\vec{\ell}|||j'\rangle$ (0)		2.00	0.54	-0.54	0.00	1.61
$(\frac{3}{16\pi})^{\frac{1}{2}}\langle j|||\vec{\ell}|||j'\rangle$ (1)		3.47	0.93	-0.93	0.00	2.78
$(\frac{3}{16\pi})^{\frac{1}{2}}(1.759)\langle j|||\vec{s}|||j'\rangle$ (0)		0.88	-0.94	0.94	0.74	-0.47
$(\frac{3}{16\pi})^{\frac{1}{2}}(9.412)\langle j|||\vec{s}|||j'\rangle$ (1)		8.16	-8.72	8.72	6.90	-4.36
$\langle j|||\vec{M1}|||j'\rangle$ 0		2.88	-0.41	0.41	0.74	1.14
$\langle j|||\vec{M1}|||j'\rangle$ 1		11.63	-7.80	7.80	6.90	-1.58

The essential characteristics of the M1 operator arise from
the values of the orbital and intrinsic spin g-factors of the neutron and proton as they enter into expression 3. Since the g^s
values are significantly larger than the g^ℓ values, the "s" terms
tend to dominate the M1 values. The "ℓ" contribution is most significant for the 3/2 — 3/2 matrix element, for which the two contributions are of opposite sign and of the same order of magnitude.
The rough similarity between the relative matrix elements of the
s and the M1(ΔT = 1) operator which is evident in Table 6 is the
basis for the numerous heuristic attempts to correlate the strengths
of Gamow-Teller beta decay with those of magnetic dipole decay.
Obviously, the correspondence will be closer for decays dominated
by 5/2 and 1/2 one-body transitions than for decays in which the
$d_{3/2}$ and $s_{1/2}$ orbits play the dominant roles.

The final significant characteristic of the entries in Table 6
arises from opposite signs of the magnetic moments of the neutron
and proton. The g^s factors combine so that in the isovector case
the coefficient of the s component of M1 is five times larger than
it is in the isoscalar case. Except for the 3/2 — 3/2 matrix elements, this enhancement of isovector over isoscalar carries through
to the full M1 value. The consequence of this enhancement is that
M1 transitions between many-nucleon nuclear states which can proceed
via ΔT = 1 one-body operators are greatly favored over those which
are constrained to ΔT = 0 paths only.

The nuclear processes of magnetic dipole decay and/or excitation and of Gamow-Teller beta decay constitute the most convenient sources available of experimental information on $\Delta J = 1$ excitations because such data can be reduced to yield a single number, the "reduced strength", characterizing each transition. Moreover, as was outlined above, the form of the operators is easily calculable, in the free-nucleon assumption, from the known properties of the neutron and proton. There are, moreover, other general aspects of the s, ℓ, and M1 operators, beyond those mentioned, which also aid the understanding and interpretation of observations of M1 and Gamow-Teller phenomena. The operators assumed for these processes have no explicit dependence upon the radial form $\rho(r)$ of the single particle wave functions (although there is the implicit assumption that the radial form is identical in the initial and final state). Hence, the values calculated for such matrix elements are independent of whether harmonic oscillator, Saxon-Woods or whatever are assumed for the radial functions. The s and ℓ operators do not connect states with different principal quantum numbers n, so that the M1 and Gamow-Teller processes do not mix configurations from different major oscillator shells. This has the consequence that what is experimentally observed as the M1 or Gamow-Teller "giant resonance" is, theoretically, an intra-sd-shell-space phenomenon for the nuclei under consideration, and hence, unlike the case for quadrupole excitations, should have a correspondence in this model. Finally, the inability of s and ℓ to connect $s_{1/2}$ and $d_{3/2}$ states means that even though wave functions may be related via significant one-body-transition-density values for $(3/2 - 1/2)$, the simple physical processes with which such connections can be probed are unable to detect these relationships. This comment is obviously most germane to the nuclei $A \geq 30$.

COMPARISON OF PREDICTIONS TO EXPERIMENT

The reduced strength for an electromagnetic transition such as an M1 transition is calculated from shell-model wave functions and single-particle matrix elements according to the formula

$$B(M1) = (2J' + 1)^{-1}(2T + 1)^{-1}[<\psi^{NJT}|||\vec{0}_P|||\psi^{NJ'T'}>_{\Delta T=0}\delta_{T\,T'} +$$

$$<T'T_z'10|TT_z><\psi^{NJT}|||\vec{0}_P|||\psi^{NJ'T'}>_{\Delta T=1}]^2\delta_{T_z T_z'} \tag{4}$$

where, as developed in the preceding sections,

$$<\psi^{NJT}|||\vec{0}_P|||\psi^{NJ'T'}>_{\Delta T} = \sum_{j,j'}[\text{SPME } 0_P(\Delta J,\Delta T);\ j,j']D^{N,JT,J'T'}_{\Delta J,\Delta T}(j,j'). \tag{5}$$

The "D" values of Tables 1 and 5 and the SPME of Table 6 have been combined according to formulae (4) and (5) to obtain predictions of the B(M1) values for magnetic dipole excitations of the $J^{\pi} = 0^{+}$ ground states of $18 \leq A \leq 38$ nuclei which lead to $J^{\pi} = 1^{+}$, T = 1 states. These calculated values are presented in Table 7, together with the calculated excitation energies of the 1^{+} states, and are shown in comparison to available experimental data.[9-14]

Before embarking upon a discussion of the comparisons presented in Table 7 we comment upon two general aspects of the calculated results. One concerns the intrinsic strength of the typical transitions and the other concerns the importance of the phase relationships between the various one-body transition routes.

In discussing electromagnetic transition strengths it is both customary and instructive to at least implicitly compare them to the strength of a pure single-particle transition. While there is some ambiguity in defining what is an appropriate value for the single-particle strength, a good working assumption follows from assuming a pure $d_{5/2} \rightarrow d_{3/2}$ transition, with D = 0.707 and SPME = 7.8. This leads to a B(M1) of 5 μ_n^2 for $J^{\pi} = 0^{+} \rightarrow J^{\pi} = 1^{+}$, $\Delta T = 1$ transitions. In this context, inspection of Table 7 reveals that even the strongest calculated (and experimentally observed) transitions are of the order or less than the strength of a single-particle transition. Hence, while the magnitudes of the strong transitions dealt with here are quite large relative to the usual values for M1 transitions, they are still, at their strongest, only fractions of a single-particle unit.

In obtaining the B(M1) values from the combination of one-body-transition densities and single-particle matrix elements, the magnitudes of the elements of each group are important, of course, but in cases for which the D values are individually of significant magnitudes, the determining factor governing the resultant transition strength is that of the phase relationships between the various one-body-transition paths. Of the phase relations noted for the SPME in the preceding section, the most systematically important one is that the $d_{5/2} \rightarrow d_{3/2}$ and $d_{3/2} \leftarrow d_{5/2}$ terms are of opposite sign. The importance of this can be gathered by inspection of the signs of the corresponding D values. The entries in Tables 1, 3, 4, and 5 systematically show values of D(5/2,3/2) and D(3/2,5/2) which have the same sign for a given transition. It follows that there is a systematic cancellation of strength between these two one-body-transition paths for those groups of transitions. For the transitions in Table 2 this phase relationship is reversed, so that the 5/2 ← 3/2 and 3/2 ← 5/2 contributions are constructive. (The same type of phase relation holds, in the same instances, between the D(3/2,1/2) and D(1/2,3/2) terms.)

Table 7. Comparison of calculations for the energies of $J^\pi = 1^+$, T = 1 states of $18 \leq A \leq 38$ nuclei and their M1 excitation strengths from the ground state to experimental values.

energy		B(M1) (μ_n^2)		energy		B(M1) (μ_n^2)	
theory	exp.	theory	exp.	theory	exp.	theory	exp.
(^{18}O:	J=0,T=1 $\rightarrow$ J=1,T=1)			(^{26}Mg:	J=0,T=1 $\rightarrow$ J=1,T=1)		
9.45		0.73		6.12		0.03	
10.10		0.02		7.28		0.00	
(^{20}Ne:	J=0,T=0 $\rightarrow$ J=1,T=1)			8.82		0.01	
				8.89		0.29	
11.29	11.249	1.82	2.04±0.36	9.78		0.15	
13.02		0.01		10.03		0.53	
14.43		0.00		10.31	10.20	0.11	1.3±0.6
15.41		0.00		10.91	10.65	0.21	1.6±0.4
15.60		0.03		(^{28}Si:	J=0,T=0 $\rightarrow$ J=1,T=1)		
17.13		0.04					
(^{22}Ne:	J=0,T=1 $\rightarrow$ J=1,T=1)			12.05 $\{+\begin{smallmatrix}10.60\\10.73\end{smallmatrix}\}$		0.33	0.5±0.2
				12.29	10.90	2.04	1.1±0.2
5.24	5.33	0.18	0.45±0.06	13.28	11.44	4.20	3.6±0.5
6.54	6.85	0.64	1.12±0.15	14.00	12.33	0.48	1.0±0.3
8.70	8.55	0.16	0.54±0.21	14.91	12.79	0.01	
9.27	9.16	2.53	1.86±0.18	15.46	(13.4)	0.00	
9.80	10.08	0.12	0.57±0.10	(^{30}Si:	J=0,T=1 $\rightarrow$ J=1,T=1)		
10.81	10.84	0.33	0.27±0.10				
11.81		0.38		4.30	3.77	0.07	
11.93		0.01		6.85		0.03	
(^{24}Mg:	J=0,T=1 $\rightarrow$ J=1,T=1)			7.34		0.00	
				7.67		0.00	
10.22	9.97	0.57	1.3±0.3	8.25		0.00	
11.14	10.71	3.51	3.2±0.6	8.65		4.07	
13.30	(12.9)	0.31		8.98		0.62	
13.59	(13.1)	0.23		9.24		0.03	
14.37		0.24		(^{32}S:	J=0,T=0 $\rightarrow$ J=1,T=1)		
14.77		0.01					
14.96		0.02		7.12	7.00	0.04	
15.33		0.13		8.24	8.13	1.32	1.4±0.7
16.03		0.08		9.18	9.21	0.35	
16.48		0.09		10.02	9.66	0.64	0.4±0.3
16.68		0.03		10.61		0.52	
16.96		0.00					

Table 7 Continued.

energy		$B(M1)$ (μ_n^2)		energy		$B(M1)$ (μ_n^2)	
theory	exp.	theory	exp.	theory	exp.	theory	exp.
(^{32}S: J=0,T=0 → J=1,T=1; cont.)				(^{36}Ar: J=0,T=0 → J=1,T=1)			
10.96	11.14	2.36	3.6±1.2	8.01	7.71	0.03	
11.59	11.62	0.31	1.6±0.8	8.55	8.13	0.44	
12.03		0.03		9.70	9.22	0.02	(0.6±0.3)
12.48		0.71		10.27	10.05	1.13	1.6±0.5
12.78		0.04		11.76	11.25	0.00	1.6±0.5
13.11		0.00		12.33	(12.09)	1.29	(0.7±0.5)
13.30		0.05		12.85		0.00	
				13.36		0.28	
(^{34}S: J=0,T=1 → J=1,T=1)				14.34		0.00	
				14.50		0.04	
4.00	4.07	0.14		15.44		0.10	
5.88	5.38	0.04		15.57		0.02	
7.15		0.11					
7.35		0.19		(^{38}Ar: J=0,T=1 → J=1,T=1)			
8.18		2.93					
8.57		0.02		5.08	(5.55)	0.00	
9.31		1.04		8.44		3.15	
9.59		0.01					

The relative phases of the SPME and the D values for the 5/2 − 5/2 to 3/2 − 5/2 contributions are such that, for the special class of the strongest transitions at least, these two paths tend to contribute constructively. The strongest transition in each of the systems A = 20—28 can be characterized as the one in which the 5/2 ← 5/2 and 3/2 ← 5/2 contributions are both "large" and of the same sign. Above A = 28, the situation is more complex, with the strongest transitions tending to be those in which 1/2 ← 1/2 and 3/2 ← 5/2 (particle notation) paths constructively interfere. The 3/2 ← 3/2 path is relatively unimportant because of the smallness of the corresponding SPME.

The comparison in Table 7 of calculated to experimentally identified excitation energies of $J^\pi = 1^+$, T = 1 states suggests that, with the exception of ^{28}Si, the calculated 1^+, T = 1 spectra are accurate facsimilies of the actual ones both in terms of relative locations in the total spectra and in terms of density of levels. In the case of ^{28}Si, the mass furtherest removed from

the energy level data used to fix the Hamiltonians, the spectrum
is distorted by an excessively depressed ground state energy, so
that calculated excitations uniformly are about 1.5 MeV too high.
For ^{20}Ne, ^{24}Mg, and ^{28}Si the lowest 1^+, T = 1 state occurs at 10—11
MeV excitation. For ^{32}S and ^{36}Ar this excitation energy drops
to 7—8 MeV. For ^{22}Ne and ^{26}Mg, the lowest 1^+ states occur at 5—6 MeV
excitation, while for ^{30}Si and ^{34}S this excitation energy drops to
4 MeV. The lowered excitation energies of the 1^+, T = 1 states
above ^{28}Si seem understandable from the consideration of the two-
particle states of identical nucleons. Two neutrons in the two
most energetically favored orbits for A $\leq$ 28, namely $d_{5/2}$ and $s_{1/2}$,
cannot combine to form a 1^+ state. The less energetically favored
pair $d_{5/2} - d_{3/2}$ is necessary, followed by the $s_{1/2} - d_{3/2}$ pair.
Above A = 28, the most energetically favored pair of orbits, $s_{1/2}$
and $d_{3/2}$, can form a 1^+ state and the excitation energy is corre-
spondingly lowered.

The M1 properties of the $J^\pi = 1^+$, T = 1 states relative to
the ground states also appear to accurately reflect the corresponding characteristics of the observed states. The observed characteristics for ^{20}Ne, ^{24}Mg, and ^{28}Si are that the M1 strength is concentrated into a single state for ^{20}Ne, into two states for ^{24}Mg and
into four states (summing over the two fragments of the lowest
1^+, T = 1 state which is split by isospin mixing with a 1^+, T = 0
state) for ^{28}Si. The predicted M1 properties of these systems
correctly reproduce this observed increasing fragmentation with
increasing mass and, in each mass individually, correctly match
the distribution of strength over the states of the spectrum. Thus,
the lowest 1^+, T = 1 state of ^{20}Ne is correctly predicted to dominate
the spectrum, the second 1^+, T = 1 state to be strongest in ^{24}Mg,
and the third such state to be strongest in ^{28}Si, some strength
appearing the lower states in both of the latter cases, however.
In addition, the total predicted strength for the lowest several
states for each system is in good agreement with the corresponding
observed total strengths.

In ^{32}S four states are definitely assigned M1 strength. The
second, fourth, sixth, and seventh 1^+, T = 1 states calculated
in the model have M1 properties consistent with these experimental
data. The corresponding energies are also in reasonable accord,
so it appears that while the M1 spectrum of ^{32}S is considerably
more complex in its characteristics than is the case for the lighter
N = Z system, the present calculations continue to quantitatively
replicate the observed feature. The increased complexity of ^{32}S
is not unexpected, since the $s_{1/2} - d_{3/2}$ type of excitation, energetically favored in this region as noted above, cannot give rise
to M1 strength because of the characteristics (ℓ-conservation)
of the M1 operator.

Four states in ^{36}Ar are experimentally assigned as M1 excitations, two strong transitions definitely, two weaker ones tentatively. The predicted spectrum of 1^+, T = 1 excitations also shows
four states appreciably excited, in the experimentally observed
order of weak, strong, strong, weak. However, as best as the energies of the observed and predicted 1^+, T = 1 spectra can be recon-
ciled, there are mismatches in the one-to-one correspondence of
M1 characteristics; e.g., the model predicts that the fourth and
sixth 1^+, T = 1 states have the largest M1 strength, while experimentally it appears that the strongest states are the fourth and
fifth. In any case, as continued to be true also for ^{32}S, the
predicted energy centroid and summed total strength of the M1 excitation in the first fifteen MeV of excitation in ^{36}Ar agrees with
the experimental result.

Of the T_z = −1 nuclei, a thorough comparison between prediction
and experiment is possible only for ^{22}Ne. The observed distribution
of M1 strength for ^{22}Ne is quite complex but seems replicated in
detail as well as in average properties by the calculations. The
calculations for ^{26}Mg 1^+, T = 1 states unfortunately were not
extended high enough in excitation energy to encompass the dominant
M1 strength. The predicted summed strength below 11 MeV excitation
(encompassing a total of eight 1^+ states) is only a small fraction
of the total calculated ground-state strength and is significantly
smaller than the observed strength between 10 and 11 MeV excitation.
The prediction for the M1 spectrum of ^{30}Si is of a single dominant
1^+ state (the sixth) at 8.5 MeV excitation. For ^{34}S two 1^+ states,
at 8 and 9 MeV excitation, are predicted to dominate the M1 spectrum.

There are only two observed Gamow-Teller decays which can
be directly compared to the M1 results of Table 7. The decay
of the lowest 1^+ state of ^{24}Na to the ground state of ^{24}Mg has
an experimental Gamow-Teller matrix element of 0.75 ± 0.018, compared to the theoretical value, assuming the free neutron decay
rate, of 0.151. This yields a ratio of theoretical to experimental
matrix elements $<GT>_{th}/<GT>_{exp}$ = 2 ± 0.5. On the other hand, the
comparison of M1 strengths in Table 7 yields $<M1>_{th}/<M1>_{exp}$ = 0.66 ±
0.20. Some of the apparent inconsistancy between these two results
(an inconsistancy based on the assumption of the analogy between
the M1 and the Gamow-Teller processes) disappears under the detailed
examination of the $\vec{D}$ values for this transition and the SPME of
$\vec{s}$ versus those of $\vec{M1}$. The two largest d values, for 5/2 − 5/2
and 5/2 − 3/2, are of the approximate same size and sign. For
$\vec{s}$, the cancelation of these two terms (see Table 6) is almost exact,
where for $\vec{M1}$, the 5/2 − 5/2 to 5/2 − 3/2 magnitude difference con-
stitutes the largest contribution to the total matrix element.
In essence, the ℓ part of the M1 operator contributes two thirds
of the total M1 matrix element, thus making the Gamow-Teller to
M1 analogy rather flimsy. While this observation does not yield

an obvious resolution of why the two predictions both deviate from
experiment, and in opposite directions, there are a large number
of possibilities in which relatively small changes in the small
and canceling D values could push the Ml value up and the Gamow-
Teller value down.

There are no Ml data for the other conjunction of Ml and Gamow-
Teller decay, that corresponding to the decay of ^{32}P to ^{32}S. The
theoretical Gamow-Teller matrix element for this transition incorporates very large cancellations (the resultant is about one tenth
of a single-particle unit) but it is still twenty times larger
than the almost completely inhibited experimental transition. It
can be expected that the Ml strength differs in detail from the
beta decay strength. In contrast to these two weak J^π, $T = 0^+, 0$
to $1^+, 1$ beta transitions, the other Gamow-Teller decays involving
our chosen set of $J^\pi = 0^+$ ground states for which data exist are
relatively strong. These are transitions proceeding from J^π,
$T = 0^+, 1$ to $1^+, 0$ states, corresponding to the D values of Table 2.
Comparisons are presented in Table 8. While agreement for aggregate
strengths in these decays is as good as for the Ml phenomena of
Table 7, the individual results perhaps show more scatter. The
accurate characterization of the wave functions of $T = 0$, $J^\pi = 1^+$
states is traditionally one of the most difficult challenges to
a shell-model calculation and our failure to meet this challenge
with complete success probably accounts for the discrepancies.

Predictions of transition strengths based on Tables 3 and 4
(and most of 2) are not presented, although of course the B(Ml)
values follow immediately by combining the D-values with the SPME
of Table 6. To generate theoretical excitation spectra the relevant excitation energies of the $T = 0$ and $T = 2$ states are also
necessary. These are given in Table 9.

Comparisons of predictions based on the D values of Tables 1—3
and data involving hadronic interactions or non-zero momentum transfer take us outside the scope of the present study, which was
directed towards phenomena for which the physical probes are osten-
sibly simple in order to facilitate concentration on the nuclear
many-body aspects of the problem. The combination of the one-body-
transition density matrices with appropriate reaction and effective
operator theories will, of course, extend the utility of Tables 1—3
to a much wider range of experimental studies than just B(Ml) and
log ft measurements.

DISCUSSION

The present status of the ongoing comparison of theory to
experiment in the field of $\Delta J^\pi = 1^+$, $\Delta T = 1$ excitations can, on
the basis of the preceding, be summarized as following. The

B. H. WILDENTHAL AND W. CHUNG

Table 8. Comparison of calculated and measured matrix elements[7]
for Gamow-Teller beta decays connecting J^π, T = 0^+,1
states with 1^+,0 states.

decay	final state index	matrix elements	
		measured	calculated
$^{18}Ne \rightarrow {}^{18}F$	1	2.219 ± 0.008	2.293
$^{22}Mg \rightarrow {}^{22}Na$	1	1.148 ± 0.021	1.574
$^{26}Si \rightarrow {}^{26}Al$	1	1.331 ± 0.029	1.620
	2	0.960 ± 0.034	0.916
	3	0.449 ± 0.056	0.986
$^{30}Si \rightarrow {}^{30}P$	1	0.507 ± 0.014	0.233
	2	0.115 ± 0.023	0.425
	3	1.375 ± 0.048	0.829
$^{34}Ar \rightarrow {}^{34}Cl$	1	0.190 ± 0.015	0.274
	2	0.318 ± 0.009	0.107
	3	0.656 ± 0.045	1.019
	4	1.446 ± 0.073	2.012
$^{38}Ca \rightarrow {}^{38}K$	1	<0.319	0.934
	2	1.637 ± 0.104	2.291
	3	0.861 ± 0.144	0.809

theoretical replication of observed Ml and Gamow-Teller phenomena -
in specific details, in "gross" or summed properties, and in systematic trends as a function of neutron and proton number - is
successful enough to justify proceeding to work further on the
assumption that the microscopic structures of such excitations
are, in the typical case, given correctly. This assumption con-
stitutes the foundation for expanded research on several fronts.
Theoretically, it should be instructive to analyze the present
one-body-transition densities in detail in order to understand
the structure of the various $\Delta J^\pi = 1^+$ excitations. New calculations,
which increase the average accuracy of predicted properties and
remove the present clear-cut discrepancies with observation, should
be feasible at the next level of improvement in effective Hamiltonians. Experimentally, more extensive and more precise Ml excitation data for the targets considered here would be extremely useful
in providing a complete and unambiguous catalog of 1^+, T = 1 states
together with their accurate B(Ml) values. Such data, in conjunction

Table 9. Calculated excitation energies of $J^\pi,T = 1^+,0$ states
in A = 20, 24, 28, 32, and 36 and of $J^\pi,T = 1^+,2$ states
in A = 22, 26, and 34.

mass	ground state (J^π,T)	excitation energy	mass	ground state (J^π,T)	excitation energy
20	$0^+,0$	13.19	36	$0^+,0$	7.06
		14.33			8.23
		14.63			10.88
		15.18			11.25
24	$0^+,0$	7.78	22	$0^+,1$	15.51
		10.03			16.46
		12.64			17.17
		13.20			18.26
28	$0^+,0$	10.00	26	$0^+,1$	13.93
		10.73			15.39
		12.70			16.39
		13.32			16.70
32	$0^+,0$	4.92	34	$0^+,1$	10.99
		7.14			12.06
		8.50			13.33
		8.76			14.75

with present and future theory for the many-body aspects of nuclear
structure, should allow delineation of the degree to which the
effective M1 operator in finite nuclei differs from the free nucleon
form of eq. (3). The study of M1 excitations in odd-mass targets
of the sd-shell should also become fruitful now that it appears
that the increased complexity of such data, resulting from the
multiplicity of final state spins, can be countered by the inter-
pretive assistance provided by nuclear structure predictions.

The study of $\Delta T = 1$ excitations initiated by probes presumably
more complex than the M1 and Gamow-Teller operators should be fruitful both in that new groups of excitations are encompassed, such
as with the (p,n) and (n,p) or (^{3}He,t) and (t,^{3}He) types of reactions leading to T = 0 or T = 2 states, and in that the same T = 1
states populated with the M1 process can be comparatively studied
with probes which, while similar, differ in details to degrees
as yet uncertain. Such comparisons of different effective operators

as probes are, of course, much more productive when the many-body aspects of the nuclear excitations are theoretically understood.

In the foregoing we have attempted to develop the thesis that $\Delta J^\pi = 1^+$ excitations constitute a field of exceptional opportunity for nuclear research. This opportunity arises from the conjunction of the development of a comprehensive and accurate theory for the many-body aspects of the nuclear excitation, as outlined here, a growing theoretical endeavor to understand the details of the nucleon transition operators in the finite-nucleus environment, which we have not discussed here at all, and great advances in the capabilities for experimentally studying these phenomena. These advances include new high-intensity, high-resolution electron accelerators and spectrometers, high-resolution, high-energy facilities for studying (p,p) and (p,n) reactions, and a variety of improved studies with mesonic probes made possible with the new meson factories. The focussing of these diverse approaches onto a compact field of concentration, where they interact with and amplify each other, should yield a significant improvement in our understanding of nuclear structure.

ACKNOWLEDGMENTS

This material is based upon work supported by the National Science Foundation under Grant No. Phy-7822696.

REFERENCES

1. W. Chung, Thesis, Michigan State University, 1976.
2. J.B. French, E.C. Halbert, J.B. McGrory, and S.S.M. Wong, Advances in Nuclear Physics, Vol. 2, ed. M. Baranger and E. Vogt, Plenum Press, 1969.
3. R.R. Whitehead, A. Watt, B. Cole, and I.J. Morrison, Advances in Nuclear Physics, Vol. 9, ed. M. Baranger and E. Vogt, Plenum Press, 1977.
4. T.T.S. Kuo, Nucl. Phys. A103, 71 (1967).
5. B.H. Wildenthal, Elementary Modes of Excitation in Nuclei, ed. R. Broglia and A. Bohr, Soc. Italiana de Fisica, 1977.
6. B.H. Wildenthal, NUKLEONIKA 23, 459 (1978).
7. B.H. Wildenthal and W. Chung, Mesons in Nuclei, ed. M. Rho and D.H. Wilkinson, North-Holland Publ., 1979.
8. B.A. Brown, W. Chung, and B.H. Wildenthal, Phys. Rev. Lett. 40, 1631 (1978) and to be published.
9. L.W. Fagg, Rev. Mod. Phys. 47, 683 (1975).
10. P.M. Endt and C. van der Leun, Nucl. Phys. A310, 1 (1978).
11. U.E. Berg, K. Wienhard, and H. Wolf, Phys. Rev. C11, 1851 (1975).
12. X.K. Maruyama et al., Phys. Rev. C10, 2257 (1974).
13. U.E.P. Berg and K. Wienhard, private communication.
14. R. Schneider and A. Richter, private communication.

ELEMENTS OF COMPARISON BETWEEN ELECTRON-NUCLEUS AND NUCLEON-

NUCLEUS SCATTERING

F. Petrovich

Department of Physics
Florida State University
Tallahassee, Florida 32306

ABSTRACT

The relationship between electron-nucleus scattering and
nucleon-nucleus scattering is reviewed. Specific features of
this relationship are used to obtain information on the effective
interaction for nucleon-nucleus scattering. Emphasis is on the
100-200 MeV incident energy region which is currently being investigated experimentally at the Indiana University Cyclotron
Facility. It is shown that a local t matrix interaction with
central, spin-orbit, and tensor parts gives a good description
of many features of the experimental data. Normal parity transitions of approximate isoscalar character (including elastic
scattering), isovector dipole transitions in the (p,n) reaction,
and transitions to high spin states of 'stretched' configuration
are all discussed in some detail. The manner in which different
parts of the interaction are manifested in the cross section is
illustrated in the examples. Some new (π,π') and proton analyzing
power data for 'stretched' states in ^{28}Si is discussed briefly.

INTRODUCTION

Electrons and nucleons have been available as probes of the
nucleus for a number of years now; however, most of the nucleon-
nucleus scattering data has been limited to incident energies in
the 20-60 MeV range and the quality of the electron-nucleus
scattering data has often been limited by poor energy resolution.
It is only recently that high quality proton beams with energies
greater than 100 MeV have become available and that significant

advances in the resolution of electron scattering measurements
have been made. The Indiana University Cyclotron Facility and
the MIT Bates Laboratory are examples of experimental installation
with these new capabilities. The first does nucleon work, of
course, while the second does electron work.

In electron scattering the coupling between the incident
electron and the target nucleus is electromagnetic in origin and
thus understood in principle. Rather definite information about
many aspects of nuclear structure can be obtained from this reaction provided the experimental data are of good quality. In nucleon
scattering at incident energies in the 20-60 MeV range the precise
nature of the effective interaction which provides the coupling
between the incident nucleon and the target nucleus is uncertain.
There has been progress in this energy region, and some connection between the effective interaction and the free two-nucleon
data has been achieved in calculations based on bound state G
matrix interactions. Some of this work is reviewed in Ref. 1 and
in another contribution these these proceedings.[2] In nucleon
scattering at incident energies greater than 100 MeV the impulse approximation[3] provides a definite prescription for the
effective interaction, namely the free two-nucleon t matrix which
can be obtained directly from the free two-nucleon scattering
data. In view of the above, there is much to be gained at this
time, from the simultaneous study of electron-nucleus and nucleon-
nucleus scattering.

In the present paper, the relationship between electron-
nucleus and nucleon-nucleus scattering is reviewed. Certain
features of this relationship are exploited to obtain information
about the effective interaction for nucleon-nucleus scattering.
The emphasis is on the 100-200 MeV incident energy region, although a few applications at lower energies are discussed. The
local t and G matrix interactions described in Ref. 2 are used
in these calculations. These have central, spin-orbit, and
tensor parts. The results to be presented serve to illustrate
the role of the different parts of the effective interaction in
the nucleon scattering cross sections. They also allow for an
assessment of the validity of the t matrix and G matrix descriptions of nucleon-nucleus scattering. Elastic scattering, normal
parity transitions of approximate isoscalar character, dipole
transitions in the (p,n) reaction, and abnormal parity transitions
to 'stretched' high spin states are specific examples which are
discussed in some detail. An interesting comparison between pion-
nucleus and nucleon-nucleus scattering is made for the case of
'stretched' excitations in ^{28}Si and some forthcoming proton
analyzing power data for these same excitations is mentioned.

THE CORRESPONDENCE BETWEEN ELECTRON AND NUCLEON-NUCLEUS SCATTERING

The essential relations for the microscopic descriptions of electron-nucleus and nucleon-nucleus scattering are given in this section. The purpose is to clearly illustrate the manner in which information may be taken from electron scattering to nucleon scattering. The relations also serve to elucidate the interplay of the various components of the effective interaction in the nucleon scattering cross sections. Initially, Born approximation is assumed and the non-local knockout exchange amplitudes associated with antisymmetrization in nucleon-nucleus scattering are ignored. This Born description is followed by a discussion of methods for including distortion and knockout exchange amplitudes. Some general comments concerning the calculations of this work close the section. The presentation here is necessarily brief due to space limitations. A more complete exposition on these matters, based on momentum space methods discussed elsewhere,[1,4] is currently in preparation.[5]

Electron-Nucleus Scattering

In Born approximation the differential cross section for electron-nucleus scattering is given by[6]

$$\frac{d\sigma}{d\Omega} = \frac{4\pi\sigma_M}{\eta} \left\{ \left| F_L(q) \right|^2 + (\tfrac{1}{2} + \tan^2 \tfrac{1}{2}\theta) \left| F_T(q) \right|^2 \right\} \tag{1}$$

where σ_M is the Mott cross section, η is a recoil factor, θ is the scattering angle, q is the momentum transfer, and $\left| F_L(q) \right|^2$ and $\left| F_T(q) \right|^2$ are the longitudinal and transverse form factors which contain the essential information about nuclear structure. The difference in the angular dependence of the longitudinal and transverse terms in the cross section allows for the separate determination of the longitudinal and transverse form factors. At $\theta=\pi$ only the transverse form factor survives.

At this point the discussion will be limited to transitions in nuclei with spin-zero ground states for which there is a unique value, J, for the total angular momentum transfer in both the electron and nucleon-nucleus scattering reactions. This restriction on the amplitudes contributing to the cross sections is essential for making sensible comparisons between the two reactions. One distinguishes between two types of transitions,

1.   normal parity → $\Delta\pi = (-1)^J$
2.   abnormal parity → $\Delta\pi = (-1)^{J+1}$

where $\Delta\pi$ is the parity change in the transition. Normal parity transitions include elastic scattering and inelastic transitions to collective states such as 2^+ and 3^- levels. Abnormal parity transitions include inelastic transitions to magnetic dipole states (1^+ levels) and to 'stretched' states, e.g. 6^- levels in s-d shell nuclei.

For normal parity transitions the longitudinal electron scattering form factor is just the Bessel transform of the ground state proton density in the case of elastic scattering or the proton transition density in the case of inelastic scattering. Specifically,

$$|F_L(q)|^2 = (2J + 1)|\rho_p^J(q)|^2 \tag{2}$$

$$\rho_p^J(q) = \int_0^\infty j_J(qr)\rho_p^J(r)r^2 dr \tag{3}$$

$$\rho_p^J(r) = <J_f||\sum_i \frac{\delta(r-r_i)}{r_i^2} Y_J(\hat{r}_i)||J_i> \tag{4}$$

where the operator appearing in the reduced matrix element in Eq. (4) is the Jth multipole of the density operator

$$\hat{\rho}_{op}(\bar{r}) = \sum_i \delta(\bar{r} - \bar{r}_i) \tag{5}$$

and the sum on i is over protons only. Strictly speaking, electron-nucleus scattering measures the proton charge density distribution which differs from the proton point density distribution because of the finite size of the proton.[6] It is the latter distribution which enters the description of nucleon-nucleus scattering. To relate the two reactions it is necessary to correct for the finite size of the nucleon. This is understood, although it is not indicated explicitly in the formulae. There is also a transverse form factor in the case of normal parity transitions. This involves coupling to spin and current distributions in the target. Although these transverse form factors are interesting in themselves, the corresponding amplitudes for the nucleon-nucleus scattering reaction are rather small. This is mainly the result of collective correlations in the target nucleus.[7] These are not important for any of the present applications and will not be discussed in detail here.

For abnormal parity transitions the longitudinal form factor vanishes. The transverse form factor is given by

$$|F_T(q)|^2 = |\frac{q\hbar}{2Mc} \sum_{\lambda,\alpha} \{c_\lambda^\ell g_\ell^\alpha \rho_{J\lambda}^{\ell\alpha}(q) + c_\lambda^s g_s^\alpha \rho_{J\lambda}^{s\alpha}(q)\}|^2 \qquad (6)$$

where C_λ are constants,[6] λ is the orbital angular momentum transfer which is restricted to J-1 and J+1, α can assume the values p and n which distinguish between proton and neutron quantities, g_ℓ^α and g_s^α are the orbital and spin g-factors, and $\rho_{J\ell}^{\ell\alpha}$ and $\rho_{J\lambda}^{s\alpha}$ are the Bessel transforms

$$\rho_{J\lambda}^{\ell\alpha}(q) = \int_0^\infty j_\lambda(qr)\rho_{J\lambda}^{\ell\alpha}(r)r^2 dr \qquad (7)$$

$$\rho_{J\lambda}^{s\alpha}(q) = \int_0^\infty j_\lambda(qr)\rho_{J\lambda}^{s\alpha}(r)r^2 dr \qquad (8)$$

of the orbital current and spin transition densities defined by

$$\rho_{J\lambda}^{\ell\alpha}(r) = <J_f||\sum_i \frac{\delta(r-r_i)}{r_i^2} [Y_\lambda(\hat{r}_i) \times \bar{\ell}_i]^J||J_i> \qquad (9)$$

$$\rho_{J\lambda}^{s\alpha}(r) = <J_f||\sum_i \frac{\delta(r-r_i)}{r_i^2} [Y_\lambda(\hat{r}_i) \times \bar{\sigma}_i]^J||J_i> \qquad (10)$$

with the sums on i running over protons or neutrons as α = p or n, respectively. The operators appearing inside the reduced matrix elements in Eq. (9) and Eq. (10) should be recognized as the Jλ multipoles of the current and polarization operators:

$$\hat{\bar{J}}_{op} = \{\sum_i \delta(\bar{r}-\bar{r}_i)\bar{v}_i\}_{sym} \qquad (11)$$

$$\hat{\bar{P}}_{op} = \sum_i \delta(\bar{r}-\bar{r}_i)\bar{\sigma}_i \qquad (12)$$

Abnormal parity transitions in electron-nucleus scattering are more complicated than normal parity transitions in the sense that they do not generally provide information on a unique density distribution.

Nucleon-Nucleus Scattering

The differential cross section for nucleon-nucleus scattering is given by[8]

$$\frac{d\sigma}{d\Omega} = (\frac{\mu}{2\pi\hbar^2})^2 \frac{k_f}{k_i} \frac{1}{2J_i+1} \sum_{\substack{M_f M_i \\ m_f m_i}} |T_{M_f m_f, M_i m_i}|^2 \tag{13}$$

where μ and k are the relativistic reduced energy and wave number, respectively, T is the transition amplitude, and m and M denote projectile and target spin projections. In Born approximation the transition amplitude is

$$T_{M_f m_f, M_i m_i} = \int e^{-i\overline{k}_f \cdot \overline{r}_p} <m_f|<J_f M_f|\sum_i v_{ip}|J_i M_i>|m_i>e^{i\overline{k}_i \cdot \overline{r}_p} d^3 r_p \tag{14}$$

where v_{ip} is the effective interaction which is assumed to be of the form

$$v = v_0^C(r) + v_\sigma^C(r)\overline{\sigma}_1 \cdot \overline{\sigma}_2 + v_\tau^C(r)\overline{\tau}_1 \cdot \overline{\tau}_2 + v_{\sigma\tau}^C(r)\overline{\sigma}_1 \cdot \overline{\sigma}_2 \overline{\tau}_1 \cdot \overline{\tau}_2$$

$$+ \{v_0^{LS}(r) + v_\tau^{LS}(r)\overline{\tau}_1 \cdot \overline{\tau}_2\}\overline{L} \cdot \overline{S} \tag{15}$$

$$+ \{v_0^T(r) + v_\tau^T(r)\overline{\tau}_1 \cdot \overline{\tau}_2\}S_{12}$$

where r is the relative nucleon-nucleon separation, C, LS, and T refer to central, spin-orbit, and tensor, and $\overline{\sigma}$, $\overline{\tau}$, $\overline{L} \cdot \overline{S}$, and S_{12} are the usual spin, isospin, spin-orbit, and tensor operators.[2,7,9] The matrix element

$$U_p(\overline{r}_p) = <J_f M_f|\sum_i v_{ip}|J_i M_i> \tag{16}$$

is recognized as the scattering potential seen by the projectile. Strictly speaking this depends on $\overline{\sigma}_p$, $\overline{\tau}_p$, and $\overline{p}_p$ as well as $\overline{r}_p$.

A brief consideration of the tensors obtained in a multipole decomposition of the effective interaction shows that the central component of the interaction provides coupling between the projectile and the density and spin density distributions of the

target. The tensor component of the interaction is bilinear in
the spin operators and only provides coupling between the projectile and spin density distributions of the target. The spin-orbit
component of the interaction provides coupling between the
projectile and the density, spin density, current, and spin-
current distributions of the target. These points are evident in
the final Born expressions[5] for the cross sections for $0 \to J$
normal parity transitions

$$\frac{d\sigma}{d\Omega} = 4\pi \left(\frac{\mu}{2\pi\hbar^2}\right)^2 \frac{k_f}{k_i}(2J+1)\left\{\left|\sum_\alpha v_\alpha^C(q)\rho_\alpha^J(q)\right|^2 + \left|\sum_\alpha v_\alpha^{LS}(q)\rho_\alpha^J(q)\right|^2\right\}$$

(17)

and for $0 \to J$ abnormal parity transitions

$$\frac{d\sigma}{d\Omega} = 4\pi \left(\frac{\mu}{2\pi\hbar^2}\right)^2 \frac{k_f}{k_i}(2J+1)\left\{\left|\sum_{\lambda,\alpha} v_\alpha^{LS}(q)B_\lambda \rho_{J\lambda}^{s\alpha}(q)\right|^2 + \sum_\lambda \left|\sum_\alpha [v_\alpha^C(q)\rho_{J\lambda}^{s\alpha}(q)\right.\right.$$

(18)

$$+ v_\alpha^T(q)\sum_{\lambda'} Z_{\lambda\lambda'}^J \rho_{J\lambda'}^{s\alpha}(q) + v_\alpha^{LS}(q)A_\lambda \rho_J^{\ell\alpha}(q)]\Big|^2\Bigg\}$$

in the nucleon-nucleus scattering reaction. Here the $v(q)$ are
Bessel transforms of the interaction components defined in Eq.
(15). In writing Eq. (17) the small terms involving the target
spin and current distributions that were mentioned previously
have been ignored. In Eq. (18) A_λ, B_λ, and $Z_{\lambda\lambda'}^J$ are coefficients
which can be obtained elsewhere[5,10,11] and $\rho_J^{\ell\alpha}(q)$ is a particular
linear combination of the $\rho_{J\lambda}^{\ell\alpha}(q)$. Equation (17) and Eq. (18) are
to be compared with the electron scattering expressions of Eq.
(2) and Eq. (6), respectively. It should also be noted that the
products $v(q)\rho(q)$ are the essential parts of the scattering potential $U(q)$ in momentum space.

Distortion

With the possible exception of electron scattering from very
light nuclei, Born approximation is not adequate for the description of electron and nucleon-nucleus scattering.[6,8] Elastic scattering is strong because the projectile "sees" the whole nucleus.
To describe the scattering properly it is necessary to construct a
scattering potential and solve the appropriate wave equation. In the
case of electron scattering the elastic scattering potential is just
the Coulomb potential of an extended charge distribution, i.e.
the average of the Coulomb interaction over $\rho_p(r)$. For nucleon
scattering the nuclear part of the scattering potential is just

Eq. (16) with $J_i = J_f$. Inelastic scattering is generally weaker than elastic scattering and is most often described in the distorted wave approximation. In this approximation the plane waves of Born approximation are replaced by exact elastic scattering wave functions. In Born approximation for each scattering angle and asymptotic momentum transfer $q = q(\theta)$ a definite momentum component of the densities $\rho(q)$ or scattering potentials $U(q)$ is sampled. When distortion is present, the essential effect is that a range of momentum components of the densities or scattering potentials contribute to the cross section at any particular scattering angle. As an example note that the central term in Eq. (17)

$$|\sum_\alpha v_\alpha^C(q)\rho_\alpha^J(q)|^2 \to |\sum_\alpha \int k^2 dk D(k,\theta) v_\alpha^C(k)\rho_\alpha^J(k)|^2 \qquad (19)$$

when distortion is present.[4,5] Here $D(k,\theta)$ is a distortion function which typically peaks at $k = q(\theta)$, at least for scattering angles where the cross section is large.

Distortion effects are treated properly in all of the calculations of this work. The main point of the above discussion is that distortion effects do not destroy the basic utility of the Born relations for electron and nucleon-nucleus scattering given in the preceding sections. Often it is possible to make meaningful comparisons directly from the Born expressions (like that on the left in Eq. (19)) by simply introducing scale factors and/or momentum shifts to approximate the effect of the distortion integrals (like that on the right in Eq. (19)). More information on the momentum space treatment of distortion effects is available in Refs. 4,5.

Knockout Exchange Amplitudes

Knockout exchange amplitudes can be included in Eq. (14) by making the replacement

$$v_{ip} \to v_{ip}(1-P_{ip}) \qquad (20)$$

where P_{ip} is the permutation operator which exchanges all coordinates of the ith and pth particles. This replacement adds an integral to Eq. (14) which is non-local in the coordinates appearing in the scattering wave functions. For the central and spin-orbit parts of the interaction, there are accurate approximate prescriptions for[1,12,13] including this non-local knockout

exchange amplitude in the local direct term by adding an energy
dependent term to the interaction. No corresponding approximation is available for the tensor interaction. There are also
computer codes for evaluating these integrals exactly.[14] These
knockout exchange amplitudes have been properly included in all
of the calculations of this work. The approximate methods have
been used in some cases. In other cases, particularly where the
tensor interaction is important, these have been treated exactly.
In all figures displaying the effective interaction in the text
below, the energy dependent modifications of the central and spin-
orbit interactions corresponding to knockout exchange are included.

General Comments

Aside from the details associated with distortion and knockout exchange, the basic idea of the calculations of this work is
to deduce $\rho(q)$ from the electron scattering data, construct
$U(q) = v(q)\rho(q)$, use these to generate the nucleon-nucleus scattering cross section, and compare the results with the experimental data. The normal parity case is the most straightforward.
To the extent that $\rho_p^J(q)$ and $\rho_n^J(q)$ are simply related only a single
density enters Eq. (2) and Eq. (17). For elastic scattering the
ground state density distributions are involved. For most nuclei
these satisfy $\rho_n(q) \approx N/Z\, \rho_p(q)$ to a good approximation.[15] For
the inelastic excitation of collective levels $\rho_p^J(q) \approx \rho_n^J(q)$ even
in nuclei with N≠Z.[16,17] Simplifying assumptions such as these
are made in the calculations to follow. At another level one
might hope to use electron-nucleus and nucleon-nucleus scattering
to study the small differences between proton and neutron density
distributions. For the case of abnormal parity transitions,
multiple densities are involved in Eq. (6) and Eq. (18) and, in
general, there is no unambiguous connection between electron-
nucleus and nucleon-nucleus scattering. For magnetic dipole excitations and for transitions to states of 'stretched' configuration, certain simplifications occur and a meaningful comparison
between the two reactions can be made. These simplifications
will be discussed at the appropriate points in the text below.

GENERAL PROPERTIES OF THE EFFECTIVE INTERACTIONS

Several local representations of the complete G matrix interaction have been presented in Ref. 9. In addition, two local
versions of the complete t-matrix interaction have appeared
recently in the literature.[18,19] These have all been discussed
in some detail in Ref. 2. The lower energy calculations of this
work make use of the version of the G-matrix referred to as M3Y

in Ref. 2. The higher energy calculations make use of the central
and spin-orbit components of the t matrix interaction of Ref. 18
and the M3Y tensor interaction. The latter is quite close to
the tensor part of the t matrix.[2] A brief discussion of the
general properties of the complete t matrix interaction follows.
The specific properties of the G matrix interaction that are of
interest here, will be introduced at the appropriate points in
the later text.

    The moduli of the Bessel transforms of the components of
the t matrix interaction are shown in Fig. 1. These have been
grouped into sets of components which contribute to nucleon-
nucleus scattering amplitudes with no spin transfer to the target
(S=0) and with spin transfer to the target (S=1). The former are
the most important in normal parity transitions and only the

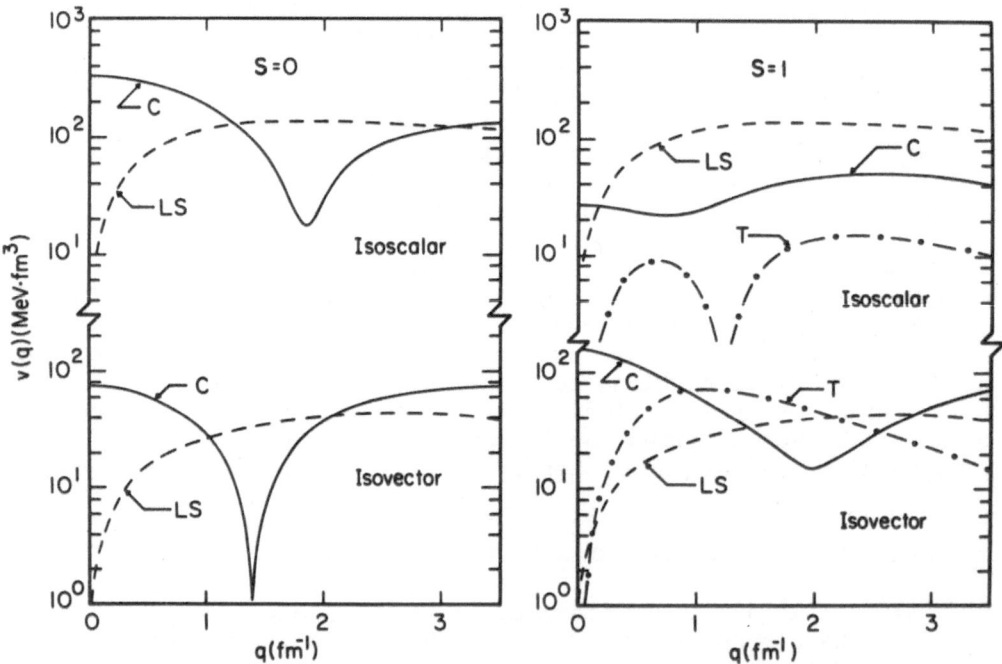

Fig. 1.   The moduli of the Bessel transforms of the complex t
          matrix interaction of Refs. 2,18. C, LS, and T cor-
          respond to central, spin-orbit, and tensor, respectively.

latter contribute to abnormal parity transitions. The components
are also separated' into isoscalar and isovector parts. These
contribute to transitions with T=0 and T=1 transfer to the target,
respectively. The central and spin-orbit parts of the interaction have been modified to include the contributions from knockout exchange at E = 135 MeV according to the approximation of
Refs. 1,12,13. This has not been done for the tensor parts.

For normal parity transitions the central parts of the
interaction are most important at low q and the spin-orbit components are most important at high q. The minima in the moduli
of the S=0 central parts of the interaction are sign changes in
the t matrix associated with the short-range repulsion in the
nucleon-nucleon force. For isoscalar normal parity transitions
which sense momentum components of t with q $\approx$ 2fm^{-1}, the spin-
orbit parts of t will be dominant. The isovector terms in the
S=0 parts of the interaction are considerably weaker than the
isoscalar terms. Because of this S=0 cross sections will be
minimally sensitive to differences between proton and neutron
target densities, whenever they are small.

For isovector abnormal parity transitions the central part
of t is dominant at low q and the spin-orbit and tensor parts
become more important with increasing q. For isoscalar abnormal
parity transitions the central interaction is weak and the spin-
orbit interaction is strong. For both isoscalar and isovector
abnormal parity transitions, the tensor interaction is more
important than Fig. 1 implies because of knockout exchange and
the relative sizes of the coefficients appearing in Eq. (18).
This will be seen clearly in the results to be discussed below.

APPLICATIONS

Elastic Scattering

Accurate experimental cross sections for the elastic scat-
tering of 80-180 MeV protons from ^{28}Si, ^{40}Ca, ^{90}Zr, and ^{208}Pb
have been determined at IUCF.[20,21] The distorted waves used in
most of the inelastic scattering calculations of this work have
been generated from phenomenological optical potentials deduced
from this data.[20,21] Consistency demands that the t matrix
interaction used in these microscopic inelastic scattering calculations also gives a reasonable description of elastic scattering.
An impulse approximation study of this elastic scattering data
is underway, but is still in its early stages.[22]

In these calculations the central and spin-orbit parts of
the optical potential are obtained by folding the t matrix inter-

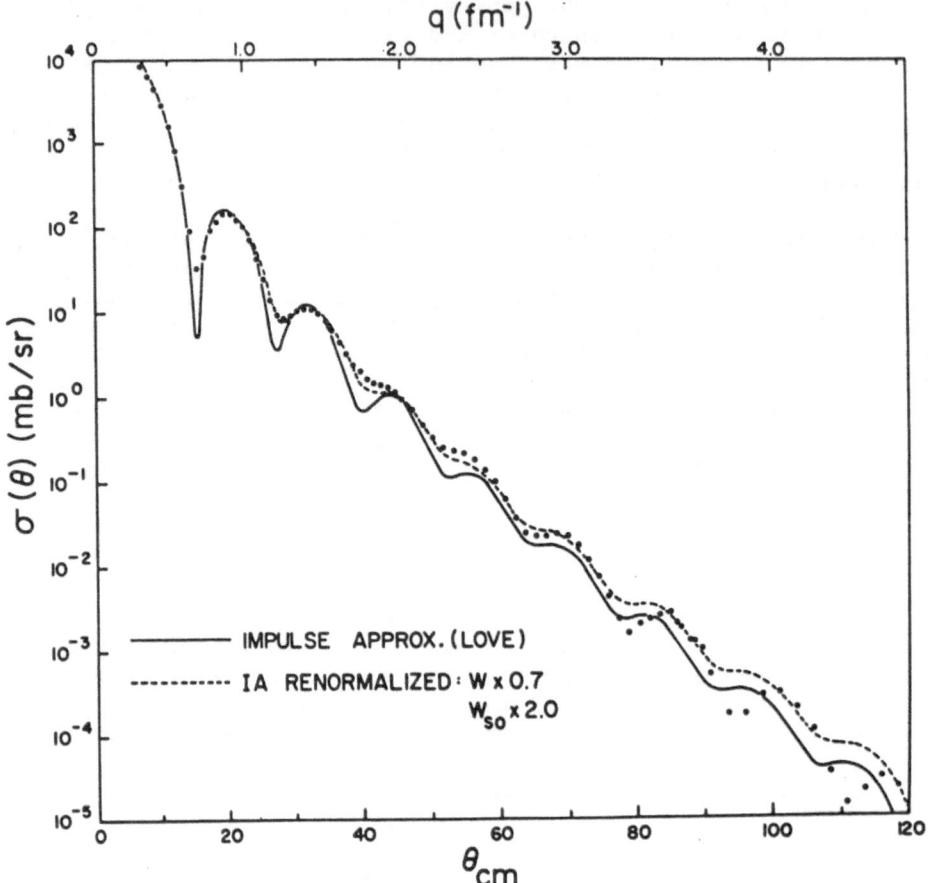

Fig. 2. The results of impulse approximation calculations for
         the elastic scattering of 135 MeV protons from ^{90}Zr. SO
         refers to spin-orbit.

action of Fig. 1 with density distributions describing the ground
states of the target nuclei. In momentum space folding refers to
the operation $U(q) = v(q)\rho(q)$. The particular density distributions
used were taken from recent impulse approximation studies[23-25] of
nucleon-nucleus elastic scattering at 800 MeV. The elastic cross
sections are obtained by solving the Schrödinger equation containing
these calculated potentials with knockout exchange terms included
via the approximations of Refs. 1,12,13. Typically it is found
that the impulse approximation underestimates the radius of the real
part of the central optical potential and overestimates the strength
of the imaginary part of the central optical potential. These

deficiencies, which are not large, are symptomatic of the need for
including Pauli blocking corrections to the t matrix interaction.

Figure 2 shows that result obtained for the elastic scattering
of 135 MeV protons from ^{90}Zr. Even without corrections the agreement between theory and experiment is quite reasonable. By reducing
the strength of the central imaginary potential by 30% and increasing the weak spin-orbit imaginary potential by about 100%, the

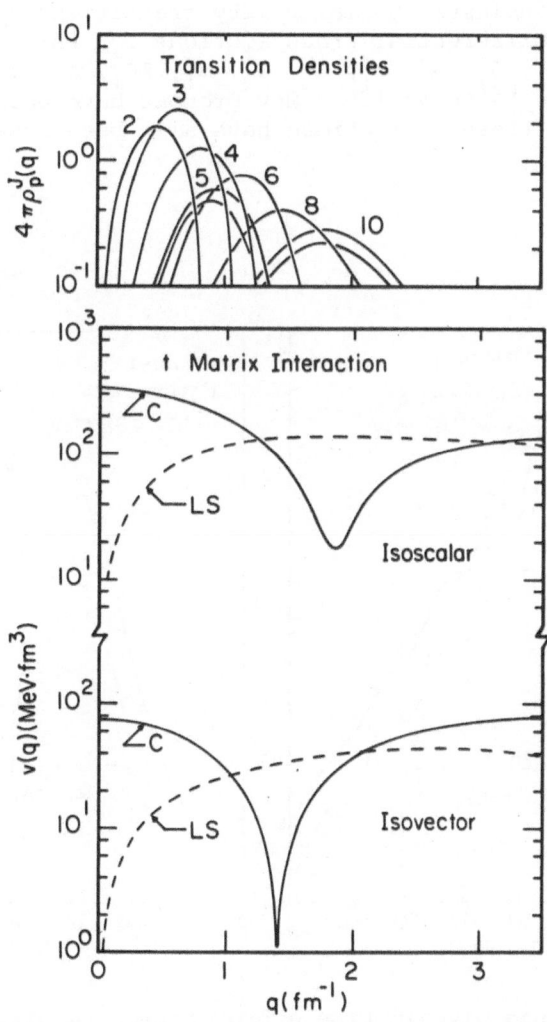

Fig. 3.   The proton transition densities from electron scattering
          for the J=2 to 10 transitions in ^{208}Pb and the moduli of
          the S=0 components of the t matrix interaction of Fig. 1.

theoretical results are substantially improved. Another result for
the elastic scattering of 182 MeV protons from ^{208}Pb is shown in
Fig. 7 of Ref. 2. Overall the results obtained to date for elastic
scattering seem satisfactory.

Inelastic Normal Parity Transitions

The elastic scattering cross sections are primarily sensitive
to the low momentum components of the S=0 part of the t matrix interaction. This information can be extended to a wider range of q
by considering inelastic normal parity transitions to states of
various spin. Differential cross sections for the inelastic exci-
tation of the 3_1^-, 5_1^-, 5_2^-, 2_1^+, 4_1^+, 6_1^+, 8_1^+, 10_1^+, and 10_2^+ normal
parity levels in ^{208}Pb by 135.2 MeV protons have been measured at
IUCF.[26] All of these excitations have also been observed in

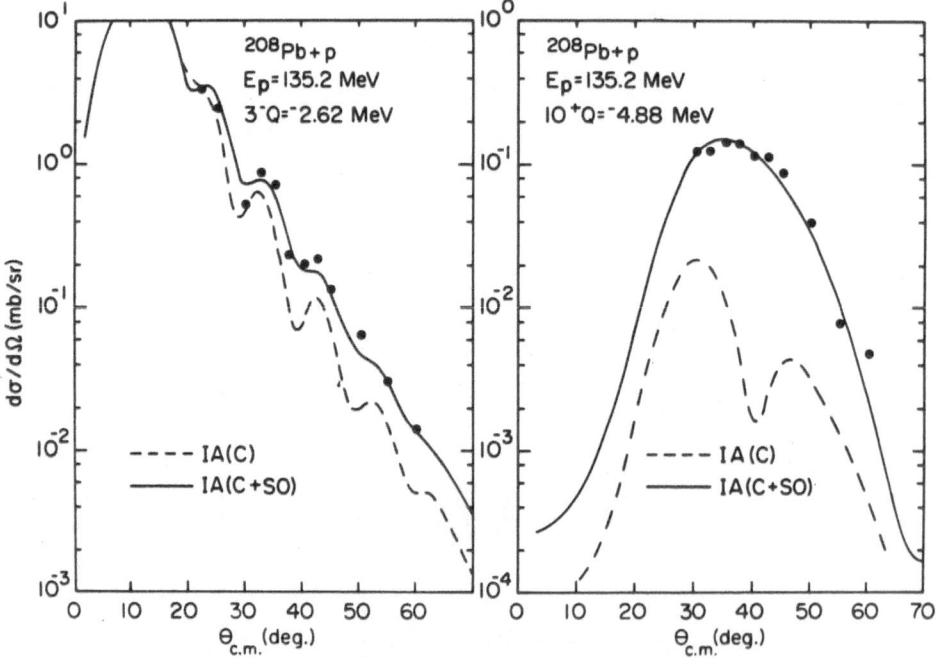

Fig. 4. Distorted wave impulse approximation results for the in-
         elastic excitation of the 3_1^- and 10_1^+ levels in ^{208}Pb by
         135.2 MeV protons.

inelastic electron scattering.[27-31] Distorted wave impulse approximation calculations have been made for these transitions.[32] The prescription for constructing the scattering potentials in these calculations is essentially the same as described above for elastic scattering. The proton transition densities were taken from the electron scattering studies[27-31] and it was assumed that $\rho_n(q) = \rho_p(q)$ throughout.

The proton transition densities extracted from the electron scattering data for the nine transitions in ^{208}Pb are compared in the S=0 parts of the t matrix interaction in Fig. 3. From Fig. 3 it is clear that the 2^+ through 10^+ transitions in ^{208}Pb will sample the interaction over the momentum range $q \approx 0.3-2.2$ fm^{-1}. The nucleon-nucleus scattering cross sections should show a transition from a central interaction shape to a spin-orbit shape over this range of q since the isoscalar central interaction virtually disappears at $q \approx 2$ fm^{-1}. This is clearly seen in the distorted wave impulse approximation results for the 3^-_1 and 10^+_1 transitions shown in Fig. 4. It also shows as a break between the peak magnitudes of the (e,e') and (p,p') cross sections as a function of J which is illustrated in Fig. 5. The break occurs at J=8. The fact that the experimental cross sections follow closely the irregular behavior of the S=0 isoscalar part of the t matrix interaction is particularly encouraging. Electron scattering data corresponding to a possible 12^+ state at 6.06 MeV observed in the proton scattering experiment of Ref. 26 would help to shed more light on this interesting point. It is also important to extend the calculations described here to isoscalar normal parity transitions in N=Z nuclei where the assumption $\rho_n(q) = \rho_p(q)$ is on firmer ground.

Magnetic Dipole Excitations in the (p,n) Reaction

The initial (p,n) experiments at IUCF have emphasized the determination of forward cross sections mainly for transitions in N=Z targets.[33-35] Under these conditions one expects to see abnormal parity transitions to the analogs of the 1^+ T=1 levels of the target. These data contain information on the low momentum components of the isovector S=1 part of the effective interaction. OPEP is effective in this spin-isospin channel and it is of particular interest to know how close the effective interaction is to OPEP at low q.

The first experiment[33,34] was carried out at an incident energy of 62 MeV and included the targets of ^{12}C, ^{24}Mg, and ^{28}Si. Although this energy is somewhat below the energy region of primary interest in this paper, it is an instructive situation to consider. Low q electron scattering data is available for the 1^+ T=1 excitations in these three nuclei.[36-38] An experimental β-decay

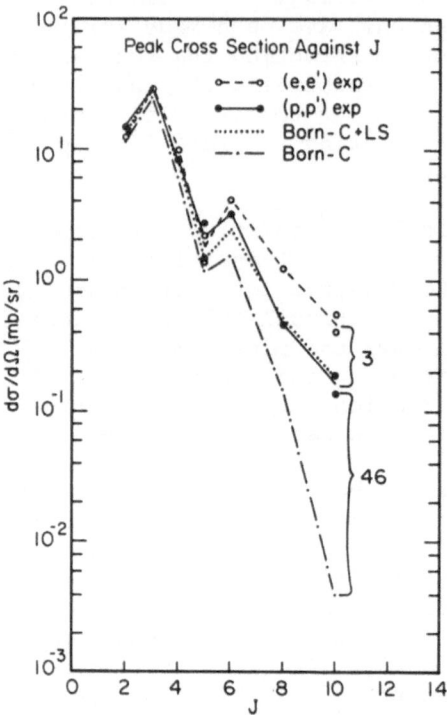

Fig. 5. The peak experimental (p,p') cross sections are shown as
         a function of J and compared with the peak experimental
         (e,e') form factors and the results of the theoretical
         (p,p') calculations. The (e,e') form factors have been
         normalized to the (p,p') cross section for the 3⁻ cross
         sections. Note the break between the (e,e') and (p,p')
         data for J=8 which is associated with the rapid decrease
         in the central contribution to the (p,p') cross sections.

matrix element is also available for the single transition that
has been observed in ^{12}C.[39] A theoretical study of the 1^+ T=1
transitions in ^{12}C, ^{24}Mg, and ^{28}Si has been made and a complete
set of results for the (p,n) reaction at 62 MeV have been obtained.[40]
This study is based on the shell model wave functions of Cohen and
Kurath (^{12}C)[41] and Chung and Wildenthal (^{24}Mg, ^{28}Si).[42]

    The results of the study of Ref. 40 are summarized in Table
1 and Fig. 6. The table lists the theoretical and experimental

Table 1. Comparison of Shell Model Results with Experiment for Isovector M1 Transitions in ^{12}C, ^{24}Mg, and ^{28}Si

Nucleus	Final State	E_x^{exp} (MeV)	E_x^{th} (MeV)	$B(M1\uparrow)^{exp}$ ($e^2 fm^2$)	$B(M1\uparrow)^{th}$ ($e^2 fm^2$)	R_{tr}^{exp} (fm)	R_{tr}^{th} (fm)	$\rho_{10}^s(10)$	$\rho_{10}^{\ell}(10)$
^{12}C	$1_1^+ T=1$	15.11	15.08	$(3.22\pm0.33)\times10^{-2}$	2.54×10^{-2}	$2.70\pm.20$	2.67	-.221	.026
^{24}Mg	$1_1^+ T=1$	9.97	10.22	$(1.30\pm0.21)\times10^{-2}$	6.31×10^{-3}	$3.05\pm.44$	3.10	-.035	-.339
	$1_2^+ T=1$	10.72	11.14	$(4.10\pm0.77)\times10^{-2}$	3.85×10^{-2}	$2.94\pm.14$	2.94	-.221	-.210
	$1_3^+ T=1$		13.30		3.41×10^{-3}		3.45	-.156	.361
	$1_4^+ T=1$		13.59		2.53×10^{-3}		2.75	.090	-.104
^{28}Si	$1_1^+ T=1$	10.48	12.05	$(5.99\pm2.75)\times10^{-3}$	3.86×10^{-3}	$3.90\pm.40$	2.78	.022	.293
	$1_2^+ T=1$	10.86	12.29	$(1.28\pm0.27)\times10^{-2}$	2.23×10^{-2}	$2.98\pm.25$	2.95	.186	.080
	$1_3^+ T=1$	11.41	13.28	$(4.01\pm0.77)\times10^{-2}$	4.64×10^{-2}	$2.58\pm.23$	2.97	-.292	.001
	$1_4^+ T=1$	12.27	14.00	$(1.13\pm0.29)\times10^{-2}$	5.26×10^{-3}	$2.93\pm.36$	2.87	-.098	-.001

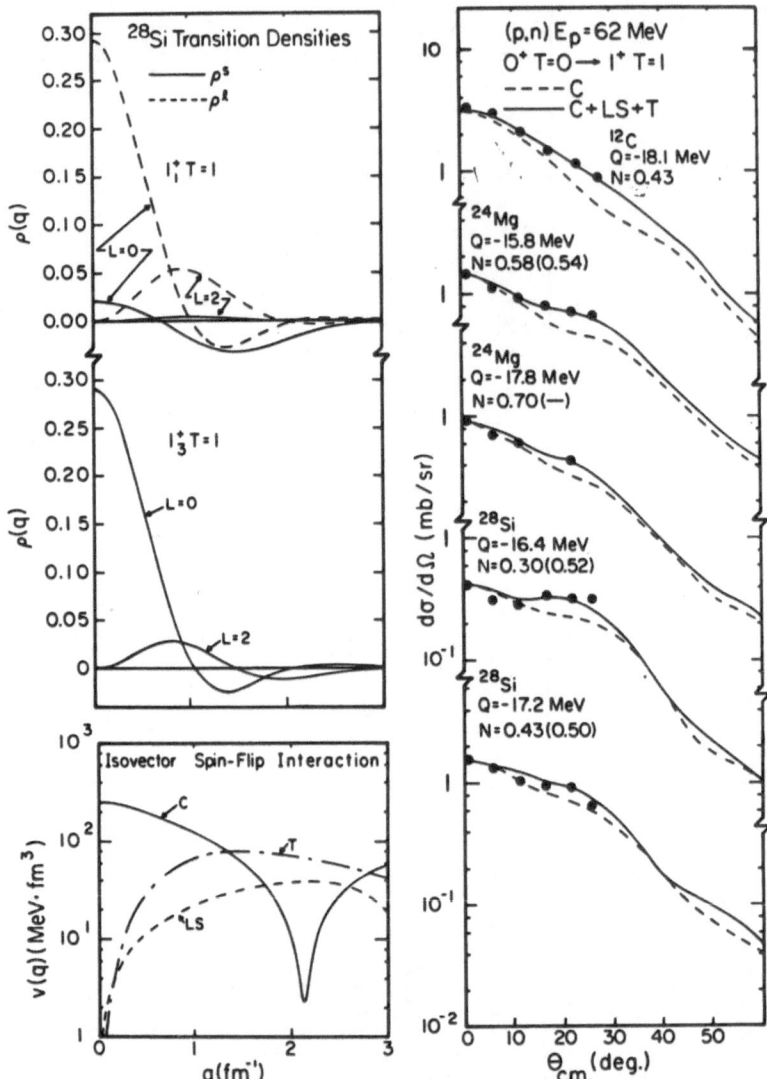

Fig. 6.   On the left, spin and orbital current transition densi-
          ties for 1_1^+ and 1_3^+ T=1 excitations in ^{28}Si compared with
          isovector S=1 parts of G matrix interaction displayed in
          the same format as in Fig. 1. L in figure corresponds
          to λ in text. On the right, results of distorted wave
          calculations normalized to data as described in the text.
          The levels observed in ^{24}Mg and ^{28}Si correspond to 1_2^+
          and 1_3^+ T=1 excitations listed in Table 1.

excitation energies, reduced transition probabilities B(M1↑), and transition radii R_{tr} for the first 1^+ T=1 excitation in ^{12}C and the first four 1^+ T=1 levels in ^{24}Mg and ^{28}Si. The B(M1↑) and R_{tr} are the two parameters required to describe the low q electron scattering data.[37,38] The last two columns in the table contain the values of the theoretical λ=0 spin and orbital current transition densities at q=0. The complete set (λ=0 and λ=2) of theoretical spin and orbital current transition densities for the 1_1^+ and 1_3^+ T=1 excitations in ^{28}Si are compared with the isovector S=1 components of the G matrix interaction on the left of Fig. 5. The results of theoretical distorted wave calculations are compared with the experimental (p,n) data on the right in Fig. 6.

From Table 1 it is clear that the theoretical wave functions give a good qualitative account of the experimental electromagnetic data. One can distinguish between two types of states. One type might be called current correlated, i.e. $\rho_{10}^\ell >> \rho_{10}^s$. The 1_1^+ T=1 levels in both ^{24}Mg and ^{28}Si are of this type. These are weak in electron scattering because $g_\ell^1 \approx .2 g_s^1$ where $g^1 = \frac{1}{2}(g^p - g^n)$. They are weaker still in the (p,n) reaction at low q because ρ_{10}^ℓ couples to the projectile through the spin-orbit interaction which is essentially zero at q=0. These levels were not seen in the 62 MeV experiment of Ref. 33,34. The other type might be called spin dominated, i.e. $g_s^1 \rho_{10}^s > g_\ell^1 \rho_{10}^\ell$. All of the other excitations in the table fall in this group. Using the fact that the tensor interaction, the spin-orbit interaction, and the λ=2 transition densities are small at q=0, Eq. (6) and Eq. (18) reduce approximately to

$$\left| F_T(q) \right|^2 \approx \left| \frac{q\hbar}{2Mc} \left(C_0^\ell g_\ell^1 \rho_{10}^{\ell 1}(q) + C_0^s g_s^1 \rho_{10}^{s1}(q) \right) \right|^2 \tag{21}$$

$$\frac{d\sigma}{d\Omega} \approx 8\pi \left(\frac{\mu}{2\pi\hbar^2} \right) \frac{k_f}{k_i} 3 \left| v_{\sigma\tau}^C(q) \rho_{10}^{s1}(q) \right|^2 \tag{22}$$

for the spin dominated transitions at low q. These relations clearly show that the scaling of the forward (p,n) cross sections for spin dominated transitions measures the scaling of the λ=0 spin transition densities at low q and departures in the scaling of M1 rates and forward (p,n) cross sections are a direct consequence of the λ=0 orbital current transition densities at low q.

Equation (21) and Eq. (22) do not allow for a model independent determination of $v_{\sigma\tau}^C(q)$ at low q since there are two experimental values, $\left| F_T(q) \right|^2$ and dσ/dΩ, and two structure unknowns, ρ_{10}^{s1} and $\rho_{10}^{\ell 1}$, for each transition. β-decay is more useful in this regard because allowed Gamow-Teller decays provide a direct

measure of $\rho_{10}^{s1}(0)^{43,44}$ which can be used in Eq. (22). The
$^{12}N(1^+) \to {}^{12}C(0^+)$ β^+ rate[36,39] implies that $\rho_{10}^{s1}(0) = .223$ for the
$0^+ \to 1_1^+$ T=1 transition in ^{12}C. This is quite close to the value of
.221 obtained from the Cohen and Kurath wave functions. It is
concluded that the discrepancy between the experimental and
theoretical B(M1↑) for this transition is associated with the orbital current transition density. Multiplication of $\rho_{10}^{\ell}(0)$ in
Table 1 by about -2 is sufficient to remove the discrepancy.

The theoretical (p,n) cross sections shown in Fig. 5 have
been calculated using the wave functions of Ref. 41,42 and the G
matrix interaction. The theoretical results give a good reproduction of the shape of the experimental data, but have to be
renormalized by factors N = .3-.7 to reproduce the magnitude of
the experimental cross sections. Much less scatter is obtained in
these normalization factors if the theoretical wave functions are
first normalized to the experimental B(M1↑). This is reflected in
the normalization factors shown in **parentheses** in Fig. 6. This
prescription was not applied to the ^{12}C transition because of the
β-decay information discussed above. These final renormalization
factors range from N = .43 to N = .54.

It is concluded from this study that the isovector S=1 component of the G matrix interaction is too strong by a factor of
approximately **1.4 at low q**. It is interesting that the corrected
$v_{\sigma\tau}^C(0) = 180$ MeV·fm^3 is close to the value 119 MeV·fm^3 corresponding
to OPEP. It is further concluded that the wave functions of Cohen
and Kurath give a good description of ρ_{10}^S, but not ρ_{10}^ℓ, for the
$0^+ \to 1^+$ T=1 transition in ^{12}C while the Chung and Wildenthal wave
functions for ^{24}Mg and ^{28}Si exhibit about the correct values for
the ratio $\rho_{10}^\ell(0)/\rho_{10}^S(0)$, but miss slightly on the overall magnitude of these quantities. Comparative studies of dipole excita-
tion in the (p,n) reaction, electron scattering, and β-decay are
a promising source of information on nuclear structure.

The calculations of Ref. 40 are currently being extended to
the newer experimental data[33,34] for incident energies of 120 MeV.
The result of a theoretical distorted wave calculation for the
excitation of the 1_1^+ T=1 level of ^{12}C in the (p,p') reaction at
122 MeV is shown in Fig. 9 of Ref. 2. The Cohen and Kurath
wave functions and the t matrix interaction of Fig. 1 have been
used in this calculation. Unlike the G matrix calculations at 62
MeV, no normalization factor is required to bring theory and experiment into agreement in this case. Since the only differences
between the (p,p') and (p,n) calculations are a factor of two in
cross sections associated with the charge exchange of the projectile and some small Coulomb effects, it seems safe to assume that
the impulse approximation will do well in describing the low q (p,n)
data at 120 MeV. More complete results should be available shortly.

Energy Dependence of v_τ^C and $v_{\sigma\tau}^C$

One of the more striking features of the (p,n) reaction is the sharp decrease in the strength of analog transitions relative to transitions which are presumably of J=1 character that has been noted in comparing (p,n) spectra for incident energies of 120 MeV with older data at lower energies.[33,34] Although accurate cross sections are not yet available to quantify this effect, it is interesting to see if current prejudices concerning the effective interaction are consistent with the observed trend. Figure 7 shows the values of v_τ^C and $v_{\sigma\tau}^C$ at q=0 for the G matrix interaction as a function of energy from 25–135 MeV. $v_\tau^C(0)$ is effective in analog transitions while $v_{\sigma\tau}^C(0)$ is effective in the J=1 transitions.

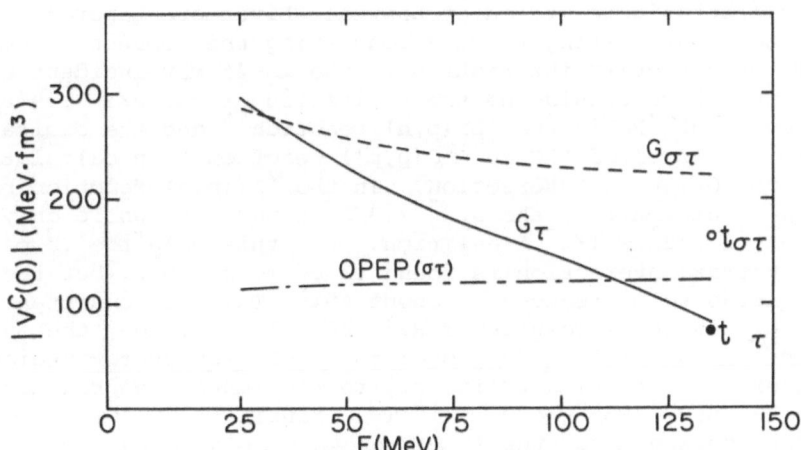

Fig. 7.   Values of $v_\tau^C(0)$ and $v_{\sigma\tau}^C(0)$ for G matrix interaction as a function of energy. OPEP values for $v_{\sigma\tau}^C(0)$ and impulse approximation values for $v_\tau^C(0)$ and $v_{\sigma\tau}^C(0)$ at 135 MeV are also shown for comparison.

The values of $v^C(0)$ have been modified to include the effect of knockout exchange according to the approximation of Ref. 1,12,13. The values of $v_\tau^C(0)$ and $v_{\sigma\tau}^C(0)$ for the t matrix interaction of Fig. 1 at 135 MeV are also shown in Fig. 7 along with the OPEP $v_{\sigma\tau}^C(0)$ values over the energy range from 25–135 MeV.

The main point to notice is that for the G matrix interaction $v_{\sigma\tau}^C(0)$ varies slowly with energy in comparison with $v_\tau^C(0)$. The

former is about 35% larger than the impulse approximation value at
135 MeV. This fact, coupled with the results of the last section,
suggests that the entire G matrix $v_{\sigma\tau}^C(0)$ curve should be reduced
uniformly by about 1.4. The G matrix $v_\tau^C(0)$ curve goes smoothly over
to the impulse approximation value – the two differ by less than 10%
at E = 135 MeV. The slow energy variation of $v_{\sigma\tau}^C(0)$ for the G
matrix interaction is similar to the behavior of the OPEP $v_{\sigma\tau}^C(0)$.

These curves suggest an increase by a factor of about 8 in the
strength of J=1 cross sections relative to the cross sections for
analog transitions in going from 25-135 MeV in incident energy.
This number is subject to uncertainties in the approximate treatment of knockout exchange which has been assumed and which is known
to be only marginally accurate at low energies.[1,5,6] Nonetheless,
a large effect is expected and it will be very interesting to have
the difference in energy dependence determined from accurate data.

A collaboration centered at Lawrence Livermore Laboratory has
produced some interesting results concerning the isovector components of the effective interaction in the 25-45 MeV incident energy
region. They have considered the excitation of the $3/2^-$ T=½ and
$½^-$ T=½ levels of ^{7}Be in the ^{7}Li(p,n) reaction[45] and the excitation
of the 1^+ T=1 level of ^{12}C in ^{12}C(p,p') reaction[46] in calculations
based on the G matrix interaction. In the ^{7}Li(p,n) reaction both
v_τ^C and $v_{\sigma\tau}^C$ contribute to the $3/2^- \rightarrow 3/2^-$ transition while only $v_{\sigma\tau}^C$
enters in the $3/2^- \rightarrow 1/2^-$ transition. For this case the v_τ^C part
of the G matrix interaction is found to be reasonable, but the
$v_{\sigma\tau}^C$ part needs to be reduced by about 40%. This is consistent
with the 62 MeV (p,n) results of Ref. 40. In addition, the energy
dependence of the ratio $v_\tau^C/v_{\sigma\tau}^C$ over the 25-50 MeV energy region is
found to extrapolate almost linearly to the impulse approximation
value at 135 MeV. For the $0^+ \rightarrow 1^+$ T=1 transition in ^{12}C reasonable
results are obtained for the 27 MeV cross section without reducing
$v_{\sigma\tau}^C$. This is in direct contradiction with the 62 MeV (p,n) results
of Ref. 40. This might be explained by the presence of possible
resonant contributions to the $0^+ \rightarrow 1^+$ T=1 cross sections in the
25-45 MeV energy region.[47] There is no data for the dipole excitations in ^{12}C over the 45-62 MeV energy range.

Abnormal Parity Transitions to 'Stretched' T=1 States

Particle-hole configurations $(j_p j_h^{-1})^{J_{max}}$, where $j_p = \ell_p + ½$,
$j_h = \ell_h + ½$, and $J_{max} = \ell_p + \ell_h + 1 = j_p + j_h$, are said to be
'stretched'. Transitions of the type $0^+ \rightarrow (j_p j_h^{-1})^{J_{max}}$ are
necessarily abnormal parity in character because

$$\Delta\pi = (-1)^{\ell_p + \ell_h} = (-1)^{J+1}$$

The maximum orbital angular momentum that can be transferred in this kind of transition is $\lambda = \ell_p + \ell_h = J - 1$. Since the operators involved in the construction of the $\rho^\ell_{JJ\pm1}(q)$ orbital current transition densities, Eq. (9), and the $\rho^s_{JJ+1}(q)$ spin transition density, Eq. (10), have rank J and J+1 in orbital space, it follows that

$$\rho^s_{JJ+1}(q) = \rho^\ell_{JJ+1}(q) = \rho^\ell_{JJ-1}(q) = 0 \qquad\qquad (23)$$

for transitions to 'stretched' configurations. Equivalently, there is a unique spin transition density, $\rho^s_{JJ-1}(q)$, which describes these transitions. These points were first emphasized by Moffa and Walker.[48]

Transitions to states of 'stretched' configuration where J_{max} is unique in a $1\hbar\omega$ basis will satisfy Eq. (23) even if the true wave functions are not pure closed shell and pure particle-hole in nature provided $3\hbar\omega$ excitations do not mix in strongly. This follows simply from the fact that below $3\hbar\omega$, only these particular j_p and j_h orbits can be connected by tensor operators of the form given in Eq. (9) and Eq. (10) with rank J_{max}. More specifically, Eq. (23) will be satisfied and

$$\rho^s_{JJ-1}(q) = S \, \rho^s_{ph}(q) \qquad\qquad (24)$$

where $\rho^s_{ph}(q)$ is the $\lambda = J-1$ spin transition constructed from pure closed shell and pure particle-hole waves functions and S is a spectroscopic amplitude which measures the deviation of the true wave functions from this simple model.

Several transitions to states of this unique 'stretched' character have now been seen in the (e,e') reaction[49-57] and in the (p,p') reaction at 135 MeV.[58-60] Most of the targets which have been studied in both reactions are N=Z nuclei which yield simultaneous data only for T=1 excitations. The T=0 excitations are not seen in the (e,e') reaction because the isoscalar spin g-factor, i.e. $g_s^0 = \frac{1}{2}(g_s^p + g_s^n)$, is more than 5 times weaker than the isovector spin g-factor. ^{208}Pb is an example of a nucleus with N$\neq$Z where these transitions have been observed in both reactions.[55,57,60]

The data of Ref. 49-60 has been studied[61] under the assumptions of Eq. (23) and Eq. (24), which allow the following reductions in Eq. (6) and Eq. (18),

$$\left| F_T(q) \right|^2 = \left| \frac{q\hbar}{2Mc} C_{J-1}^S \, g_s^l \, S \, \rho_{ph}^S(q) \right|^2 \tag{25}$$

$$\frac{d\sigma}{d\Omega} = 4\pi \left(\frac{\mu}{2\pi\hbar^2}\right) \frac{k_f}{k_i} (2J+1) \left\{ \left| v_\tau^{LS}(q) B_{J-1} S \rho_{ph}^S(q) \right|^2 \right.$$

$$\left. + \sum_\lambda \left| v_{\sigma\tau}^C(q) S \rho_{ph}^S(q) \delta_{\lambda, J-1} + v_\tau^T(q) Z_{\lambda, J-1}^J S \rho_{ph}^S(q) \right|^2 \right\} \tag{26}$$

respectively. Equation (25) and Eq. (26) have been written for the special case of an isovector transition. The modifications required for other cases should be obvious. In the calculations of Ref. 61 the geometrical parameters needed to construct $\rho_{ph}^S(q)$ and the spectroscopic amplitudes have been fixed by fitting the (e,e') data. The (p,p') calculations were carried out using these same $\rho_{ph}^S(q)$ and the interaction of Fig. 1. Separate spectroscopic amplitudes were determined from the (p,p') calculations.

Table 2 summarizes the available experimental data for transitions to states of 'stretched' character. The first column lists the target, the 'stretched' particle-hole configuration, and the

Table 2. A tabulation of excitation energies and (e,e') and (p,p') spectroscopic amplitudes for unnatural parity states of 'stretched' configurations.

Transition	E_{exp} (MeV)	E_{thy} (MeV)	$S^2_{(e,e')}$	$S^2_{(p,p')}$
$^{12}C(d_{5/2}p_{3/2}^{-1})4^-$	19.5	17.6	0.27	
$^{16}O(d_{5/2}p_{3/2}^{-1})4^-$	18.9	17.7	0.45	0.32
$^{24}Mg(f_{7/2}d_{5/2}^{-1})6^-$	15.14	17.6	0.27	0.30
$^{28}Si(f_{7/2}d_{5/2}^{-1})6^-$	14.36	16.9	0.59	0.29
$^{58}Ni(g_{9/2}f_{7/2}^{-1})8^-$	10.30	10.4	0.30	
$^{208}Pb(\pi i_{13/2}h_{11/2}^{-1})12^-$	7.06	7.18	0.56	0.20
$^{208}Pb(\nu j_{15/2}i_{13/2}^{-1})12^-$	6.42	6.49	0.54	0.80
$^{208}Pb(\nu i_{15/2}i_{13/2}^{-1})14^-$	6.75	6.49	0.56	0.46

spin of the state. The assignments are based on the comparison of the observed energy of the states, E_{exp}, with the pure particle-hole excitation energy, E_{th}, and on the shapes of the experimental angular distributions. The last two columns list the spectroscopic amplitudes obtained in the calculations of Ref. 61.

Specific results of the calculations of Ref. 61 for the 6^- T=1 excitation in ^{24}Mg are compared with the experimental data in Fig. 8. The theoretical (p,p') cross section has been decomposed into the separate contributions from the central, tensor, and spin-orbit interaction. The tensor contribution is dominant. This is easily understood by noting that $v_\tau^{LS}(q) \approx v_\tau^T(q) \gg v_{\sigma\tau}^C(q)$ in the important region of momentum transfer $q \approx 2\,\text{fm}^{-1}$. Combining this with the fact that $Z_{J+1,J-1}^J > Z_{J-1,J-1}^J > B_{J-1}$ in Eq. (26)[5,10,11] it is clear that the reaction is dominated by the tensor amplitude with J+1 orbital angular momentum transfer to the projectile and J-1 orbital angular momentum transfer to the target. Another result for the 14^- excitation in ^{208}Pb is shown in Fig. 11 of Ref. 2. The comparison between the (e,e') and (p,p') reaction thus contains information mainly on the isovector part of the tensor interaction. This is true even for cases when the iso-scalar part of the tensor interaction enters. The contribution of the isovector part of the tensor interaction to isoscalar ampli-tudes due to knockout exchange outweighs, by far, the direct contribution associated with the weak isoscalar tensor interaction.

Returning to Table 2, for ^{16}O, ^{24}Mg, and ^{28}Si, the three self-conjugate nuclei where the comparison is possible, there is good agreement between the (e,e') and (p,p') spectroscopic amplitudes for ^{24}Mg and reasonable agreement for ^{16}O. There is reason to believe that the disparity with the proton results for ^{28}Si[55] is due to overestimation of the (e,e') experimental cross section due to poor energy resolution, which is in the process of being remeasured.[62]

There is also reasonable agreement between the electron and proton results for the 14^- excitation in ^{208}Pb. The 12^- excita-tions in ^{208}Pb are not of the unique 'stretched' type we have been discussing. They have been included in Table 2 because their nearly separate proton and neutron character allows the possibility of gaining information about the isospin mixture of the nucleon-nucleon force once the extent of configuration mixing has been determined. This problem is being considered elsewhere.[60] Scat-tering from ^{58}Ni provides another opportunity for gaining informa-tion on the isospin mixture of the nucleon-nucleon force since several 8^- states, including both T_0 and $T_0 + 1$ states, are ob-served in electron scattering.

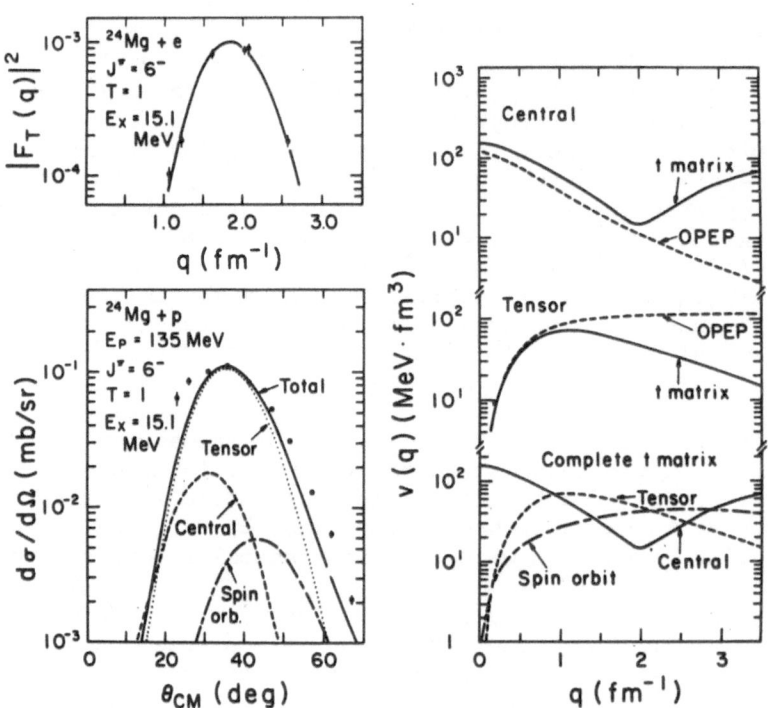

Fig. 8.   (a) Inelastic electron form factor for the 6^- (15.14 MeV) state in ^{24}Mg; (b) inelastic proton cross sections for the 6^- (15.14 MeV) state in ^{24}Mg. The theoretical cross section is decomposed into central, tensor, and spin-orbit contributions; (c) Isovector S=1 parts of the t matrix interaction of Fig. 1 again. The central and tensor parts of OPEP are shown for comparison.

In summary, the overall comparison between the inelastic electron and proton results is encouraging. The results further affirm that it is reasonable to use the free nucleon-nucleon t matrix to describe the (p,p') reaction at 135 MeV. More specifically they suggest that the isovector part of the tensor interaction has about the right strength and momentum dependence near

$q=2$ fm^{-1}. Its departure from OPEP, which is indicated in Fig. 8, is consistent with the viewpoint that short range contributions associated with heavier meson exchanges are important.

Abnormal Parity Transitions to 'Stretched' T=0 States

Although there is no electron scattering data available for comparison, the results of (p,p') calculations for the excitation of 'stretched' T=0 states are still of interest. There is experimental (p,p') data available for the T=0 companions of the 4^-, 6^-, and 6^- T=1 excitations in ^{16}O, ^{24}Mg, and ^{28}Si, which were discussed in the last section.[58,59] Theoretical (p,p') calculations have been carried out for all three transitions. Again the calculations were made with the t matrix interaction of Fig. 1 and it was assumed that the geometry for the T = 0 $\rho_{ph}^s(q)$ was the same as the T=1 case. The results of these calculations for the 6^- T=0 excitation in ^{28}Si are compared with the results for the 6^- T=1 excitation in this same nucleus in Fig. 9.

As anticipated in the discussion of Fig. 1, the spin-orbit interaction comes in more strongly for the T=0 transition. The tensor contribution to the T=0 cross section is still quite important even though the isoscalar tensor interaction is weak. This is the effect of the T=1 tensor interaction through knockout exchange which was also mentioned earlier. It is important to emphasize that both the T=0 spin-orbit interaction and the T=1 tensor interaction were shown to have about the right strengths in the calculations for T=0 normal parity transitions and T=1 'stretched' excitations, respectively. There is a definite shift in the peak of the theoretical T=0 cross section relative to the peak of the theoretical T=1 cross section. This shift is seen in the experimental data in all cases.

The squared spectroscopic amplitude for the 6^- T=0 transition in ^{28}Si is about 3 times smaller than for the 6^- T=1 excitation. This result is typical. This suggests that the wave functions for the 'stretched' T=0 states are quite different from those for the 'stretched' T=1 states. Differences of this size are difficult to achieve in standard particle-hole models - even when extended to $3\hbar\omega$ bases.

Inelastic Analyzing Powers for High Spin States

In an experiment just completed a week ago,[63] inelastic analyzing powers for the excitation of the 5^- T=0, 6^- T=0, and 6^- T=1 levels in ^{28}Si by 135 MeV protons were measured. Although the data are not available to show, it is worthy of note that the

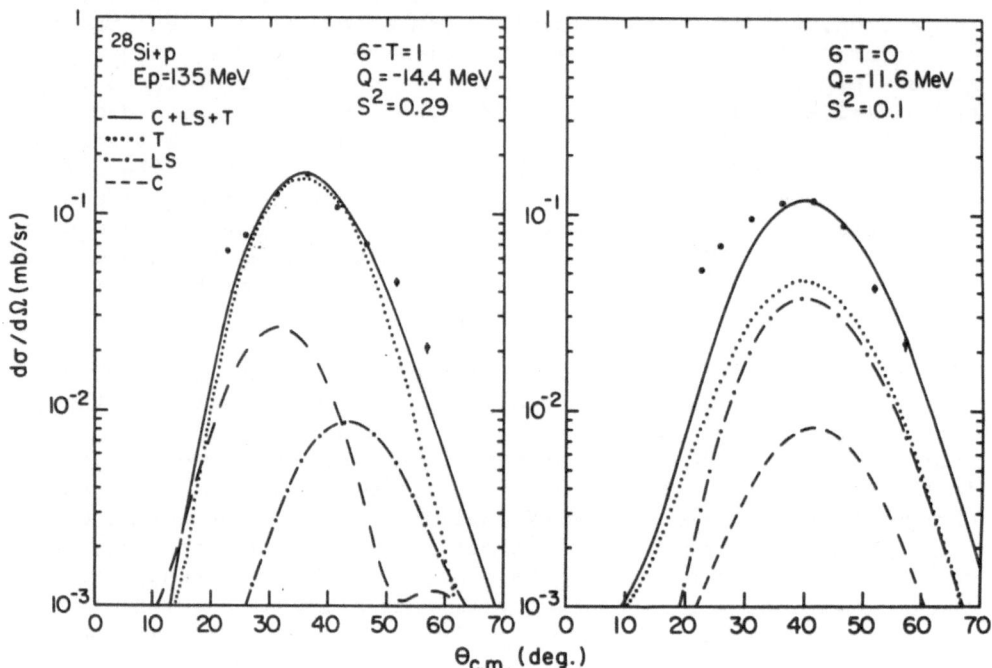

Fig. 9. Comparison of distorted wave impulse approximation re-
         sults for the excitation of the 6⁻ T=0 and 6⁻ T=1
         'stretched' states in ^{28}Si by 135 MeV protons. C, LS, and
         T refer to central, spin-orbit, and tensor contributions
         to the cross sections.

analyzing powers for the three states have distinctly different
shapes as a function of angle. The differences in shape are in
essentially complete agreement with the theoretical analyzing
powers generated along with the theoretical cross sections described in the preceding two sections. The time lag between the
theoretical calculations and the experimental data was approximately 9 months.

Observation of 'Stretched' States in Inelastic Pion Scattering

The 6^- T=0 and T=1 levels in ^{28}Si have also been seen in inelastic pion-nucleus scattering.[64,65] In the most simplistic model
the ratio of inelastic π^+ cross sections for the 6^- T=0 and T=1
excitations in ^{28}Si would be given by[66]

$$\frac{\sigma_{\pi^+}(T=0)}{\sigma_{\pi^+}(T=1)} = \left(\frac{t_{\pi N}^0}{t_{\pi N}^1}\right)^2 \left(\frac{S^0}{S^1}\right)^2 \tag{27}$$

where $t_{\pi N}$ denotes the free π-nucleon t matrix and S denotes the
spectroscopic amplitudes introduced above. Using $(t_{\pi N}^0/t_{\pi N}^1)^2 = 4$
as obtained from the free π-nucleon data near the (3,3) resonance
and assuming $S^0 = S^1$ leads to the value 4 for the cross section
ratio in Eq. (27). This is to be compared with the ratio of the
experimental cross sections at peak which is $1.5\pm.2$.[64,65] Using
the value $(S^1/S^0)^2 = 2.9$ deduced from the (p,p') calculations
leads to the value 1.38 for the ratio of Eq. (27). This agrees
quite nicely with the experimental $(\pi^+,\pi^+{}')$ ratio.

Before this result can be taken seriously possible differences
in shape between the spin transition densities for the T=0 and T=1
excitations must be carefully considered. The differences between
absorption in the (π,π') and (p,p') reactions must also be examined.
Nonetheless, the correlation between the (p,p') and (π,π') reactions is tantalizing and further work along these lines should
prove rewarding.

CONCLUSION

Parallels between electron scattering and nucleon scattering
have been exploited to gain reliable information on the effective
interaction for nucleon scattering. It has been shown that the
impulse approximation gives quite a reasonable description of
nucleon-nucleus scattering in the 100-200 MeV energy region. The
results shown test most thoroughly the S=0, T=0 and S=1, T=1 parts
of the t matrix interaction over the momentum transfer range
$q \approx 0\text{-}2$ fm^{-1}. Evidence for the non-central parts of these interaction components shows up clearly in the experimental cross sections. It is found that isoscalar spin-flip transitions of high
q are dominated by the isoscalar spin-orbit interaction and the
isovector tensor interaction through knockout exchange. Since
these interaction components have been verified in other transitions, the theoretical results for the isoscalar spin-flip transitions, at high q might be considered reliable. The results of new
analyzing power measurements support this assertion. Definite

information of the S=0, T=1 part of the t matrix interaction awaits new data from the (p,n) reaction.

The correspondence between the isovector components of the impulse approximation t matrix interaction and the low energy G matrix interaction has also been investigated. Recent 62 MeV (p,n) data suggests that the central part of the S=1, T=1 component of the G matrix interaction is about 40% too strong. With this correction it is found that the G matrix interaction extrapolates smoothly to the impulse approximation t matrix interaction at 135 MeV. A striking feature of the isovector components of the effective interaction is a decrease in the ratio $(v_T^C(0)/v_{\sigma T}^C(0))^2$ by a factor of 8 over the energy region from 25–135 MeV. A definite drop in the strength of analog cross sections relative to the cross sections for dipole transitions has been observed in the new (p,n) data at 120 MeV although accurate cross sections are not yet available.

It is clear that a new level of reliability in the microscopic interpretation of nucleon-nucleus scattering has been achieved by moving to incident energies greater than 100 MeV. For maybe the first time some definite understanding of abnormal parity transitions has been obtained. The possibilities for gaining new information on nuclear structure through comparative studies of nucleon, electron, and pion-nucleus scattering are extremely encouraging. The high energy nucleon scattering studies also promise to provide useful benchmarks for investigations of reaction mechanisms which complicate the interpretation of low energy nucleon scattering data.

ACKNOWLEDGMENTS

The author acknowledges the close assistance of W. G. Love who has been a major collaborator in this work. Additional thanks go to R. A. Lindgren for information on recent developments in electron scattering and to the theorists, experimentalists, and users at the Indiana University Cyclotron Facility who have supplied much of the information of nucleon scattering shown within. This work was supported in part by the National Science Foundation.

REFERENCES

1. F. Petrovich, "Microscopic Optical Potentials," H. V. v. Geramb, ed., Springer-Verlag, New York (1979), 89:155 and references contained therein.
2. W. G. Love, these proceedings.

3. A. K. Kerman, H. McManus, and R. M. Thaler, Ann. Phys. (N.Y.)
    8:551 (1959).
4. F. Petrovich, Nucl. Phys. A251:143 (1975).
5. F. Petrovich and W. G. Love, to be published.
6. T. de Forest, Jr. and J. D. Walecka, Adv. Phys. 15:1 (1966).
7. F. Petrovich, H. McManus, J. Borysowicz, and G. R. Hammerstein,
    Phys. Rev. C16:839 (1977).
8. G. R. Satchler, Nucl. Phys. 55:1 (1964).
9. G. Bertsch, J. Borysowicz, H. McManus, and W. G. Love, Nucl.
    Phys. A284:399 (1977).

10. W. G. Love and L. J. Parish, Nucl. Phys. A157:625 (1970).
11. W. G. Love, Nucl. Phys. A192:49 (1972).
12. F. Petrovich, H. McManus, V. A. Madsen, and J. Atkinson, Phys.
    Rev. Lett. 22:895 (1969).
13. W. G. Love, Nucl. Phys. A312:160 (1978).
14. J. Raynal and R. Schaeffer, Computer code DWBA 70 as modified
    by W. G. Love.
15. W. D. Myers and W. J. Swiatecki, Ann. Phys. (N.Y.) 55:395
    (1969).
16. E. C. Halbert and G. R. Satchler, Nucl. Phys. A233:265 (1974).
17. A. Scott, N. P. Mathur, and F. Petrovich, Nucl. Phys. A285:222
    (1977).
18. W. G. Love, A. Scott, F. Todd Baker, W. P. Jones, and J. D.
    Wiggins, Jr., Phys. Lett. 73B:277 (1978).
19. A. Picklesimer and G. Walker, Phys. Rev. C17:237 (1978).
20. P. Schwandt, A. Nadasen, P. P. Singh, M. D. Kaitchuk, W. W.
    Jacobs, J. Meek, A. D. Bacher, and P. T. Debevec, "IUCF
    Technical and Scientific Report," unpublished, p. 79 (1977-
    1978).
21. A. Nadasen, P. Schwandt, P. P. Singh, A. D. Bacher, P. T.
    Debevec, W. W. Jacobs, M. D. Kaitchuk, and J. T. Meek,
    submitted to Phys. Rev. C.
22. P. Schwandt, F. Petrovich, and A. Picklesimer, "IUCF Technical
    and Scientific Report," unpublished, p. 27 (1978-1979).
23. A. Chamaux, V. Layly, and R. Schaeffer, Phys. Lett. 72B:33
    (1977).
24. L. Ray, G. W. Hoffmann, G. S. Blanpied, W. R. Coker, and R. P.
    Liljestrand, Phys. Rev. C18:1756 (1978).
25. L. Ray, W. R. Coker, and G. W. Hoffmann, Phys. Rev. C18:2641
    (1978).
26. G. S. Adams, A. D. Bacher, G. T. Emery, W. P. Jones, D. W.
    Miller, W. G. Love, and F. Petrovich, submitted to Phys.
    Lett.
27. J. F. Ziegler and G. A. Peterson, Phys. Rev. 165:1337 (1968).
28. J. H. Heisenberg and I. Sick, Phys. Lett. 32B:249 (1970).
29. M. Nagao and Y. Torizuka, Phys. Lett. 37B:383 (1971).
30. J. Friedrich, K. Voegler, and H. Euteneuer, Phys. Lett. 64B:
    269 (]976).

31. J. Lichtenstadt, J. Heisenberg, C. N. Papanicolas, C. P.
     Sargent, A. N. Courtemanche, and J. S. McCarthy, Phys.
     Rev. Lett. 40:1127 (1978).
32. F. Petrovich, W. G. Love, G. S. Adams, A. D. Bacher, G. T.
     Emery, W. P. Jones, and D. W. Miller, submitted to Phys.
     Lett.
33. B. D. Anderson, D. E. Bainum, A. Baldwin, C. C. Foster, C. D.
     Goodman, C. A. Goulding, M. B. Greenfield, J. N. Knudson,
     R. Madey, J. Rapaport, and T. R. Witten, "IUCF Technical
     and Scientific Report," unpublished, p. 117 (1977-1978).
34. C. D. Goodman, these proceedings.
35. C. L. Moake, P. T. Debevec, L. J. Gutay, P. A. Quinn, and
     R. P. Scharenberg, "IUCF Scientific and Technical Report,"
     unpublished, p. 114 (1977-1978).
36. B. T. Chertok, C. Sheffield, J. W. Lightbody, Jr., S. Penner,
     and D. Blum, Phys. Rev. C18:23 (1973).
37. L. W. Fagg, W. L. Bendel, R. A. Tobin, and H. T. Kaiser, Phys.
     Rev. 171:1250 (1968).
38. L. W. Fagg, W. L. Bendel, E. C. Jones, Jr., and S. Numrich,
     Phys. Rev. 187:1378 (1969).
39. F. Ajzenberg-Selove, Nucl. Phys. A248:1 (1975).
40. F. Petrovich, W. G. Love, and R. J. McCarthy, submitted to
     Phys. Rev. Lett.
41. S. Cohen and D. Kurath, Nucl. Phys. 73:11 (1965).
42. W. Chung and B. H. Wildenthal, to be published.
43. A. Bohr and B. Mottelson, "Nuclear Structure, Benjamin, New
     York (1969) p. 345.
44. J. D. Anderson, C. Wong, and V. A. Madsen, Phys. Rev. Lett.
     24:1074 (1970).
45. F. Petrovich, R. H. Howell, S. M. Grimes, C. Wong, C. H.
     Poppe, G. M. Crawley, and S. M. Austin, to be published.
46. R. H. Howell, F. S. Dietrich, D. W. Heikkinen, and F.
     Petrovich, to be published.
47. H. V. Geramb, K. Amos, R. Sprichmann, K. T. Knöpfle, M.
     Rogge, D. Ingham, and C. Mayer-Böricke, Phys. Rev. C12:
     1697 (1975).
48. P. J. Moffa and G. E. Walker, Nucl. Phys. A222:140 (1974).
49. T. W. Donnelly, Jr., J. D. Walecka, I. Sick, and E. B.
     Hughes, Phys. Rev. Lett. 21:1196 (1968).
50. L. W. Fagg, P. B. Ertel, H. Crannell, R. A. Lindgren, J. B.
     Flanz, and G. A. Peterson, Bull. Am. Phys. Soc. 23:583
     (1978).
51. I. Sick, E. B. Hughes, T. W. Donnelly, Jr., J. D. Walecka, and
     G. E. Walker, Phys. Rev. Lett. 23:1117 (1969).
52. R. S. Hicks, private communication.
53. H. Zarek, B. O. Pich, T. E. Drake, D. J. Rowe, W. Bertozzi,
     C. Creswell, A. Hirsch, M. V. Hynes, S. Kowalski, B. Norum,
     F. N. Rad, C. P. Sargent, C. F. Williamson, and R. A.
     Lindgren, Phys. Rev. Lett. 38:750 (1977).

54. T. W. Donnelly, Jr., J. D. Walecka, G. E. Walker, and I. Sick, Phys. Letts. 32B:545 (1970).
55. R. A. Lindgren, C. W. Williamson, and S. Kowalski, Phys. Rev. Lett. 40:594 (1978).
56. J. Lichtenstadt, J. Heisenberg, C. N. Papanicolas, C. P. Sargent, A. N. Courtemanche, and J. S. McCarthy, Phys. Rev. Lett. 40:1127 (1978).
57. J. Lichtenstadt, J. Heisenberg, C. N. Papanicolas, J. S. McCarthy, and A. N. Courtemanche, Bull. Am. Phys. Soc. 24: 53 (1979).
58. R. S. Henderson, B. M. Spicer, J. D. Svalbe, B. C. Officer, G. G. Shute, W. W. Devons, W. L. Friesel, W. P. Jones, A. C. Attard, to be published.
59. G. S. Adams, A. D. Bacher, G. T. Emery, W. P. Jones, R. T. Kouzes, D. W. Miller, A. Picklesimer, and G. E. Walker, Phys. Rev. Lett. 38:1387 (1977).
60. A. D. Bacher, G. T. Emery, W. P. Jones, D. W. Miller, G. Adams, F. Petrovich, and W. G. Love, Bull. Am. Phys. Soc. 23:945 (1978) and to be published.
61. R. A. Lindgren, W. J. Gerace, A. D. Bacher, W. G. Love, and F. Petrovich, Phys. Rev. Lett. 42:1524 (1979).
62. S. Yen and T. E. Drake, private communication.
63. S. Yen, T. E. Drake, and A. D. Bacher, private communication.
64. H. A. Thiessen, private communication.
65. C. Olmer, B. Zeidman, D. F. Geesaman, R. E. Segel, R. L. Boudrie, C. L. Morris, and H. A. Thiessen, Bull. Am. Phys. Soc. 24:687 (1979).
66. F. Petrovich, W. G. Love, R. A. Lindgren, W. J. Gerace, G. Walker, A. D. Bacher, E. Siciliano, H. A. Thiessen, B. Zeidman, D. Geesaman, and C. Olmer, Bull. Am. Phys. Soc. 24:677 (1979).

THE (p,n) REACTION AT INTERMEDIATE ENERGY

Charles D. Goodman

Oak Ridge National Laboratory*
Oak Ridge, Tennessee 37830

INTRODUCTION

If the effective interaction for the (p,n) reaction were the free N-N force, and the free N-N force were the one pion exchange force, then the $^{12}C(p,n)^{12}N$ spectrum at 0° would look just like what is actually seen in a recent 120 MeV experiment (see fig. 1). Inverting the statement to claim that fig. 1 shows that the effective force is that of the one pion exchange potential would be overstating the case. It is, however, fair to say that new data in the 100-200 MeV energy range suggest an OPEP-like interaction with its characteristic spin-isospin operator, while at lower energy (p,n) spectra look qualitatively different with the spin independent IAS transition dominating the spectrum.

Until recently high energy (p,n) data with resolved levels were not available because of the difficulty of performing the experiments. However, such data are now becoming available with the installation of a beam swinger[1] at the Indiana University Cyclotron Facility and the development of large, time compensated detectors.[2]

In this paper I shall review the use of the (p,n) in exploring effective interactions. I shall show some new data from IUCF, and shall emphasize the differences between low and high energy data. I shall also briefly describe the experimental problems and discuss the techniques we are using at IUCF.

---

*Operated by Union Carbide Corporation for the U.S. Department of Energy, under contract W-7405-eng-26.

THE (p,n) REACTION AS AN INTERACTION PROBE

The (p,n) reaction was the discovery reaction for isobaric
analog states. The IAS discovery spectrum[3] is shown in fig. 2.
The dominance of the IAS peak is typical of (p,n) spectra in the
10-40 MeV range.

The earliest model,[4] due to Lane, assumed that a symmetry
term was present in the nucleon-nucleus optical potential. In
isospin notation this term had the form $(1/A)V_\tau(\tau \cdot T)$, where τ is
the isospin operator for the nucleon and T is the total isospin
operator for the nucleus. V_τ is a strength. Treated as a per-
turbing potential, this term gives rise to the IAS excitation in
a (p,n) reaction through the τ^+T^- component.

In the Lane model the dominance of the IAS peak comes about
through the cooperative interaction of reaction and structure
effects. The $\tau \cdot T$ term in the interaction connects the target
state to any final state that has a structure component differing
from the target state only in isospin projection. At the same
time the nuclear structure is such that such a component is found
almost exclusively in a single state, the IAS.

In this purely macroscopic form the model accounts for the
excitation only of the IAS. No other state is connected to the
target state by the τ^+T^- component. The magnitude of the IAS
cross section can be used as a sensitive measure of the value of
V_τ. In addition, since the total T^- operator is the sum of
single particle t^- operators for the individual nucleons, the
IAS cross section can also be analyzed to yield an effective
two-body v_τ in a microscopic model. Many experimental studies
have been motivated by these ideas.[5] The extraction of the Lane
potential is covered in other contributions to this conference.

It was recognized shortly after the first IAS experiments
that other states connected to the target state by a spin-isospin
operator could also be excited through the (p,n) reaction. In
beta decay language the IAS transition is between states connected by the allowed Fermi beta decay operator. The (p,n) reaction can also excite states connected by the Gamow-Teller
operator. The Fermi and the Gamow-Teller states are excited
through different pieces of the effective interaction.

The macroscopic term $(\tau \cdot T)$ can be thought of as the sum of
two-body, nucleon-nucleon interaction terms $\Sigma(\tau \cdot \tau_i)$ where the
subscripted operator refers to nucleons in the nucleus. It happens for this Fermi-like term that the sum of the individual
nucleon isospin operators is the total isospin operator for the
nucleus. The Gamow-Teller like operator, $\Sigma(\sigma \cdot \sigma_i)(\tau \cdot \tau_i)$, does not
correspond to a total nuclear operator, but it does have a simple

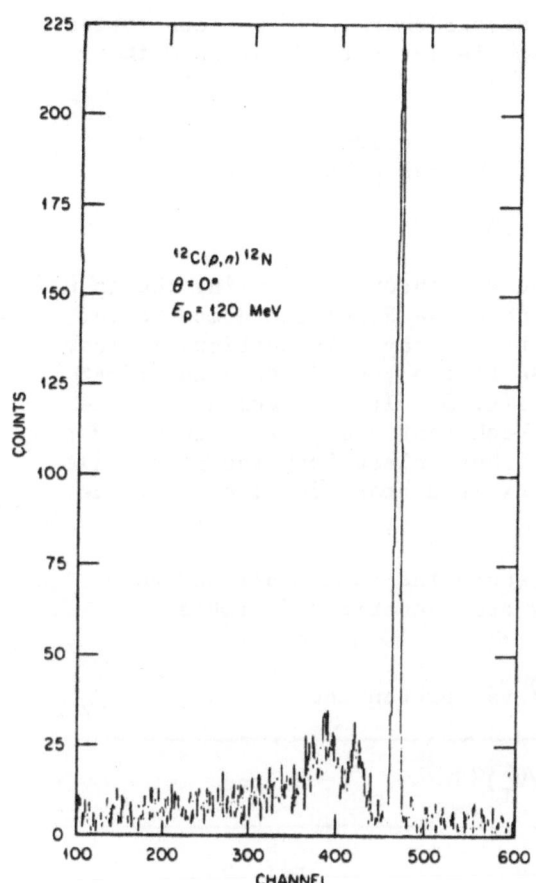

$^{12}C(p,n)^{12}N$
$\theta = 0°$
$E_p = 120$ MeV

Fig. 1. Time-of-flight spectrum for $^{12}C(p,n)^{12}N$, 0°, $E_p = 120$ MeV. Flight path is 62 m.

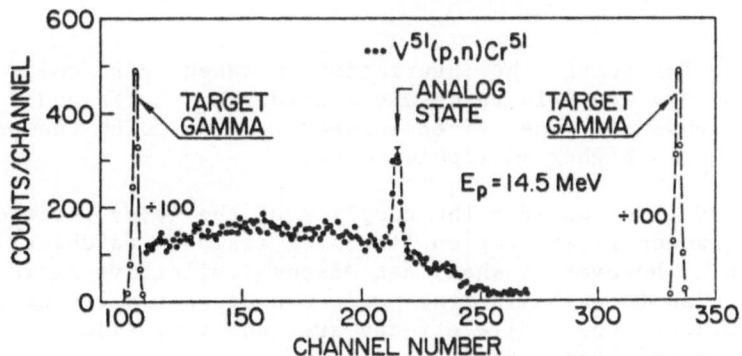

Fig. 2.   The analog state discovery spectrum (from ref. 3).

connection with well known nuclear structure; it connects states between which there would be Gamow-Teller beta decay were it energetically possible.

The Livermore group[6] exploited these ideas using a model in which the (p,n) interaction is the sum of two parts,

$$V_\tau \Sigma(\tau \cdot \tau_i) \text{ and } V_{\sigma\tau} \Sigma(\sigma \cdot \sigma_i)(\tau \cdot \tau_i). \tag{1}$$

They determined values for V_τ and $V_{\sigma\tau}$ through measuring the (p,n) cross sections for transitions where the Fermi and Gamow-Teller matrix elements were known from beta decay. In particular, they exploited the case of $^7\text{Li}(p,n)^7\text{Be}$ to the ground state and first excited state as illustrated in fig. 3. It is assumed that the ground state transition is the incoherent sum of a V_τ term and a $V_{\sigma\tau}$ term. (The interference vanishes unless both the projectile and target are polarized.) The excited state involves only the $V_{\sigma\tau}$.

A Michigan State group[7] extended the same model and analysis up to 45 MeV. The results are summarized in Table I.

TABLE I. $(V_{\sigma\tau}/V_\tau)^2$ vs. proton energy.

E_p (MeV)	$(V_{\sigma\tau}/V_\tau)^2$	Source
10-20	0.44	ref. 6
25	0.48	ref. 7
35	0.61	ref. 7
45	0.76	ref. 7

In the foregoing, the interaction is taken to be quite simple and only monopole terms are considered. Moffa and Walker[8] discuss the generalized interaction including tensor, spin-orbit, and higher multipole terms.

One may also consider the coupling of the (p,n) reaction to collective modes of excitation (see, for example, Satchler[9]). In this talk, however, I shall not discuss collective excitations. I wish to emphasize the nucleon-nucleon aspects of the (p,n) reaction. Collective effects are, however, discussed elsewhere in this conference.

The interactions derived from the (p,n) studies already mentioned have been essentially phenomenological. Currently there

is considerable effort to derive interactions based on the free
N-N force. Following the work of Kuo and Brown[10] for bound
states, Bertsch et al.[11] have derived effective interactions
for reactions that are parameter free in the sense that they are
constrained to agree with the free N-N force at some point in
their derivations and are not adjusted to fit reaction data.

Love[12] and Petrovich[13] have used the MSU G-matrix in
nucleon-nucleus reaction calculations. Love[14] has also carried
out DWIA calculations using a t matrix generated by fitting free
N-N phase shifts. Picklesimer and Walker[15] have created a
generalized interaction from a global fit over a wide energy
range of the free N-N phase shifts. Comparisons of these various
calculations with data are shown in papers in these proceedings
by Love, Petrovich, and Picklesimer, and in this paper.

Inasmuch as the reaction cross section depends on both the
interaction and the structure, one can emphasize the exploration
of one or the other aspects by the choice of targets and final
states. I wish now to emphasize the use of known structure to
filter out selectively certain components of the interaction.

In free N-N scattering the natural observables do not make a
representation that picks out a force of the form of OPEP. As
Signell has said "It might be of interest to note that the one
pion exchange process is not a prominent one in the experimental
data; finding its effect is much like looking for the proverbial
needle in a haystack."[16] However, in nuclear scattering the
structure relationships between the target state and the final
excited states may act as selective filters for components of the
interaction. In $^{12}C(p,n)^{12}N$ the ground state acts as a filter
for the OPEP component of the nuclear force, and the sharp peak
in fig. 1 is, so to speak, the needle clearly standing out of the
haystack.

The long range part of the N-N interaction is believed to be
the one pion exchange potential which may be written as

$$V_{OPEP} = f^2(mc^2)\tau_1 \cdot \tau_2 [\sigma_1 \cdot \sigma_2 + S_{12}(1+3/x+3/x^2)]e^{-x}/3x \qquad (2)$$

$$x = mcr/\hbar .$$

To discern the qualitative aspects of the signature of OPEP
let us consider the way in which the isospin and spin operators
enter into the central term. (We shall ignore the tensor part
temporarily and say only that S_{12} is a function of the relative
orientation of the two nucleon spins with respect to each other
and with respect to the radius vector between the nucleons.

There is no term in the OPEP that does not contain both spin and isospin operators.)

We are concerned with the operator product $(\tau_1 \cdot \tau_2)(\sigma_1 \cdot \sigma_2)$. Suppose particle 1 is the projectile and particle 2 is a neutron in the nucleus. This may be expanded out to the form:

$$[\frac{1}{2}(\tau_1^+ \tau_2^- + \tau_1^- \tau_2^+) + \tau_1^0 \tau_2^0][\frac{1}{2}(\sigma_1^+ \sigma_2^- + \sigma_1^- \sigma_2^+) + \sigma_1^0 \sigma_2^0]. \qquad (3)$$

For the direct part in which particle 1 emerges as a neutron, only the terms with τ_1^+ as a factor contribute. (Calculations show that the direct part dominates over the exchange part at forward angles.[17]) The operator of interest is then

$$\frac{1}{2}\tau_1^+ \tau_2^- [\frac{1}{2}(\sigma_1^+ \sigma_2^- + \sigma_1^- \sigma_2^+) + \sigma_1^0 \sigma_2^0]. \qquad (4)$$

If the projectile interacts with a single bound nucleon, the resultant excitation will in general project onto several of the possible final states, and the projections differ according to how the spin, isospin and momentum are changed by the interaction.

Consider the schematic picture shown in fig. 4 representing a ^{12}C nucleus. The boxes represent the single particle states in an L-S coupling representation. The occupancy arrangement represents one term in a 0^+, T=0 state.

The spin independent isospin operator, $\sum_i (\tau_1^+ \tau_i^-)$ which is the (p,n) part of $\sum_i (\tau_1 \tau_i)$, will yield zero for a T=0 target. From the diagram this can be seen in that if a neutron is changed to a proton the corresponding state that the proton must occupy is already filled. (This need not be true for every term individually, but the result will be zero when the summation is carried out.)

$$| \sum_i (\tau_1^+ \tau_i^-) | \psi_{target}(t=0)p > = 0 \qquad (5)$$

The spin dependent isospin operator, $\sum_i (\sigma_1 \cdot \sigma_i)(\tau_1 \cdot \tau_i)$ will, in general, not yield zero. From the diagram this can be seen also since the corresponding spin flipped state need not be occupied unless the total S=0.

The spin independent operator is similar to a Fermi beta
decay operator and the spin dependent operator is similar to a
Gamow-Teller beta decay operator.

THE (p,n) REACTION ON SELF CONJUGATE TARGETS

As seen in fig. 1, the ^{12}C(p,n)^{12}N spectrum for 120 MeV at
0° is dominated by a single peak, the ground state transition.
The analog in ^{12}C of the ground state of ^{12}N is the $J^{\pi} = 1^+$, T=1
state at 15.1 MeV.

It has been known that in back angle inelastic electron
scattering on carbon the 1^+, T=1 state at 15.1 MeV is preferentially excited.[18] Back angle electron scattering emphasizes M1
transitions. The enhanced excitation was explained by the combined work of Morpurgo[19] and Kurath.[20] Morpurgo pointed out
that on a T=0 nucleus there is a near cancellation of the isoscalar part of the M1 operator and the isovector part is large. This
isovector part contains an orbital current term and a magnetic
moment (hence spin) term. Kurath showed that the spin-isospin
strength for carbon in particular, and more generally in T=0 nuclei
is concentrated in a single state or very few close states for the
case of an approximately half filled major shell. This is an
interesting finding when one considers, for example, that even
the restricted space of eight particles in the p shell (^{12}C)
allows eight T=1, 1^+ states. The spin-flip isospin-flip strength
from the ground state of ^{12}C is not distributed at all uniformly
among them, but is concentrated almost entirely in a single
state. This implies that $|\tau^-(\sigma \cdot \sigma)\psi^0_{g.s.}>$ is approximately an
eigenfunction of the nuclear Hamiltonian.

For the ^{12}C(p,n)^{12}N reaction, the nuclear operator in the
interaction term $(\sigma \cdot \sigma_i)(\tau \cdot \tau_i)$ is just a different isospin projection of the same operator that connects the ground state of ^{12}C
to the 15.1 MeV state. Thus the concentration of strength seen
for the 15.1 MeV state in 180° electron scattering on ^{12}C is
another manifestation of the same effect that produces a concentration of strength in the ground state in ^{12}C(p,n)^{12}N. In
general the relationship between 180° electron scattering and 0°
(p,n) reactions will be simple only for states for which the
transition density is primarily the spin density. In selecting
the 0° point of observation for (p,n) we have emphasized the
monopole part of the isovector spin-spin interaction. The 180°
electron scattering emphasizes magnetic transitions for which the
magnetic moment part is equivalent to a spin interaction. There
is also a current interaction term. In general a comparison of
electron scattering with (p,n) should provide more structure
information than either probe can provide separately.

There is also a simple relationship between (p,n) and (p,p')
according to an idea due to Adair.[22] The cross section for
$^{12}C(p,p')^{12}C(15.1)$ is simply related to $^{12}C(p,n)^{12}N(g.s.)$ by an
isospin coupling coefficient. $\sigma(p,n) = 2\sigma(p,p')$ for this case.
This result is borne out by experiment, see figs. 5 and 6. This
result also gives us the opportunity to study the reaction as a
charged particle reaction with good resolution. In (p,p'),
however, a background of T=0 states is present, so the effect of
the $\sigma\tau$ operator is not striking in the raw data. A more detailed
discussion of comparisons between (p,n) and (p,p') spectra is contained in a paper by C. C. Foster in this conference.

The solid curves in figs. 5 and 6 are DWIA calculations by
Picklesimer and Walker[22] using an interaction t matrix that is a
global fit to N-N phase shifts over a wide energy range. Figure 7
shows a calculation by Love[14] using a t matrix fit to N-N phase
shifts near the actual energy. The long range part of that interaction is expressly matched to OPEP and the fit to N-N phase
shifts is made by adding shorter range components also.[15]

In $^{28}Si(p,n)^{28}P$ the situation is quantitatively similar to
the carbon case. The strong M1 state in ^{28}Si is at 11.4 MeV and
the 2.1 MeV 1^+ state in ^{28}P is its analog and is strongly excited
in the (p,n) reaction. Figure 8 shows that the 2.1 MeV state is
persistently dominant in the spectrum over a large angular range.

In $^{24}Mg(p,n)^{24}Al$ the spin-isospin strength seems to be split
between two states, at 1.1 and 3.1 MeV, as seen in fig. 9.

Figures 10 and 11 show the differential cross sections
for some 1^+ levels. The solid curves are calculations by F.
Petrovich using the MSU G-matrix interaction[25] and Wildenthal-
Chung wave functions.[26]

COMPARISON WITH BETA DECAY

As we have seen, the nuclear matrix element for a $(\sigma\cdot\sigma)(\tau\cdot\tau)$
interaction is the same as that which appears in Gamow-Teller
beta decay, and the nuclear matrix element for the $(\tau\cdot\tau)$ interaction is the same as that which appears in Fermi beta decay.
Thus, we should expect simple relationships between (p,n) cross
sections and beta decay matrix elements. We have also seen that
such a relationship has already been exploited by the Livermore
and MSU groups in the study of $^7Li(p,n)^7Be$. In particular, in
that case, since the two levels that are compared are only a few
hundred keV apart it is argued that the distortion effects are
quite similar for the two levels, so the interaction information
can be extracted reliably even without precisely accurate DWBA
calculations.

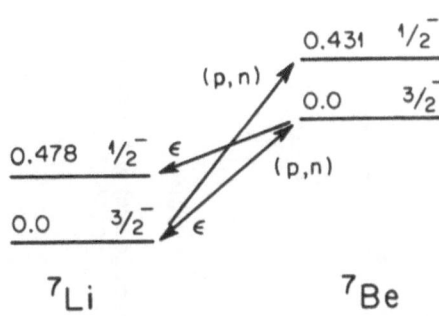

Fig. 3. Energy level scheme for A=7.

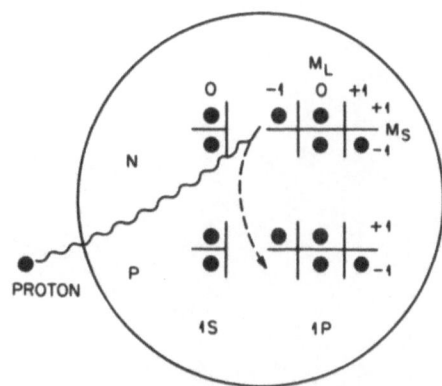

Fig. 4. Occupation diagram in L-S coupling representation for eight particles in the p shell. This would, for example, represent one term in the ground state wave function of ^{12}C.

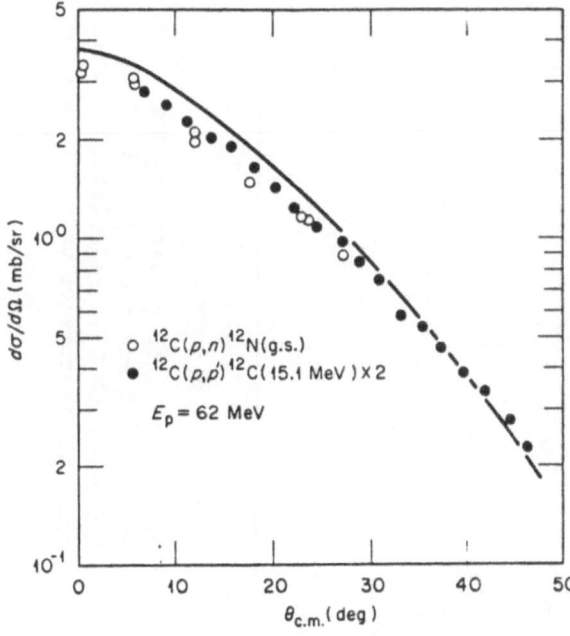

Fig. 5. Differential cross sections for ^{12}C(p,n)^{12}N(g.s.) and ^{12}C(p,p')^{12}C(15.1 MeV) at E_p = 62 MeV with DWIA calculation. See text for sources of data and calculation.

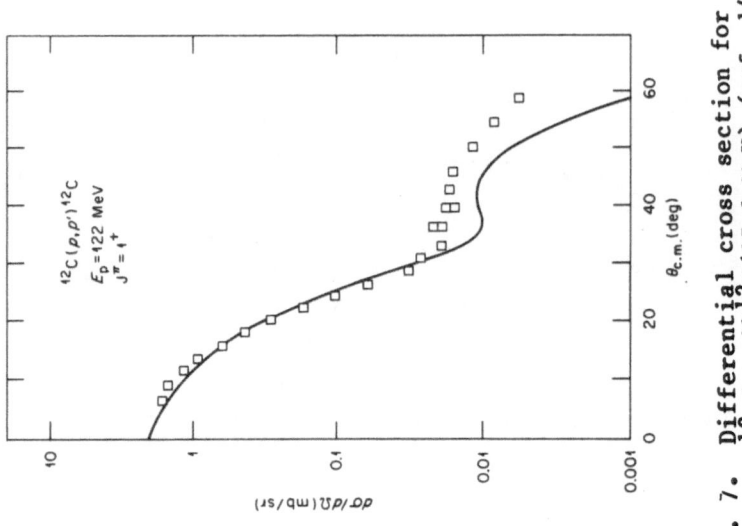

Fig. 7. Differential cross section for $^{12}_{6}C(p,p')^{12}C(15.1 \text{ MeV})$ (ref. 14).

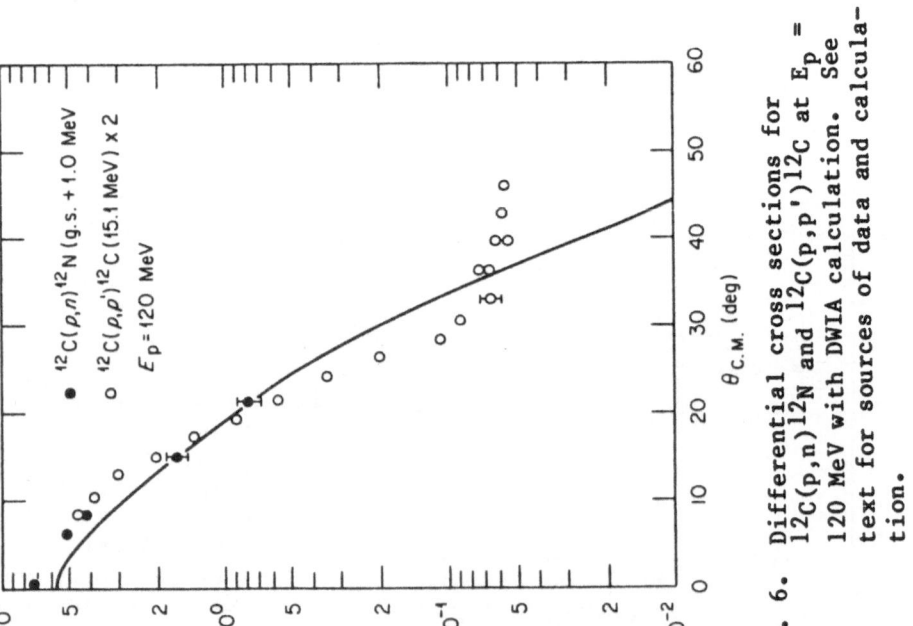

Fig. 6. Differential cross sections for $^{12}C(p,n)^{12}N$ and $^{12}C(p,p')^{12}C$ at $E_p = 120$ MeV with DWIA calculation. See text for sources of data and calculation.

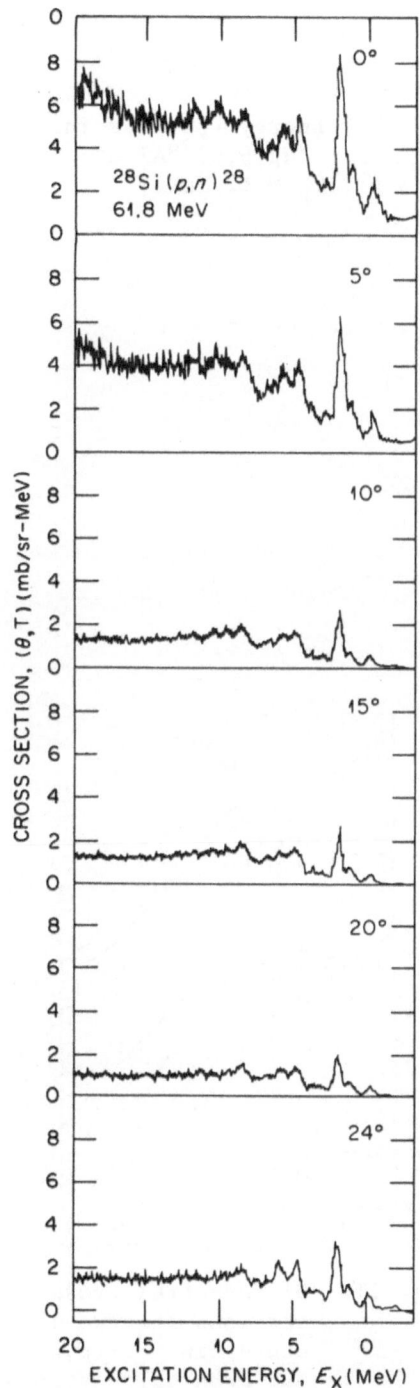

Fig. 8.   Energy spectra for
$^{28}Si(p,n)^{28}P$ at
several angles for
E_p = 62 MeV.

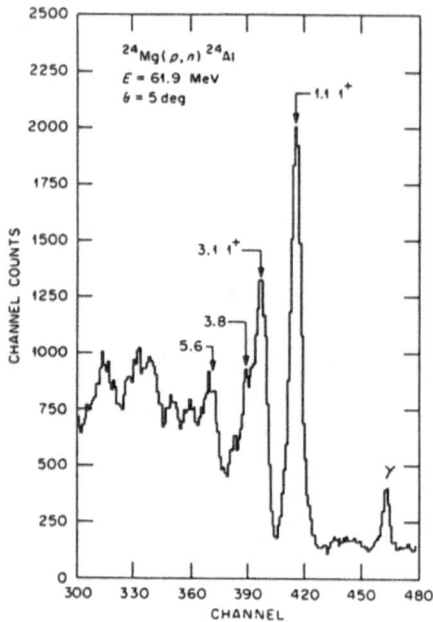

Fig. 9. Energy spectrum for
^{24}Mg(p,n)^{24}Al at
E_p = 62 MeV.

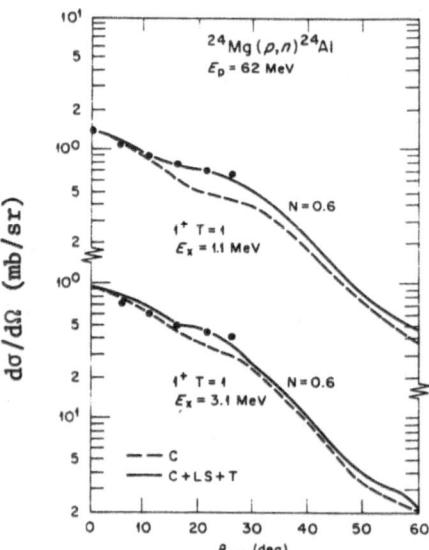

Fig. 10. Differential cross
sections for two 1^+
states in ^{24}Mg(p,n)^{24}Al
with DWBA calculations
(ref. 25).

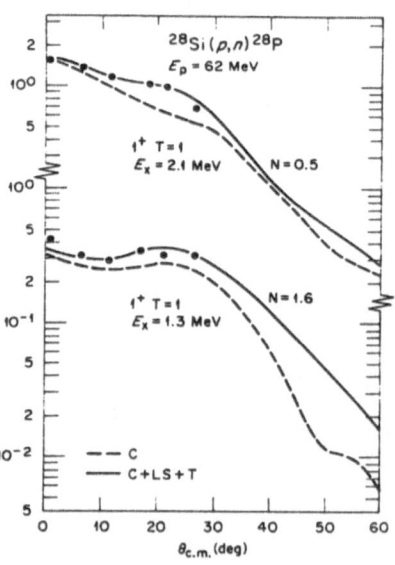

Fig. 11. Differential cross
sections for two 1^+
states in ^{28}Si(p,n)^{28}P
and DWBA calculations
(ref. 25).

In Born approximation the expression for the reaction cross section has the form:

$$\frac{d\sigma}{d\Omega} = \left(\frac{\mu^2}{2\pi\hbar^2}\right) \quad \frac{k_f}{k_i} \; 8\pi \; \frac{2J_f+1}{2J_i+1} \quad |v(q)\rho(q)|^2 \tag{6}$$

where $v(q)$ represents an interaction strength as a function of momentum transfer and $\rho(q)$ is a transition density as a function of momentum transfer.

For beta decay a similar expression applies but we are dealing only with zero momentum transfer and only $\rho(0)$ is relevant. In the reaction, if the Q value is not too different from zero the momentum transfer at 0° is approximately zero. Also we can reason that the effect of distortion is to reduce the cross section by some factor, N_d, that depends on the target and on the projectile energy. The 0° reaction cross section for Q=0 ($k_i = k_f$) is then approximately

$$\frac{d\sigma(0°)}{d\Omega} = N_d \left(\frac{\mu}{2\pi\hbar^2}\right)^2 \frac{2J_f+1}{2J_i+1} \quad |v(0)\rho(0)|^2 \tag{7}$$

For Fermi transitions ρ is the isospin density, $|\rho(0)|^2 = \langle F \rangle^2$, and for Gamow-Teller transitions ρ is the spin-isospin density $|\rho(0)|^2 = \langle GT \rangle^2$. For a general mixed transition

$$\frac{d\sigma(0°)}{d\Omega} = N_d \; \text{Const.} \; \frac{(2J_f+1)}{(2J_i+1)} \left(\langle F \rangle^2 + \left(\frac{v_{\sigma T}}{v_\tau}\right) \; \langle GT \rangle^2 \right) \tag{8}$$

This simple expression is valid only if tensor and exchange contributions can be ignored. Petrovich has shown that this is so for small momentum transfers.[13] For beta decay

$$\frac{C_o}{ft} = \frac{(2J_f+1)}{(2J_i+1)} \left(\langle F \rangle^2 + \left(\frac{C_A}{C_V}\right)^2 \; \langle GT \rangle^2 \right) \tag{9}$$

where we use values of the constants from Kavanagh,[27] $(C_A/C_V)^2 = 1.5 \pm 0.03$ and $C_o = 6200 \pm 60$ sec.

Figure 12 shows the 0° (p,n) cross sections for ^{12}C and ^{28}Si at E_p = 62 MeV plotted against the $\langle GT \rangle^2$. There is no proportionality. In fig. 13 a distortion correction is made and there is a proportionality. The factor N_d was arrived at by comparing plane wave calculations with a distorted wave calculation using an appropriate optical potential and the correct charges and Q values. It should be noted that the distortion at 62 MeV is appreciable.

Figures 14 and 15 show similar plots at E_p = 120 MeV, without and with distortion corrections. Here the distortion correction is very small suggesting the possibility of obtaining reliable GT matrix elements for transitions that cannot be observed through beta decay. In the case of mixed F and GT transitions what is plotted is the 0° cross section vs. $\langle GT \rangle^2 + (V_\tau / V_{\sigma\tau})^2 \langle F \rangle^2$. Details on how values for the matrix elements were found are given in the section Odd Mass Mirror Nuclei below.

$V_{\sigma\tau}$ AND V_τ

Figure 16 shows a spectrum for ^{26}Mg(p,n)^{26}Al. With the help of a fitting program the high energy end of the spectrum can be decomposed into peaks assigned to the 0^+ analog state at 0.23 MeV, and 1^+ states at 1.06 MeV and 1.85 MeV. The differential cross sections are shown in fig. 17. Beta decay to these states from the ground state of ^{26}Si has been observed and log ft values are given.[28] We assume that the analogous transitions from ^{26}Mg have the same matrix elements. Using the beta decay information to obtain the matrix elements and comparing the (p,n) cross sections for the 0.23 MeV and the 1.06 MeV states, we arrive at $(V_{\sigma\tau}/V_\tau)^2$ = 4.6.

ODD MASS MIRROR NUCLEI

In general the beta decay transitions between odd mass mirror nuclei are mixed F and GT transitions. For ^{7}Li(p,n)^{7}Be the first excited state of ^{7}Be at 0.48 MeV is not resolved, so the differential cross section shown in figs. 14 and 15 is assumed to be made of the F and GT parts of the mirror transition and the GT transition to the 0.48 MeV states. The beta decay from the ground state of ^{7}Be to the ground state and first excited states of ^{7}Li have been studied and log ft values are known.[29] The equivalent matrix element used in figs. 14 and 15 is $\langle GT \rangle^2_{0.48} + \langle GT \rangle^2_0 + (V_\tau/V_{\sigma\tau} \langle F \rangle_0)^2$.

In ^{9}Be(p,n)^{9}B it is not possible to measure beta decay. Figure 18 shows the distribution of calculated GT strength using

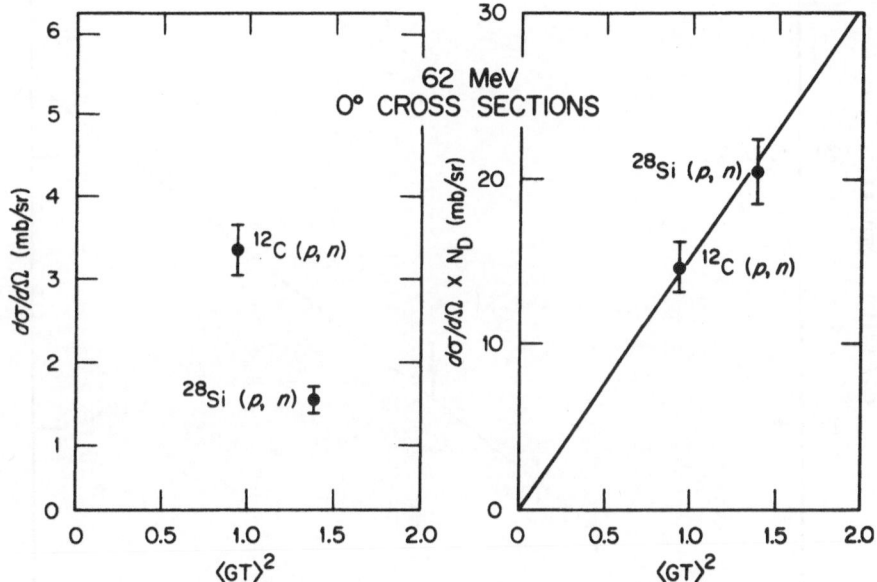

Fig. 12. Two (p,n) cross
sections at 0° for
E_p = 62 MeV plotted
vs. the squares of
the corresponding
GT matrix elements.

Fig. 13.   The data shown in
fig. 12 corrected
approximately for
distortion effects.

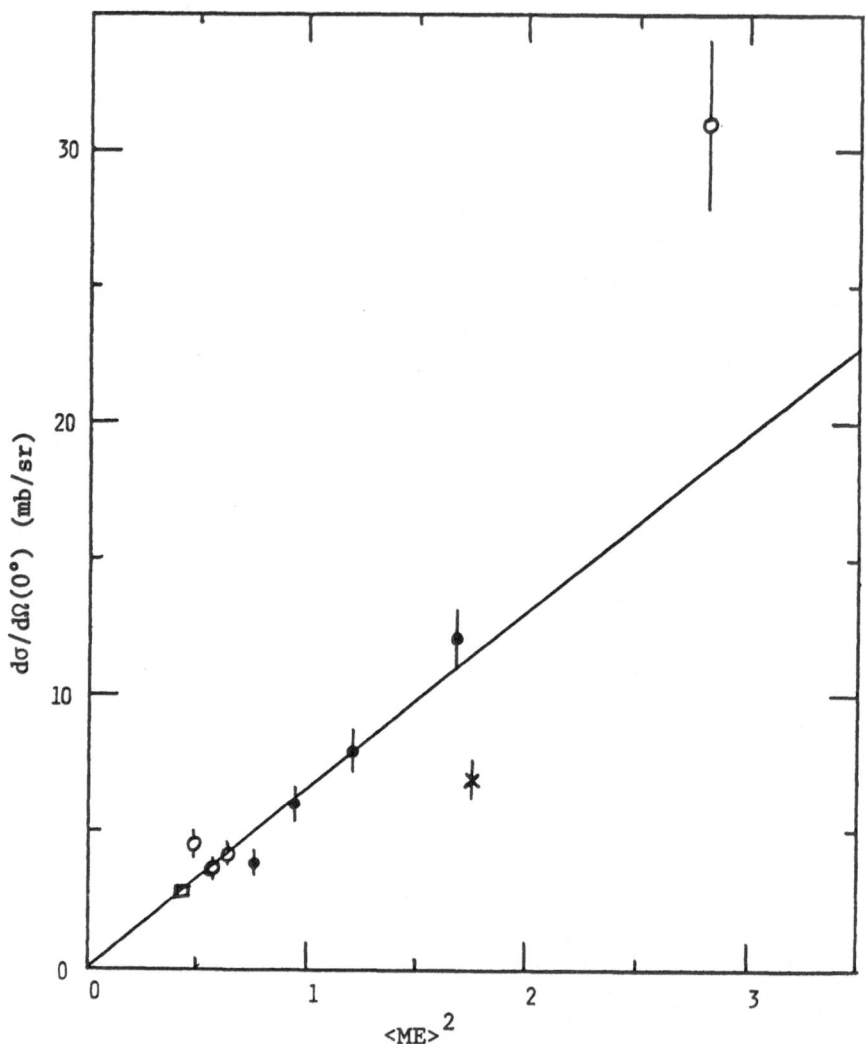

Fig. 14. Zero degree (p,n) cross sections at 120 MeV plotted vs. beta decay matrix elements where $\langle ME \rangle^2 = \langle GT \rangle^2 + (V_\tau/V_{\sigma\tau})^2 \langle F \rangle^2$. The solid dots represent direct comparisons with GT matrix elements. The line is drawn through the $^{26}Mg(p,n)^{26}Al(1.06)$ point. The choice of $(V_{\sigma\tau}/V_\tau)^2 = 4.6$ places the ^{26}Mg IAS point on the line (open square). The circles are mixed F and GT transitions plotted with $(V_{\sigma\tau}/V_\tau)^2 = 4.6$. The cross is for $^{28}Si(p,n)$ plotted vs. the calculated sum of two matrix elements for a pair of unresolved states.

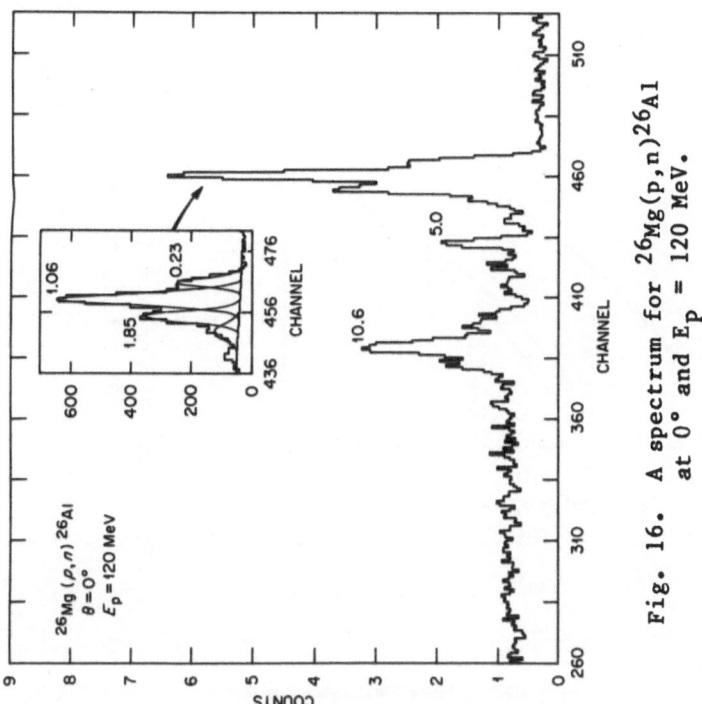

Fig. 16. A spectrum for ^{26}Mg(p,n)^{26}Al at 0° and E_p = 120 MeV.

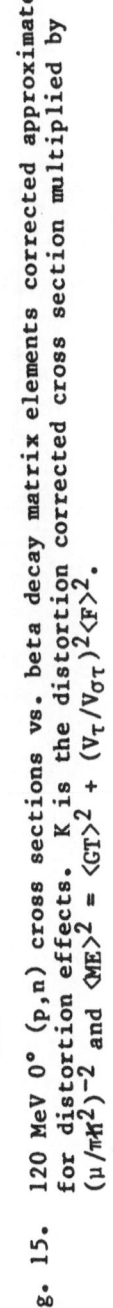

Fig. 15. 120 MeV 0° (p,n) cross sections vs. beta decay matrix elements corrected approximately for distortion effects. K is the distortion corrected cross section multiplied by $(\mu/\pi K^2)^{-2}$ and $\langle ME \rangle^2 = \langle GT \rangle^2 + (V_\tau/V_{\sigma\tau})^2 \langle F \rangle^2$.

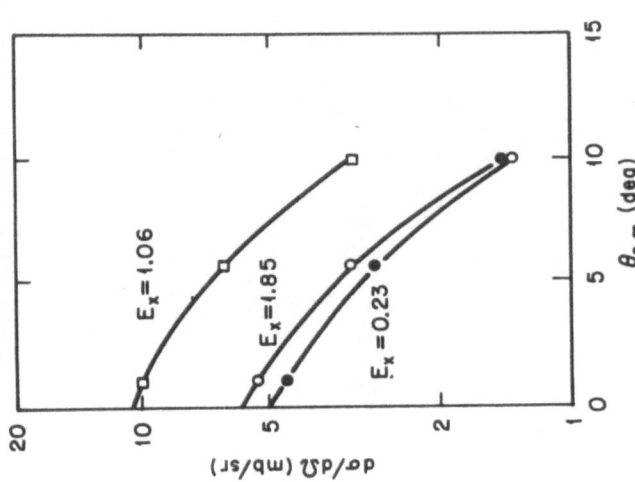

Fig. 17. Differential cross sections for $^{26}Mg(p,n)^{26}Al$.

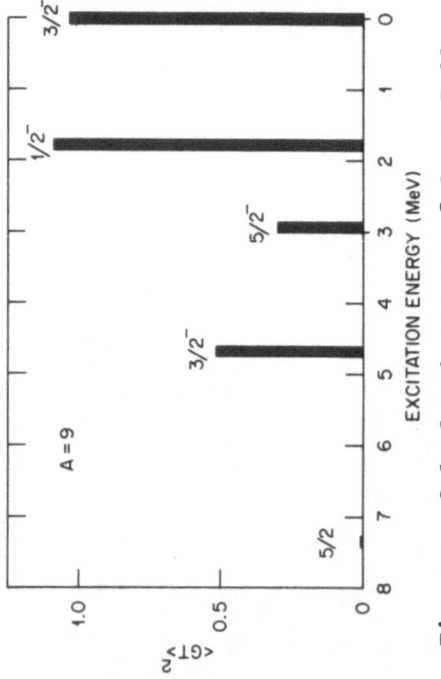

Fig. 18. Calculated squares of Gamow-Teller matrix elements for $^9Be \rightarrow {}^9B$.

the Cohen and Kurath interaction.[30] For E_p = 120 Mev the
spectrum appears to contain three peaks, the ground state, a
state at about 3.6 MeV and a weak shoulder on that peak
corresponding to about 4.7 MeV of excitation. The differential
cross sections for the ground state and the 3.6 MeV state track
each other quite well, fig. 19, and it seems reasonable to
assume that the 3.6 MeV peak represents a GT transition.

Figure 20 shows a $^{13}C(p,n)^{13}N$ spectrum. Figure 21 shows a
calculation of GT strength. The ground state beta decay has been
measured and the deduced value of $\langle GT \rangle^2_{exp}$ = 0.274. The calculated
value of 0.323 is in reasonable agreement with the beta decay
value. The ^{13}C g.s. points on figs. 14 and 15 are based on the
experimental $\langle GT \rangle^2$.

The $^{27}Al(p,n)^{27}Si$ spectrum is shown in fig. 22. Beta decay
rates are known to a few of the low lying levels in ^{27}Al from the
ground state of ^{27}Si. The ground state transition is of course
mixed F and GT. We assume that the second peak in the spectrum
is due to the sum of the GT transitions to the levels at 2.2
MeV, 2.7 MeV, and 3.0 MeV.

The expected ratio of the second peak to the ground state
from the beta decay information is 0.52. Peak fitting results on
the (p,n) spectrum range from 0.47 to 0.67, these limits
corresponding to two extreme assumptions about the background
under the unresolved peaks.

ENERGY DEPENDENCE OF THE (p,n) SPECTRA

The most striking difference between low energy (p,n) spectra
($E_p \lesssim$ 45 MeV) and high energy (p,n) spectra ($E_p \gtrsim$ 100 MeV) is
that the Fermi transitions dominate the low energy spectra and the
Gamow-Teller transitions dominate the high energy spectra. To
some extent this can be understood in terms of a changing ratio of
$V_{\sigma\tau}$ to V_τ.

Love has analyzed this question and has calculated volume
integrals of different components of the effective interaction at
various energies.[12] Figure 23 shows a representation of Love's
results together with some experimental data. Only the central
terms in the interactions are used in this comparison. The low
energy experimental data come from analyses of $^7Li(p,n)^7Be$,[6,7]
and the 120 MeV experimental datum is from ^{26}Al as discussed
above.

At high energy the concept of exciting Fermi and Gamow-Teller
transitions through different terms in the effective N-N interaction is on reasonably firm ground. The proton wave length is

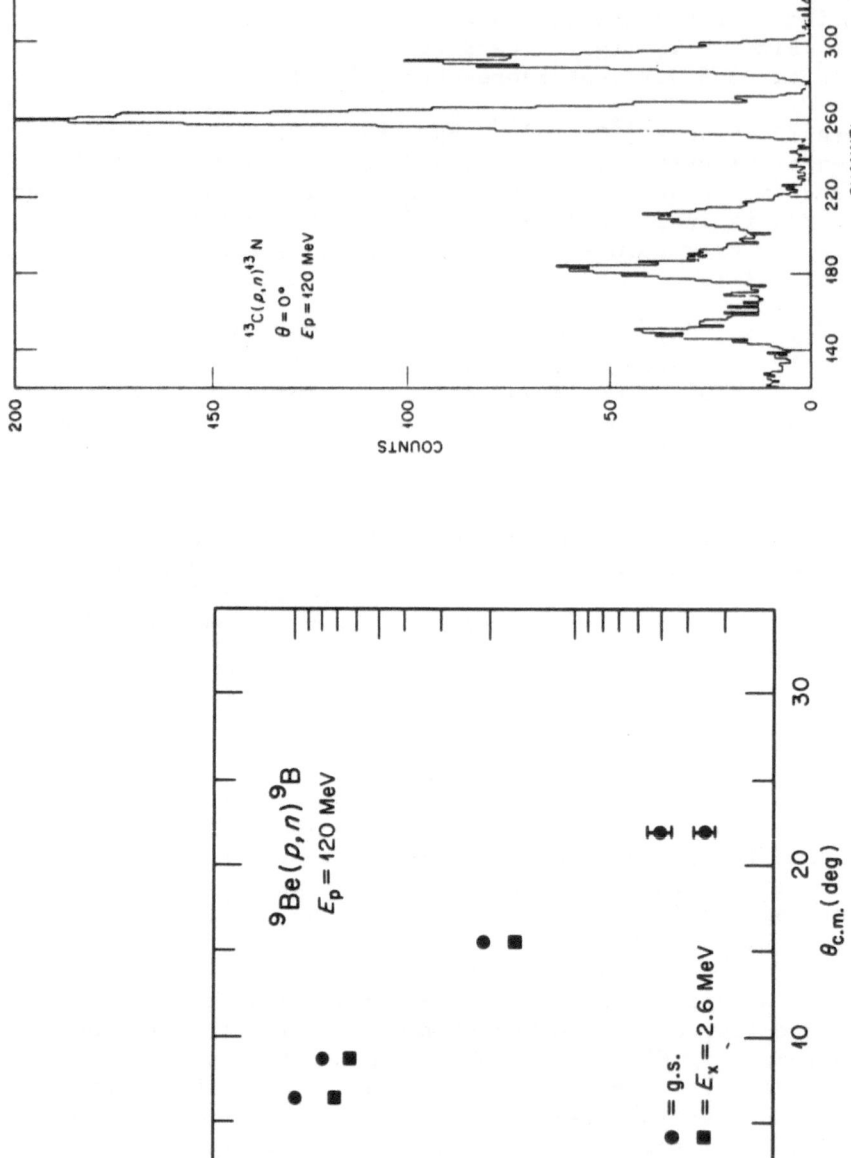

Fig. 20. Time-of-flight spectrum for $^{13}C(p,n)^{13}N$ at 0° $E_p = 120$ MeV.

Fig. 19. Differential cross sections for the ground state and excited state in $^9Be(p,n)^9B$.

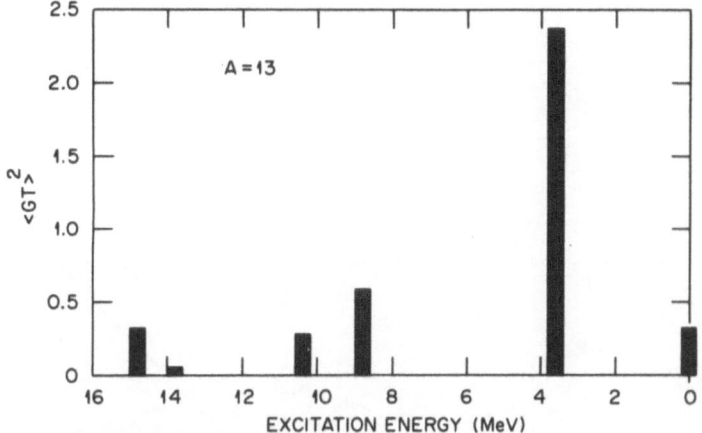

Fig. 21.   Calculated squares of Gamow-Teller matrix elements for
$^{13}C \rightarrow ^{13}N$.

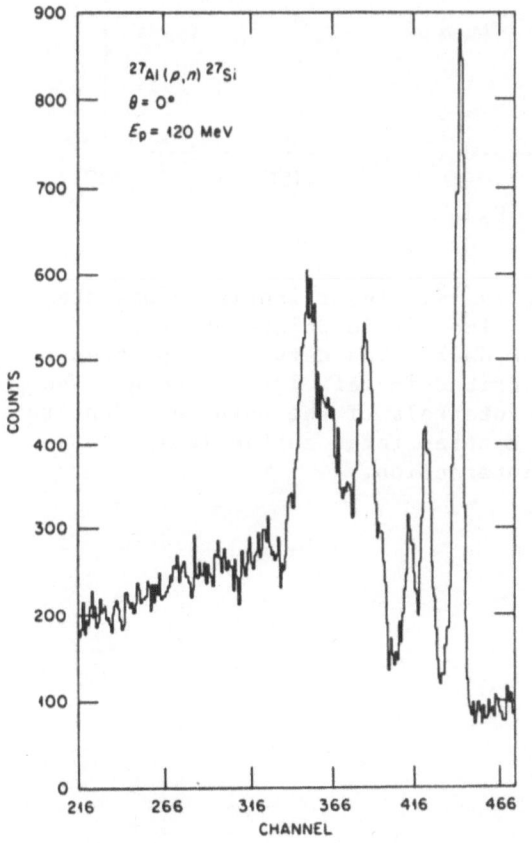

Fig. 22.   Time-of-flight
spectrum for
$^{27}Al(p,n)^{27}Si$ at
$0°$, E_p = 120 MeV.

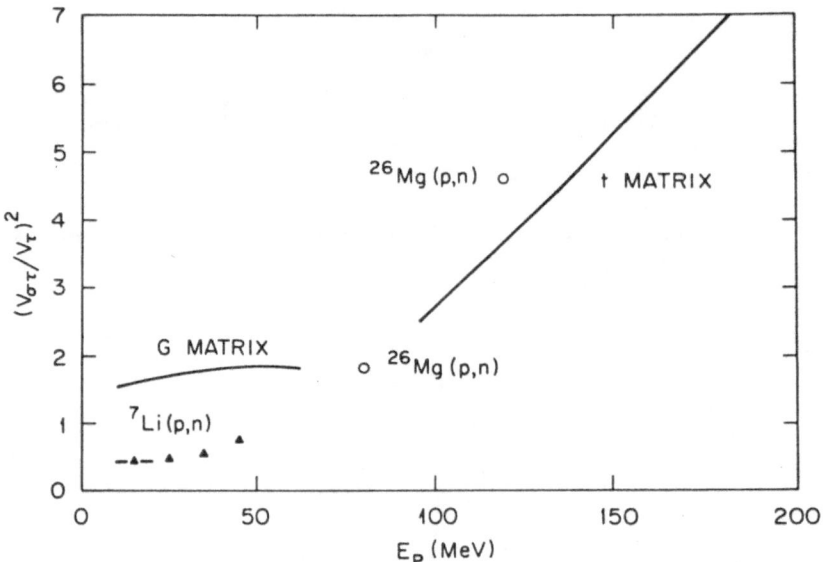

Fig. 23. Determinations of $(V_{\sigma\tau}/V_{\tau})^2$. The triangle points are
plotted from Table I. The circle points are new
measurements from IUCF data. The curves are plotted
from calculations described in ref. 12. These are the
ratios of the volume integrals of the spin dependent to
the spin independent central terms in the isospin
dependent effective interaction.

smaller than the nucleus; the N-N cross section is near a minimum
and it should be probable for the proton to traverse the nucleus
with no more than a single collision; distortion is small; and
the momentum transfer at 0° is small. These three statements are
not true in the low energy range where the IAS transitions dominate the spectra.

At low energy the Gamow-Teller peaks seem to disappear almost
completely. Figures 24 and 25 show the differential cross sections for the ground state and excited state at about 3.4 MeV in
$^9Be(p,n)^9B$ for E_p = 11 and 23 MeV. Compare this with fig. 19.
In the spectrum for $^{27}Al(p,n)^{27}Si$ for E_p = 32 MeV only the ground
state (IAS) transition is prominent.[33] In fig. 22 most of the
strength is in other peaks which we assume are Gamow-Teller transitions. For $^{90}Zr(p,n)^{90}Nb$, published spectra at E_p = 35 and E_p =
45 MeV look qualitatively different from each other[34] in that in
the higher energy spectrum presumed 1^+ states above and below the
IAS are relatively more prominent.

Perhaps we can understand the low energy spectra by working
downward from high energy where the interpretation seems clearer.

EXPERIMENTAL TECHNIQUES

Even though the (p,n) reaction was recognized many years
ago as a tool for the exploration of effective interactions and
nuclear structure, its use in the intermediate energy range has
been virtually nil owing to lack of suitable apparatus.

Because the neutron is uncharged, the precision tools of
charged particle spectroscopy, the magnetic spectrograph and the
solid state detector, are not applicable. Neutrons are easily
detected by the proton recoils they produce in hydrogenous
material and by the reactions they produce upon striking nuclei.
The signals from these interactions, however, do not precisely
reflect the energies of the causing neutrons. Time of flight
where the detection signals are used as time markers rather than
energy signals is the most successful method for measuring
neutron energies.

A time of flight facility[1] has now been set up at the
Indiana University Cyclotron Facility where a proton beam up to
200 MeV is available with subnanosecond bursts.

Even with subnanosecond time resolution, flight paths
greater than 100 m are required for some experiments as can be
surmised from fig. 26 which shows how the resolution depends on
neutron energy. Long flight paths bring on two attendant

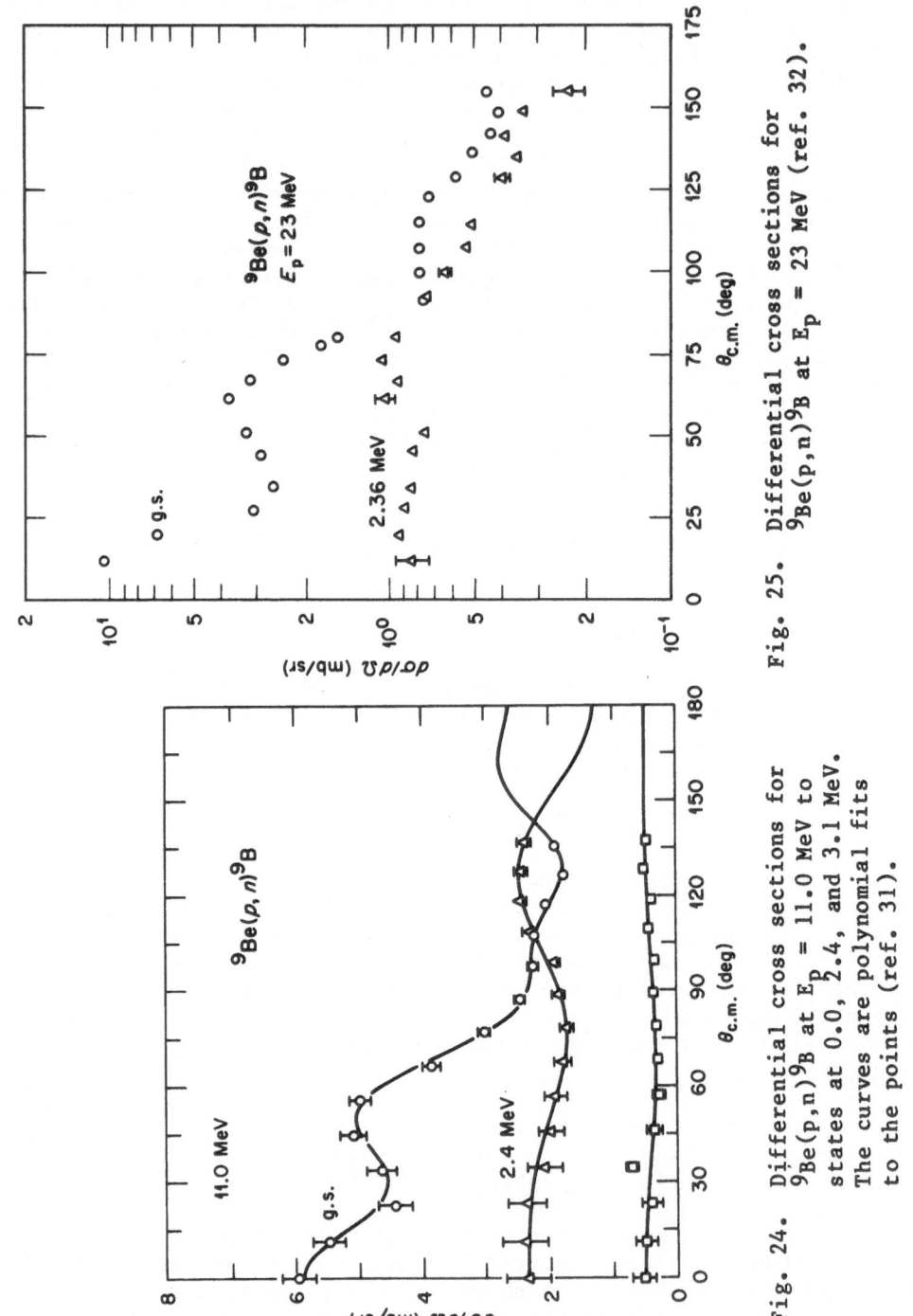

Fig. 25. Differential cross sections for $^9Be(p,n)^9B$ at E_p = 23 MeV (ref. 32).

Fig. 24. Differential cross sections for $^9Be(p,n)^9B$ at E_p = 11.0 MeV to states at 0.0, 2.4, and 3.1 MeV. The curves are polynomial fits to the points (ref. 31).

problems: it is not practical to swing the detector about the
target for angular distribution measurements, and the solid
angle subtended by reasonable sized detectors becomes miniscule.

The beam swinger solves the angular distribution problem by
varying the angle of incidence of the beam rather than the angle
of observation.

The solid angle problem demands large detectors, and, since
the dimensions are so large that the variations in transit time
of neutrons and light within the scintillator exceed the
required time resolution, some form of time compensation is
essential.

We are using a compensation scheme illustrated in fig. 27.
The origin times of the light arrival curves for scintillations
taking place in different parts of the scintillator can be made
approximately independent of the position of the scintillation
by tilting the scintillator appropriately. The variation of
rise time with position can be compensated by using quadratic
extrapolated zero timing.[2] The tilt system is simple and allows
the scintillator to be used in an orientation in which it is
thicker than the recoil track length, so in a diagram of pulse
height vs. time from the detector, high pulse height corresponds
to high neutron energy. This feature is particularly important
when one deals with the spectrum wrap-around that occurs when the
flight time differences between the high and low ends of the
spectrum are greater than the separation of beam bursts.

Other approaches to the detector problem have been developed by other groups at IUCF: a thin trigger detector followed
by a thick detector to stop the proton recoils,[35] and a thin,
large area transverse detector with mean timing.[36]

SUMMARY AND CONCLUSIONS

The forward angle (p,n) spectra for $E_p \gtrsim 100$ MeV are dominated by Gamow-Teller transitions, in contrast to low energy
spectra $(E_p \lesssim 45$ MeV) where the Fermi transitions (IAS
transitions) dominate.

Prominent GT transitions are expected from a pion exchange
interaction, and it is expected that OPEP is the dominant component of the interaction in the 100-200 MeV range.

The (p,n) reaction is a valuable probe of the effective
interaction in that the reaction cross sections to selected states are sensitive to particular components of the interaction.

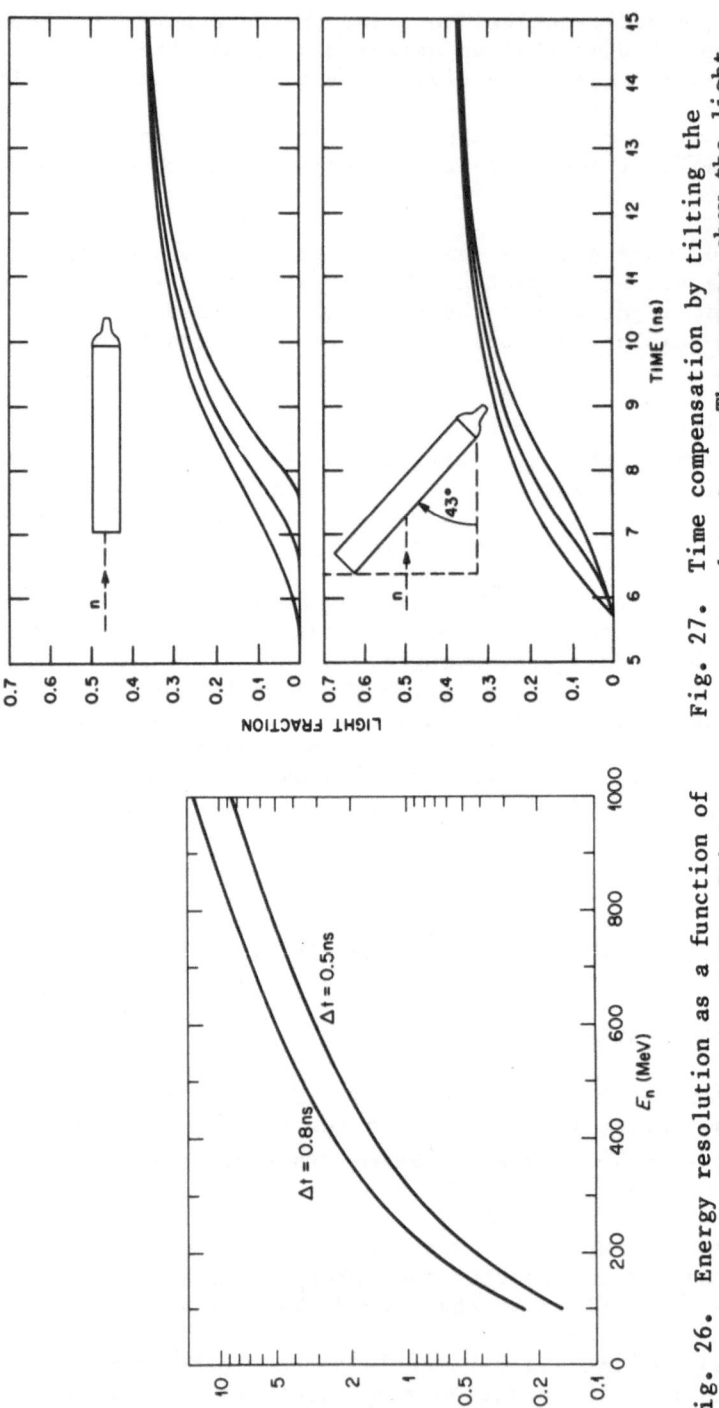

Fig. 27. Time compensation by tilting the detector. The curves show the light arrival profiles for scintillations at the front, middle, and back of a 1 m long scintillator for E_n = 100 MeV, n = 1.6, and phosphor decay time constant of 2.4 ns.

Fig. 26. Energy resolution as a function of neutron energy for a 100 m flight path and the time resolutions indicated on the curves.

This is being used to explore the monopole spin dependent and spin independent parts of the interaction and will be used to explore higher multipole components.

The (p,n) reaction is also a valuable structure probe and can be used to explore the distribution of Gamow-Teller strength. If we gain enough confidence in reaction calculations we can deduce GT matrix elements for nuclear excitations totally inaccessible to beta decay. Since distortion effects are seen to be small around 120 MeV, it seems possible to learn to extract structure information reliably.

ACKNOWLEDGEMENTS

The experimental work on (p,n) reactions at the Indiana University Cyclotron Facility has involved the efforts of a great many people and it is difficult to distribute credit justly. Table II lists the experimenters who have worked on any (p,n) experiment at IUCF. Special credit is due C. C. Foster whose heroic efforts got the beam swinger system installed.

TABLE II. Participants in (p,n) experiments.

B. Anderson	Kent State
D. Bainum	Ohio U.
A. Baldwin	Kent State
P. Debevec	IUCF (Illinois)
C. Foster	IUCF
C. Goulding	Florida A and M
M. Greenfield	Florida A and M
L. Gutay	Purdue
D. Horen	Oak Ridge
J. Knudson	Kent State
D. Lind	Colorado
R. Madey	Kent State
G. Moake	Wisconsin
P. Quin	Wisconsin
J. Rapaport	Ohio U.
R. Scharenberg	Purdue
S. Schery	Texas A and M, Galveston
E. Sugarbaker	Colorado
J. Watson	Kent State
T. Witten	Kent State
C. Zafiratos	Colorado

R. S. Lord, E. D. Hudson, M. J. Saltmarsh, and J. Rylander are to be thanked for much of the early magnet design work on the beam swinger when it was intended to be installed in Oak Ridge. E. Kowalski is to be thanked for assistance with pulse selection and phase stabilization. We thank the IUCF management and staff for their support and a major commitment of resources to this project.

REFERENCES

1. C. D. Goodman, C. C. Foster, M. B. Greenfield, C. A. Goulding, D. A. Lind and J. Rapaport, IEEE Trans. Nucl. Sci. NS-26, 2248 (1979).
2. C. D. Goodman, J. Rapaport, D. E. Bainum and C. E. Brient, Nucl. Instrum. Methods 151, 125 (1978). C. D. Goodman, J. Rapaport, D. E. Bainum, M. B. Greenfield and C. A. Goulding, IEEE Trans. Nucl. Sci. NS-25, 577 (1978).
3. J. D. Anderson, C. Wong and J. W. McClure, Phys. Rev. 126, 2170 (1962).
4. A. M. Lane, Phys. Rev. Lett. 8, 171 (1962).
5. See for examples: J. D. Carlson, C. D. Zafiratos and D. A. Lind, Nucl. Phys. A249, 29 (1975), and R. R. Doering, D. M. Patterson and A. Galonsky, Phys. Rev. C 12, 378 (1975).
6. S. D. Bloom, J. D. Anderson, W. F. Hornyak and C. Wong, Phys. Rev. Lett. 15, 264 (1965) and J. D. Anderson, C. Wong and V. A. Madsen, Phys. Rev. Lett. 24, 1074 (1970).
7. R. R. Doering, L. E. Young, R. K. Bhowmik, S. M. Austin, S. D. Schery and R. DeVito, Michigan State University Cyclotron Laboratory Annual Progress Report, Sept. 1976.
8. P. J. Moffa and G. E. Walker, Nucl. Phys. A222, 140 (1974).
9. G. R. Satchler, Nucl. Phys. A100, 497 (1967).
10. T. T. S. Kuo and G. E. Brown, Nucl. Phys. 85, 40 (1966).
11. G. Bertsch, J. Borysowicz, H. McManus and W. G. Love, Nucl. Phys. A284, 399 (1977).
12. W. G. Love, proceedings this conference.
13. F. Petrovich, proceedings this conference.
14. Indiana University Cyclotron Facility Progress Report Feb. 1977-Jan. 1978, pp. 91-93.
15. W. G. Love, A. Scott, F. T. Baker, W. P. Jones and J. D. Wiggins, Jr., Phys. Lett. 73B, 277 (1978). See also W. G. Love and G. R. Satchler, Nucl. Phys. A159, 1 (1970).
16. A. Picklesimer and G. E. Walker, Phys. Rev. C 17, 237 (1978).
17. Peter Signell, Advances in Nuclear Physics II, 223 (1969).
18. F. Petrovich, proceedings this conference.
19. W. C. Barber, F. Berthold, G. Fricke and F. E. Gudden, Phys. Rev. 120, 2081 (1960).

20. G. Morpurgo, Phys. Rev. 110, 721 (1958).
21. D. Kurath, Phys. Rev. 130, 1525 (1963).
22. R. K. Adair, Phys. Rev. 87, 1041 (1952).
23. C. C. Foster, proceedings this conference.
24. We thank A. Picklesimer and G. E. Walker for providing us with these calculations.
25. We thank F. Petrovich for providing us with these calculations.
26. B. H. Wildenthal and W. Chung, s-d shell wave functions communicated to F. Petrovich prior to publication.
27. R. W. Kavanagh, Nucl. Phys. A129, 172 (1969).
28. P. M. Endt and C. van der Leun, Nucl. Phys. A310, 1 (1978).
29. F. Ajzenberg-Selove and T. Lauritsen, Nucl. Phys. A227, 1 (1974).
30. We thank J. B. McGrory for calculating these GT matrix elements with the Oak Ridge shell model code.
31. R. Byrd, Ph.D. Thesis, Duke University, Dept. of Physics, 1978.
32. R. D. Bentley, Ph.D. Thesis, University of Colorado, Dept. of Physics and Astrophysics, 1972.
33. R. K. Jolly, T. M. Amos, A. Galonsky, R. Hinrichs and R. St.Onge, Phys. Rev. C 7, 1903 (1973).
34. R. R. Doering, A. Galonsky, D. M. Patterson and G. F. Bertsch, Phys. Rev. Lett. 35, 1691 (1975).
35. G. L. Moake, P. T. Debevec, L. J. Gutay, P. A. Quin and R. P. Scharenburg, Indiana University Cyclotron Facility Technical and Scientific Report 1977 to 1978, p. 114.
36. R. Madey, private communication.

EXPERIMENTAL TEST OF ONE-PION EXCHANGE AND PCAC IN PROTON-NUCLEUS CHARGE EXCHANGE REACTIONS AT 144 MeV.[*]

G. L. Moake, L. J. Gutay, R. P. Scharenberg, P. T. Debevec, and P. A. Quin

Department of Physics, Purdue University, West Lafeyette, IN 47907, Department of Physics, University of Illinois at Urbana-Champaign, Urbana, IL 61801, Department of Physics, University of Wisconsin, Madison, WI 53706

The inelastic excitation of nuclei by intermediate energy nucleons is a topic of long standing experimental and theoretical interest. At intermediate energy it has been argued, the inelastic scattering amplitudes should bear some more or less transparent relation to the free nucleon-nucleon scattering amplitudes. Although this relation would be difficult to apply to an arbitrary complex final state, at the very least the relation should be evident in a judiciously chosen set of discrete final states. Not only has the latest generation of accelerators and experimental techniques fostered a renewed interest in this problem, but the nucleon-nucleon amplitudes themselves are now rather well determined.

The nucleon-nucleon interaction arises from the exchange of mesons. A persistent feature of neutron-proton elastic scattering is the charge exchange process, i.e. the rapid variation in the differential cross section for $n+p \rightarrow p+n$ at small momentum transfer, typically, $q^2 \lesssim M_\pi^2$. This distinctive feature of the differential cross section is generally regarded as arising from the interference of the one-pion exchange amplitude, which varies rapidly over the range $0 \leq q^2 \leq M_\pi^2$, and a number of other amplitudes, which cumulatively have a much less rapid dependence on momentum transfer. It has been shown, for example, by Ashmore et al[1] that the pion-nucleon coupling constant, $g_{p\pi n}(-M_\pi^2)$, that can be obtained from an analysis of neutron-proton scattering is consistent with the

[*]This work has been supported by the U.S. Department of Energy and the National Science Foundation.

coupling constant that can be obtained from an analysis of pion-nucleon scattering. It should be noted that the process by which this interference arises is still a point of some controversy; however, the one-pion exchange amplitude is so firmly established that it has been incorporated into all nucleon-nucleon phase shift analyses.

It is due to the small mass of the pion that it is afforded special consideration; yet it is not the only component of the nucleon-nucleon interaction. The exchange of a pseudo scalar meson provides a natural motivation for the spin-isospin term and the tensor term in a nucleon-nucleon potential. A comparison of the static one-pion exchange potential and a realistic nucleon-nucleon potential, e.g. Hamada-Johnston, shows that at distances greater than a pion Compton wavelength the tensor force is dominantly due to one-pion exchange. The various components of the central force even at large distances require in addition to one-pion exchange the exchange of isoscalar and isovector scalar mesons which represent the effective contribution of non-resonant two-pion exchange. Given this degree of complexity it is in general of little use to consider the nucleon induced inelastic scattering as arising from the exchange of mesons. Nevertheless there is a class of inelastic excitations for which we wish to argue that this point of view is not without merit. These are the so-called Gamow-Teller transitions.

It is a common feature of the three reactions, $p+{}^6Li \rightarrow n+{}^6Be$, $p+{}^{12}C \rightarrow n+{}^{12}N$, and $p+{}^{14}N \rightarrow n+{}^{14}O$, that the change in nuclear angular momentum (J), isospin (I), and parity (P) is the same; namely, $\Delta J=1$, $\Delta I=1$, $\Delta P=0$. These nuclei undergo the same change in their quantum numbers in the (p,n) reaction as in an allowed Gamow-Teller beta decay. This change in quantum numbers is also that which can be induced by the exchange of a virtual p-wave pion. Recalling the contribution of one-pion exchange to neutron-proton charge exchange and the relation between the axial vector matrix element and the pion-nucleus coupling constant[2], we conjectured that the reactions are dominated at intermediate energy by one-pion exchange for small momentum transfer, $q^2 \lesssim M_\pi^2$. This appears to be the case for ${}^6Li$ and ${}^{12}C$ where the axial vector matrix element is large (although in A=6 it is, to be accurate, ${}^6He$ for which the beta decay is known), but it does not appear to be the case for ${}^{14}N$ where the matrix element is vanishingly small.

Our measurement was carried out at the Indiana University Cyclotron Facility using an incident proton energy of 144 MeV. This energy was chosen to take advantage of existing elastic scattering data[3] and was at the time of our experiment near the maximum available energy. The experimental setup is shown in Fig. 1. (This beam line was available for neutron time-of-flight experiments from the fall of 1976 to the spring of 1978. It has since been dis-

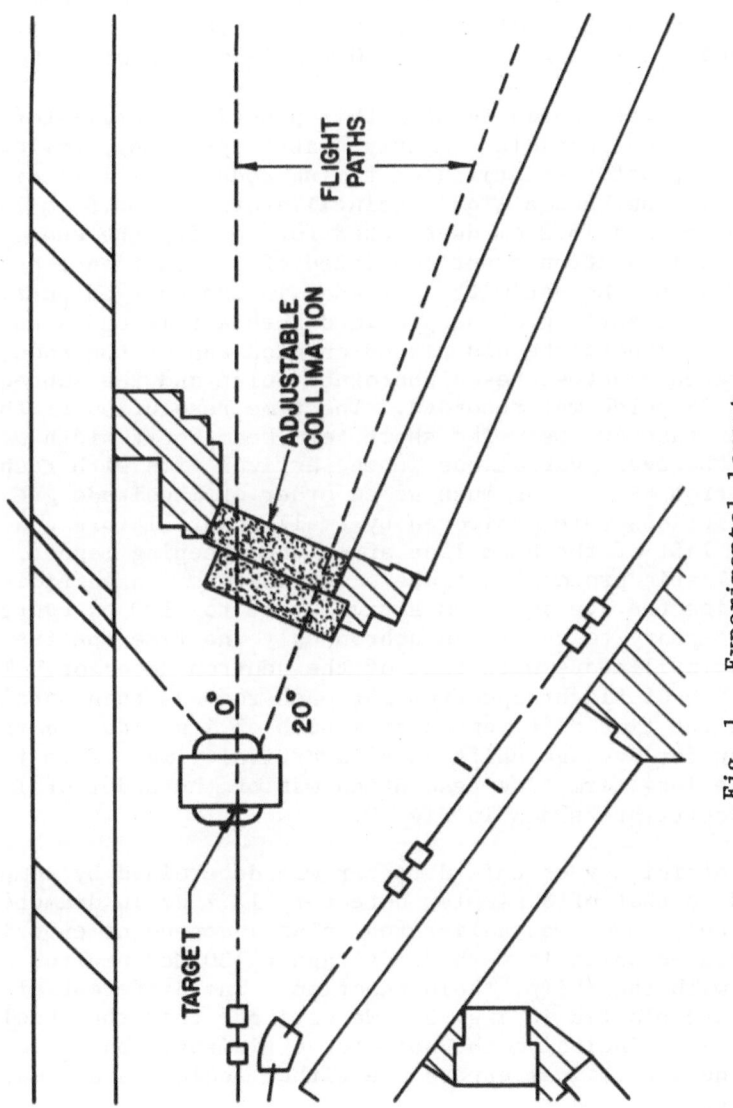

Fig. 1. Experimental layout.

assembled.) The target, with an areal density of ~50 mg/cm^2, was located at the entrance of a large gap dipole magnet, which swept the proton beam into a well shielded beam dump where the total charge was integrated. Neutrons passed between the pole faces of the magnet and exited the vacuum chamber through a 4 mil Kapton window. Adjustable lead and concrete collimators permitted the remote detector to view only the target. Background from scattering off of the collimators was negligible. A small correction for attenuation in air along the 20 m flight path was made.

The detector consisted of a thin plastic scintillator to identify charged particles (mostly elastic protons), followed by two independent plastic scintillator timing rods, 4.4 cm in diameter and 45.7 cm long, and a liquid scintillator vat, 50.8 cm long, 25.4 cm high, and 15.2 cm deep, used for setting the energy threshold. A valid neutron event consisted of a coincidence between one of the rods and the vat with no signal in the charged particle identifier. Elastic protons provided both a time and a pulse height reference. A phototube was placed on each end of the rods, and the time difference between each phototube pulse and the subsequent cyclotron RF pulse was recorded. The time resolution of the rods was better than 400 ps. The short term beam burst width was ≲ 300 ps; however, variations in the arrival time with respect to the cyclotron RF were as much as an order of magnitude greater. These variations were monitored by small scintillators placed to the right and left of the beam line after the sweeping magnet. The rate of elastic protons in these scintillators was sufficiently high to determine the average beam arrival time to ~100 ps every second. It was necessary to record asynchronously the time spectra of the monitor scintillators with that of the neutron detector. The neutron time-of-flight spectrum for each rod was then obtained by averaging the time differences from each of its ends and then correcting for average shift in beam arrival time. With this procedure the long term time resolution was of the order of 1 ns. Sample spectra are shown in Fig. 2.

The efficiency of this detector was determined by comparing its yield to that of a simpler detector, 12.7 cm in diameter and 12.7 cm deep, which was calibrated in an independent experiment using the associated particle method. A tagged 130 MeV neutron beam was produced with the ^{7}Li(p, ^{7}Be)n reaction. The differential cross sections are plotted in Fig. 3. We estimate that the absolute normalization, including the detector efficiency, has an accuracy of ±7%; the statistical errors are either indicated or smaller than the data points.

The theoretical calculations were done in the framework of an absorption modified one-pion exchange model.[4] The six independent pion exchange Born helicity amplitudes for a ΔJ=1, ΔI=1, ΔP=0 can

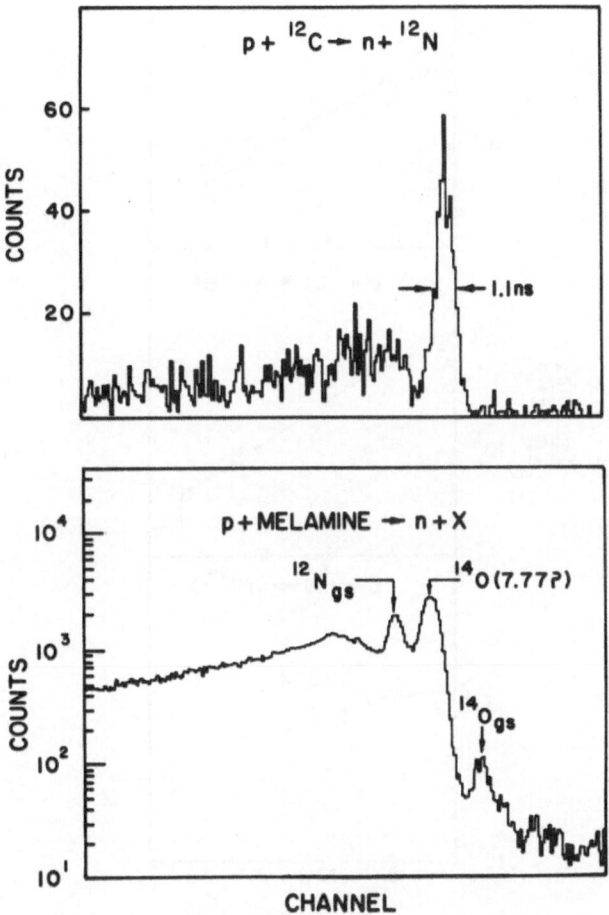

Fig. 2. Sample neutron time of flight spectra.

184   G. L. MOAKE ET AL.

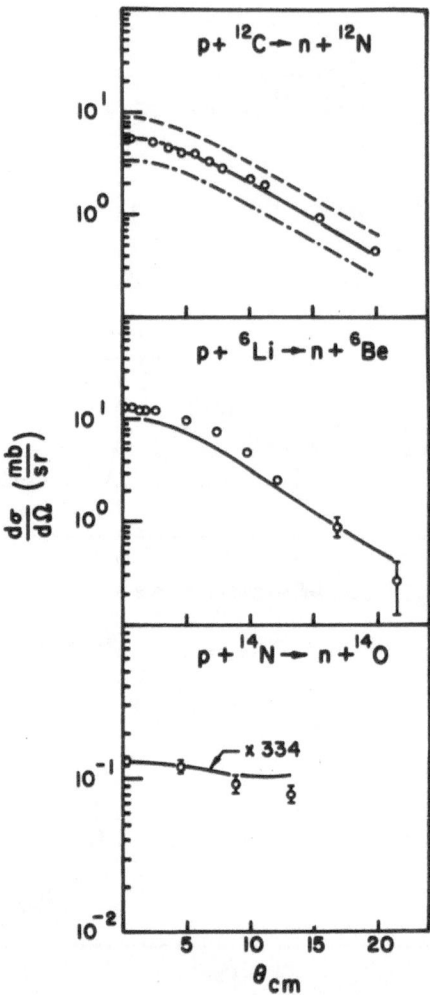

Fig. 3. Differential cross sections. For ^{12}C the three calculations are, as described in the text, plane wave Born approximation (dashed curve), the usual absorption model (dot-dashed curve), and our approximation for distortions (solid curve). For ^{6}Li and ^{14}N only our approximation is indicated.

be written in the form[5]

$$T_{\lambda'\lambda}(x) = (\tfrac{1+x}{2})^{|\lambda+\lambda'|} (\tfrac{1-x}{2})^{|\lambda-\lambda'|} [\frac{A_{\lambda'\lambda}(z)}{z-x} - B_{\lambda'\lambda}(x)]$$

where λ and λ' are the magnitude of the differences of the particle
helicities in the initial and final states, $x=\cos\theta$, and z is a
variable depending only on the masses and momenta of the particles.
The term $B(x)$ is known as the "extraordinary" term in the literature,
and for the cases considered here it is simply a rotation function
of order 1/2, which corresponds to S-wave scattering. It is
necessary to modify these amplitudes in three ways before a sensible
comparison with experiment is made. First of all, the extraordinary
term is removed. By this procedure, the Born amplitudes are made
to have a smoothly varying partial wave expansion[6], an expansion
which is made with the technique given by Hogaasen and Hogaasen.[5]
This removal is analogous to the neglect of the contact term in the
one-pion exchange potential. Secondly, vertex functions are chosen
to provide the correct dependence on momentum transfer. These two
modifications are customary in all relativistic one boson exchange
calculations.[7] We discuss the choice of the vertex functions below.
The third modification takes cognizance of the distortions from
initial state and final state interactions.

In high energy applications of the absorption model, the
process under consideration was peripheral in the sense that the
initial and final state interactions were of short range compared to
one-pion exchange. The prescription used to modify the Born
amplitudes to account for distortions was to multiply each partial
wave by the product of the elastic scattering phase shifts of the
initial and final channels,

$$T_{\lambda'\lambda}^{j} \rightarrow (S_f^{j})^{1/2} \; T_{\lambda'\lambda}^{j} \; (S_i^{j})^{1/2}$$

This form is justified for the case where the elastic scattering is
highly diffractive. In our application both the elastic scattering
and the exchange amplitudes have a range of the order of nuclear
dimensions, invalidating this argument. Consequently, we propose
the modification

$$T_{\lambda'\lambda}^{j} \rightarrow \tfrac{1}{2}[(S_i^{j})^{1/2} + (S_f^{j})^{1/2}] \; T_{\lambda'\lambda}^{j}.$$

This form preserves the transparent separation of the Born amplitudes
and the distortions, and can be motivated by Green function techniques. The elastic scattering S-matrix elements are generated by
an optical model potential fit to the appropriate elastic scattering

data. We show in Fig. 3 for the case of ^{12}C that the (plane wave)
Born approximation is too large, that the usual absorption model
is too small, and that our approximation provides a rather accurate
representation of the data.

The magnitude and momentum dependence in our calculation are
largely governed by the vertex functions $g_{p\pi n}(q^2)$ and $g_{N\pi N'}(q^2)$.
Following Kim and Primakoff[2] we note that at $q^2=0$ these vertex
functions are given by application of a modified Goldberger-
Trieman relation[8] to nuclei.

$$g_{N\pi N'}(0) = ((M_N M_{N'})^{1/2}/a_\pi)\ F_A(0)$$

In this equation M_N, $M_{N'}$, and $a_\pi=0.131$ GeV denote the masses of the
initial and final nuclear states, and the pion decay constant,
respectively. The axial vector matrix element $F_A(0)$ is obtained
from the ft value of the nuclear beta $N' \rightarrow N + e + \nu$. (For A=6
we use the ^{6}He decay.)

$$F_A^2(0) = \frac{2\pi^3 \ln 2 (2J_{N'}+1)}{3G^2 \cos^2\theta_C\ m_e^5\ ft}$$

For the nucleon it is, of course, the physical coupling constant
$g_{p\pi n}(-M_\pi^2)$ which is measured independently and which appeared in the
original Goldberger-Trieman relation. The modified relation follows
from the Partially Conserved Axial Current hypothesis (PCAC)[9]
which assumes that the divergance of the axial current satisfies an
unsubtracted dispersion relation. From this relation it is clear
that $g_{N\pi N'}^2(0)$, and hence the Born cross sections, are inversely
proportional to the beta decay ft value.

Using the approximation that the momentum dependence of the
axial vector matrix element, $F_A(q^2)$, is identical to that of the
weak magnetism matrix element, $F_M(q^2)$, which, assuming CVC, is
measured in inelastic electron scattering, Kim[10] and Kim and
Townsend[11] find that

$$g_{N\pi N'}(q^2) = \frac{1 - q^2/25m_\pi^2}{(1 + q^2/M_A^2)^2}\ g_{N\pi N'}(0)$$

for A=6 and A=12, where $M_A=2M_\pi^2$ and $2.6\ M_\pi^2$, respectively.[12] Nucleon
beta decay and neutrino scattering yield the equation[13]

$$g_{p\pi n}(q^2) = g_{p\pi n}(0)/(1+q^2/47m_\pi^2)^2.$$

For A=14, $F_A(0) \tilde{} 0$, and the usual approximation required to obtain $F_A(q^2)$ from inelastic electron scattering is not valid. By making a plausible separation of the spin and orbital contributions to the inelastic electron scattering, Goulard et al[14] suggest that

$$F_A(q^2) \quad \alpha \quad G_{N\pi N'}(q^2) \quad \alpha \quad \frac{q^2/m_\pi^2}{(1+q^2/2.4m_\pi^2)}$$

Although this approximation is of value in the extrapolation of $F_A(q^2)$ to $q^2=m_\mu^2$ (muon capture), in our application the 0° (p,n) cross section is then zero.

The results of application of one-pion exchange and PCAC to these (p,n) reactions are shown in Fig. 3 and in Table I.

Table I.

Target	$g_{N\pi N'}(0)$ [PCAC]	$g_{N\pi N'}(0)$ [experiment]	$d\sigma/d\Omega\vert 0^o$ (mb/sr)
^{6}Li	69.2	60.9	13.1
^{12}C	59.5	59.9	5.56
^{14}N	.994	18.0	0.13

The absolute magnitude of $g_{N\pi N'}(0)$ obtained from our 0° cross section is within 15% of the value predicted by PCAC from nuclear beta decay for A=6 and A=12. It is gratifying that this well defined relationship between the weak interaction and pion exchange holds at this level.

With regard to A=14, as $F_A(0) \approx 0$, in our calculation the (p,n) cross section must essentially vanish at 0°. The measured cross section is too large by a factor of ~300. Although the cross section is small compared to ^{6}Li and ^{12}C, it cannot be due largely to one-pion exchange. As the tensor component of the nucleon-nucleon interaction is largely due to one-pion exchange, it may not be appropriate to constrain an effective interaction, suitable for nucleon inelastic scattering impulse approximation calculations, by the measured ^{14}N cross section.

It is a pleasure to acknowledge the assistance of Dr. C. C. Foster in the design and construction of the neutron beam line and shielding arrangements. The support of Professors R. E. Pollock and G. T. Emery throughout this work has been appreciated. We would like to thank Professor K. Vasavada for theoretical assistance during

the proposal and analysis stage. Professors H. Primakoff and B.
Goulard were kind enough to illucidate the problems in ^{14}N.
Finally, we thank Dr. H. Foelsche of the BNL Accelerator Division
for arranging the loan of the sweeping magnet without which our
experiment would not have been possible.

REFERENCES

1. A. Ashmore, W. H. Range, R. T. Taylor, B. M. Townes, L.
    Castillejo and R. F. Peierls, Nucl. Phys. 36, 258 (1962).
2. C. W. Kim and H. Primakoff, Phys. Rev. 139B, 1447 (1965).
3. O. N. Jarvis, C. Whitehead, and M. Shah, Nucl. Phys. A 184,
    615 (1974).
4. K. Gottfried and J. D. Jackson, N. Cim. 34, 5783 (1964) L.
    Durand and Y. T. Chiu, Phys. Rev. 139, 646 (1965).
5. H. Hogaasen and J. Hogaasen, N. Cim. 39, 941 (1965).
6. G. Fox, CALT-68-335.
7. K. Enkelenz, Phys. Rep. C 13, 191 (1974).
8. M. Goldberger and S. B. Trieman, Phys. Rev. 110, 1178 (1958).
9. M. Gell-Mann and M. Levy, N. Cim 16, 705 (1960).
10. C. W. Kim, Rev. d. N. Cim. 4, 189 (1974).
11. C. W. Kim and J. S. Townsend, Phys. Rev. D11, 656 (1975).
12. Our equation is obtained by combining the nuclear PCAC relation,
     Ref. 11, Eq. 6, with the impulse approximation for the pseudo-
     scalar form factor, Ref. 10, Eq. 56, and Ref. 11b, Eq. 16, with
     the q^2 dependence of the axial vector matrix element, Ref. 10,
     Eq. 59 and 60.
13. Ref. 10, Eq. 8. The numerical constants have been chosen such
     that $g_{p\pi n}(-M_\pi^2) = 1.43$ and $g_{p\pi n}(0) = 1.32$.
14. B. Goulard, B. Lararo, H. Primakoff, and J. D. Vergados, Phys.
     Rev. C16, 1999 (1977).

DISCUSSION

    Debevec: After the oral presentation of the above paper,
Drs. Love, McManus, and Petrovich expressed disagreement with the
interpretation of the discrepancy between the ^{14}N(p,n)^{14}O experiment and the calculation. On the basis of our discussion I
should like to correct the last sentence of the above paper. In
the calculation presented, the 0° cross section is essentially
zero as $F_A(0)$ is essentially zero. The treatment of distortions,
however, through the eikonal approximation, is inadequate for
^{14}N(p,n)^{14}O. Although there is no model-independent determination of the momentum dependence of the vertex function for
^{14}N, it must be similar to that suggested by Goulard et al. (for
$q^2 \lesssim M^2\pi$), i.e. $q^2/(1+q^2/M^2)^2$. Although the asymptotic momenta
of the incident and outgoing particles are colinear for 0°, the
distorting potentials can change the direction as well as the
amplitude and phase of the incident and outgoing waves. Thus the

distorting potentials permit the sampling of the vertex function, which, although zero at $q^2=0$, obviously rises. Therefore, (even) in the model presented -- one-pion exchange -- there will be a finite 0° cross section. The treatment of distortions is adequate for the ^{6}Li and ^{12}C 0° comparison simply because the vertex functions peak at $q^2=0$. The comparison then in Table I for ^{14}N is rather without merit. As the 0° ^{14}N cross section depends (even) in the one-pion exchange model on the distortions and momentum dependence of the vertex function, I can make no quantitative statement of the physical phenomena that contribute to the cross section. The inhibition of the beta decay is dramatic, but the (p,n) cross section is only small. The DWIA calculation presented by Dr. Love shows its magnitude to be the coherent contribution of many amplitudes. It is therefore unreasonable to expect any high level of quantitative agreement with experiment for ^{14}N.

EVIDENCE FOR GAMOW-TELLER STRENGTH IN BROAD BUMPS IN (p,n)

AND (^{3}He,t) SPECTRA[*]

Aaron Galonsky

Cyclotron Laboratory,
Michigan State University
East Lansing, MI 48824

INTRODUCTION

Following the discovery by Anderson and Wong of isobaric analog states[1] in (p,n) reactions (equivalent to superallowed or giant Fermi transitions in β decay) a giant Gamow-Teller (G-T), or spin-flip, transition in charge-exchange reactions was predicted.[2] Subsequently, a broad, structured peak in the ^{90}Zr(p,n)^{90}Nb reaction was interpreted as verifying the predicted transition,[3] a $0^+ \to 1^+$, T = 5 → T = 4 transition from the ground state of ^{90}Zr to an excitation energy in ^{90}Nb centered at 8.4 MeV. Some of the neutron spectra of Doering, et al.[4] are shown in Fig. 1. In each spectrum the isobaric analog of the ground state (IAS) of ^{90}Zr is represented by the narrow peak at E_n = 33 MeV and the 1^+ state by the broad peak around 30 MeV. If generalized to other nuclei, concentration of G-T strength at such excitation energies could account for the generally low rates of G-T decay from the ground states of radioactive nuclei.

The same transition has been seen with the (^{3}He,t) reaction at 80 MeV at Grenoble[5] and at 130 MeV at Jülich.[6] The Grenoble work separated the 8.4-MeV peak into two groups, one at 7.2 MeV having the angular distribution of a $0^+ \to 1^+$ transition, and another of unknown multipolarity at 9.7 MeV. The Jülich work revealed another broad bump, also present in re-scrutinized (p,n) spectra,[4] at 18.5 MeV in ^{90}Nb. Before examining in detail work on Zr-that already cited plus some current unpublished data--other cases of broad bumps in (p,n) and (^{3}He,t) spectra will be briefly noted.

[*] This material is based upon work supported by the National Science Foundation under Grant No. Phy-7822696.

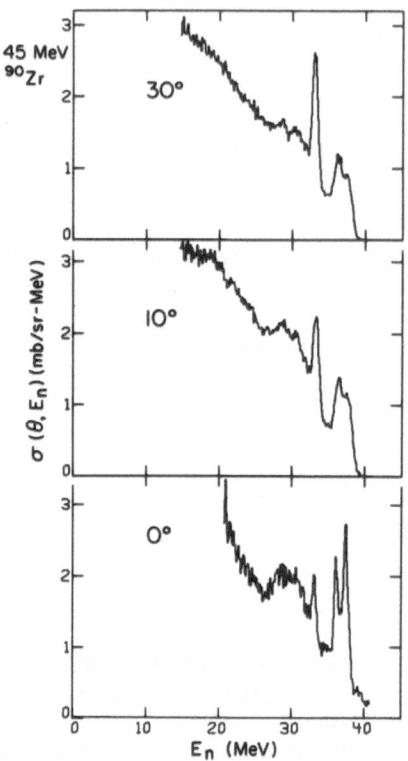

Fig. 1.   Neutron spectra (Ref. 4) from ^{90}Zr + p at 45 MeV showing
          the sharp IAS at 33 MeV and a broad bump $\simeq$ 30 MeV.

SURVEY

     A low-resolution spectrum of the ^{58}Ni(p,n)^{58}Cu reaction is
shown in Fig. 2.[7] The analogs and antianalogs of 1^+ states seen
in ^{58}Ni(t,^{3}He)^{58}Co[8] and ^{58}Ni(e,e')[9] are at excitation energies
around 10 and 6 MeV, respectively, and indeed these regions of
Fig. 2 (above background) have the angular distribution of a 1^+
state. In a high-resolution experiment[7] the 1^+ identification
has been verified for several individual states between 9.8 and
11.0 MeV in ^{58}Cu.

     The third and fourth figures give neutron spectra from (p,n)
reactions on ^{120}Sn and ^{208}Pb, respectively.[4] The sharp peak (of
the IAS) in each tin spectrum is superimposed upon a broad peak
which is particularly manifest at 0°. This juxtaposition is similar
to that seen in Fig. 1 for ^{90}Zr, but in this case no test has been
made of a possible 1^+ assignment for the broad peak. In the 0°
spectrum for lead one can almost believe that the IAS peak is once
again on, in this case squarely on, a broad peak, but that idea
is not reinforced at the other angles.

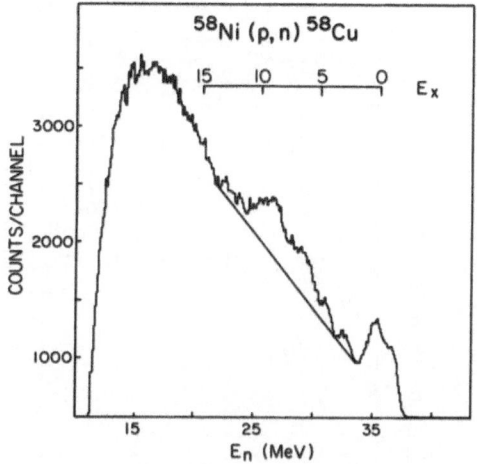

Fig. 2. Neutron spectrum (Ref. 7) at 20° from ^{58}Ni + p at 45 MeV.

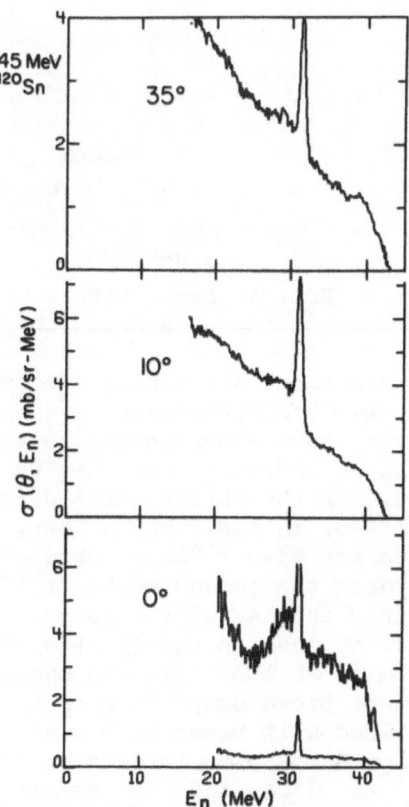

Fig. 3. Neutron spectra (Ref. 4) from ^{120}Sn + p at 45 MeV showing the sharp IAS at 31 MeV and a broad bump $\simeq$ 30 MeV.

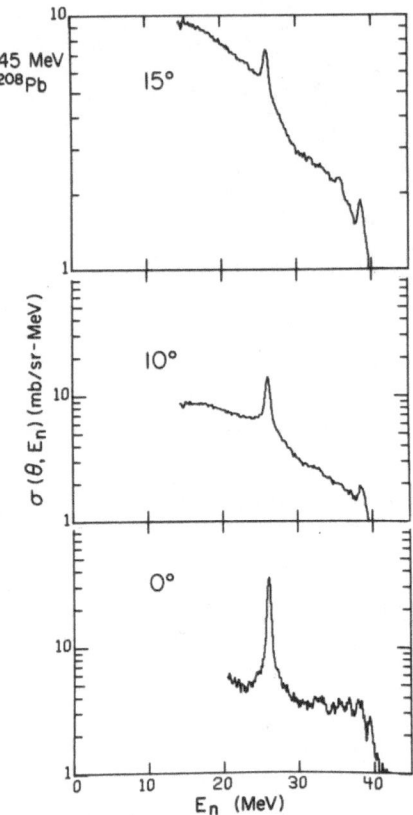

Fig. 4. Neutron spectra (Ref. 4) from ^{208}Pb + p at 45 MeV. The
         IAS appears to be on top of a broad bump ≃ 26 MeV.

        Although (^{3}He,t) reactions are subject to multistep complexi-
ties, they induce the same transitions as (p,n) reactions, and
in some laboratories they have experimental advantages over (p,n).
Experience will tell us the relative sensitivity of the two reactions to new structure. In the Jülich work[6] with ^{3}He at 130 MeV,
targets of ^{120}Sn and ^{208}Pb, in addition to ^{90}Zr, were employed.
One of the spectra obtained with ^{120}Sn is shown in Fig. 5. Except
for a few small peaks near the ground state of ^{120}Sb, in 85 MeV
of excitation energy only the IAS of the target appears. Absent
is a bump near the IAS, as seen in the (p,n) spectra of Fig. 3.
The Jülich spectra from ^{208}Pb show only one peak, that due to the
IAS. An investigation of broad bumps is currently underway at
the University of Maryland with beams of ^{3}He at 130 and 180 MeV.[10]
There is no evidence for bumps with targets of ^{12}C, ^{46,48}Ti, ^{58}Ni,
^{59}Co, and ^{120}Sn, but ^{27}Al, ^{40}Ca, and ^{90}Zr do yield such. One of
the spectra from ^{40}Ca is given in Fig. 6. The center of the bump
is at an excitation energy ≃ 12 MeV (Q ≃ -26 MeV). At Groningen
a study of ^{48}Ca(^{3}He,t)^{48}Sc at 70 MeV has just been completed.[11]

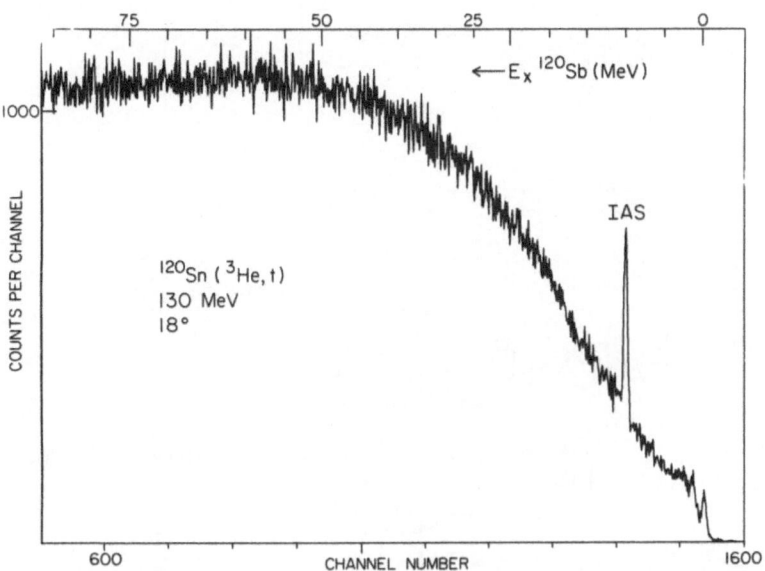

Fig. 5. Triton spectrum (Ref. 6) from ^{120}Sn + ^{3}He at 130 MeV.
At high excitation the IAS is the only discernable peak.

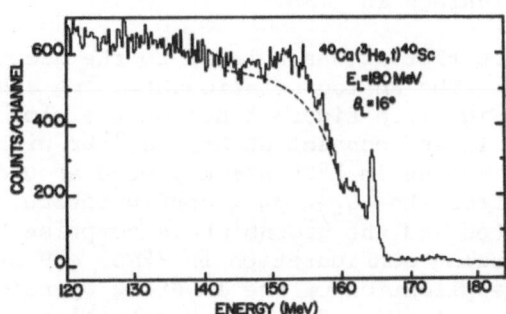

Fig. 6. Triton spectrum (Ref. 10) from ^{40}Ca + ^{3}He at 180 MeV.

By their angular distributions, broad peaks at 7.8, 10.6, and
13.3 MeV are interpreted as envelopes of groups of 1^{+} states carrying a significant part of the G-T strength.

THE CASE OF ^{90}Zr

Returning now to the case of ^{90}Zr, it was proposed[3] that the
bump $\simeq$ 8.4 MeV in ^{90}Nb($E_n \simeq$ 30 MeV in Fig. 1) was not only a giant

A. I. GALONSKY

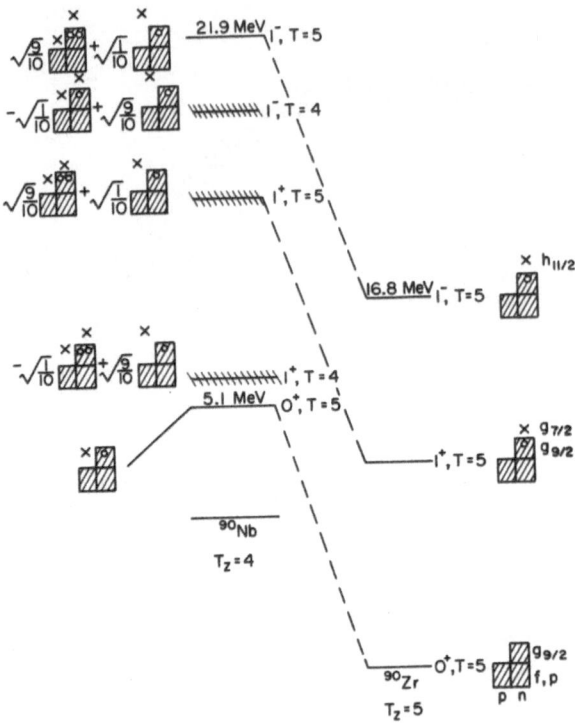

Fig. 7. Giant resonance states of ^{90}Zr and their isobaric analogs
         and antianalogs in ^{90}Nb.

G-T transition, but that it was related to the giant magnetic dipole
resonance in ^{90}Zr. The analog relationships are easy to follow
with the help of Fig. 7, a figure based on the closed-shell model
of ^{90}Zr. Dashed lines connect states in ^{90}Zr with their isobaric
analogs in ^{90}Nb. States in ^{90}Zr are expected around 9 MeV with
the 1^+ character from the $\nu g_{9/2}^{-1} g_{7/2}$ configuration. Transitions
between these states and the ground state comprise the giant M1
resonance. The analog configuration in ^{90}Nb, $\simeq 9$ MeV above the
IAS, obtained by application of the lowering operator, has the
two components shown in Fig. 7. Only the smaller component can
be reached in a direct (p,n) transition from the ground state of
^{90}Zr. An orthogonal linear combination of the two components having
T = 4, an antianalog state, should be isospin split[12] downward
by $\simeq 100(N-Z)/A \simeq 9$ MeV. In other words, it should be close to
the IAS (at 5.1 MeV). As shown in Fig. 7, the part of this state
which can be reached in a direct reaction has the larger amplitude,
$\sqrt{9/10}$. Population of this state is presumably responsible for
at least part of the broad bump found near the IAS in the neutron[3,4]
(Fig. 1) and triton[5,6] spectra. The discovery of a second bump[6]
about 10 MeV higher in excitation energy was taken as confirmation
of this description wherein the analog and antianalog of the giant
M1 were seen in charge-exchange reactions.

There are several loose parts of the above picture. For one
thing, there has been no calculation with a detailed ^{90}Zr wave
function which might predict equal populations of the two isospin
components. The factor of nine implied by the oversimplified closed
shell model is certainly not observed. Another point to remember
is that the bump around the IAS has structure. The French group[5]
has concluded that only the component centered at 7.2 MeV is a
1^+ state. What is the origin of the other component? And finally,
we must note that the energies are not quite in agreement with
the best estimates. The two bumps are separated by more than 10
or 11 MeV (depending on which energy we take for the T = 4 bump),
whereas the Bertsch and Mekjian estimate[12] is a bit lower, 9 MeV.
The largest energy discrepancy involves the energy of the 1^+ resonance in ^{90}Zr and the energy of its analog, presumably of the upper
bump $\simeq$ 18.5 MeV in ^{90}Nb. Since the IAS is at 5.1 MeV, the 1^+ analog
is 13.4 MeV higher, which is more than 4 MeV higher than expected.
The expected value of 9 MeV comes from a study of 180° electron
scattering by Cecil, et al.[13] One of their spectra is shown in
Fig. 8. However, subsequent studies[14] with much higher resolution
by the Darmstadt group revealed only $\simeq$ 15% of the possible M1
strength. Most of the strength they find $\simeq$ 9 MeV is M2 rather
than M1. Perhaps there is not good evidence for a "giant" M1 in
^{90}Zr, to say nothing of its excitation energy.

Another possibility for the character of the upper bump is
that it is connected with the famous giant E1 resonance in ^{90}Zr.
This resonance, at 16.8 MeV,[15] is included in Fig. 7. The excitation energy of the E1 state and of its analog in ^{90}Nb are too high
by $\simeq$ 3 1/2 MeV to be identified with the upper bump. However,

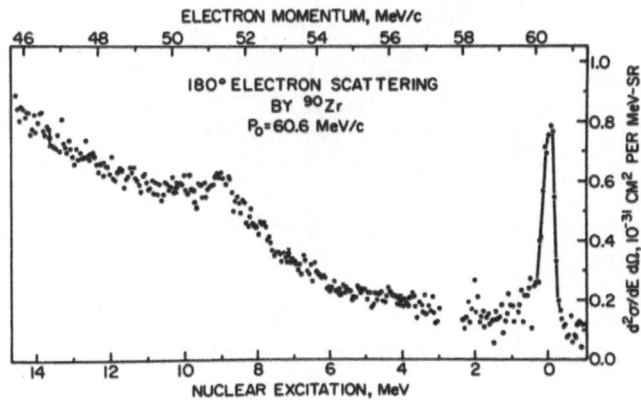

Fig. 8.   Spectrum of electrons scattered by ^{90}Zr through 180°
         (Ref. 13). The anomaly around 9 MeV was presumed M1.

as suggested by N. Marty, the isospin splitting of the giant El, because of its collectivity, is much less than that of the M1. Applied to ^{90}Nb, estimates of the T = 4, T = 5 splitting are $\simeq$ 3 1/2 MeV,[15] just the right amount to make the T = 4 component agree with our upper bump. Perhaps the original idea of the antianalogs being populated more abundantly than the analogs is correct, and we are seeing the antianalogs of the M1 and the El.

OTHER Zr ISOTOPES

In an attempt to clarify the situation, a (p,n) study on targets near ^{90}Zr has begun.[16] Thus far neutron spectra from ^{90,91,92,94}Zr have been obtained from proton bombardment at 45 MeV. The first result is that both bumps appear in spectra from all four targets. Figure 9 presents the spectra at 15°. The location of the ground state is indicated by a vertical bar under a zero. At E_n = 33 MeV are the IAS peaks; at 16.5 MeV target gamma rays; and around 18 and 30 MeV the upper and lower bumps, respectively.

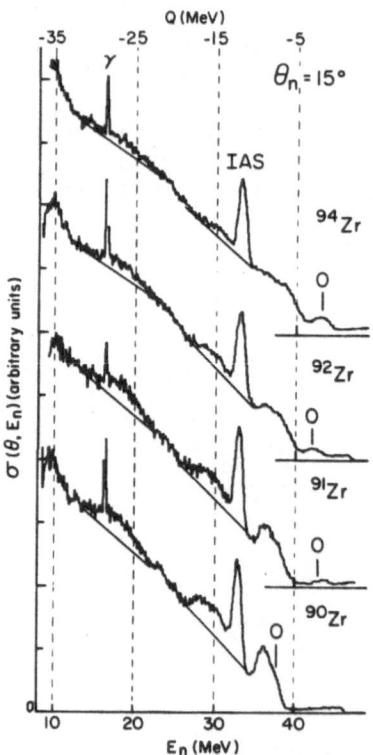

Fig. 9. Neutron spectra (Ref. 16) at 15° from bombardment of ^{90,91,92,94}Zr with 45-MeV protons.

As we go from ^{90}Zr to ^{94}Zr the bumps become less prominent. This
is the second result. Finally, the bump energies are clearly not
related to the ground states of their respective nuclei; the energy
behavior is more one of fixed Q values, much as for the IAS. A
fixed Q value is consistent with the interpretation of the upper
bump as the analog of a state at fixed excitation energy in the
four targets, perhaps a 1^+ state at 14 to 15 MeV. The lower bump
does not, however, follow the systematics expected of the antianalog
of such a state. If it did, isospin splitting, $\simeq 100(N-Z)/A$, would
shift it $\simeq$ 4 MeV lower in ^{94}Nb than in ^{90}Nb.

The observed decreasing strength of the bump transitions from
^{90}Zr to ^{94}Zr is also difficult to reconcile with $\nu g_{9/2} \rightarrow \pi g_{7/2}$
transitions. In the simple view, the addition of neutrons to the
$g_{7/2}$ shell begins with ^{91}Zr. This addition should have the effect
of reducing the antianalog-state cross section. Referring to Fig. 9
that cross section, for ^{90}Zr, should be proportional to the square
of the amplitude $\sqrt{9/10}$. In general for the Zr isotopes this amplitude is $\sqrt{9/(N-Z)}$. The observed trend (see Fig. 9) with (N-Z) is
somewhat stronger than predicted. The real discrepancy is with
the 1^+, T = 5 state. Its cross section should be proportional to
$[1 - 9/(N-Z)]$, an increase with (N-Z)! If the addition of neutrons
beyond the N = 50 shell creates $\nu g_{9/2}$ holes and/or $\pi g_{7/2}$ particles,
the observed decrease with N could conceivably follow for both
states. But this may not be the case.

Spectra from the four Zr targets have been obtained at 5°
intervals from 10° to 70°. The angular distributions for the IAS
and the two bumps are shown in Fig. 10 for ^{90}Zr. The other targets
give similar results. Previous determinations[3] of the IAS and
lower-bump (open points) distributions agree with these and have
been shown to satisfy 0^+ and 1^+ assignments. Within experimental
error the shape of the angular distribution for the upper bump
(solid points) is the same as for the lower bump. Preliminary
calculations with DWBA-70[17] of angular distributions utilizing
a realistic effective nucleon-nucleon interaction[18] are in agreement
with the data for 1^+ but not for 1^-.

We intend to seek similar systematic effects in the Mo isotopes, in ^{93}Nb, and perhaps in other targets. Studying these phenomena at Indiana University, where higher proton energies are
available, might be very interesting, particularly in light of
the reported[19] increase with energy of the spin-isospin-flip interaction in comparison to the pure isospin-flip interaction.

REFERENCES

1. J.D. Anderson and C. Wong, Phys. Rev. Lett. 7, 250 (1961).
2. K. Ikeda, S. Fujii, and J.I. Fujita, Phys. Lett. 3, 271 (1963).

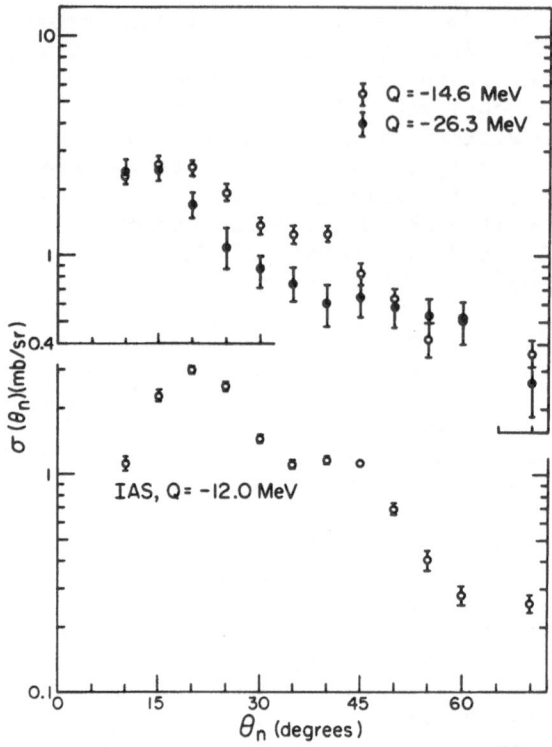

Fig. 10. Angular distributions (Ref. 16) from ^{90}Zr + p at 45 MeV.
         The Q values plus Fig. 9 identify the three distributions
         plotted here.

3. R.R. Doering, A. Galonsky, D.M. Patterson, and G.F. Bertsch,
    Phys. Rev. Lett. 35, 1691 (1975).
4. R.R. Doering, A. Galonsky, D.M. Patterson, and F.E. Serr,
    Michigan State University report MSUCP-29 (1976).
5. D. Ovazza, A. Willis, M. Morlet, N. Marty, P. Martin, P. de
    Saintignon, and M. Buenerd, Phys. Rev. C18, 2438 (1978).
6. A. Galonsky, J.-P. Didelez, A. Djaloeis, and W. Oelert, Phys.
    Lett. 74B, 176 (1978).
7. L.E. Young, Sam M. Austin, R.R. Doering, and Aaron Galonsky,
    Bull. Am. Phys. Soc. 22, 28 (1977).
8. E.R. Flynn and J.D. Garrett, Phys. Rev. Lett. 29, 1748 (1972).
9. R. A. Lindgren, W.L. Bendel, E.C. Jones, Jr., L.W. Fagg, X.K.
    Maruyama, J.W. Lightbody, Jr., and S.P. Fivozinsky, Phys.
    Rev. C14, 1789 (1976).
10. S.L. Tabor, C.C. Chang, M.C. Collins, G.J. Wagner, and J.R.
    Wu, Bull. Am. Phys. Soc. 23, 953 (1978) and private communica-
    tion from S.L. Tabor.
11. C. Gaarde, J.S. Larsen, M.N. Harakeh, S.Y. van der Werf, M.
    Igarishi, and A. Müller-Arnke, submitted to Nucl. Phys.
12. G.F. Bertsch and A. Mekjian, Ann. Rev. Nucl. Sci. 22, 25 (1972).

13. F.E. Cecil, W.L. Bendel, L.W. Fagg, and E.C. Jones, private
     communication from L.W. Fagg and L.W. Fagg, Rev. Mod. Phys.
     47, 683 (1975); p. 702.
14. W. Knüpfer, R. Frey, A. Friebel, W. Mettner, D. Meuer, A.
     Richter, E. Spamer, and O. Titze, Phys. Lett. 77B, 367 (1978).
15. B.L. Berman and S.C. Fultz, Rev. Mod. Phys. 47, 713 (1975).
16. W.A. Sterrenburg, Sam M. Austin, U.E.P. Berg, R. DeVito, and
     A.I. Galonsky, Bull. Am. Phys. Soc. 24, 649 (1979).
17. J. Raynal and R. Schaeffer, CEN Saclay, unpublished.
18. G. Bertsch, J. Borysowicz, H. McManus, and W.G. Love, Nucl.
     Phys. A284, 399 (1977).
19. C.L. Goodman, reported at this conference.

AN EMPIRICAL EFFECTIVE INTERACTION[*]

Sam M. Austin

Cyclotron Laboratory and Physics Department
Michigan State University
East Lansing, MI   48824   U.S.A.

INTRODUCTION

Cross sections for charge exchange or inelastic scattering reactions depend upon a nuclear matrix element which contains both the wavefunctions of the nuclear states involved and the effective two-body interaction V_{eff} mediating the transition between these states. Consequently one can use these reactions as a spectroscopic tool only if one has a priori knowledge of the effective interaction. Two approaches have been taken to obtaining V_{eff}. One of these is entirely theoretical[1] in nature, beginning with a more or less accurate potential model for the bare two nucleon force. At energies below about 50 MeV, one can argue that the interaction must basically resemble that used in the nuclear shell model and a real interaction derived from the shell model G matrix has commonly been used. At energies above about 100 MeV, the impulse approximation provides a good approximation to V_{eff}. This approach and applications of it have been discussed in detail at this conference by Love[1] and by Petrovich.[2]

The second approach, that discussed in this lecture, is entirely empirical in nature.[3,4] One chooses a simple form for the effective interaction and a transition for which the wavefunctions are well known. The strength of the interaction is then adjusted to fit the experimental cross section. If, after a representative sample of cases has been considered, a relatively consistent value of V_{eff} emerges, this interaction is suitable for spectroscopic studies.

[*] This material is based in part upon work supported by the National Science Foundation under Grant No. Phy-7822696.

One may question why one should pursue an empirical approach when rather sophisticated theoretical estimates of V_{eff} are available. Indeed, I expect that the theoretical approach will eventually prevail, but at present substantial difficulties arise, especially at energies below 100 MeV where G matrix interactions are used. Of course, the bare two nucleon interaction itself is not perfectly known. Further uncertainties are introduced by the procedures used to obtain first the G matrix and then representations of it in terms of functions convenient for computation. These uncertainties seem to be largest for the triplet even (TE) part of the interaction. For example, the volume integral of the TE interaction extracted[5] from the Reid potential is 43% larger than that from the Hamada-Johnston potential. A further problem is that the shell model G matrix is real while V_{eff} must be complex. An effective interaction has been derived[6] from a complex G matrix appropriate to one particle in the continuum but has been little used in calculations. The effects of the imaginary part of V_{eff}, as judged by macroscopic-model calculations, are not dominant but neither are they negligible.

It is difficult to estimate the errors which various approximations and uncertainties introduce into the theoretical V_{eff}. For the empirical interactions, on the other hand, the consistency of the results provides an immediate check on the accuracy of the procedure. In addition, systematic uncertainties in the reaction mechanism are automatically corrected by a renormalization of the empirical effective interaction. Effects which might cause such renormalizations are neglected density dependences and contributions of the imaginary part of V_{eff} and of sequential transfer processes[7] such as (p,d)(d,n).

The results of empirical fits to V_{eff} were last reviewed in 1972.[4] Recent experimental work allows one to determine, much better than at that time, the isospin transfer parts of the interaction responsible for charge exchange reactions such as (p,n). We will confine this survey, in most cases, to the bombarding energy range from 20 to 50 MeV. At lower energies, compound nucleus effects cannot be excluded with certainty for many of the light nuclei most useful in these studies. At higher energies too little information is as yet available to permit useful consistency checks. Furthermore, at energies above 100 MeV the impulse approximation appears to provide a reasonable approximation to V_{eff} although there are indications that its imaginary part is too large.[1] Further information in the 50-100 MeV range would be highly desirable and will likely be forthcoming from Indiana. Finally, we will consider only nucleon induced reactions. The use of complex projectiles requires additional assumptions about reaction mechanisms and about the relationship of the two-nucleon and cluster-nucleon interactions.

In order to make this lecture self contained, I will describe briefly the theoretical background necessary for the description of the reactions involved and outline the assumptions involved in extraction of an empirical V_{eff}. A review of available information on various components of the effective interaction then follows.

We find that in the 20-50 MeV range the interaction between unlike nucleons (V^{pn}) is 3 to 4 times stronger than that between like nucleons (V^{pp} or V^{nn}). Inelastic proton (neutron) scattering reactions are then sensitive mainly to the neutron (proton) components of the nuclear wavefunction, so that information from inelastic proton scattering complements that from electron scattering. Furthermore, comparisons of (p,p') and (n,n') to the same states can provide information about the isospin structure of core excitations, much as do comparisons of π^+ and π^- scattering, but in a setting where the reaction mechanism is presently better understood. Studies to date[8] support this conclusion.

We also find that as the energy increases the interaction mediating isospin transfer weakens relative to that mediating simultaneous spin and isospin transfer. At higher energies[1,2,9,10] this effect becomes obvious, implying that the (p,n) reaction above 100 MeV will be extremely useful in searches for spin flip strength (Gamow-Teller strength) in nuclei.

THEORETICAL BACKGROUND

In the distorted wave approximation (DWA) neglecting exchange, the cross-section for a reaction A(a,b)B is proportional to the square of a transition amplitude

$$T_{ba} = \int \chi_b^{(-)*} \langle \psi_f | V_{eff} | \psi_i \rangle \chi_a^{(+)} d\vec{r} \ , \tag{1}$$

where the χ's are distorted waves generated in an appropriate optical potential. All nuclear structure information is contained in the form factor $\langle \psi_f | V_{eff} | \psi_i \rangle$ which depends on the wavefunctions of the nuclear states involved and on the effective interaction. One assumes that V_{eff} is the sum of the two body interactions V_{ip} between the projectile "p" and the valence nucleons "i" of the target, where

$$V_{ip} = V_{ip}^{central} + V_{ip}^{tensor} + V_{ip}^{spin-orbit} .$$

In more detail, we have

$$V_{ip}^{cent} = V_{00}(r) + V_{10}(r)\vec{\sigma}_i \cdot \vec{\sigma}_p + V_{01}(r)\vec{\tau}_i \cdot \vec{\tau}_p + V_{11}(r)(\vec{\sigma}_i \cdot \vec{\sigma}_p)(\vec{\tau}_i \cdot \vec{\tau}_p) \tag{2}$$

$$V_{ip}^{ten} = (V^{ten} + V_{\tau}^{ten}\vec{\tau}_i \cdot \vec{\tau}_p)S_{ip} \tag{3}$$

$$V_{ip}^{s.o.} = (V^{s.o.} + V_{\tau}^{s.o.}\vec{\tau}_i \cdot \vec{\tau}_p)\vec{L} \cdot \vec{S}. \tag{4}$$

The subscripts on the central V_{ST} are the spin and isospin transferred in the reaction; S and T are limited to 1 or 0 for nucleons, since their spin and isospin are 1/2. For mnemonic purposes we will often write $V_o = V_{00}$, $V_{\sigma} = V_{10}$, $V_{\tau} = V_{01}$, $V_{\sigma\tau} = V_{11}$, since for example, V_{11} is responsible for a reaction involving both spin and isospin flip. For the tensor and spin orbit forces only S = 1 matrix elements are nonzero and we simply indicate by a subscript τ whether or not there is isospin transfer of 1. S_{12} is the usual tensor operator.

The selection rules for the scattering process determine which parts of V_{eff} contribute to a particular reaction. These are

$$\vec{J} = \vec{J}_f - \vec{J}_i \qquad\qquad \vec{T} = \vec{T}_f - \vec{T}_i$$

$$\vec{S} = \vec{s}_i - \vec{s}_f \qquad\qquad \pi_i \pi_f = (-1)^L \tag{5}$$

$$\vec{L} = \vec{J} - \vec{S}.$$

Here $\vec{J}$, $\vec{S}$, and $\vec{L}$ are the total, spin, and orbital angular momenta transferred in the reaction and T is the transferred isospin. The subscripted quantities refer to initial (i) and final (f) states, π denoting the parity of these states. As an example, consider a (p,n) reaction from a $J_i^{\pi} = 0^+$, $T_i = 1$ state to a $J_f^{\pi} = 1^+$, $T_f = 1$ state (denoted as $(0^+,1) \rightarrow (1^+,1)$). We find S = 1, T = 1, so that only $V_{\sigma\tau}$, V_{τ}^{ten} and $V_{\tau}^{s.o.}$ can contribute to the reaction.

The selection rules outlined above apply strictly only to the direct part of the transition amplitude for central forces. However, if the interaction acts <u>only</u> in even states of relative motion (or only in odd states) the effective exchange amplitude is proportional to the direct amplitude. In the present energy range an even effective interaction provides a good description of most cross sections, the destructive interference of direct and exchange amplitudes yielding small net odd-state amplitudes, so simple selection rule considerations remain useful. Exchange amplitudes which do not satisfy $\pi_i \pi_f = (-1)^L$ are also possible, but these are normally small except for large L transfer. Finally, non-central forces (tensor and spin-orbit) are large only for large momentum transfer[5] and are not usually significant at energies below 50 MeV.

A final point concerns the relationship of the operators of Eq. 2 and those mediating electromagnetic (EM) or β decay transitions. All are one body operators and one expects that states strongly connected by EM or β transitions will also be strongly connected by inelastic scattering or charge exchange provided the spin-isospin parts of the operator involved are equivalent. For example, E2 transitions between natural parity collective states, e.g. $(0^+ \rightarrow 2^+)$, are strongly enhanced and it is well known that inelastic scattering between such states, mediated by V_o, is also enhanced and (if N = Z) by about the same factor. A similar relationship exists for transitions mediated by allowed Gamow-Teller β decay and $V_{\sigma\tau}$. Implicit in these statements is the fact that the wavefunctions ψ_i and ψ_f in Eq. 1 must closely reproduce EM or β transitions if they are satisfactorily to describe inelastic scattering and charge exchange.

PROCEDURES

At present, the DWA description of direct reaction data is far from perfect. To keep the "noise" in the fitted parameters at an acceptable level, the fitting procedure must therefore involve the smallest possible number of parameters, and experience quickly indicates that fits with more than one parameter are often ambiguous. This has a number of implications. First, to obtain information about the individual V_{ST} one needs to find transitions which isolate a single V_{ST}, else a subtraction is involved. Consequently, (p,p') studies are confined to N = Z nuclei, since otherwise both T = 0 and T = 1 are allowed, and to 0^+ initial (or final) states, since otherwise both S = 0 and S = 1 are allowed. In the case of the (p,n) reaction T = 1 always, because $|\Delta T_z| = 1$, and one need not require N = Z, but the restrictions on spin remain. There are, of course, situations where it might be more appropriate to use a p-n formalism rather than isospin. For example, to describe a (p,p') interaction on a nucleus with proton valence orbitals only, one needs the proton-proton interaction $V^{PP} = V_o + V_\tau$, rather than V_o and V_τ separately. Unfortunately it is difficult to fix V^{PP} with empirical procedures, since, for N ≠ Z nuclei, one needs to know the core-polarization charges for both neutrons and protons. This information is seldom available. The only practical approach then, is to use the isospin formalism and to concentrate on N = Z nuclei whenever core polarization effects are likely to be important, as is the case for natural parity transitions between collective states. The effects of core polarization for both neutrons and protons can then be determined from electron scattering or EM transitions.

A second limitation is that the interaction must have a very simple form. We use a real interaction with a Yukawa shape:

$$V_{ST}(r) = V_{ST} \frac{e^{-r/\mu_{ST}}}{r/\mu_{ST}} \tag{6}$$

with all ranges $\mu_{ST}=1.0$ fm. This central interaction has about
the same mean square radius as the commonly used G-matrix interactions. Moreover, for low energies and $L \lesssim 3$, the shapes of the
predicted angular distributions are relatively insensitive to μ_{ST}.
The restriction to a real interaction is a more serious approximation. Effects of the imaginary part of the force are not negligible, as one can show by including an imaginary term obtained from
the phenomenological optical model in a macroscopic model DWA calculation. However, the overall magnitude of the total cross section
is often not greatly changed, the major effects being on the detailed
angular structure. For this reason we usually fit total cross
sections or the average of measured cross sections rather than
peak cross sections. Of course, the resulting real interaction
may be renormalized to some extent by the effects of the imaginary
interaction.

The wavefunctions of the nuclear states are also sensitively
involved as mentioned at the end of the last section. We summarize
only those transitions for which the wavefunctions are believed to
describe rather well the transitions between these states. This
is most important for collective transitions mediated by V_0. On
occasion effective charge techniques were used to make an approximate correction when the wave functions were known to be inadequate.

Most of the calculations described here were made with the code
DWBA 76[11] which can include the effects of central, tensor and
spin-orbit interactions and the knock-on exchange amplitudes.
V_{eff} must be expressed as sums of Yukawa shapes. In some cases
effective interactions taken from calculations without exchange
were corrected for the average effects of the exchange amplitude
and/or strengths for non-Yukawa V_{eff} were converted to Yukawa
strengths by matching volume integrals.

Treatment of elastic scattering is a special case. The real
part of the optical model potential for a self conjugate nucleus
is given in first order[12] by folding the nuclear matter density
$\rho(r')$ with the two body interaction V_0:

$$U(r) = \int \rho(r') V_0(|\vec{r} - \vec{r'}|) d\vec{r'}. \tag{7}$$

It follows from Eq. 7 that

$$\int V_0(r) d\vec{r} = \frac{1}{A} \int U(r) d\vec{r} \tag{8}$$

where A is the atomic weight. Thus from the volume integral of
$U(r)$ one obtains the volume integral of V_0 and hence V_0 itself.

A similar relationship (see Eq. 10) connects the optical model symmetry potential and V_τ. An energy dependent exchange correction must be applied to values of V_0 and V_τ so extracted.

To summarize, the main points of procedure are

1. Transitions are chosen which isolate a given V_{ST} and for which the available wavefunctions satisfactorily describe the related EM or β transition. When necessary, effective charge techniques are used to correct for wavefunction inadequacies.

2. Calculations include knock-on exchange or are corrected for its effects. In many cases the strength of the dominant V_{ST} was adjusted with other components of the interaction held near their empirical values. In other cases different assumptions were made, for example, that V_{eff} was a Serber interaction.[4] These various assumptions yield different exchange amplitudes, affecting the extracted V_{ST}. Such effects are usually not large, but contribute to the spread of resulting values.

3. The strength of a real Yukawa interaction with a 1.0 fm range is adjusted to fit the total or average cross section.

CENTRAL INTERACTION

The results obtained for the central interaction are summarized in Figs. 1–4, with the bombarding energy on the abscissa and the strength of the real 1.0 fm Yukawa interaction on the ordinate. The relative signs of the V_{ST} are chosen to agree with the universal predictions of the theoretical interactions, namely that V_0 is attractive, while V_σ, V_τ and $V_{\sigma\tau}$ are repulsive. Lines shown on each of these figures and labeled by R, HJ, KK, or KKD are the predictions of various theoretical interactions. Specifically, they are the strengths of 1.0 fm Yukawas with the same volume integral as interactions based on the Hamada-Johnston potential (HJ),[5] the Reid soft core potential (R),[5] the Kallio-Koltveit potential (KK)[13] and the density dependent KK potential (KKD)[14] averaged over the lead nucleus. The numbers near the points are the mass numbers of the targets. Results of a given experiment or analysis are denoted by a common symbol and/or connected by lines. Comments on the individual V_{ST} follow.

$\underline{V_0(r)}$

This part of the interaction operator is related to the electric multipole operator Eλ and mediates natural parity transitions

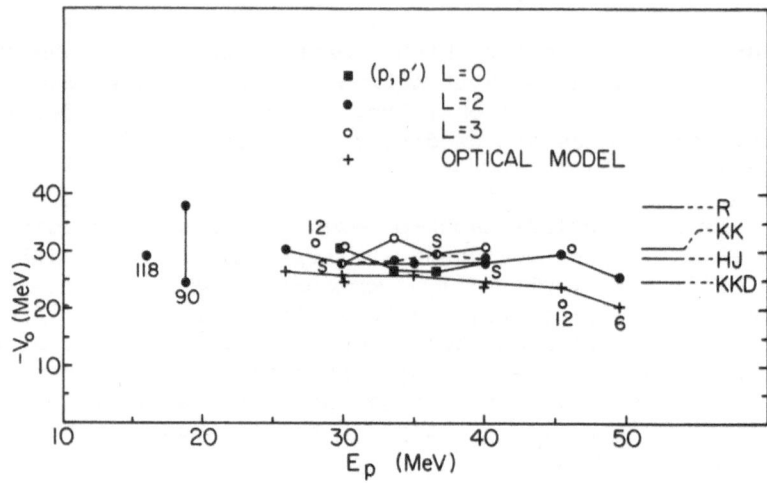

Fig. 1. Values of V_o for a real 1.0 fm range Yukawa interaction.
The mean (± standard deviation) of the points is −27.9 ±
3.5 MeV. Points not numbered are from ^{16}O. S means two
points are superimposed.

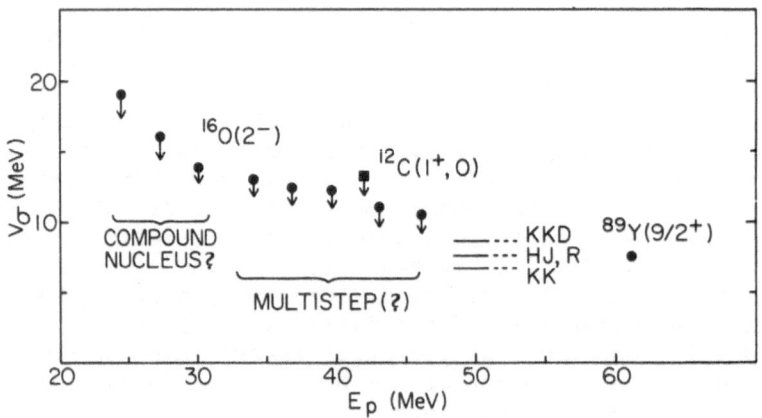

Fig. 2. Values of V_σ for a real 1.0 fm range Yukawa interaction.
The downward pointing arrows indicate upper limits.

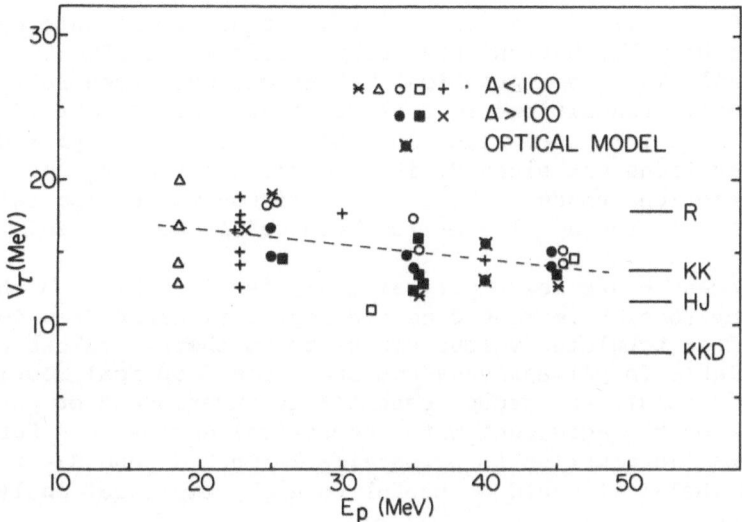

Fig. 3.   Values of V_τ for a real 1.0 fm range Yukawa interaction.
The mean (± standard deviation) of the points is 15.2 ±
2.2 MeV. The dashed line is drawn through the points
by eye and has a slope dV_τ/dE_p = -0.1.

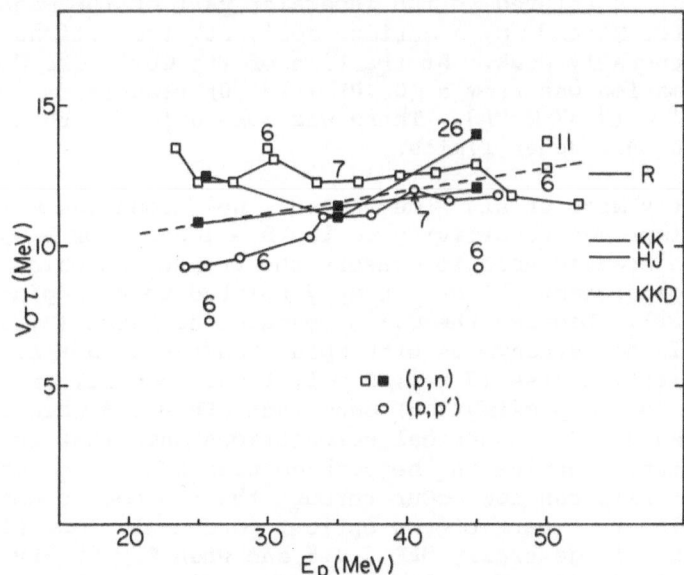

Fig. 4.   Values of $V_{\sigma\tau}$ for a real 1.0 fm range Yukawa interaction.
For space reasons a $^{11}B(p,n)^{11}C$ point, $V_{\sigma\tau}$ = 16.9 MeV at
30 MeV, has not been plotted. The mean (± standard devia-
tion) of the points is 11.7 ± 1.7 MeV. The dashed line
is drawn through the points by eye and has a slope
$dV_{\sigma\tau}/dE_p$ = 0.075. The solid points are from the data on
$^7Li(p,n)$ and $^{26}Mg(p,n)$ described here.

with $T = 0$. It might seem that $0^+ \to 0^+$ transitions would be ideal for determining V_o, but unfortunately, reliable wavefunctions are seldom available. As a practical matter one must then rely on natural parity transitions to collective states: $0^+ \to 2^+, 3^-$, etc. The V_σ part of the force also contributes its $S = 1$ amplitude to these transitions but since V_σ is much smaller than V_o and collective correlations enhance the $S = 0$ amplitude while depressing that for $S = 1$, the $S = 1$ amplitude can safely be ignored.

In recent years few empirical analyses which bear on this part of the force have been done and Fig. 1 is taken from Ref. 4. References to tabulated values can be found there. Values of V_o from available (p,p') analyses and from folded optical model analyses form a consistent group, essentially independent of energy and in reasonable agreement with theoretical estimates. Further experimental information is now available on ^{40}Ca and $N = Z$ nuclei in the sd shell; it would be useful to apply empirical analyses to these data.

$V_\sigma \vec{\sigma}_i \cdot \vec{\sigma}_p$

$V_\sigma \vec{\sigma}_i \cdot \vec{\sigma}_p$ is related to the isoscalar part of the magnetic multipole operator Mλ and, by a curious accident, transitions it mediates are also generally weak. At the time of the Gull Lake Conference, most information was from a $(0^+, 0) \to (2^-, 0)$ transition in ^{16}O $[^{16}O(p,p')^{16}O$ (8.87 MeV)]. There was some evidence that these values were only upper limits.

Recently Moss et al.[15] developed a new technique which promised a greatly improved sensitivity to the $S = 1$ parts of V_{eff}. A polarized proton beam is used to measure the transverse polarization transfer coefficient which is simply related to the spin-flip probability (SFP). Because the $\vec{\sigma}_1 \cdot \vec{\sigma}_2$ operator contains $(\sigma_{1z}\sigma_{2z})$ terms, spin-flip is not synonymous with spin transfer or $S = 1$. But in a simple limiting case ($J = 1$, $S = 1$, $L = 0$, central interactions, plane wave Born approximation) one finds SFP $= 2/3$ when $S = 1$ and SFP $= 0$ when $S = 0$. Numerical calculations show that these qualitative features survive in the more complex DWA environment, even though spin flip can now occur through the two body tensor force and the spin-orbit part of the optical potential. One finds[15] that when $S = 1$, generally SFP $\gtrsim 0.5$ and when $S = 0$, SFP $\lesssim 0.1$ except at very backward angles where there is a peak due to spin flip by the optical model spin-orbit potential.

It had been thought that this method of obtaining SFP for unnatural parity states would be useful in obtaining estimates of the $S = 1$ part of V_{eff}. This hope has not been realized for

reasons that will become clear. Fig. 5 shows the results[16] of
measurements of SFP for $^{16}O(p,p')^{16}O$ (2[−], 8.87 MeV), a transition
which must be S = 1 in the DWA. Yet one finds SFP $\simeq$ 0.2.   The
most reasonable explanation of this result is that multistep processes with small spin flip dominate the cross section. But whatever the explanation one clearly cannot use simple DWA calculations
to obtain V_σ from this cross section. It has been argued[16] that
the spin flip cross section σ_{SF} extracted from the data according
to

$$\sigma_{SF}(\theta) \equiv SFP(\theta)\sigma(\theta)$$

can be compared with the results of one-step DWA calculations.
This cannot be precisely correct, since the multistep processes
do not have zero spin flip, it's only small, and interference
effects may be important. Nevertheless it seems reasonable to
pursue this approach, though results so far are mostly tentative
because of wavefunction uncertainties. One result of these analyses
leads[17] to the upper limit for $^{12}C(p,p')^{12}C$ (1[+],0; 12.7 MeV) shown
along with other upper limits in Fig. 2. Perhaps an additional
word of warning is in order concerning the $^{16}O(p,p')^{16}O$ (2[−]) limits
since there is some evidence,[16] in my opinion not yet convincing,

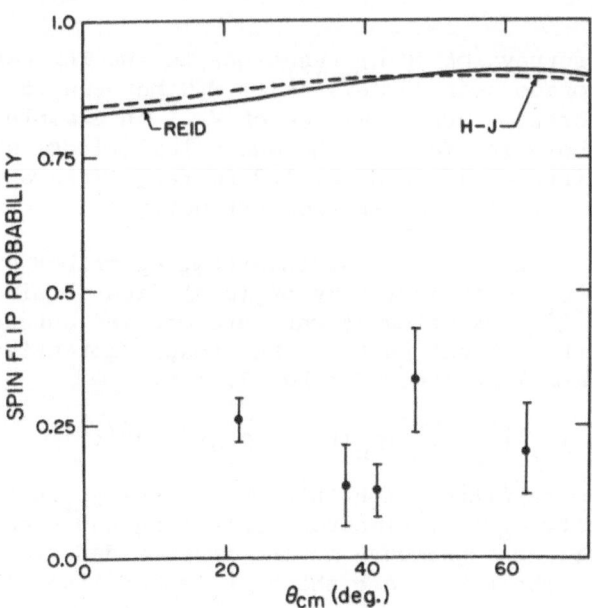

Fig. 5.   Spin flip probability for the 2[−], T = 1 state at 8.87 MeV
          in ^{16}O (Ref. 16) measured at E_p = 40·MeV. The curves
          are results of DWA calculations based on the effective
          interactions of Ref. 5.

that the Gillet-Vinh Mau wavefunctions[18] used in these calculations are indadequate.

To summarize, only upper limits are presently available, but these limits lie near theoretical estimates of V_σ. There is some hope that the spin-flip measurements will eventually yield values of V_σ but it is clear that the effects of multistep processes must first be better understood.

$V_\tau \vec{\tau}_i \cdot \vec{\tau}_p$

This operator is related to that for Fermi β_+ decay and in the case of charge exchange reactions mediates $(0^+,T) \to (0^+,T)$ transitions to the isobaric analog state (IAS). Furthermore, its strength relative to V_0 determines the ratio of the p,p and p,n interactions since

$$V^{pp} = V_0 + V_\tau \qquad V^{pn} = V_0 - V_\tau. \tag{9}$$

We note that V_0 is attractive, i.e. $V_0 < 0$ so that V_τ must be repulsive, i.e. $V_\tau \gtrsim 0$, if the unlike-nucleon interaction is to be stronger than that for like nucleons, as it must be in this energy range.

Extensive surveys of (p,n) reactions to the IAS carried out recently at Colorado near 23 MeV and at MSU between 25 and 45 MeV have greatly increased our knowledge of V_τ. An example of the MSU data[19] is shown in Fig. 6. The DWA calculations, shown both for a G matrix interaction and for 1.0 fm range Yukawas, describe the data fairly well though far from perfectly.

A technique rather different than that described previously has also been used to treat a body of (p,n) data obtained at MSU and Colorado.[20,21] The calculations were carried out in a macroscopic model with the real part of the isospin potential obtained from a folding model analogous to Eq. 7, i.e.

$$V_1 = \frac{-A}{(N-Z)} \int V_\tau(|\vec{r} - \vec{r}'|)[\rho_n(r') - \rho_p(r')]d\vec{r}'. \tag{10}$$

Cross sections were fitted with this purely real potential by fixing the proton density ρ_p from electron scattering and then adjusting three parameters: the geometry of the neutron density ρ_n and the strength of V_τ. These fits are shown as dashed lines (labeled A) in Fig. 7.

The isospin potential is expected to have an imaginary part, W_1, corresponding to an imaginary part of the two nucleon interaction V_τ. A parameter-free estimate of W_1 was made[20] using the

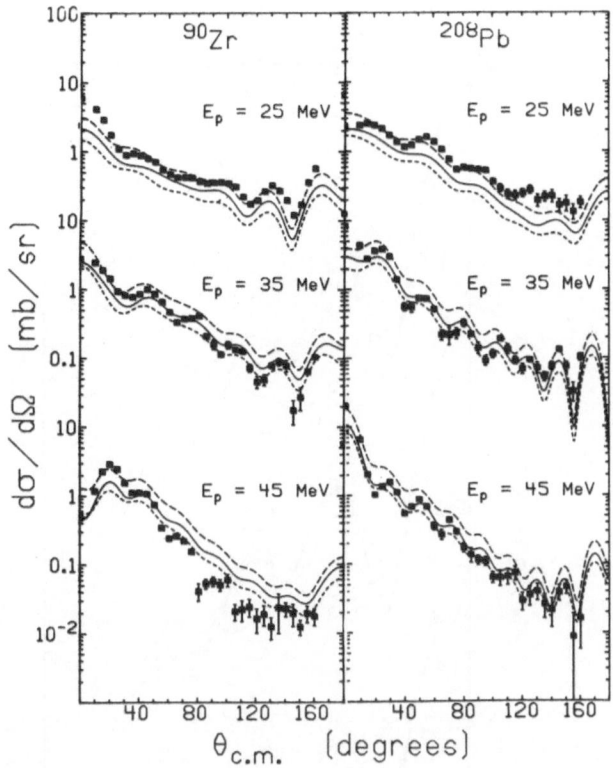

Fig. 6.   Comparison of measured ^{90}Zr, ^{208}Pb(p,n) IAS cross sections
with microscopic model DWA calculations.   The solid curve
is based on an early version of the G-matrix interaction
of Ref. 5 and the dashed curves on 1.0 fm range Yukawa
interactions with strengths of 12.0 and 18.0 MeV.

forward scattering amplitude approximation, basically a high energy
approximation, but nevertheless reasonably successful in describing
the imaginary potential for elastic proton scattering near 30 MeV.
For present purposes we regard it as simply a reasonable estimate
of W_1 which will allow an evaluation of its effects on extracted
values of V_τ.   Three parameter fits, again involving V_τ, led to
the solid curves labeled C in Fig. 7.   These fit the data much
better than those calculated with a purely real interaction.   In
Fig. 8 we show the values of V_τ resulting from real and real-plus-
imaginary fits to a variety of nuclei.   Inclusion of W_1 reduces
the values of V_τ by about 13% on the average.

A summary of results (Refs. 3, 19, 20, 22-26) for V_τ is shown
in Fig. 3.   Some analyses have been ignored, a few for $E_p < 15$ MeV
on light nuclei, where compound nucleus effects may be important,

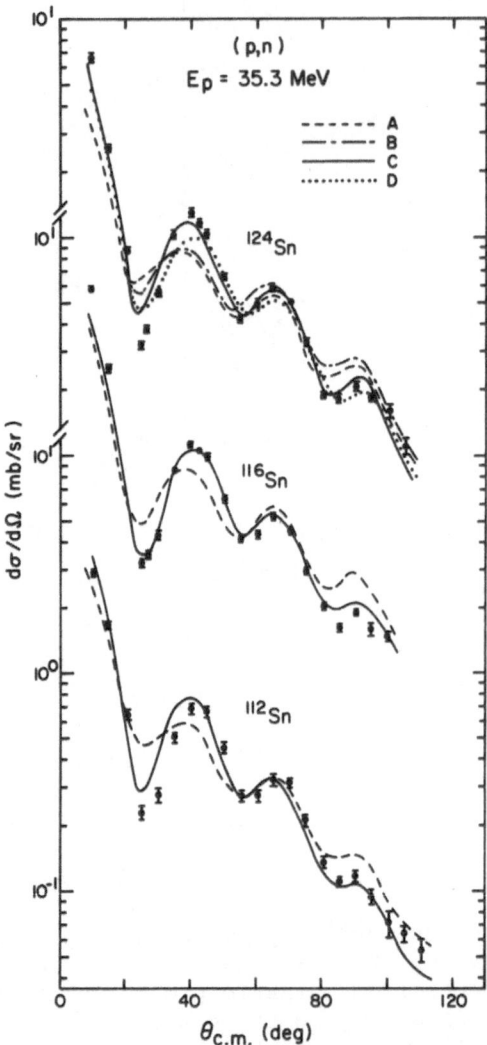

Fig. 7.   Comparison of measured ^{112,116,124}Sn(p,n) cross sections
          leading to the iosbaric analog states with macroscopic
          DWA calculations based on a folding model (Ref. 20). The
          curves were calculated with: A, a real potential; B and
          C a complex potential whose imaginary part is based on
          the forward scattering amplitude approximation, C including
          a finite range correction; and D a complex potential with
          a phenomenological imaginary part.

and others between 30 and 50 MeV done prior to 1970. I believe
the standard of both data and analysis has improved sufficiently
in the past 10 years to warrant this omission, but the omitted
results are plotted in Ref. 4 for the skeptical reader. Rapaport[27]
has recommended that V_τ be plotted as a function of $E-E_{coulomb}$.
We have not done so here and it appears this would not much change
the results, since any energy dependence is small and there is
no clear trend in the ratio of V_τ(light)/V_τ(heavy) at a given
energy. Furthermore, V_τ is an effective interaction and may be
renormalized by the effects of ignored processes with a different
dependence on A and Z.

The most important of such processes are those for sequential
transfer (ST), e.g. (p,d)(d,n), which apparently interfere destructively with the DWA one step (OS) amplitudes of our analyses.
Kunz[7] shows that simple calculations of OS + ST require V_τ = 33 MeV
to fit the data, the large value a result of the destructive interference between the two amplitudes. Including continuum states
increases the ST amplitude while finite range and Pauli principle
effects reduce it. But calculations including all of these effects
have not been done, so definite conclusions are not possible.
Since the results of one step calculations yield values of V_τ close
to theoretical expectations, it seems there are two possibilities
(assuming that OS,ST interference remains destructive): either
the ST amplitude is small or it is approximately twice the size
of the OS amplitude, so that $|OS + ST| \simeq |OS|$. Furthermore, if
the ST amplitudes are large, they must have an angular dependence

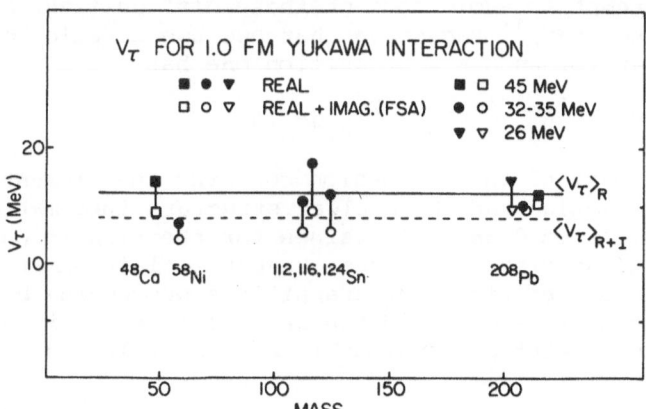

Fig. 8.   Comparisons of real V_τ resulting from macroscopic DWA
           fits with purely "real" and with "real + imaginary" poten-
           tials. The imaginary part is taken from the forward scat-
           tering amplitude approximation. The solid (dashed) line
           is the average value of V_τ for real (complex) interactions.

very similar to that for OS, since the OS cross sections reproduce the data fairly well. In any case, values of V_τ obtained by the empirical procedure yield a rather well defined value of V_τ which can be used to predict (p,n) cross sections via a one-step DWA calculation.

$$\underline{V_{\sigma\tau}(\vec{\sigma}_i \cdot \vec{\sigma}_p)(\vec{\tau}_i \cdot \vec{\tau}_p)}$$

This operator corresponds to that for Gamow-Teller β decay. Strong Gamow Teller transitions should therefore be strong in (p,n) reactions[28] and conversely, one can use the (p,n) reaction to search for β decay strength. Because direct observation of the major part of the β decay strength is energetically forbidden in most heavier nuclei, substantial attention has been devoted to the use of charge exchange for this purpose. Studies of isolated final states in sd shell nuclei have shown that relative strengths in (p,n) reactions are close to those in electromagnetic or β transitions[29] and there is some evidence that concentrations of GT strength have been found in heavier nuclei.[30] To make these conclusions quantitative requires a knowledge of the "nuclear coupling constant" $V_{\sigma\tau}$. Unfortunately empirical evidence on $V_{\sigma\tau}$ is relatively weak. Most transitions studied previously[4] have been in the lower part of the lp shell. While the wavefunctions employed are probably adequate, the optical models are uncertain and DWA predictions are unreliable.

In an attempt to avoid some of these difficulties, Anderson, Wong and Madsen (AWM)[31] suggested that one use a ratio technique. They noted that for an L = 0 transition one has

$$\sigma(\theta) \simeq [V_\tau^2 <1>^2 \delta_{S0} + V_{\sigma\tau}^2 <\sigma>^2 \delta_{S1}]\sigma_o(\theta). \tag{11}$$

The bracketed part of the expression contains the elements of the effective interaction and the nuclear structure information, with $<\sigma>^2$ obtainable from β decay ft values for the transition, while $\sigma_o(\theta)$ contains the distorted waves, kinematical information, etc. The ratio of cross sections for transitions proceeding by L = 0 to nearby states "a" and "b" in the same nucleus is given by the ratio of the bracketed expressions in Eq. 11 evaluated for the two states

$$\frac{\sigma_a(\theta)}{\sigma_b(\theta)} = \frac{[\ ]_a}{[\ ]_b} \frac{\sigma_{oa}}{\sigma_{ob}} \simeq \frac{[\ ]_a}{[\ ]_b}. \tag{12}$$

The second equality follows from $\sigma_{oa} \simeq \sigma_{ob}$, i.e. the distorted wave dependence vanishes in the ratio. AWM applied this approximation to the ^7Li(p,n) reaction at $E_p \lesssim 25$ MeV to obtain values of the ratio $V_{\sigma\tau}/V_\tau$.

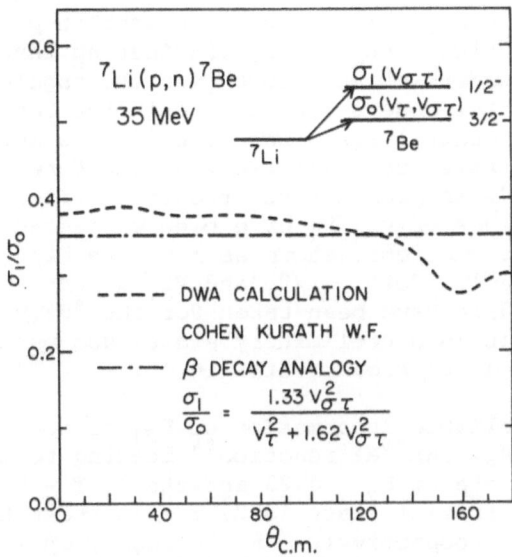

Fig. 9. A test of the AWM approximation (Ref. 31) for the ^{7}Li(p,n)^{7}Be reaction.

To determine the validity of the AWM approximation we performed DWA calculations for this reaction leading to the ground (3/2$^-$) and first excited (1/2$^-$) states of ^{7}Be as shown in Fig. 9. According to the AWM model

$$\frac{\sigma_1}{\sigma_0} = \frac{v_{\sigma\tau}^2 <\sigma>_1^2}{v_{\tau}^2 + <\sigma>_0^2 v_{\sigma\tau}^2} \tag{13}$$

reflecting the fact that only S = 1 amplitudes contribute to 3/2$^-$ → 1/2$^-$ while S = 0, 1 for 3/2$^-$ → 3/2$^-$. The ratio of Eq. 13, with $<\sigma>^2$ evaluated from Cohen-Kurath[32] wavefunctions [(6—16)2BME], and the results of central force DWA calculations with the same wavefunctions and V_τ, $V_{\sigma\tau}$, are shown in Fig. 9. The agreement is surprisingly good, although a detailed evaluation of contributing amplitudes shows that this is somewhat fortuitous (J = 3 amplitudes contribute 9% of the cross section for the ground state but are not allowed for the 1/2$^-$ state). Nevertheless, given this close agreement, ratio measurements should provide reliable estimates of $V_{\sigma\tau}/V_\tau$. As V_τ is rather well determined, values of $V_{\sigma\tau}$ immediately follow.

Results of cross section measurements[33] carried out at 25 MeV are shown in Fig. 10; the cross sections for the two states are indeed similar. Detailed plots of the measured ratios at 25, 35,

and 45 MeV shown in Fig. 11 reflect a remarkable proportionality
of these cross sections, especially considering that they vary
by more than two orders of magnitude over the angular range shown
at 35 MeV. There is, however, structure in the ratios which is
not reproduced by central force DWA calculations such as those
shown in Fig. 9. Later we shall see that the forward angle peak,
near 35° at 35 MeV, is probably related to contributions of the
tensor force. Values of $V_{\sigma\tau}/V_\tau$ have been extracted from ratios
of the total cross sections, shown as lines in Fig. 10, and are
tabulated in Table 1. Both $V_{\sigma\tau}/V_\tau$ and $V_{\sigma\tau}$ appear to increase with
energy. Similar data have been taken for the ^{9}Be(p,n) reaction
but the analysis is in a preliminary state, due partly to ambiguities in assigning J^π to broad states in ^{9}B.

Relatively reliable information on $V_{\sigma\tau}$ is now available from
studies of the ^{26}Mg(p,n)^{26}Al reaction[34] leading to the 0^+ T = 1
isobaric analog state at E_x = 0.23 and the 1^+, T = 0 state at 1.06
MeV. Because the initial state is 0^+, T = 1, these transitions
isolate V_τ and $V_{\sigma\tau}$ respectively. Furthermore, optical model potentials are relatively reliable and DWA calculations indicate that
the tensor force is unimportant. Data have been obtained at 25,
35, and 45 MeV. Values of V_τ and $V_{\sigma\tau}$ were extracted by fitting
cross sections using spectroscopic amplitudes obtained from the
wavefunctions of Chung and Wildenthal.[35] Results at 35 MeV are
shown in Fig. 12. In this case the ratios are not reproduced more
reliably than the individual cross sections.

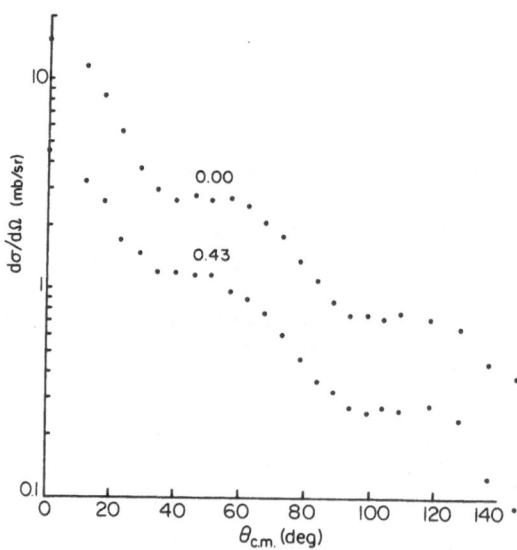

Fig. 10.   Cross sections at E_p = 25 MeV, for the ^{7}Li(p,n)^{7}Be reac-
           tion leading to the ground and first excited states of ^{7}Be.

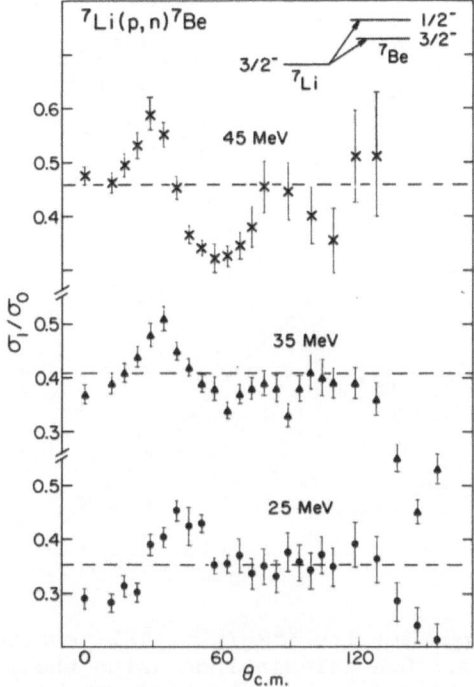

Fig. 11. Ratios of cross sections σ_0 and σ_1, for the ^{7}Li(p,n)^{7}Be
reaction leading to the ground (3/2$^-$) and first excited
(1/2$^-$) states of ^{7}Be, respectively.

Table 1. Values of $V_{\sigma\tau}/V_\tau$ and $V_{\sigma\tau}$ from ^{7}Li(p,n)^{7}Be cross section
ratio measurements.

E_p (MeV)	σ_1/σ_0	$V_{\sigma\tau}/V_\tau$[a]	$V_{\sigma\tau}/V_\tau$ corr[b]	V_τ (MeV)[c]	$V_{\sigma\tau}$ (MeV)
24.8	0.353	0.727	0.677	15.9	10.8
35	0.408	0.830	0.766	14.9	11.4
45	0.461	0.939	0.858	14.0	12.0

[a]Obtained from Eq. 13 using experimental values of ^{7}Be β decay
ft values which lead to $V_{\sigma\tau}/V_\tau = [1.14\sigma_0/\sigma_1 - 1.34]^{-1/2}$.

[b]Corrected to account for the error in the AWM approximation discussed earlier.

[c]From Fig. 3.

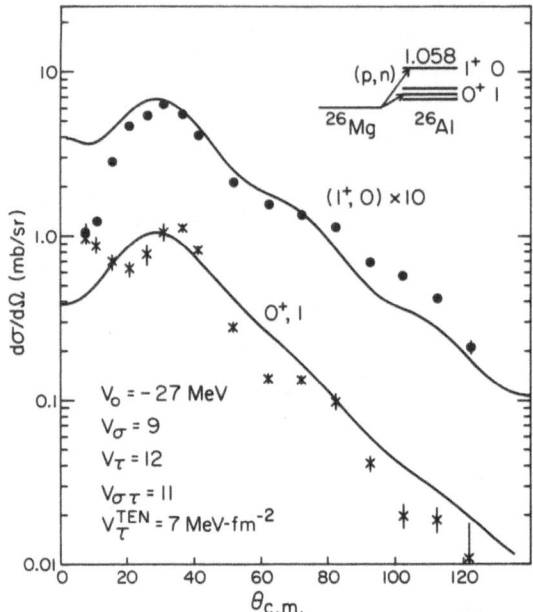

Fig. 12. Cross sections for the ^{26}Mg(p,n)^{26}Al reaction at 35 MeV.
The curves are DWA calculations using the 1.0 fm range
Yukawa central interactions shown. The tensor interaction
has an r^2 × Yukawa form; it is not very important for
either of these cross sections.

        Available data for $V_{\sigma\tau}$ (Refs. 33, 34, 36-39) are summarized in
Fig. 4. There is weak evidence for an increase of $V_{\sigma\tau}$ with E_p.

TENSOR INTERACTION

        Because the tensor force is important only for large momentum
transfer, at the relatively low energies considered here its effects
are usually not large and one must seek out special circumstances
which emphasize them. A favorable case involves transitions which are
isospin analogs of the greatly inhibited β decay of ^{14}C($0^+1 \rightarrow 1^+0$).
For ^{14}C(p,n)^{14}N(g.s.), for example, the relationship between β decay
and (p,n) reactions guarantees that the L = 0 amplitude from $V_{\sigma\tau}$
is small,[40] and hence that the total contribution of the central
force is small. The selection rules for the tensor interaction,[41]
however, allow the angular momentum transfer to the nucleus and
projectile to differ by two units (they must be the same for central
forces) and the contribution of the tensor interaction to the cross
section can then be important. Studies of this transition were
pioneered by the Livermore group.[40,42]

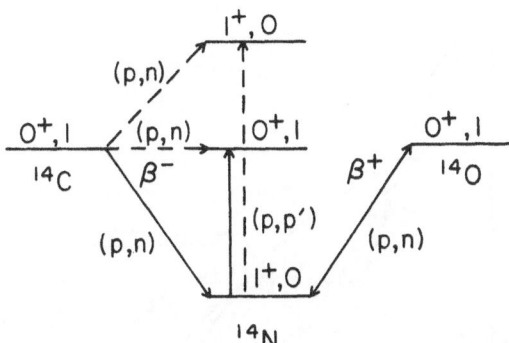

Fig. 13. Energy levels of the mass-14 system. Solid lines show
the greatly inhibited β decay of ^{14}C and analogs of this
transition. All of these transitions as well as those
shown by dashed lines have been measured in (p,p') or
(p,n) reactions.

The isospin analogs of ^{14}C(p,n), shown in Fig. 13, have also
been exploited in the same fashion. Cross sections for ^{14}N(p,p')^{14}N
(2.31) obtained by Fox and Austin[43] are shown in Fig. 14, along
with the results of DWA calculations. A central interaction alone
produces an angular distribution which is too flat; addition of
a tensor force of the form

$$v_\tau^{ten}(r) = v_\tau^{ten} r^2 \frac{e^{-r/\mu_{ten}}}{r/\mu_{ten}} \tag{14}$$

yields a much more accurate reproduction of the cross section.
Here μ_{ten} = 0.816 fm, obtained from matching the low momentum transfer components of the Hamada-Johnston tensor force[4] and v^{ten} is
omitted because G matrix interactions indicate that it is much
smaller than v_τ^{ten}.

Unfortunately, at higher energies, 36.6 and 40 MeV,[43] the
nature of the angular distribution changes in a way not easily
reproduced by DWA calculations. The fits shown in Fig. 15 are
much better than fits without a tensor interaction but are not
of excellent quality. It is perhaps worth noting that the back angle
peak in the cross section at 36.6 and 40.0 MeV is seen at the same
momentum transfer in 122 MeV data.[45] In summary, this data yields
definite evidence for the importance of the tensor interaction
but one cannot determine its strength with precision. Similar
fits and similar conclusions are obtained in the studies at lower
energy,[42] and Love[1] has noted that there are also difficulties
reproducing the data at E_p = 122 MeV.

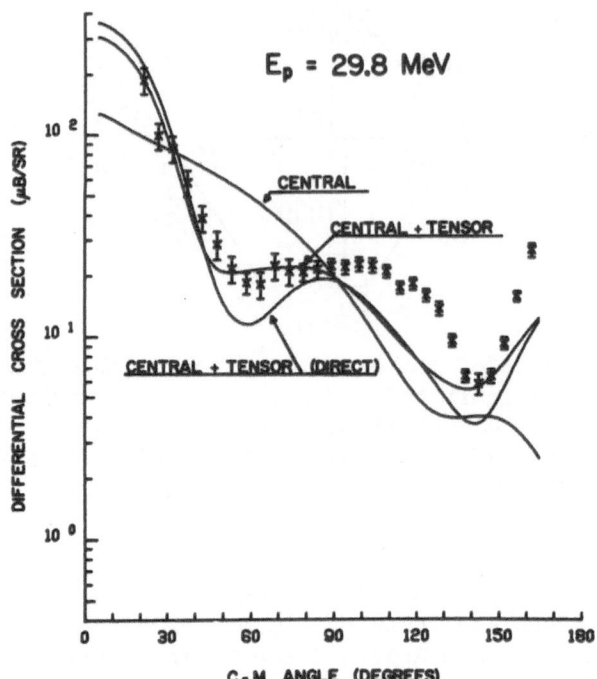

Fig. 14. Cross sections for the $^{14}N(p,p')^{14}N$ (2.311 MeV, 0^+), T = 1)
reaction. The DWA calculations shown include empirical
tensor and central interactions adjusted to fit the data.
The curve labeled "direct" contains only the direct amplitude (no exchange). This figure is from Ref. 4.

It is not clear why the DWA calculations are inadequate to
describe the mass 14, $1^+0 \leftrightarrow 0^+1$ transitions. One might expect
multistep processes to dominate when the one-step amplitude is so
small, but a measured[46] spin flip probability ≥ 0.8 for the transition apparently precludes this possibility. Still, relatively small
coupled channel amplitudes might have serious effects and are shown
to do so in calculations[47] at lower energies. Cross sections at 35
MeV are now available[48] for all the transitions shown in Fig. 13
and may permit a rigorous test of the coupled channel approach.

The measurements of the $^7Li(p,n)^7Be$ reaction discussed earlier
may also yield an estimate of V_τ^{ten}. It has not been possible to
reproduce the peak in the ratios near 35° at 35 MeV without including a tensor force. When it is included, the peak appears naturally and this result is stable against changes in optical potentials and wavefunctions. Fig. 16 gives the results[49] for the
values of $V_{\sigma\tau}$ and V_τ^{ten} shown, with other parts of V_{eff} fixed at
values from Figs. 1-3.

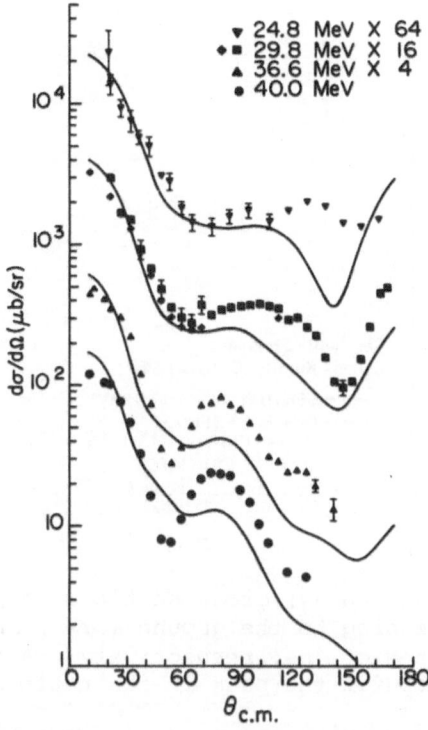

Fig. 15. Cross sections for the ^{14}N(p,p')^{14}N (2.31 MeV, 1^{+}, T = 0)
reaction between 24.8 and 40.0 MeV (Ref. 43). The DWA
calculations include both central and tensor forces, with
their <u>ratio</u> adjusted for best overall fit at all energies.

Calculations by various authors have been performed with radial
forms of the tensor force different from that given in Eq. 14. To
compare these results, we use the volume and r^2 weighted integrals

$$J_o \equiv 4\pi \int_o^\infty V_\tau^{ten}(r) r^2 dr \qquad J_2 \equiv 4\pi \int_o^\infty r^2 V_\tau^{ten}(r) r^2 dr. \qquad (15)$$

J_2 is most appropriate if small momentum transfers are most important for the reaction,[4] while J_o seems empirically appropriate
at least for the mass-14 case. I.e. if one compares two interactions with different radial shapes and then adjusts the strength
of V_τ^{ten} to produce the same cross sections, J_o is approximately
the same for the two. Presumably this reflects the fact that high
momentum transfers are important.[50] Results are shown for J_o and J_2
in Fig. 17. Comparing via J_o, the mass-14 results are in good
agreement with G-matrix estimates but the ^{7}Li(p,n) results are
not. Comparing via J_2 the converse is true. In any case, the

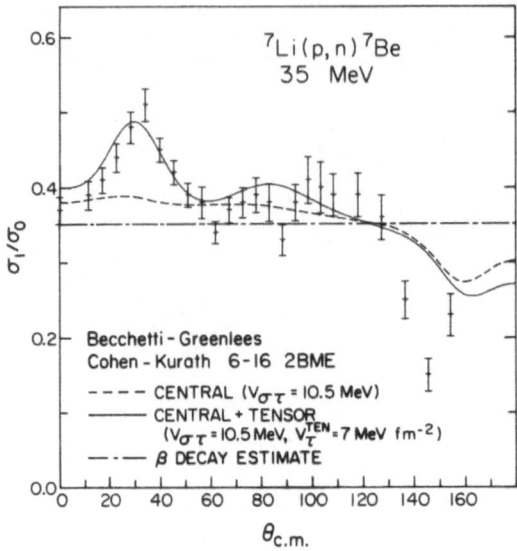

Fig. 16. Ratios of σ_0 and σ_1, cross sections for the $^7Li(p,n)^7Be$ reaction leading to the ground (3/2$^-$) and first excited (1/2$^-$) states of ^7Be, respectively. A tensor force seems necessary to fit the peak in the ratio near 35°.

mass 14 and $^7Li(p,n)^7Be$ results are inconsistent. Clearly, additional studies are required before firm conclusions can be drawn. Some additional detail on the lower energy work is given in Ref. 4.

SPIN-ORBIT INTERACTION

Very little new information is available on the spin orbit interaction $V^{s.o.}$ in the 20—50 MeV bombarding energy range. Because it is large only at large momentum transfers, $V^{s.o.}$ is of dominant importance mainly for excitation of high spin states (J = 6 or 8, for example). Unfortunately, in these cases the effects of collective enhancements are not as yet well known so reliable empirical $V^{s.o.}$ cannot be obtained. Ref. 4. discusses much of the available data. However, the possibility discussed there, that $^{16}O(p,p')^{16}O$ (2$^-$,0; 8.87 MeV) might be dominated by $V^{s.o.}$ appears now to be ruled out by the low spin flip probability for this state.[16]

The spin orbit interaction is important for inelastic scattering to high spin states at Indiana energies[1,2] and such studies with polarized and unpolarized protons should greatly clarify the nature of the collective enhancements for high J states. It may then be possible to obtain information on $V^{s.o.}$ at lower energies.

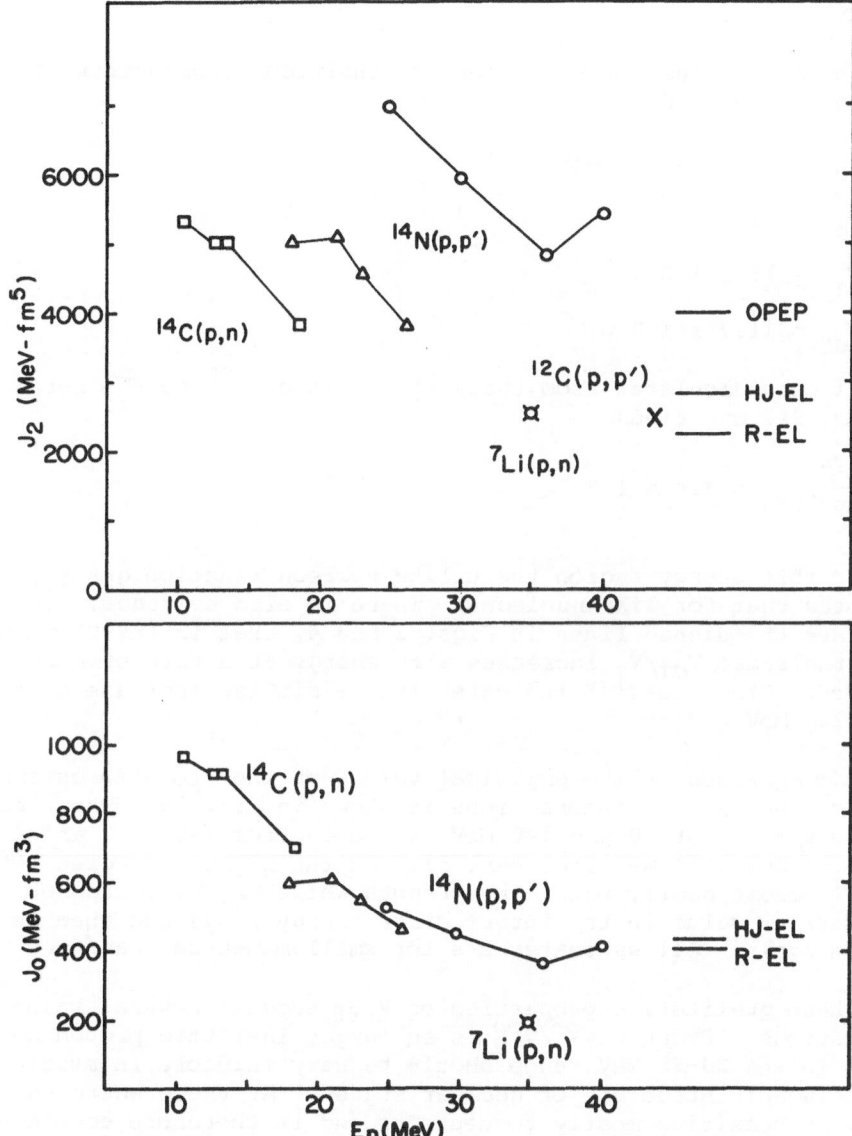

Fig. 17. Values of J_0 (lower graph) and J_2 (upper graph) for the
tensor force. Only the analyses leading to points denoted
(o⌗x) include exchange. The theoretical values labelled
OPEP (the one-pion exchange potential), HJ-EL and R-EL (both
from Ref. 5 where even parts of the interaction are based
on the Hamada-Johnston(HJ) and Reid (R) potentials
and odd parts on the effective oscillator G-matrix elements of Elliott (EL)) are shown as short horizontal lines.
Data are from Refs. 41-44, 49.

SUMMARY

The mean values obtained for the individual components of the central force are

$$V_o = -27.9 \pm 3.5 \text{ MeV}$$

$$V_\sigma \leq 11.5$$

$$V_\tau = 15.2 \pm 2.2$$

$$V_{\sigma\tau} = 11.7 \pm 1.7 .$$

If one calculates from these the ratio of V^{pp} to V^{pn} for S = 0 (see Eq. 9), one finds

$$\frac{V^{pn}}{V^{pp}})_{S=0} = 3.4 \pm 1.2$$

Thus in this energy region the unlike nucleon reaction greatly dominates that for like nucleons. There is also evidence, for which see the dashed lines in Figs. 3 and 4, that in the 25—50 MeV range the ratio $V_{\sigma\tau}/V_\tau$ increases with energy at a rate of about 0.01/MeV. Other work[1,9] indicates that a similar increase continues up to 120 MeV.

A comparison of the empirical values of the V_{ST} with estimates based on theoretical interactions is shown in Fig. 18. The theoretical estimates at 40 and 140 MeV are taken from Tables 2 and 4 of Ref. 1 and at 1 GeV from Ref. 51. At the higher energies V^{pn} and V^{pp} become nearly equal in strength while $V_{\sigma\tau}/V_\tau$ increases to a maximum value in the intermediate energy range and then decreases again. All estimates are for small momentum transfers only.

These qualitative properties of V_{eff} suggest several valuable applications. Because V^{pn}/V^{pp} is so large, inelastic proton scattering in the 20-50 MeV range should be very valuable in studies of the isospin structure of nuclear states. At these energies (p,p') is sensitive mostly to neutrons and it therefore complements electron scattering or electromagnetic data, permitting one to obtain information on both proton and neutron excitations. When it is technically feasible, comparisons of proton and neutron inelastic scattering will better serve the same purpose, as (p,p') and n,n') are sensitive mostly to neutrons and protons respectively and the reaction mechanisms are very similar. As an example, Fig. 19 shows cross sections for ^{138}Ba(p,p')^{138}Ba leading to the low lying 2^+, 4^+, 6^+ states.[52] DWA calculations were done with large basis shell model wavefunctions[53] but for proton valence orbitals only. The

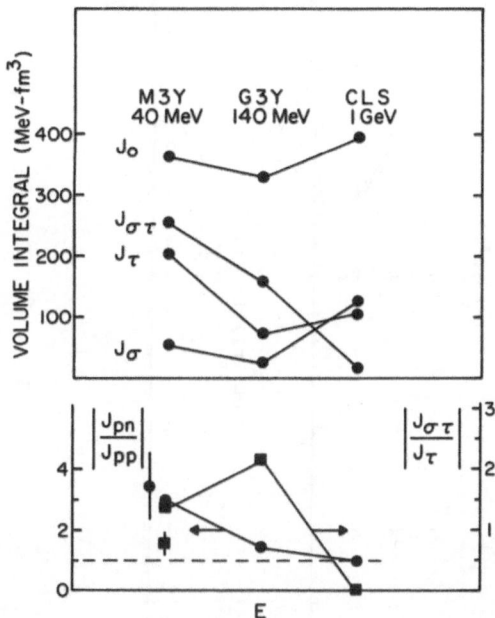

Fig. 18. Upper graph: volume integral estimates of the V_{ST}. Those labeled M3Y and G3Y were taken from Ref. 1. Tables 2 and 4 and that labeled CLS were taken from Chaumeaux, Layly, and Schaeffer, Ref. 51. Lower graph: ratios obtained from upper graph compared with results of empirical analyses reported here (shown with error bars).

actual cross sections are enhanced by a factor of 37 over the calculations compared to an enhancement factor of about 3.2 for the E2 transition $0^+ \rightarrow 2_1^+$. The enhancement is so great because of the large ratio V^{pn}/V^{pp} and corresponds (for $0^+ \rightarrow 2_1^+$) to a reasonable value of the neutron polarization charge, $\delta_n \simeq 1.3$. (Note: this calculation is done by the model of Ref. 52, but using the present ratio V^{pn}/V^{pp}.)

A second application follows from the large value of the ratio $V_{\sigma\tau}/V_\tau$ above 100 MeV.[1,2,9] Work at 45 MeV[29,30,54] suggests that (p,n) reactions are useful for locating concentrations of spin-flip strength in nuclei. The strong increase with energy of $V_{\sigma\tau}/V_\tau$ means that searches for the giant Gamow-Teller resonance with (p,n) reactions above 100 MeV will be much more sensitive and selective than work at lower energies.

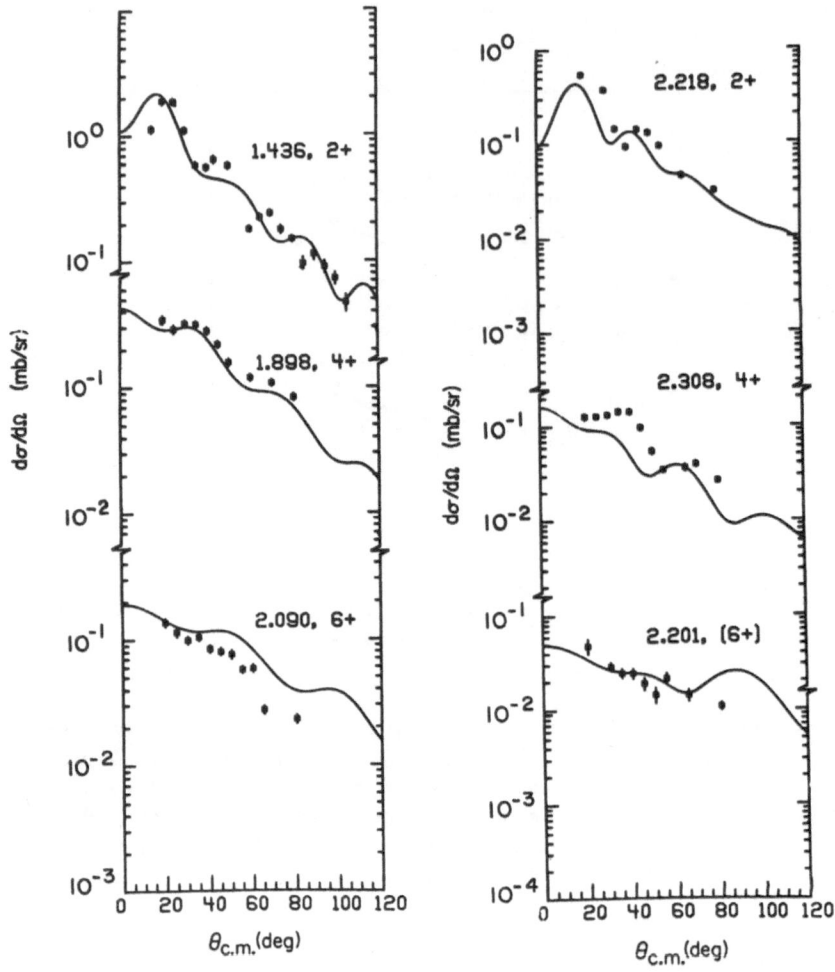

Fig. 19. Cross sections at E_p = 30 MeV for ^{138}Ba(p,p') leading to the two lowest lying states of spin-parity 2^+, 4^+, and 6^+. The DWA calculations have all been multiplied by a factor of 37.

ACKNOWLEDGMENTS

    This survey is obviously based on the work of many people. I wish to thank them for their assistance and in many cases for providing information prior to publication. I have also had the benefit of conversations concerning interpretation of data with G. Bertsch, W.G. Love, and F. Petrovich.

REFERENCES

Except for summarized work (Figs. 1-4, 17), no attempt has been made to be complete. Rather, I have given illustrative references and have referred where possible to other talks at this conference in the form "Author, Telluride Conference, 1979." Ref. 4 provides a more complete discussion (with references) of much of the earlier work.

1. W.G. Love, Telluride Conference, 1979.
2. F. Petrovich, Telluride Conference, 1979.
3. G.R. Satchler, Nucl. Phys. A95, 1 (1967).
4. S.M. Austin, p. 285 in The Two-Body Force in Nuclei, ed. S.M. Austin and G.M. Crawley, Plenum, New York (1972).
5. G. Bertsch, J. Borysowicz, H. McManus, and W.G. Love, Nucl. Phys. A284, 399 (1977).
6. F.A. Brieva, H.V. Geramb, and J.R. Rook, Phys. Lett. 79B, 177 (1978).
7. P.D. Kunz, Telluride Conference, 1979.
8. D.E. Bainum, R.W. Finlay, J. Rapaport, J.D. Carlson, and J.R. Comfort, Phys. Rev. Lett. 39, 443 (1977).
9. C.D. Goodman, Telluride Conference, 1979.
10. C.C. Foster, Telluride Conference, 1979.
11. R. Schaeffer and J. Raynal, unpublished. Modified by W.G. Love.
12. D. Slanina and H. McManus, Nucl. Phys A116, 271 (1968).
13. A. Kallio and K. Koltveit, Nucl. Phys. 53, 87 (1964).
14. A.M. Green, quoted in A. Lande and J.P. Svenne, Phys. Lett. 25B, 91 (1967).
15. J.M. Moss, W.D. Cornelius, and D.R. Brown, Phys. Lett. 69B, 154 (1977).
16. J.M. Moss, W.D. Cornelius, and D.R. Brown, Phys. Rev. Lett. 41, 930 (1978).
17. J.M. Moss, W.D. Cornelius, and D.R. Brown, Phys. Lett. 71B, 87 (1977) and private communication.
18. V. Gillet and N. Vinh Mau, Nucl. Phys. 54, 321 (1964).
19. R.R. Doering, D.M. Patterson, and A. Galonsky, Phys. Rev. C12, 378 (1975).
20. S.D. Schery, S.M. Austin, A. Galonsky, L.E. Young, and U.E.P. Berg, Phys. Lett. 79B, 30 (1978).
21. S.D. Schery, Telluride Conference, 1979.
22. L.D. Rickertsen and P.D. Kunz, Phys. Lett. 47B, 11 (1973).
23. S.D. Schery, D.A. Lind, and H. Wieman, Phys. Rev. C14, 1800 (1976).
24. J.D. Anderson, S.D. Bloom, C. Wong, W.F. Hornyak, and V.A. Madsen, Phys. Rev. 177, 1416 (1969).
25. W. Sterrenburg, S.M. Austin, U.E.P. Berg, R. DeVito, and A. Galonsky, to be published.
26. F.D. Becchetti and G.W. Greenlees, Phys. Rev. 182, 1190 (1969).
27. J. Rapaport, Telluride Conference, 1979.

28. K. Ikeda, S. Fujii, and J.I. Fukita, Phys. Lett. $\underline{3}$, 271 (1963).
29. U.E.P. Berg, Telluride Conference, 1979.
30. A. Galonsky, Telluride Conference, 1979.
31. J.D. Anderson, C. Wong, and V.A. Madsen, Phys. Rev. Lett. $\underline{24}$, 1074 (1970).
32. S. Cohen and D. Kurath, Nucl. Phys. $\underline{73}$, 1 (1965).
33. R.R. Doering, L.E. Young, R.K. Bhowmick, S.M. Austin, S.D. Schery, and R. DeVito, B.A.P.S. $\underline{21}$, 978 (1976).
34. W. Sterrenburg, S.M. Austin, U.E.P. Berg, R. DeVito, and A. Galonsky, to be published.
35. B.H. Wildenthal and W. Chung, Telluride Conference, 1979.
36. A.S. Clough, C.J. Batty, B.E. Bonner, C. Tschalär, L.E. Williams, and E. Friedman, Nucl. Phys. $\underline{A137}$, 222 (1969); A.S. Clough, C.J. Batty, B.E. Bonner, and L.E. Williams, Nucl. Phys. $\underline{A143}$, 385 (1970).
37. K.H. Bray, M. Jain, K.S. Jayaraman, G. Lobianco, G.A. Moss, W.T.H. Van Oers, and D.O. Wells, Nucl. Phys. $\underline{A189}$, 35 (1972).
38. S.M. Austin, P.J. Locard, W. Benenson, and G.M. Crawley, Phys. Rev. $\underline{176}$, 1227 (1968); S.M. Austin and G.M. Crawley, Phys. Lett. $\underline{27B}$, 570 (1968).
39. P.J. Locard, S.M. Austin, and W. Benenson, Phys. Rev. Lett. $\underline{19}$, 1141 (1967).
40. C. Wong, J.D. Anderson, J. McClure, B. Pohl, V.A. Madsen, and F. Schmittroth, Phys. Rev. $\underline{160}$, 769 (1967).
41. W.G. Love and L.J. Parish, Nucl. Phys. $\underline{A157}$, 625 (1970).
42. C. Wong, J.D. Anderson, V.A. Madsen, F.A. Schmittroth, and M.J. Stomp, Phys. Rev. C$\underline{3}$, 1904 (1971).
43. S.H. Fox and S.M. Austin, to be published; S.M. Austin and S.H. Fox, p. 388 in Proc. Int'l. Conf. on Nuclear Physics, ed. J. de Boer and H.J. Mang, North Holland, 1973.
44. H.F. Lutz, D.W. Heikkinen, and W. Bartolini, Nucl. Phys. $\underline{A198}$, 257 (1972).
45. J.R. Comfort, S.M. Austin, P. Debevec, G. Moake, R.W. Finlay, and W.G. Love, to be published.
46. J.M. Moss, Int'l. Symp. on Direct Nuclear Reaction Mechanisms, Fukuoka, 1978.
47. L.F. Hansen, S.M. Grimes, J.L. Kammerdiener, and V.A. Madsen, Phys. Rev. C$\underline{8}$, 2072 (1973).
48. T.N. Taddeucci, R.R. Doering, L.C. Dennis, A. Galonsky, and S.M. Austin, B.A.P.S. $\underline{24}$, 594 (1979).
49. S.M. Austin, R. DeVito, L.E. Young, and R.R. Doering, B.A.P.S. $\underline{24}$, 610 (1979).
50. T. Taddeuchi and W.G. Love, private communication.
51. A. Chaumeaux, V. Layly, and R. Schaeffer, Phys. lett. $\underline{72B}$, 33 (1977); R. Schaeffer, private communication.
52. D. Larson, S.M. Austin, and B.H. Wildenthal, Phys. Rev. C$\underline{11}$, 1638 (1975).
53. B.H. Wildenthal, Phys. Rev. Lett. $\underline{22}$, 1118 (1969).
54. W.A. Sterrenburg, S.M. Austin, U.E.P. Berg, R. DeVito, and A.I. Galonsky, B.A.P.S. $\underline{24}$, 649 (1979).

DETERMINATION OF THE MACROSCOPIC ISOVECTOR POTENTIAL FROM NUCLEON-NUCLEUS SCATTERING*

J. Rapaport

Physics Department, Ohio University

Athens, Ohio 45701

INTRODUCTION

The nucleon-nucleus elastic scattering represents the largest of all partial cross-sections. It is usually described in terms of the optical model potential (OMP). In the past few years much work has been done in the understanding and parameterization of the OMP.

In a mixed audience such as the one we have today, some confusion may arise when one speaks about the OMP. One reason is that experimental and theoretical physicists describe by the word OMP different but somewhat related things. In my talk, I will concentrate on the "empirical OMP", the one used and obtained from phenomenological analyses. We also have the "theoretical OMP" which I understand may be and has been derived in various different ways.

The study of neutron and proton elastic scattering data and its analysis in terms of the OMP leads to a unique way of determining the isovector part of the empirical OMP. We will examine the energy and radial dependence of this part and predict (p,n) quasi-elastic cross sections which will be compared with experimental results.

Recently Jeukenne, Lejeune and Mahaux[1] starting from the Brueckner-Hartree-Fock approximation and Reid's hard core nucleon-nucleon interaction have calculated the isovector part of the theoretical OMP. It will be compared with some of the present phenomenological results.

*Supported in part by a National Science Foundation grant.

OPTICAL MODEL ANALYSIS

The phenomenological OMP $U(r,E)$ is a complex local potential which consists of a real term, $V(r,E)$, and an imaginary term, $W(r,E)$:

$$-U(r,E) = V(r,E) + i\,W(r,E)$$

and has features based on physical intuition. It may be written as a sum of the terms:

$$-U(r,E) = U_N(r,E) - V_c(r) - U_{so}(r,E)$$

where $V_c(r)$ is the Coulomb potential, U_{so} is the spin-orbit potential and U_N may be expressed as suggested by Lane[2]

$$U_N(r,E) = U_o(r,E) + \frac{4}{A} U_1(r,E)\vec{t}\cdot\vec{T}$$

where U_o is the isoscalar and U_1 the isovector part of the OMP; $\vec{t}$ and $\vec{T}$ are the isospins of the incident nucleon and target respectively. The isospin interaction splits the radial part of the potential into two diagonal terms which describe the proton and neutron elastic scattering while the non-diagonal or coupling term describes the (p,n) quasi-elastic scattering. The latter is given by

$$U_{pn}(r,E) = 2\sqrt{\frac{\varepsilon}{A}}\, U_1(r,E)$$

where $\varepsilon = (N-Z)/A$ is the nuclear asymmetry. The diagonal terms may be written:

$$U_n(r,E) = U_o(E)f(r) - \varepsilon U_1(E)F(r)$$

$$U_p(r,E) = U_o(E)f(r) + \varepsilon U_1(E)F(r) + \Delta U_c f_1(r)$$

for neutrons and protons respectively. In general U represents a complex quantity and the term $\Delta U_c f_1(r)$ is the so called Coulomb correction term first introduced by Lane[3]; it is usually parameterized as a single real term $\Delta V_c f_1(r) = \beta(Z/A^{1/3})f(r)$. The OMP analysis of neutron and proton scattering on T=0 (ε=0) nuclei gives a unique way to evaluate empirically the Coulomb correction term. This has been done with available nucleon scattering data on ^{40}Ca[4]; the above parameterization for the real term was used and a value $\beta = 0.46 \pm 0.07$ was derived.

The equations given above indicate that it is possible to obtain information about the isovector part of the OMP by a simple comparison of neutron and proton OMP analyses of elastic scattering data. Several attempts, based mainly on OMP analyses of proton elastic data, have been reported[5]. However, as seen from the above equations, this requires an "a priori" knowledge of $\Delta U_c f_1(r)$. The

analysis of neutron scattering data should give more reliable values
for the isovector term, because no "a priori" assumptions are
needed. An obvious way of determining $U_1(r,E)$ is by a simultaneous
OMP analysis of proton and neutron elastic scattering data at equivalent energies from the same nucleus. This has been done and
reported elsewhere[6].

To discuss the radial dependence of $U_1(r,E)$ it is useful to
write it as a sum of a real V_1, and an imaginary term, W_1:

$$U_1(r,E) = V_1(r,E) + i\ W_1(r,E) \ .$$

In empirical analysis it is generally assumed that $V_1(r,E)$ has the
same radial dependence that $V_0(r,E)$, the real term of the isoscalar
part of the OMP. This radial dependence is usually taken to be a
volume Woods-Saxon shape:

$$V_1(r,E) = V_1(E)f(r)$$

with

$$f(r) = (1 + e^x)^{-1} \text{ and } x = (r - r_R A^{1/3})/a_R.$$

Several quasi-elastic (p,n) analyses[7] have suggested a surface
peaked radial dependence for V_1. Sinha et al.[8] indicate such a
term in their empirical analysis of 30 MeV proton elastic scattering
on even Mo isotopes.

We have analyzed 7, 9, 11, 20 and 26 MeV neutron elastic scattering data on ^{208}Pb and 24, 26 and 30 MeV proton elastic scattering
on ^{208}Pb and obtained a Lane consistent OMP which was used to calculate the quasi-elastic (p,n) reaction on ^{208}Pb[9]. Volume, surface
and a mixture of volume and surface radial dependences for V_1 were
studied. The results based on the χ^2 values from the nucleon
elastic scattering data were not useful in selecting among the above
assumptions. The calculated (p,n) IAS cross sections when compared
to the data[10] indicate some preference for the pure surface-peaked
but the results were not significantly different from those reported
in ref. 9) where a pure volume form factor was assumed.

A surface peaked form factor (usually derivative Woods-Saxon)
is generally used for the radial dependence of W_1.

NEUTRON GLOBAL OMP PARAMETERS

During the last few years at Ohio University we have measured
accurate neutron elastic cross sections in the energy range 7-26 MeV.
The targets, including several enriched isotopic samples, range from

O (A=16) to Bi (A=209). From this large set of neutron data, we
have selected a sub-set of nuclei which may be considered as spheri-
cal nuclei, ^{40}Ca, ^{90}Zr, ^{92}Mo, ^{116,124}Sn, ^{208}Pb and obtained two sets
of neutron global OMP parameters[11]. These are indicated as set a)
and set b) in table 1 and either of them reproduce equally well the
neutron elastic scattering data. These parameters were used to
calculate neutron total cross sections on ^{40}Ca, ^{90}Zr and ^{208}Pb in
the energy range 1-30 MeV. The calculations[11] reproduce quite well
the available experimental results[12].

Table 1.   Neutron Global Parameters[†]

Set a)	$r_R = 1.198 \qquad r_I = 1.295 \qquad r_{so}^* = 1.01$
	$a_R = 0.663 \qquad a_I = 0.588 \qquad a_{so}^* = 0.75$
	$V_R = 54.19 - 0.33E - \varepsilon(22.7 - 0.19E)$
	$W_D = 4.28 + 0.4E - \varepsilon\ 12.8$
	$W_V^* = 0$ $\qquad\qquad\qquad\qquad\qquad$ $\left.\begin{array}{}\\ \\ \end{array}\right\}$ $E \lesssim 15$ MeV
	$W_D = 14.0 - 0.39E - \varepsilon\ 10.4$
	$W_V = -4.3 + 0.38E$ $\qquad\qquad\qquad$ $\left.\begin{array}{}\\ \\ \end{array}\right\}$ $E \gtrsim 15$ MeV
	$V_{so}^* = 6.2$
Set b)	$r_R = 1.225 - \dfrac{2.985}{A} \qquad\qquad r_I = 1.297$
	$a_R = 0.668 \qquad\qquad\qquad a_I = 0.588$
	$V_R = 54.62 - 0.30E - \varepsilon(25.3 - 0.014E)$
	$W_D = 4.29 + 0.4E - \varepsilon\ 12.6$
	$W_V^* = 0$ $\qquad\qquad\qquad\qquad\qquad$ $\left.\begin{array}{}\\ \\ \end{array}\right\}$ $E \lesssim 15$ MeV
	$W_D = 13.3 - 0.38E - \varepsilon\ 9.9$
	$W_V = 3.8 + 0.37E$ $\qquad\qquad\qquad$ $\left.\begin{array}{}\\ \\ \end{array}\right\}$ $E \gtrsim 15$ MeV

[†]Geometrical parameter values in fm; potential depths in MeV.

[*]Values kept constant during the search; same values for set b).

The energy dependence of the isovector term in the neutron global OMP parameters is contained only in the real term and the linear slopes are quite different for set a) and set b); however the geometrical parameters are also different. It is well known that the discussion of the energy dependence may be made less dependent on the choice of geometrical parameters if it is discussed within the realm of the reformulated OMP[13]. In this approach, the significant quantities to be compared are the volume integral per nucleon and the root-mean-square radius. The values for the contribution of the isovector term to the volume integral per nucleon of the OMP as a function of energy are indicated in table 2 for the two neutron global sets of OMP parameters.

Jeukenne et al.[1] have calculated the isovector volume integral per nucleon for an incident nucleon energy E = 35 MeV. Their results displayed versus A as solid lines in fig. 1 are compared with the present empirical values. A rather good agreement is observed with the values of set a).

QUASI-ELASTIC SCATTERING

Several macroscopic analyses of quasi-elastic (p,n) reactions have been reported in the literature[14-16]. These studies have focussed on obtaining an energy-dependent Lane-model consistent OMP set of parameters that describe proton elastic neutron elastic and (p,n) IAS quasi-elastic scattering. The global Becchetti and Greenlees[17], (BG), set of potentials is generally used and the parameter search is conducted[15,16] in such a way that the proton potentials should not be modified. Satchler and Perey[18] have remarked that the BG neutron potentials were based on the proton analysis and included a forced dependence on both energy and asymmetry, ε; thus it is not surprising to find that the experimental total and reaction cross sections for neutrons are poorly described by the BG potentials[18].

Table 2. Isovector Potential. Energy variation of the volume integral per nucleon in (MeV-fm^3)

	Real term: $(\varepsilon J_{V_1}/A)$		Imaginary term: $(\varepsilon J_{W_1}/A)$	
	Set a)	Set b)	(Set a) = (Set b)	
			E $\lesssim$ 15	E $\gtrsim$ 15
^{90}Zr	20.9 − 0.17E	23.0 − 0.018E	16.2	13.2
^{116}Sn	25.4 − 0.21E	28.4 − 0.022E	18.5	15.0
^{208}Pb	37.6 − 0.32E	43.1 − 0.034E	23.0	19.0

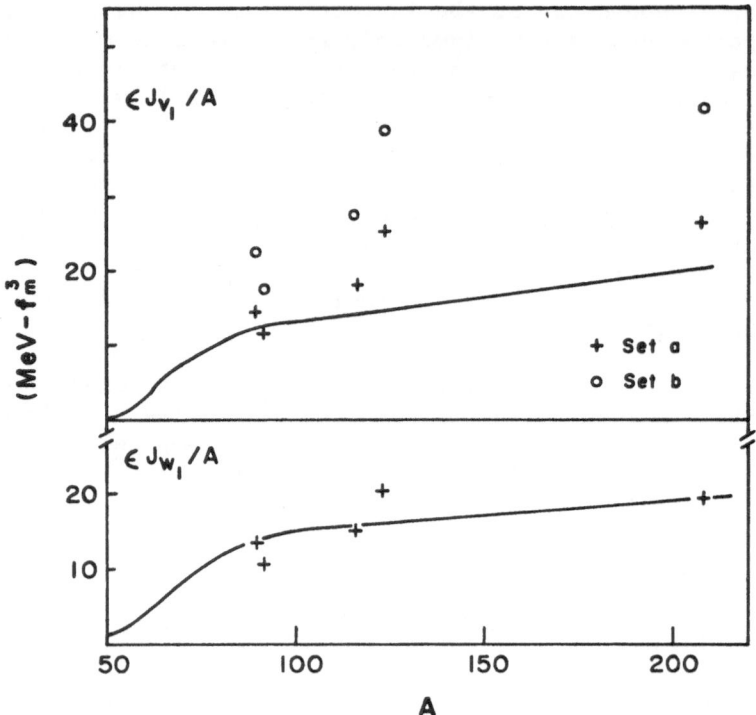

Fig. 1. Values of the volume integral per nucleon for the isovector
         potential at E=35 MeV. The solid line represents micro-
         scopic calculations (ref. 1) while the points are empirical
         values obtained using neutron global OMP parameters, set a)
         (crosses) and set b) (circles).

Patterson[16] et al. obtained several OMP parameter sets and the
one to produce best overall results, indicated as set B, assumes an
isovector potential, with an energy independent real term (V_1 = 17.7)
and an energy dependent imaginary term [$W_{D_1}(E)$ = 18.1 - 0.31 E].

The present neutron global OMP parameters may be used to
generate proton OMP parameters by reversing the sign of the isovector term and adding the empirical Coulomb correction term
ΔV_c = 0.46 $Z/A^{1/3}$. With this set of Lane consistent OMP parameters
we have calculated the $^{208}Pb(p,n)^{208}Bi$ IAS cross section for proton
energies between 20 and 50 MeV. The (p,n) IAS form factor,
$U_{pn}(r,E)$, was obtained by forming

$$U_1(r,E) = \frac{1}{2\varepsilon} [U_p(r,E_p) - U_n(r,E)]$$

where E_p is the incident proton energy, E is the outgoing neutron energy, and calculating

$$U_{pn}(r,E) = 2 \sqrt{\frac{\varepsilon}{A}} U_1(r,E)$$

This procedure which follows the Lane[2] description, differs from the one adopted by Patterson[16], where they use for U_1 the global values of V_1 and W_1. It is important to realize that the two procedures give rather different results because the radial dependence of the imaginary part of the empirical OMP is energy dependent[9,14]. Usually a surface term, radial derivative of a Woods-Saxon shape, is used to represent the radial shape for nucleon energies less than approximately 20 MeV while for energies larger than approximately 50 MeV a single volume term is used. For nucleon energies between 20 and 50 MeV a combination of surface Woods-Saxon derivative and volume Woods-Saxon are used with the same geometrical parameters r_I, a_I.

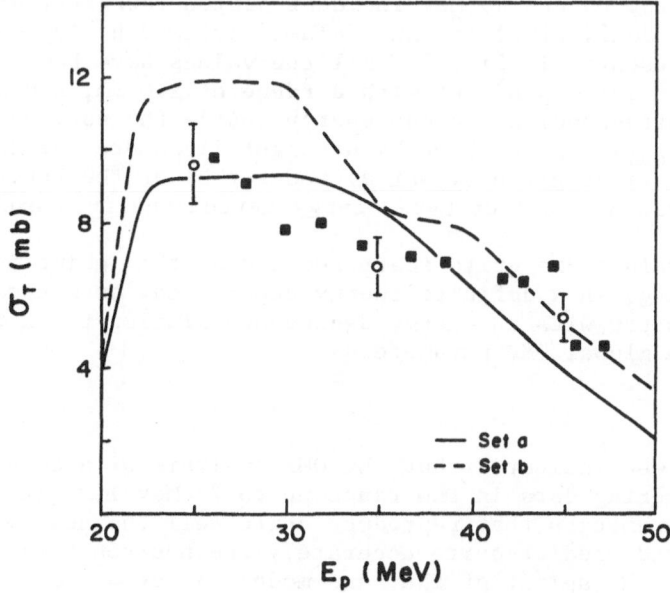

Fig. 2.   Total cross section excitation function for the $^{208}Pb(p,n)^{208}Bi$ (IAS) reaction. The DWBA predictions shown correspond to OMP parameters set a) solid line, and set b) dashed line.

The calculated excitation function for the total cross section for the ^{208}Pb(p,n)^{208}Bi IAS reaction <u>without adjustable parameters</u> is presented in fig. 2 where it is compared with available data[19,20]. The data of ref. 19 has been normalized to the absolute cross sections reported in ref. 20. The observed agreement may be considered quite good. There is also a very good agreement between the predicted (set a) and measured angular distribution at E_p = 25.8 MeV[10].

MACROSCOPIC AND MICROSCOPIC DESCRIPTIONS OF (p,n) IAS TRANSITIONS

Anderson et al.[21] have indicated that the macroscopic and microscopic descriptions of (p,n) IAS transitions are equivalent in the j-independent-amplitude approximation if the macroscopic form factor $V_1(r)$ is replaced with $(A/\pi)V_\tau g_0(r)$, a quantity which is roughly independent of nuclear size. In a naive way the energy dependence of V_τ, the isospin-flip part of the effective nucleon-nucleon interaction, should correspond to the energy dependence of the isovector strength of the OMP. The neutron global OMP parameters include an energy dependence on the real term $V_1(E)$ of the isovector potential which may be compared with energy dependent values of V_τ obtained from microscopic analyses. Values of V_τ obtained by Doering et al.[20] at 25, 35 and 45 MeV incident proton energies, at 22, 30 and 40 MeV by Jolly et al.[22] and values reviewed by Austin[23] up to 50 MeV are presented in fig. 3. All the values have been normalized to a Yukawa radial dependence with a range of 1.0 fm, and are shown versus the corresponding nucleon energy inside the nucleus, $(E = E_p - \Delta E_{Coul})$. Also shown by straight lines are the energy variations of V_1 as given by set a) and set b) of the neutron global OMP parameters. We neglect here energy dependence from exchange.

The V_τ values are quite scattered and at this point it is not easy to distinguish a definite energy dependence, even though they appear to cluster with an energy dependence similar to that given by set a) of the global OMP parameters.

CONCLUSIONS

It has been indicated that the OMP analysis of accurate neutron elastic scattering data in the range up to 26 MeV has provided a set of OMP parameters that reproduce quite well the neutron elastic scattering, and predict quite accurately the neutron total cross sections. In the spirit of the Lane-model, a set of OMP parameters for protons have been obtained which were used to calculate the total ^{208}Pb(p,n)^{208}Bi IAS cross sections in the energy range 20-50 MeV. The predicted cross sections agree rather well with reported measured values. The shape and magnitude of the ^{208}Pb(p,n)^{208}Bi angular distribution also agree well with reported

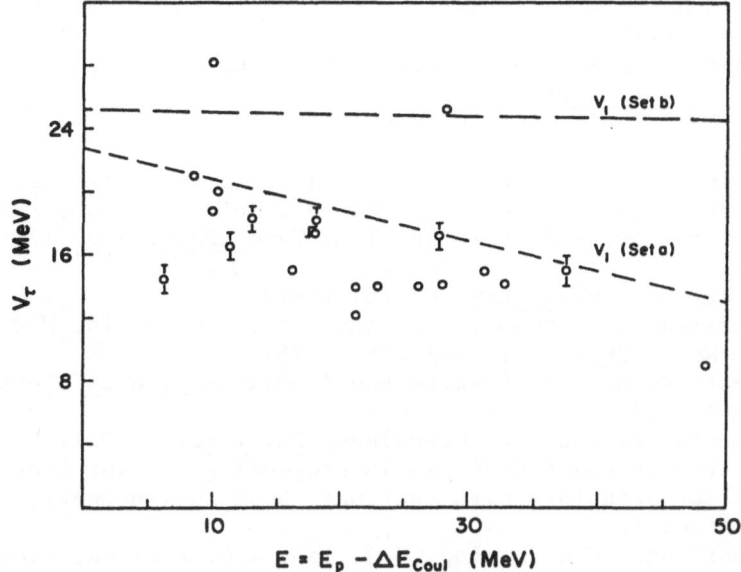

Fig. 3.   Values of V_τ for 1.0 fm Yukawa interaction. The lines
represent the energy dependence of the real term of the
isovector potential, V_1, obtained from the neutron global
OMP parameters.

measurements. Finally the energy dependence of the real part of
the isovector potential V_1, was compared with reported V_τ values
obtained from microscopic analyses at several bombarding energies.

REFERENCES

1.   J.P. Jeukenne, A. Lejeune and C. Mahaux, <u>Phys. Rev.</u> C16:80 (1970),
2.   A.M. Lane, Phys. Rev. Lett. 8:197 (1962); <u>Nucl. Phys.</u> 35:676
     (1962).
3.   A.M. Lane, <u>Rev. Mod. Phys.</u> 29:191 (1957).
4.   J. Rapaport, J.D. Carlson, E. Bainum, T.S. Cheema and R.W. Finlay,
     <u>Nucl. Phys.</u> 286:232 (1977).
5.   P.E. Hodgson, Nuclear Reactions and Nuclear Structure (Clarendon
     Press, Oxford, 1971).
6.   J.C. Ferrer, J.D. Carlson and J. Rapaport, <u>Phys. Lett.</u> 62B:399
     (1976).
7.   G.R. Satchler, Isospin in Nuclear Physics, ed. D.H. Wilkinson
     (North-Holland, Amsterdam, 1969, ch. 9).

8. B.C. Sinha and V.R.W. Edwards, Phys. Lett. 31B:273 (1970); 35B:391 (1971).

9. J. Rapaport, T.S. Cheema, D.E. Bainum, R.W. Finlay and J.D. Carlson, Nucl. Phys. A296:95 (1978).

10. S.D. Schery, D.A. Lind and C.D. Zafiratos, Phys. Rev. C9:416 (1974).

11. J. Rapaport, V. Kulkarni, R.W. Finlay, to be published.

12. D.I. Garber and R.R. Kinsey, BNL 325, Third Edition, 1976.

13. G.W. Greenlees, G.J. Pyle and Y.C. Tang, Phys. Rev. 171:1115 (1968).

14. G.W. Hoffmann, Phys. Rev. C8:761 (1973).

15. S.M. Grimes, C.H. Poppe, J.D. Anderson, J.C. Davis, W.H. Dunlop and C. Wong, Phys. Rev. C11:158 (1975).

16. D.M. Patterson, R.R. Doering and A. Galonsky, Nucl. Phys. A263:261 (1976).

17. F.D. Becchetti and G.W. Greenlees, Phys. Rev. 182:1190 (1969).

18. G.R. Satchler and F.G. Perey in Proceedings of the Conference on Nuclear Structure with Neutrons, Budapest, Hungary, 1972 (unpublished).

19. G.W. Hoffmann, W.H. Dunlop, G.J. Igo, J.G. Kulleck, C.A. Whitten, Jr., Phys. Lett. 40B:453 (1972); T.J. Woods, G.J. Igo and C.A. Whitten, Phys. Lett. 39B:193 (1972).

20. R.R. Doering, D.M. Patterson and A. Galonsky, Phys. Rev. C12:378 (1975).

21. J.D. Anderson, S.D. Bloom, C. Wong, W.F. Hornyak and V.A. Madsen, Phys. Rev. 177:1416 (1969).

22. R.K. Jolly, T.M. Amos, A. Galonsky, R. Hinrichs and R. St. Onge, Phys. Rev. C7:1903 (1973).

23. S.M. Austin, "The Two-Body Force in Nuclei", ed. by S.M. Austin and G.M. Crawley (Plenum Press, New York, 1972).

REACTION MECHANISM/STRUCTURE SENSITIVITY STUDIES OF CROSS SECTIONS, ASYMMETRIES, AND POLARIZATION IN (p,p') AND (p,n)

A. Picklesimer

Physics Department
Indiana University
Bloomington, IN 47405

INTRODUCTION

Inelastic scattering of nucleons from nuclei (in general we speak of both inelastic and charge exchange scattering as inelastic scattering) shows great promise as a source of theoretically valuable information. The recent advances in the range and quality of the available experimental cross section data and the promise of complementary asymmetry and polarization data soon to be forthcoming provides the experimental origin of this promise. In addition, the availability of data for other nuclear probes (the most important being the electron) provides an independent source of information through which the viability of proposed theoretical approaches can be checked. The final ingredient necessary is then careful theoretical analysis sufficient to obtain comparable accuracy for theory and experiment.

The promise of inelastic scattering lies in two separate areas, the possibility of obtaining both detailed nuclear structure and reaction mechanism information. The behavior of the nucleon-nucleon interaction as a function of energy is the operative factor in determining the most propitious theoretical approach to inelastic scattering. At intermediate energies the nucleon-nucleon total cross section passes through a broad, deep, flat minimum in contrast to its behavior at lower energies. It is primarily for this reason that theoretical analysis is expected to be most simple at intermediate energies. The aptness of a simple DWtA* approach can be checked by comparison with corresponding electron scattering theory and experiment and recent studies tend to indicate the viability of such an approach.[1,2] Once the DWtA mechanism has been established at intermediate energies it can be used in the extraction of nuclear

---

*Distorted wave t matrix approximation (see ref. 7).

structure information. Furthermore, reaction mechanism studies can
be carried out by examining the behavior of the same states as the
energy is decreased. In this fashion one should be able to examine
the progression from the simple DWtA theory to the more complicated
theories, for example G-matrix type calculations,[3] necessitated by
lower energies. Knowledge gained in this fashion should prove valuable in considering other strongly interacting systems, pion scattering and nuclear structure calculations for example.

In order to reliably extract the information contained in the
experimental data it is essential to have a clear understanding of
those aspects of calculations which have an important bearing on
predictions and to describe them accurately. In Section 1 we consider one such aspect, the importance for both reaction mechanism
and nuclear structure studies of appropriately describing the two-
body polarization data. It is also useful to investigate the
"boundary conditions" for the extraction of nuclear structure
information (and to compare with other nuclear probes) in order to
proceed in the most effective manner. For this purpose we investigate in Section 2 the implications of the time reversal properties
of the nuclear wavefunctions. Finally, the simple fact that the
nucleon-nucleus interaction is independent of the interaction of
other probes with the nucleus indicates that inelastic scattering
can be used to obtain new nuclear structure information. An example
of this fact is also discussed in Section 2.

SECTION 1. TWO-BODY POLARIZATION EFFECTS

It is clearly necessary to take appropriate account of effects
induced by the two-body polarization if they substantially affect
inelastic scattering predictions. In order to investigate this
possibility it is convenient to begin by defining the two-body
spin-spin amplitudes, M_{ij}, in the notation of Stapp.[4] For
singlet scattering i = j = s, whereas for triplet scattering i(j) is
the final (initial) z-projection of total spin. We suppress isospin
and adopt the coordinate system in which the z-axis is taken along
the incident direction and the azimuthal scattering angle $\varphi = 0$.
The differential cross section (I_0) in the barycentric frame is
then[4]

$$I_0 = \tfrac{1}{4} \sum_{ij} |M_{ij}|^2 \tag{1}$$

in which there is no scattering between singlet and triplet states.
Using Time Reversal Invariance[5] and the usual expression for the
polarization in the two-body system (P) it follows that, in our coordinate system,[4]

$$I_0 P = \frac{\sqrt{2}}{4} \operatorname{Im}\left[(M_{01} - M_{10})(M_{11} - M_{1-1} + M_{00})^*\right] \tag{2}$$

where 'Im' indicates that the imaginary part of the expression following is to be taken and the direction of P is along the y-axis.
It can also be shown, again in our coordinate system, that a real
transition operator leads to real amplitudes, M_{ij}. For the central
forces this statement is simply that the Fourier Transform of a real
(central) function is real, for the non-central forces it is a consequence of our choice of coordinate system.

It therefore follows from Eq. (2) that a real transition
operator corresponds to zero polarization in the two-body system and
hence cannot carry two-body polarization information into inelastic
scattering calculations. The possible ramifications of this result
for inelastic scattering polarization and asymmetry calculations are
obvious. From Eq. (1), on the other hand, the choice of a real
transition operator is not expected to be seriously deficient for
cross section calculations, although interference between different
isospin channels and interference effects due to the optical potential may affect calculations to some extent.

In order to examine the importance of using a complex transition operator (which describes the two-body polarization information) explicit distorted wave (DWtA) calculations have been carried[6]
out for two transition operators using the computer code Mephisto.
The first operator, which we refer to as complex, is the one obtained
recently by Picklesimer and Walker[7] and it provides a good description of the two-body polarization data. The other operator, which
we refer to as real, is real in all but the singlet spin channels.
The real operator was obtained at the same time as the complex one,
has the same operator and range characteristics as the complex one,
and the two operators provide essentially equivalent fits to the
magnitudes of the two-body spin and isospin dependent amplitudes.
Comparison of the distorted wave predictions of the two operators
should therefore allow the importance of describing the two-body
polarization to be assessed. The singlet spin channels were left
complex to retain the appropriate isospin interference effects and
therefore the results to be shown for the importance of using a
complex operator in cross section predictions may be weaker than
would be obtained with a purely real operator. From Eq. (2), however, this should not affect the polarization and asymmetry results.

In Figs. (1) to (3) representative examples of the results
obtained are detailed. The optical potentials used for the distorted wave calculations are as follows: ^{12}C (65 MeV) - K. Hosono,
et al[8]; ^{12}C (122 MeV) - J. Comfort, et al[9]; ^{17}O (135 MeV) -
W. Bertozzi, et al[10]; ^{28}Si (134 MeV) - G. S. Adams[11]. The
nuclear structure descriptions used in this section are very simple
and are used for purposes of illustration only. They are:
^{12}C 2^+ T=0,1 - (1p 3/2)$^{-1}$ (1p 1/2); ^{28}Si 6^- T=0 - (1d 5/2)$^{-1}$(1f 7/2);
^{17}O 1/2$^+$ - (2s 1/2).

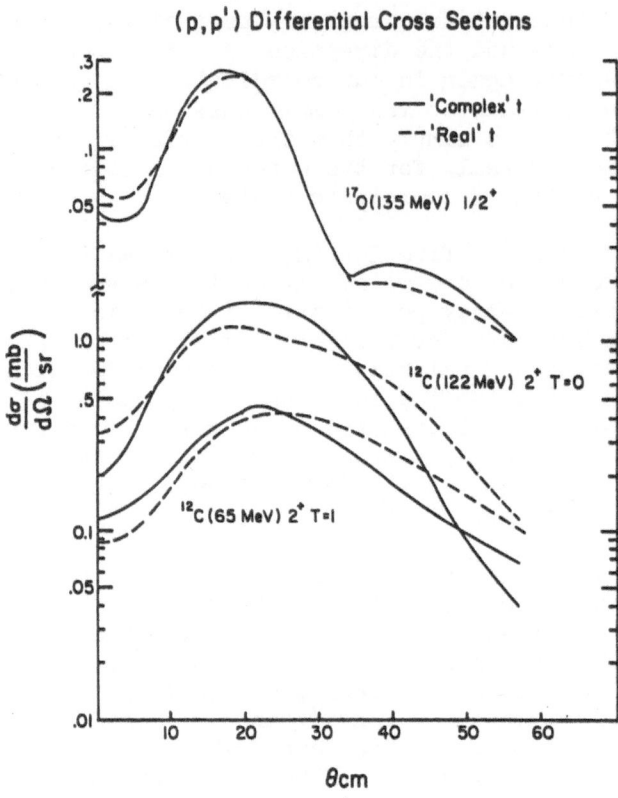

Figure 1

In Fig. (1) selected cross section predictions for the two operators are compared. Clearly the use of the complex operator causes some modification of the results although the effect is not dramatic. Predictions for the 6⁻ states in ^{28}Si (not shown) have little dependence on which transition operator is used. It therefore appears that the use of the appropriate complex form of the transition operator facilitates cross section predictions but is not a source of large effects.

Figs. (2) and (3) display the results for predicted asymmetries using the two operators. In Figs. (2) we note the very large differences in asymmetries predicted at both 65 and 122 MeV for the ^{12}C 2^+ states considered. For simplicity only asymmetries are shown although polarization predictions show the same strong dependence on the choice of the transition operator. In Figs. (3a) and (3c) we see the same feature for the ^{17}O $\frac{1}{2}^+$ and ^{28}Si 6- T = 0 states. These figures provide striking confirmation of the necessity of appropriately describing the two-body polarization data in making inelastic scattering asymmetry (and polarization) predictions.

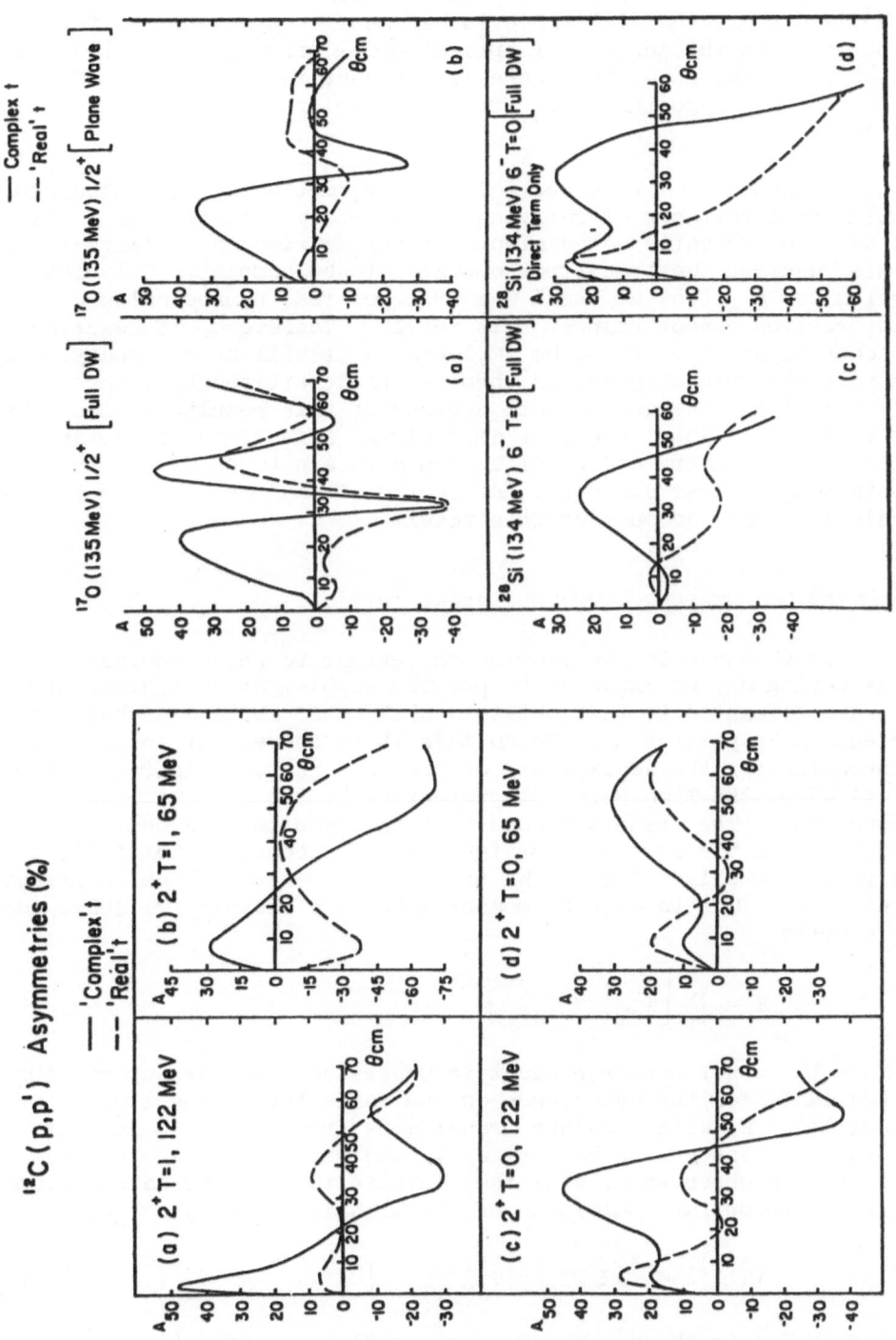

Figure 3

Figure 2

Fig. (3b) is a plane wave calculation corresponding to Fig. (3a) and shows the difference in asymmetries predicted by the two operators in the absence of optical model effects. Fig. (3d) shows that the results obtained here are present in the absence of the knock-on exchange term, and are not therefore simply exchange effects.

In summary of this section, we have motivated and shown explicitly the importance of carrying the two-body polarization information into inelastic calculations of asymmetries and polarizations. This requires the transition operator to be complex. Calculations which use a potential, real G-matrix, or real phenomenological interaction cannot respect this result. Therefore, if reaction mechanism studies are to be implemented it will be necessary to constrain the more complicated theories (necessitated by lower energies) to take appropriate account of this result, so that the theories will join together smoothly as a function of energy. Also, if asymmetry and polarization data are to be of use in obtaining nuclear structure information it is important that calculations take account of this result.

SECTION 2.   NUCLEAR STRUCTURE SENSITIVITY.

In this section we examine the extent to which inelastic proton scattering may be expected to provide supplementary nuclear structure information to that obtained with other nuclear probes, the electron in particular. To do this it is convenient to begin by investigating the sensitivity of predictions to nuclear structure implied by the Time Reversal properties of the nuclear wave functions. The transition operator is taken to be a one body operator in the nuclear coordinates and we consider initially the direct term only. Taking the transition operator to be invariant under rotations in angular momentum (J) and isospin (T) space, one can write

$$\hat{T} = \sum_{JT} \left[\hat{\hat{T}}^{(1)}_{JT} \times \hat{\hat{T}}^{(2)}_{JT} \right]^{00} \tag{1}$$

where the usual tensor product is understood and the superscript 1(2) indicates the dependence on the variables of the target (projectile). We will consider transitions from $J = T = 0$ nuclear states to states of arbitrary J, T, and parity (π). Calculations of scattering observables will then involve the nuclear matrix element of the appropriate multipole of the transition operator, viz.

$$\langle \hat{T}^{(1)} \rangle \equiv \langle JM, TT_z, \pi | \hat{\hat{T}}^{(1)}_{JM, TT_z, \pi} | 0 \rangle \tag{2}$$

where $M(T_z)$ is the z-projection of angular momentum (isospin).

Introducing the usual second quantized format and standard angular momentum coupling one finds

$$\langle \hat{T}^{(1)} \rangle = \sum_{\{\alpha\}\{\beta\}} (-1)^{j_\beta - \mu_\beta} (-1)^{\frac{1}{2} - \tau_\beta} \langle j_\alpha \mu_\alpha \tau_\alpha | \hat{T}^{(1)}_{JM, TT_z}, \pi | j_\beta \mu_\beta \tau_\beta \rangle$$

$$\times (j_\alpha \mu_\alpha j_\beta - \mu_\beta | JM)(\tfrac{1}{2}\tau_\alpha \tfrac{1}{2} - \tau_\beta | TT_z)\langle JM, TT_z, \pi | 0^+ (j_\alpha, j_\beta, JM, TT_z) | 0 \rangle \quad (3)$$

where $\{\alpha\}$ and $\{\beta\}$ are complete sets of single particle states, $j(\mu)$ denotes single particle angular momenta (z-projection), τ is the z-projection of single particle isospin, and the expressions in parentheses are the usual Clebsch-Gordon (C-G) coefficients. In Eq. (3) and the following we suppress the principle and oribtal single particle quantum numbers. The operator 0^+ is given by

$$0^+ (j_\alpha, j_\beta, JM, TT_z) = \sum_{\mu_\alpha \mu_\beta} \sum_{\tau_\alpha, \tau_\beta} (-1)^{j_\beta - \mu_\beta} (-1)^{\frac{1}{2} - \tau_\beta}(j_\alpha \mu_\alpha j_\beta - \mu_\beta | JM)$$

$$\times (\tfrac{1}{2}\tau_\alpha \tfrac{1}{2} - \tau_B | TT_z) \, a^+_\alpha a_\beta \quad .$$

In this equation the subscripts α and β on the usual creation and annihilation operators (a^+ and a) denote the full set of single particle quantum numbers. Defining the amplitude for the transition to proceed through a particular particle-hole coupling by (independent of M and T_z by the Wigner-Eckart Theorem)

$$A^{JT}_{j_1 j_2} \equiv \langle JM, TT_z, \pi | 0^+ (j_1, j_2, JM, TT_z) | 0 \rangle$$

we rewrite Eq. 3 symbolically as

$$\langle \hat{T}^{(1)} \rangle = \sum_{\{\alpha\}\{\beta\}} C_{\alpha\beta} \langle j_\alpha \mu_\alpha \tau_\alpha | \hat{T}^{(1)}_{JM, TT_z}, \pi | j_\beta \mu_\beta \tau_\beta \rangle A^{JT}_{j_\alpha j_\beta} \quad .$$

Since the interest here is in the dependence of observables on the time reversal related amplitudes $A^{JT}_{j_1 j_2}$ and $A^{JT}_{j_2 j_1}$, we will consider the contribution of

$$\langle T^{(1)} \rangle_{12} = \Sigma \left[C_{12} \, A^{JT}_{j_1 j_2} \, \langle 1 | T^{(1)}_{JM, TT_z}, \pi | 2 \rangle \right.$$

$$\left. + C_{21} \, A^{JT}_{j_2 j_1} \, \langle 2 | T^{(1)}_{JM, TT_z}, \pi | 1 \rangle \right],$$

where the sum now runs only over the z-projection variables of Eq. (3). In particular, utilizing the properties of the C-G coefficients and the anti-unitary Time Reversal Operator for the

Schrödinger theory, we find

$$\langle JM, TT_z, \pi | T^{(1)}_{JM, TT_z, \pi} | 0 \rangle_{12} = \Sigma\, C_{12} \left[\langle j_1 \mu_1 \tau_1 | \hat{T}^{(1)}_{JM, TT_z, \pi} | j_2 \mu_2 \tau_2 \rangle\, A^{JT}_{j_1 j_2} \right.$$

$$\left. + (-1)^{J + (\ell_2 + j_2) - (\ell_1 + j_1)} \langle j_1 \mu_1 \tau_1 | \sigma_y\, \hat{T}^{(1)**\dagger}_{JM, TT_z, \pi}\, \sigma_y | j_2 \mu_2 \tau_2 \rangle\, A^{JT}_{j_2 j_1} \right]$$

$$(4)$$

where the adjoint and complex conjugate operate in J-space. The relative contribution of the two amplitudes is then governed by the transformation properties of $\hat{T}^{(1)}$ under the operations in the second term of Eq. (4) and we note that the different components of $\hat{T}^{(1)}$ obey

Central/spin independent: $\sigma_y T^{(1)**\dagger} \sigma_y = T$

Central/spin dependent   : $\sigma_y T^{(1)**\dagger} \sigma_y = -T$ (5)

Tensor   : $\sigma_y T^{(1)**\dagger} \sigma_y = -T$

Using Eq. (4) and these relations it is clear that the phase relation between the contributions of the two amplitudes (of Eq. 4) is independent of the isospin of the final state.

For normal parity final states forces of both types in Eq. (5) can contribute and hence scattering observables depend on both

$$\Delta \equiv A^{JT}_{j_1 j_2} - A^{JT}_{j_2 j_1} \quad \text{and} \quad \Sigma \equiv A^{JT}_{j_1 j_2} + A^{JT}_{j_2 j_1}$$

The two particle spin-orbit interaction also contributes forces of both types to such transitions and hence depends on both Δ and Σ. A special dependence on only one amplitude (Δ or Σ), and the corresponding possibility of a large enhancement or reduction, therefore requires forces of one type to dominate the transition. For example, in the case of normal parity T = 0 states which are reached primarily by the strong central spin and isospin independent force one obtains from Eq. (4)

$$\langle T \rangle_{12} \sim A^{JT}_{j_1 j_2} + (-1)^{j_2 - j_1} A^{JT}_{j_2 j_1} \, .$$

In the case of non-normal parity states, on the other hand, only force components that are odd under the transformation in Eq. (4) can contribute (this is also true for the spin-orbit force).

So that from Eq. 4 one obtains again

$$\langle T \rangle_{12} \sim A^{JT}_{j_1 j_2} + (-1)^{j_2 - j_1} A^{JT}_{j_2 j_1} \tag{6}$$

which now does not depend on the dominance of a particular force. An example of a large effect arising from this relation is given by the 1^+ $T = 1$ (15.11 MeV) state in ^{12}C. Comparison of experimental data with calculations for a simple $(1p3/2)^{-1}$ $(1p1/2)$ structure model indicate a necessary reduction in calculations of a factor of about 4 for both inelastic proton and inelastic electron scattering.[1,12] Much of this reduction is accomplished by the mechanism of Eq. (6) as will be seen later in considering the Cohen and Kurath[13] structure amplitudes for this state.

For higher lying states which satisfy[14]

$$\text{Sign}(A^{JT}_{j_2 j_1}) = (-1)^{\ell_1 + \ell_2 + 1} \text{Sign}(A^{JT}_{j_1 j_2})$$

we find that for non-normal parity states

$$\langle T^{(1)} \rangle_{12} \sim |A^{JT}_{j_1 j_2}| + (-1)^{J + j_2 - j_1} |A^{JT}_{j_2 j_1}| . \tag{7}$$

Of interest in connection with this relation are the high spin "stretched" states seen in inelastic proton and inelastic electron scattering which require large reduction factors, for example the ^{28}Si 6^- $T = 0,1$ states. From Eq. (7) the time reversal related states are expected to interfere destructively and this mechanism may be a contributing factor in causing the large reduction observed. Note that this effect is independent of the final isospin and so is expected to be the same for both the $T = 0$ and $T = 1$ states if the amplitudes (A) are comparable. In fact, recent calculations[1] indicate that the required renormalization is very nearly the same for the two states.

For other probes, the electron, pion, and kaon, results similar to the above are obtained. These aspects will be discussed in some detail by George Walker in his talk later in this conference. The mechanism discussed here is, of course, the source of the large enhancements and reductions in cross sections that sometimes result from use of the RPA (Random Phase Approximation) rather than the TDA (Tamm-Dancoff Approximation).

In the case of the nucleon as a nuclear probe one needs to consider the role of the knock-on exchange term, in contrast to other probes. For the sake of brevity we merely list the analogous results obtained for it in a treatment similar to that given above for the direct term.

1. The exchange term can be shown to violate the relations
obtained for the direct term, either by considering the second term
of the Taylor Series Expansion of the space exchange operator or by
explicit calculations. The importance of this effect will be discussed shortly.

2. In the limit of short range forces, for which the space exchange operator may be approximated by unity, the exchange term
obeys the same relations as obtained for the direct term. For this
reason the dependence of predictions on the amplitudes Δ and Σ is
much stronger for the one upon which the direct term depends.

3. The spin-orbit exchange term satisfies the relations
obtained for the direct term to a very close degree. This fact is
demonstrated by explicit calculations and apparently arises from the
short range of the spin-orbit force. The short range can be understood from meson exchange models of the nuclear force, in which the
spin-orbit term arises from the exchange of heavy vector mesons
$(M_V \gtrsim 5M_\pi)$. This result leads, for example, to the fact that the
^{28}Si 6^- T = 0 state, which receives a large contribution from exchange, behaves as if it depended only upon the direct term. Therefore our previous discussion of the ^{28}Si 6^- T = 0,1 states remains
unchanged.

It is interesting to consider the extent to which the exchange
term, by virtue of its not obeying the relations obtained for the
direct term, affects the extraction of nuclear structure information.
In addition, we will give an example of the fashion in which inelastic proton scattering can supplement inelastic electron scattering in the determination of nuclear structure. In order to do
this it is convenient to use the ^{12}C 1^+ T = 1 (15.11 MeV) state as
an example. The use of this state is desirable due to the recent
analysis of inelastic electron scattering to it by Dubach and
Haxton.[15] The inelastic proton scattering calculations to be
presented were performed using the interaction of Picklesimer and
Walker[7] in the DWtA. In keeping with the work of Cohen and
Kurath (C-K) and of Dubach and Haxton (D-H), the nuclear structure
is confined to the lp shell and the amplitudes needed are then Δ
and Σ (for lp3/2 ↔ lp1/2), $A^{11}_{1/2\ 1/2}$, and $A^{11}_{3/2\ 3/2}$. In Fig. (4) we
show the results of calculations for the sensitivity of this transi-
tion at 61 MeV to the value of Σ, the remaining nuclear structure
parameters are those of (D-H). It is to be noted that inelastic
electron scattering to this state is independent of Σ,[15] but from
Fig. (4) it is apparent that this is not the case for inelastic
proton scattering. The exchange term, and to a lesser extent the
net prediction (Direct + Exchange) shows sensitivity to the value of
Σ, particularly at large momentum transfers. It therefore appears
that, given reliable calculations at large momentum transfers, the
nucleon may provide the opportunity to gain nuclear structure infor-

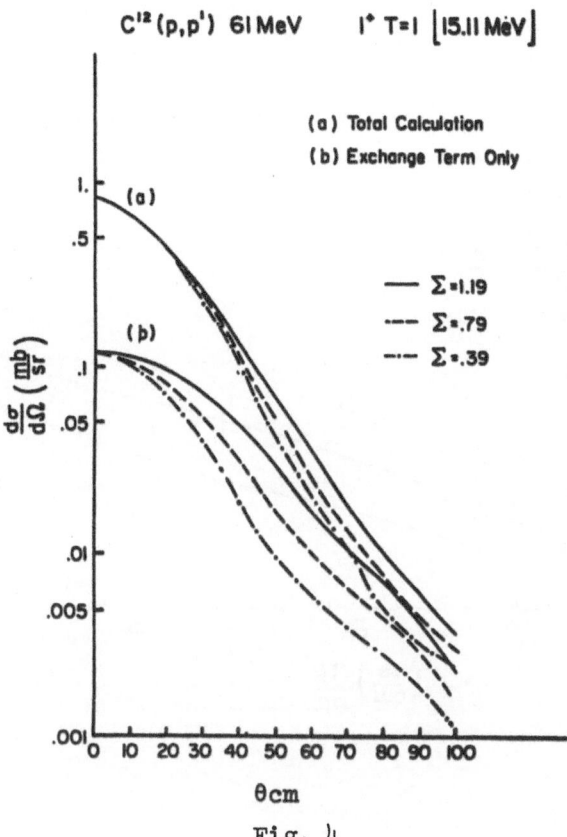

Fig. 4

mation not obtainable from inelastic electron scattering. Since the dependence on Σ is most pronounced at large momentum transfer, we note in passing that the inelastic proton scattering calculations show very little dependence on the use of exchange currents in the inelastic electron scattering results of (D-H).

In Fig. (5) we display results of inelastic proton scattering which show dramatically its usefulness as a supplement to electron scattering. Electron scattering to this state depends on only two linear combinations of the structure amplitudes[15] Δ, $A^{11}_{1/2\,1/2}$, and $A^{11}_{3/2\,3/2}$ so that it does not determine the amplitudes separately. Figure (5) shows proton scattering predictions for different choices of Δ, with the other amplitudes being chosen so that the electron scattering calculations remain unchanged. The sensitivity of the proton scattering results therefore appears to make it a promising probe of nuclear structure. In Fig. (6) proton scattering calculations using the (C-K) and (D-H) amplitudes are compared. Also shown is recent (p,n)[16] data normalized to account for the difference in (p,n) and (p,p') calculations. While the low momentum transfer electron scattering data is well described by both

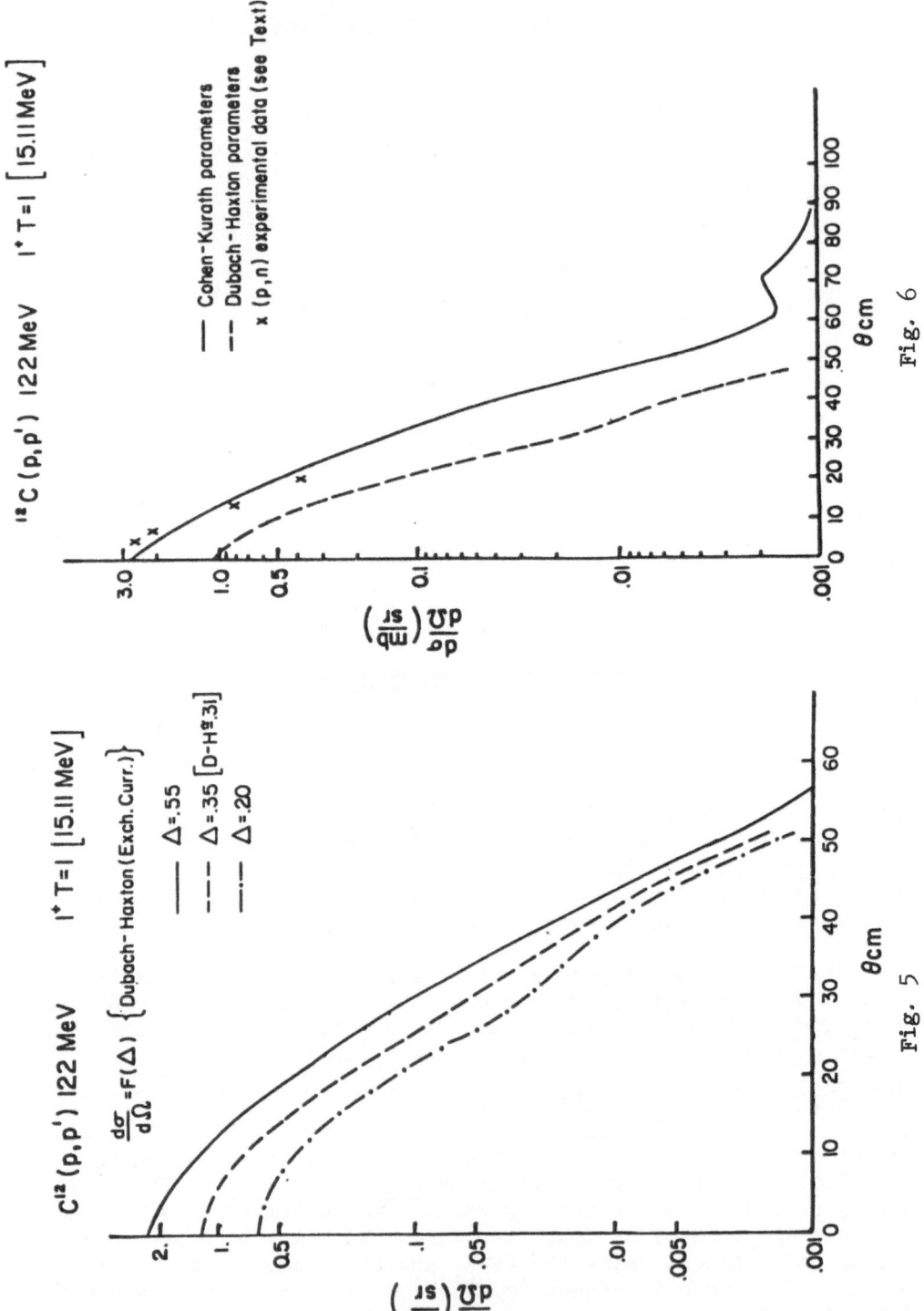

Fig. 6

Fig. 5

(C-K) and (D-H), the proton results clearly favor the (C-K) amplitudes. It therefore appears that the restriction of nuclear
structure amplitudes to the lp shell is too severe, and that inelastic proton scattering combined with electron scattering has provided evidence of this fact. In concluding it is useful to consider
how the time reversal arguments lend support to this result. The
(C-K) amplitudes $A^{11}_{1/2\ 3/2}$, $A^{11}_{3/2\ 1/2}$, $A^{11}_{1/2\ 1/2}$, and $A^{11}_{3/2\ 3/2}$ are
respectively[17] .6928, .3407, .0593 and .0764. So that in com-
paring the sizes of the smaller nuclear structure amplitudes to the
'dominant' amplitude one finds the following picture. For a simple
$(1p3/2)^{-1}$ $(1p1/2)$ approximate description the dominant amplitude has
$|A|^2 = 1$. From the C-K amplitudes one might estimate $|A^{11}_{1/2\ 3/2}|^2 \cong .5$.
Using the Time Reversal results however one finds $|\Delta|^2 = .12$. It
therefore appears that amplitudes which may be negligible when compared with a dominant amplitude of 1. or .5, may be important when
the reduction in strength due to the Time Reversal properties is
large. For the particular state considered here that this is the
case is clear. The simple $(1p3/2)^{-1}$ $(1p1/2)$ picture requires a
reduction factor of 4. From Fig. (6) the C-K amplitudes describe
this reduction very well. However $|\Delta|^2 = .12$, a reduction of the
'dominant' contribution by a factor of 8. Hence the small ampli-
tudes $|A|^2 \leq .6\%$ provide a factor of 2 enhancement! On this basis
it would not be at all surprising if s-d shell amplitudes should
prove important in describing this transition.

REFERENCES

1. A Picklesimer and G. E. Walker, to be published.
2. F. Petrovich, proceedings of this conference;
    G. Love, proceedings of this conference.
3. G. Bertsch, J. Borysowicz, H. McManus, and W. G. Love, Nucl.
    Phys. A 284, 399 (1977).
4. H. P. Stapp, T. J. Ypsilantis, and N. Metropolis, Phys. Rev.
    105, 203 (1957).
5. L. Wolfenstein and J. Ashkin, Phys. Rev. 85, 947 (1952).
6. K. Amos, V. Madsen, and I. E. McCarthy, Nucl. Phys. A 94, 103
    (1967);
    K. Amos, Nucl. Phys. 103, 657 (1967).
7. A. Picklesimer and G. E. Walker, Phys. Rev. C 17, 237 (1978).
8. K. Hosono, et al., Phys. Rev. Letts. 41, 621 (1978).
9. J. Comfort, to be published.
10. W. Bertozzi, M. Deady, M. V. Hynes, J. Kelly, B. Norum, F. N.
    Rad, M.I.T; G. E. Walker, A. Bacher, G. Baranko, G. Emery,
    C. Foster, W. Jones, D. Miller, I.U.; B. L. Berman, L.L.L.,
    private communication.
11. G. S. Adams, Indiana University Cyclotron Facility Internal
    Report 77-3 (1977).

12.   H. K. Lee and H. McManus, Phys. Rev. 161, 1087 (1967).
13.   S. Cohen and D. Kurath, Nucl. Phys. A 101, 1 (1967).
14.   G. E. Walker, Phys. Rev. C 5, 1540 (1972); and
         proceedings of this conference.
15.   J. Dubach and W: C. Haxton, Phys. Rev. Letts. 41, 1453 (1978).
16.   C. D. Goodman, private communication.
17.   K. Amos, et al., Phys. Letts. 52B, 138 (1974).

DISCUSSION

Moravcsik: Your talk was another example of the fast
growing realization both in nuclear and particle physics that
dealing only with differential cross sections ignores half of the
information (generally ignoring the phases, and, in addition,
giving only average information on the magnitudes). Doing this
one can go completely astray in guessing the correct theory.

THE ROLE OF THE N-P INTERACTION IN MICROSCOPIC

CALCULATIONS OF COLLECTIVE MOTION

J. B. McGrory

Oak Ridge National Laboratory*

Oak Ridge, Tennessee 37830

A topic of longstanding interest in nuclear structure theory has been the microscopic theory of collective nuclear behavior. There is now a large body of evidence that nuclear shell-model calculations[1] can reproduce rotational behavior which is observed in light nuclei, particularly in the sd shell. In the last few years, there has been great interest in the so-called interacting boson model[2] (IBM), which has been able to describe both rotational and vibrational behavior in medium and heavy nuclei. As I'll discuss further below, the IBM is a truncation scheme for shell-model calculations. An intrinsic assumption of the model is that the shell model is capable of describing observed collective phenomena. One question then is what are the key ingredients in a shell-model calculation which lead to rotational features? There have been several papers[3-7] in recent years on this question, and in most of them, it is suggested that the neutron-proton interaction plays a decisive role in shell-model descriptions of rotational behavior. It is the purpose of this talk to summarize these arguments on the importance of n-p interaction.

By nuclear shell model, I mean a mixed-configuration shell model. An underlying single-particle potential is assumed. All low-lying orbits up to a specific orbit are assumed to be completely filled to form an inert J=0 core. A small set of valence orbits above the core orbits is included in the active model space. All orbits above these valence orbits are explicitly excluded from

---

*Research sponsored by the Division of Physical Research, Dept. of Energy, under contract W-7405-eng-26 with the Union Carbide Corp.

the calculation. Shell-model states are formed by constructing
states of a relatively few particles in these valence orbits and
diagonalizing an effective residual Hamiltonian in the valence
space. The usual assumption is that the effective Hamiltonian is
the sum of a one-body and a two-body operator. One can write the
Hamiltonian in an n-p formalism in an obvious notation

$$H = H_{p\text{-core}} + H_{n\text{-core}} + H_{nn} + H_{pp} + H_{pn}.$$

The question of interest here is what role do these various terms
play in determining whether or not a nucleus is deformed and exhibits rotational features? I use as an operational definition of
a rotational nucleus one which exhibits one or more bands of states
where the excitation energies have a $J(J+1)$ spacing as a function
of J, and which are mutually connected by very strong B(E2)'s, and
which are not connected to the remaining states in the space by
strong B(E2)'s.

I would first like to discuss briefly the role of the single-
particle spectrum in the generation of rotations in shell-model
calculations. One of the most extensive applications[1] of the shell
models has been to the sd shell nuclei, i.e. nuclei with A = 17-39.
There, it is assumed that the first 16 particles form an inert core
and valence particles occupy the $0d_{5/2}$, $1s_{1/2}$, and $0d_{3/2}$ orbits.
The low-lying states of ^{20}Ne in this model are described as two
neutrons and two protons in these three valence orbits. In ^{20}Ne
there is a well-known ground-state rotational band. The energies
and relative B(E2)'s of these states are well reproduced by a conventional shell-model calculation. ^{44}Ti is the fp shell analog of
^{20}Ne in that in the usual model it is 2 neutrons and 2 protons outside of ^{40}Ca. The observed spectrum does not exhibit nearly so
pronounced a rotational spectrum as does ^{20}Ne. What is the difference? Here, it seems to be the one-body spin-orbit splitting which
is the decisive factor. Some insight into what is going on can be
derived from a simple Hartree-Fock argument.[3] A rotational band
of states is formed by the projections from a single intrinsic state
with a large quadrupole deformation. The state for two neutrons
and two protons with a large quadrupole moment is formed when the
particles are put in single-particle orbits with a large quadrupole
deformation. The single-particle state with the largest quadrupole
moment in a finite shell-model space can be found by diagonalizing
the single-particle quadrupole operator. In the sd shell, that
wave function is $0.82\ d_0 - 0.58\ s_0$ (d_0 is the state with $\ell = 2$, and
s_0 is the state with $\ell = 0$.) There is more d than s state in this
orbit. One might thus suspect that if the single-particle spectrum
were such that the d states were much higher in energy than the s
states, no strong deformations would occur. But in fact the $d_{5/2}$
orbit is about 1 MeV below the $s_{1/2}$ orbit, so the single-particle

spectrum is favorable for forming large deformations. The situation is rather different in the fp shell. In the fp shell, the
single-particle state with the largest quadrupole moment is
0.78 p_o - 0.63 f_o. But the single-particle spectrum has the $f_{7/2}$
orbit below the $p_{3/2}$ orbit by 2 MeV, so the single-particle spectrum
does not favor rotations. As an experiment, the spectrum of ^{44}Ti
was calculated[3] in a shell model with a ^{40}Ca core and four particles
in the fp shell. The Kuo-Brown fp shell interaction[8] was used for
the effective two-body residual interaction. Calculations were made
with several different single-particle spectra. The results are
shown in Fig. 1. Here, column two shows the realistic shell-model
calculation of the states in the ground-state band of ^{20}Ne where
the energy of the J=2$^+$ state is normalized to one MeV. The third
column shows a similar calculation for ^{44}Ti, using the experimentally

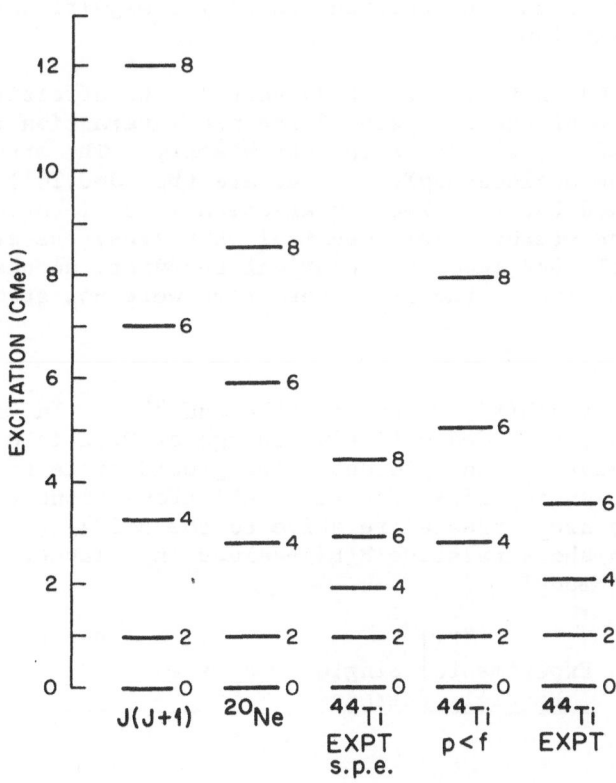

Fig. 1.   Calculated spectra of ground-state bands in ^{20}Ne and ^{44}Ti.
The first column shows a rigid rotor J(J+1) spectrum when
the energy of the 2$^+$ states is normalized to 1.0 MeV.

observed single-particle energies. The fourth column shows the same
calculation, but using a different set of single-particle energies,
in this case with the $p_{3/2}$ and $p_{1/2}$ orbits degenerate and 2 MeV below the degenerate $f_{7/2}$ and $f_{5/2}$ orbits. In this case, if our
hueristic arguments are valid, the resulting ground-state band
should become significantly more "rotational". This is indeed the
case. A similar pattern appears when one considers calculated B(E2)
values. In Table 1, the calculated B(E2) values in ^{44}Ti between
members of the ground-state band are shown. An added effective
charge of 0.5e has been used for both neutrons and protons. In
Table 1, absolute B(E2) values are shown for the $2_1^+ \rightarrow 0_1^+$ transition
and the B(E2) values relative to this $2_1^+ \rightarrow 0_1^+$ transition are shown
for the transitions. Also shown in this table are the relative
B(E2) values of the adiabatic rotational model ($\psi_M^J \propto D_{MK}^J \chi_K(\Omega)$).
From Table 1 we see that lowering the p orbits relative to the f
orbits gives a 50% enhancement of the B(E2) values, and a somewhat
more rotational behavior for the excited states. Thus, we see that
the one-body spin-orbit interaction can play a significant role in
generating deformations.

Now let me go on to the two-body part of the effective interaction. The role of the T=0 part of the n-p interaction in the
generation of deformation has a lengthy history. The strongest
components of the nucleon-nucleon force are the (J=0,T=1) pairing
matrix element and the (J=1,T=0) interaction which interacts only
in neutron-proton states. For identical particles, the strong
pairing matrix element leads to spherical behavior, so there would
be little deformation if the n-p interaction were not strong. A

Table 1. Calculated B(E2)-values in ^{20}Ne and ^{44}Ti. In all calculations, an added effective charge of 0.5e is used for
both neutrons and protons. The ground-state transitions
are given in units of e^2fm^4. All transitions to excited
states are expressed relative to the B(E2)$_{2-0}$. The last
column shows relative B(E2)-values in adiabatic rotational model.

J_i	J_f	^{20}Ne	^{44}Ti Experimental Single-Particle Energies	^{44}Ti $\varepsilon_p < \varepsilon_f$	Rotational
2	0	248	116	179	1.0
4	2	1.24	1.4	1.4	1.4
6	4	1.01	1.2	1.4	1.6
8	6	0.65	0.9	1.2	1.6

good example of this is a comparison of the ^{20}Ne and ^{20}O spectra.
Both systems are four particles in the (sd) shell. In this model
the spectrum of ^{20}O is independent of the n-p interaction. ^{20}Ne
has two neutrons and two protons, so it is a function of both the
n-p and the p-p interaction. The spectra are quite different as
seen in Fig. 2. As discussed above, ^{20}Ne is quite rotational, while
^{20}O has no rotational aspects. Federman and Pittel[5] have recently
discussed some aspects of the role of the n-p interaction in the
development of deformations in nuclei. They made some of the arguments discussed here on the importance of the n-p interaction. They
argue that for the n-p interaction to play a major role, there must
be good overlap of the single-particle wave functions. If the
single-particle functions are labeled by a principle quantum number
n and an orbital angular momentum quantum number ℓ, then there will
be a good overlap if $n_1 \approx n_2$ and $\ell_1 \approx \ell_2$. This implies that deformations will occur when neutrons and protons are filling single-

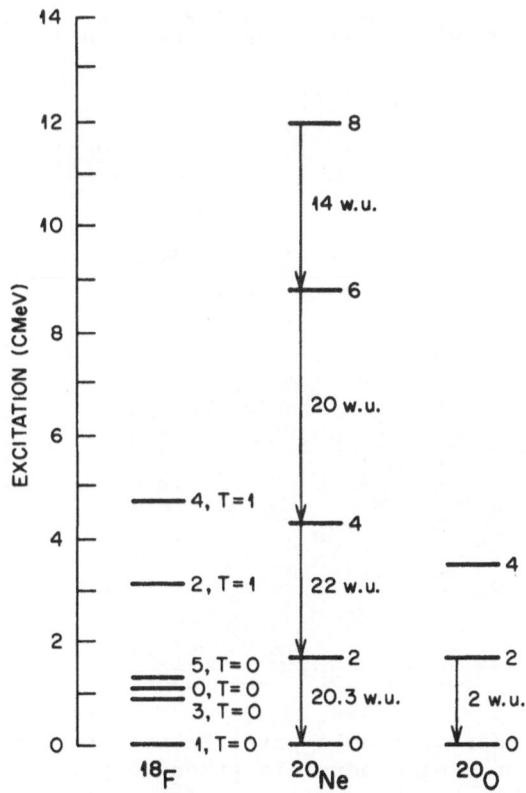

Fig. 2. Observed spectra of Yrast states in ^{18}F, ^{20}Ne, and ^{20}O.

particle orbits with approximately equal quantum numbers n and ℓ.
As an example, they discuss nuclei around the ^{88}Sr core. The
single-particle spectrum there is shown in Fig. 3. For the Mo
isotopes, four protons are in the $p_{1/2}$ and $g_{9/2}$ orbit. Their
arguments would suggest that when the $g_{7/2}$ neutron orbit starts to
fill (i.e. when 14 neutrons are added), deformation effects should
set in because the n-p interaction becomes strong. This, in fact,
is two neutrons away from where there is experimental evidence for
deformation. They make similar arguments in the rare-earth and
actinide regions. Finally, Federman and Pittel argue that these
strong correlations can come from the short-ranged $^3S_1$ neutron-
proton interaction.

I have made some explicit studies of the role of the n-p interaction in forming deformation. I have made nuclear shell-model
calculations of some nuclei in an n-p formalism in which it is easy
to see the importance of n-p correlations. Let me outline here the
procedure for the calculations. In an obvious notation, the residual shell-model interaction can be written

$$H = H_{n-core} + H_{p-core} + H_{nn} + H_{pp} + H_{pn} = H_n^{1+2} + H_p^{1+2} + H_{pn}.$$

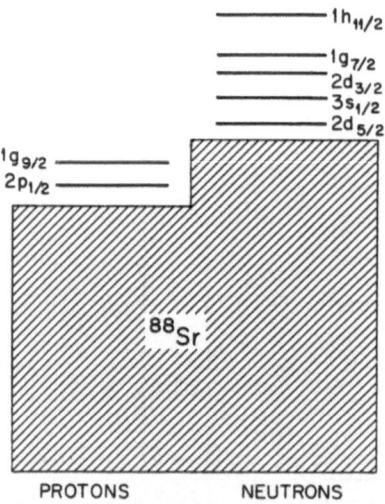

PROTONS   NEUTRONS

Fig. 3. Observed order of single-particle states near ^{88}Sr core.
Only half of single-particle strength for the $1h_{11/2}$
neutron orbit has been seen, so its position is more uncertain than other orbits.

If, initially, we ignore H_{pn}, then these eigenstates of H can be written as a direct product of eigenstates of H_p^{1+2} and H_n^{1+2}, i.e.

$$|\psi_{n,p}^{i,j}\rangle_J = [|\psi_p^i\rangle \times |\psi_n^j\rangle]^J$$

where

$$H_n^{1+2}|\psi_n^j\rangle = E_n^j|\psi_n^j\rangle$$

$$H_p^{1+2}|\psi_p^i\rangle = E_p^i|\psi_p^i\rangle.$$

If all the eigenstates of H_p^{1+2} and H_n^{1+2} are included in this direct product space, a complete (n,p) space is generated. Thus, for ^{20}Ne, we can expand the wave functions in terms of eigenstates of ^{18}O and ^{18}Ne. All the correlations of the n-n and p-p interactions are in the ^{18}O and ^{18}Ne wave functions. I have made a series of calculations of ^{20}Ne in which I successively include fewer and fewer eigenstates of the ^{18}Ne and ^{18}O spectrum. If I include only the lowest $J=0^+$, 2^+, and 4^+ eigenstates of the neutron and proton systems, I obtain the spectrum as shown in Fig. 4. The observed spectrum and the spectrum calculated in the complete $(sd)^4$ space are also shown in Fig. 4. Calculated and observed B(E2) values for transitions within the ground-state bands are shown there also for all three spectra. The calculation with only three eigenstates reproduces the exact calculation quite well. The comparison of the spectrum which results when no n-p interaction is included and the spectrum which results from the complete-space diagonalization shows there are large effects due to the n-p interaction. This is even more transparent when one analyzes the wave functions, as is done in Table 2. Here the wave functions of the ground-state rotational band are expressed as couplings of the eigenstates of ^{18}O and ^{18}Ne. If the n-p interaction were relatively weak, the ground state, $J=0^+$ state would be dominated by the $(^{18}O,J=0_1^+) \times (^{18}Ne,J=0_1^+)$ state. In fact, the state formed by coupling the two $J=2^+$ states is comparable to the $0_1^+ \times 0_1^+$ coupling in the small space calculation. In the exact wave function, the $2_1^+ \times 2_1^+$ state is actually the largest one. Thus, the n-p interaction leads to a highly correlated wave function in a case where a clear rotational band exists.

A final explicit example of the importance of the n-p interaction in producing deformation is from recent work by Nair et al.[6] They have performed major shell Hartree-Fock calculations of ^{20}Ne and ^{24}Mg. They do the calculations with a pairing plus quadrupole force. In one case, they use the full force, and a second case they use only the n-p force, and in a third case they use only the

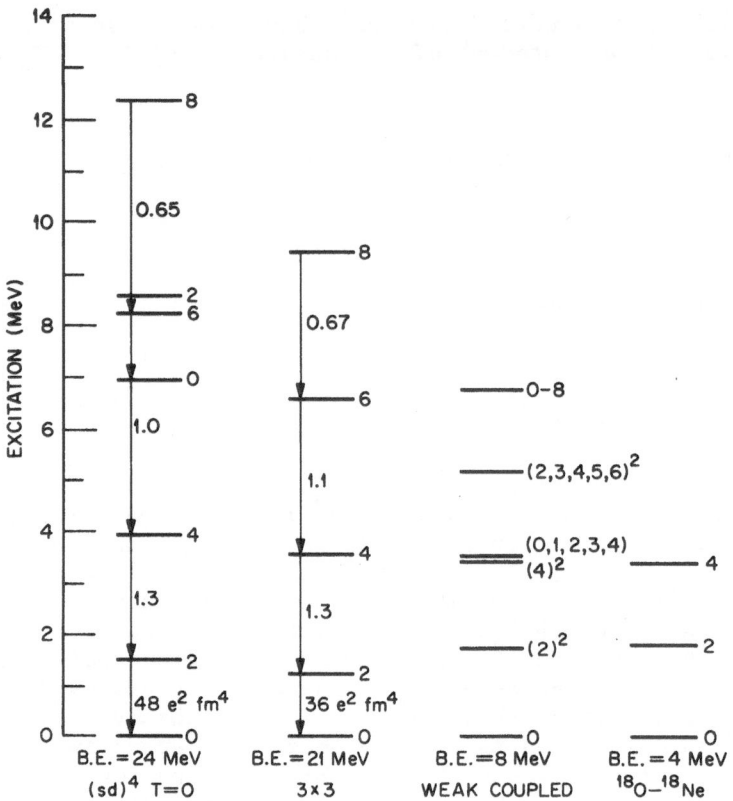

Fig. 4. Calculated spectra of ^{20}Ne in various model spaces. A
         realistic effective interaction is used in all cases. The
         first column shows results in full (sd)^4,T=0 space for
         Yrast even-J states. The last column shows the calculated
         (sd)^2,T=1 spectrum of Yrast even-J states. The next-to-
         last column shows results of calculation where the states
         in the last column are coupled to form 2n-2p states, but
         with no neutron-proton interaction. The second column
         shows results of diagonal n-p interaction in space of
         states formed by coupling the lowest three ^{18}O states to
         the lowest ^{18}Ne states. B(E2)-values in e^{2}fm^4 are shown
         for the 2→0 transitions. B(E2)-values for transitions to
         excited states are shown relative to the 2→0 transition.
         The binding energies shown exclude (approximately) the
         one-body contributions.

identical-particle force. They calculate the intrinsic quadrupole
moment of the Hartree-Fock solutions in these cases. The results
are summarized in Table 3. The calculated intrinsic quadrupole
moment arises almost completely from the n-p interaction.

Table 2.   Structure of states in ground state rotational band of
           ^{20}Ne in terms of state of form ($J^p \times J^n$), where J^p is the
           lowest eigenstate of ^{18}Ne and J^n is same for ^{18}O. Note
           that the admixtures of ($J^p \times J^n$) and ($J^n \times J^p$) are equal,
           and only one of them is shown.

	J = 0	J = 2	J = 4	J = 6
($J_p \times J_n$)				
0 × 0	0.69			
0 × 2		−0.55		
0 × 4			0.44	
2 × 2	0.67	−0.42	−0.63	
2 × 4		0.31	0.31	−0.68
4 × 4	0.27	−0.16	−0.16	0.25

Table 3.   Intrinsic quadrupole moments of Hartree-Fock Bogoliubov
           solutions for ground-state bands of ^{20}Ne and ^{24}Mg.

Residual Force	<Q> e fm^2	
	^{20}Ne	^{24}Mg
$H_{nn} + H_{pp} + H_{pn}$	7.8	9.1
H_{pn}	7.7	8.9
$H_{nn} + H_{pp}$	0.1	0.1

     Thus, there is ample evidence that microscopic wave functions
which describe rotational behavior occur because of strong correlations introduced by the neutron-proton interaction. Knowing this
is so is interesting, but how useful is the information? Is there
any way to use it to make microscopic calculations of transitional
and rotational nuclei? In an attempt to answer this, I would like
to go into a little more detail on the decomposition of the Hilbert
space into neutron states and proton states. As I pointed out
above, one can generate a complete space by solving the pure proton
and pure neutron Hamiltonians separately, then coupling the resulting eigenstates, and diagonalizing the neutron-proton interaction in that space. The matrices in the separate neutron and
proton spaces are much smaller than those in the total n-p space

because there are fewer particles, and because one must deal only
with states with maximum isospin. However, the resulting matrices
in the direct-product (n,p) space are larger than necessary because
the isospin symmetry is lost in constructing the basis. The
question then is, can one find a small enough set of neutron and
proton eigenstates so that the resulting (n,p) space remains small,
but the overlap with exact low-lying wave functions is large? This
question was studied several years ago by Hecht et al. There, the
residual interaction was assumed to be a surface-delta interaction
(V_{SDI}) with equal strengths in the n-n, p-p, and n-p systems. If
one works in the direct product space defined above, the matrix
elements of $V_{np} = V_{SDI}$

$$\langle \psi_{np}^{i,j} | V_{np} | \psi_{np}^{k,\ell} \rangle_J = -g \sum W(ijk\ell;JK) \langle \psi_n^i \| Q^K \| \psi_n^k \rangle$$

$$\times \langle \psi_p^j \| Q_p^K \| \psi_p^k \rangle \ (-)^{i+\ell-J} \ \hat{i}\hat{j}/\hat{J}$$

where $Q_M^K = Y_M^K(\Omega)$, Y^K is the usual spherical harmonic, g a coupling
constant, and W a Racah coefficient. The justification for using
the surface-delta interaction is discussed by Hecht et al. This
form of the matrix elements does suggest a means for picking out
the important neutron and proton states. If one can find a set of
states Φ_n' which is a subset of ψ_n', the complete set of neutron
states, and if for all the surface multipole operators, Q^K, the
matrix elements $\langle \Phi_n' \| Q^K \| \psi_n' \rangle$ for all ψ_n' not in Φ_n' are zero, then
eigenstates involving Φ_n are not coupled by the neutron-proton
interaction to any of the other ψ_n^i. If one can also find a set Φ_p^i
with similar properties, then the resultant direct product space
($\Phi_p^i \times \Phi_n^j$) is completely uncoupled from the rest of the neutron-
proton space. The eigenvalues of V_{np} in this space will yield
exact eigenstates of H. A space which approximately satisfies these
criteria was constructed by Hecht et al. They treated the problem
of four protons in the (fp) shell and six neutrons in the (sd)
shell, where all the single-particle orbits are degenerate. They
introduced the favored pair concept. If one diagonalizes V_{SDI} in
the space $(sd)^2$, T = 1, the spectrum shown in the first column of
Fig. 5 results. Only three states have non-zero eigenvalues, with
J = 0, 2, and 4 and J = 0. These three states are called favored
pair states. One can define creation operator for these states

$$A_J^+ = \sum_{j_1 j_2} C_{j_1 j_2}^J (a_{j_1}^+ \times a_{j_2}^+)^J$$

where a_j^+ is the usual creation operator for a particle in the orbit
j. It is possible to create four-particle states for coupling

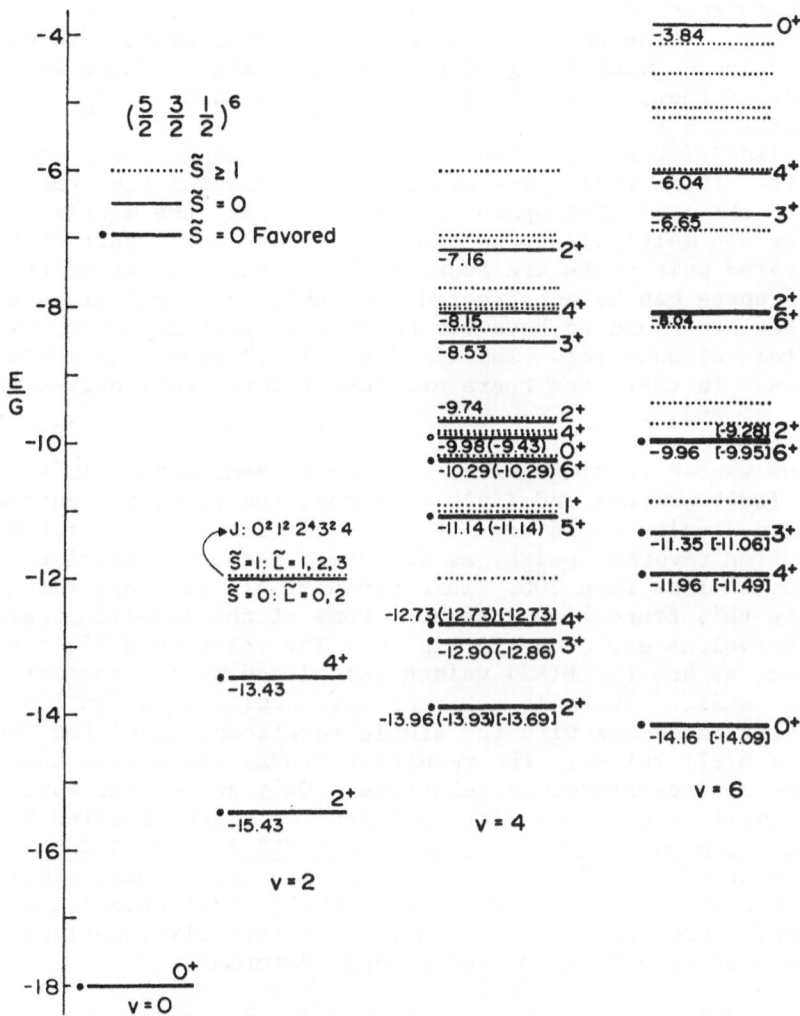

Fig. 5. Calculated spectra of eigenvalues of $(sd)^6$, T=3 states, where a surface delta interaction with unit strength in n-n, p-p, and n-p channels and degenerate single-particle energies. In full $(sd)^6$, T=3 space there are 142 states.

together two of these favored pair creation operators

$$A_4^J = (A_{J_1}^+ \times A_{J_2}^+)^J.$$

One can construct normalized states with good mixed seniority from
these states. These states created from favored pairs have many of
the properties we want for good truncation states. There are relatively few of them, ~ 20, and the expectation value of H_{np} between
these states are almost exactly equal to eigenvalues in the four-
particle $(sd)^4$, T=2 space. Finally, one can make sum rule arguments
to show that these states are essentially uncoupled from the remainder of the $(sd)^4$, T=2 space. That is to say, the matrix elements
of the surface multipole operators between a favored pair state and
a non-favored pair state are found to be small. States of the
$(sd)^6$, T=3 space can be constructed from three favored pair states.
Such states are found to have strong overlap with low-lying exact
six-particle eigenstates. The $(fp)^2$ and $(fp)^4$ spaces have similar
properties. In that case there are four favored pair states with
J = 0, 2, 4, and 6.

If we wanted to diagonalize a residual Hamiltonian in the full
space of $(sd)^6$ protons and $(fp)^4$ neutrons, the resulting matrices
for a fixed-J value would be ~ 60,000. If we use those states constructed from favored pairs, as discussed here, the matrices can
be reduced to less than 200. This ten-particle case has been
treated in this truncation scheme.[9] Some of the low-lying calculated eigenvalues are shown in Fig. 6. The relative B(E2)'s are
also shown, as are the B(E2) values calculated in the adiabatic
rotational model. There is a rather spectacular rotational spectrum
and striking agreement with the simple rotational model for the
calculated B(E2) values. The resulting ground-state wave function
has strong neutron-proton correlations. Only 36% of the wave function is comprised of J=0 seniority 0 neutron states coupled to
seniority 0 proton states. An additional 36% is composed of
neutrons coupled to J=2, seniority 2 and protons coupled similarly.
This is an explicit example of a large shell-model calculation where
neutron and proton states separately are relatively spherical, but
the correlated wave function is strongly deformed.

This model serves as a natural introduction to the now well-
known interacting boson model. The fundamental philosophy of the
IBM and the model of Hecht et al. are quite similar, i.e. microscopic shell models in terms of valence particles outside closed
shells will produce collective features, and the coupling of a
relatively few selected proton and neutron states through a strong
p-n interaction can reproduce much larger shell-model calculations.
The IBM makes other simplifying assumptions. Where Hecht et al.
constructed many particle neutron (proton) eigenstates from two-
fermion favored pair reaction operators with J = 0, 2, $(2j_{max}-1)$
(j_{max} is j of the orbit with the largest single-particle value) the
IBM uses only two two-particle creation operators, i.e. operators
analogous to the $A^+_{J=0}$ and $A^+_{J=2}$ above. Further, they map the two-
fermion operators onto boson operators, s^+ and d^+. Their many-
particle states are constructed from various appropriate couplings

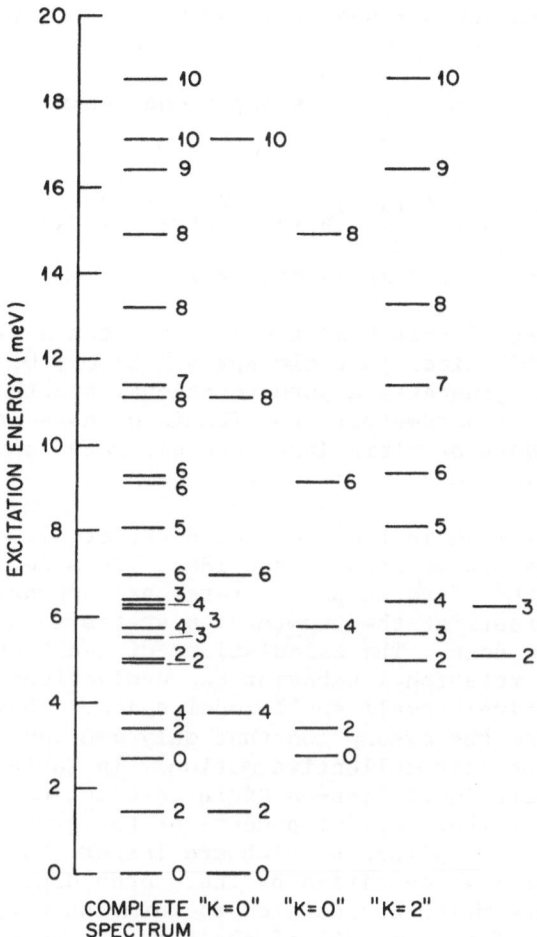

Fig. 6.   Spectrum of eigenstates in four-neutron, six-proton calculation described in the text. The first column shows
all eigenvalues below 5.0 MeV plus selected high-spin
states up to 20 MeV. The remaining columns show same
states as members of collective bands where the band
structure is determined by the fact that states in band
are connected by strong B(E2) values.

of these s^+ and d^+ bosons. In this "boson" space an effective
Hamiltonian is used which has the following form

$$H = \Sigma \ \epsilon \ (n_{\pi d} + n_{\nu d}) + K \ Q_\pi^2 \cdot Q_\nu^2$$

where $n_{\pi(\nu)d}$ is the number operator for the proton (neutron) d-bosons

  ε = energy spacing between s and d one-boson states, which is
      treated as a parameter

and $\quad Q^2_{\pi(\nu)} = (s^+d)_{\pi(\nu)} + (d^+s)_{\pi(\nu)} + \chi_{\pi(\nu)}(d^+d)_{\pi(\nu)}$

  K,χ are also treated as parameters.

The one-boson part of this Hamiltonian generates a vibrational
spectrum in the identical particle space. If ε = 0, the n-p boson-
boson interaction generates a pure rotational spectrum. Treating
ε, K, and X as free parameters, the IBM is capable of reproducing
a large body of data or vibrational nuclei, rotational nuclei, and
transitional nuclei.

  Calculations made in the model of Hecht et al. can provide
some check on the assumptions of the IBM. The model of Hecht does
not assume that the "favored pair" states are approximately bosons,
and it does not restrict the two-fermion states to J = 0^+ and
J = 2^+ as the IBM does. The calculations of Hecht et al. certainly
demonstrate that rotational behavior can evolve from the diagonalization in a relatively small shell-model space. They also provide
strong support for the assumption that only J=0^+ and J=2^+ pairs are
needed to describe this collective motion. In Table 4 an analysis
of the ground state in the ten-particle calculation described above
is given. We list there all components of the ground-state wave
function which have admixtures which are larger than 0.2 in magnitude. The states are identified by the proton eigenstate and the
neutron eigenstate which are coupled to form the given basis state.
In this J=0^+ wave function, 86% of the state is accounted for by
these "large" components. In all these large components, only
four proton (neutron) eigenstates are involved. The J = 0^+ is the
seniority state, the J = 2^+ is a seniority 2 state, while the J=2^+
and J=4^+ states are seniority 4 states constructed from coupling
two particle states with J = 2. Thus, all these components are
analogous to states which would be included in the IBM. In particular, perturbative theory would say that the J=4^+ seniority 2
state should be strongly admixed. But it is the seniority 4 state
formed from J=2^+ two-particle states which is the important one.
A similar feature exists for all the members of the ground-state
band in the ten-particle case, i.e. the states are all dominated
by states made up of J=0^+ and J=2^+ pairs.

  In summary then, I have tried to illustrate the existing
picture of the important ingredients in the microscopic origins of
nuclear collective motion. I have stressed the importance of the

Table 4. Decomposition of lowest $J=0^+$ and $J=2^+$ states in system of four pseudo-fp neutrons and six pseudo-sd protons in terms of states $(J_i^p \times J_j^n)$. Only those admixtures > 0.2 are shown. J_i^p labels the ith eigenstate in proton spectrum with given J. J_p^n is defined in an analogous fashion.

$(J_p^i \times J_n^j)$	$J = 0$	$J = 2$
$0_1 \times 0_1$	0.60	
$0_1 \times 2_1$		0.424
$0_1 \times 2_2$		0.216
$2_1 \times 0_1$		0.458
$2_1 \times 2_1$	0.594	0.266
$2_1 \times 2_2$	0.210	0.260
$2_2 \times 2_2$	0.218	
$2_2 \times 4_2$		0.255
$4_2 \times 2_1$		0.275
$4_2 \times 4_2$	0.229	
$\Sigma \, \alpha_{ij}^2$	86%	72%

one-body spin-orbit, the importance of the availability of single neutron and single proton orbits with similar n and ℓ, and, in general, the importance of the neutron-proton interaction in the development of rotational behavior. Considerations such as those discussed here have brought the nuclear structure theory significantly closer to a microscopic model of wide applicability in the mass regime which might be characterized as transitional or deformed

List of References

1. B. H. Wildenthal, p. 383 in Elementary Modes of Excitation in Nuclei, LXIX Corso, Soc. Italiana di Fisica, Bologna, Italy, 1977.
2. A. Arima and F. Iachello, Ann. Phys. (N.Y.) 99, 253 (1976).
3. K. H. Bhatt and J. B. McGrory, Phys. Rev. C3, 2293 (1971).
4. M. Danos and V. Gillet, Phys. Rev. 161, 1034 (1967).
5. P. Federman and S. Pittel, Phys. Letts. 69B, 385 (1977).
6. S. C. K. Nair, A. Ansari, and L. Satpathy, Phys. Letts. 71B, 257 (1977).
7. K. T. Hecht, J. B. McGrory, and J. P. Draayer, Nucl. Phys. A197, 369 (1972).
8. T. T. S. Kuo and G. E. Brown, Nucl. Phys. A114, 241 (1968).
9. J. B. McGrory, Phys. Rev. Lett. 41, 533 (1978).

PION SINGLE CHARGE EXCHANGE

Helmut W. Baer

Medium Energy Physics Division
Los Alamos Scientific Laboratory
University of California
Los Alamos, NM 87545

ABSTRACT

The relation of the $(\pi^{\pm},\pi^0)$ charge exchange reaction to the (p,n) reaction is discussed. The operating principles and performance of the LAMPF π^0 spectrometer are given. First results for the (π^+,π^0) reaction on ^{7}Li, ^{90}Zr, and ^{208}Pb are presented. The measured A-dependence of the 0^0 (π^+,π^0) cross sections at $T_\pi=$ 100 MeV from A=1 to A=208 are compared with the semi-classical diffraction model of Johnson.

INTRODUCTION

To understand the (p,n) reaction--the principal focus of this conference--it is important to recognize its relation to the $(\pi^{\pm},\pi^0)$ reaction. I would like to indicate some of the connections, and to describe some of the first results obtained with the LAMPF π^0 spectrometer on pion single charge exchange to individual nuclear states.

1. COMPARISON OF (p,n) AND $(\pi^{\pm},\pi^0)$ REACTIONS

For those accustomed to viewing the (p,n) reaction in terms of one-boson-exchange channels, the quantitative relation between $\pi N \rightarrow \pi N$ and $np \rightarrow np$ scattering is fundamental. [see, e.g., Ref. 1, 2] Historically, the large cross sections for 180^0 scattering of neutrons on proton targets were compelling evidence for the exchange of a charged boson in N-N scattering. The experimental cross sections at 50 MeV and at 790 MeV are shown in Fig. 1. Without the exchange of a charged boson it would be difficult to explain the sharp rise in the differential cross sections at 180^0 in the center-

of-mass system. However, if a charged meson is transferred in the
interactions, the scattering appears as forward peaking for the
incident beam particle, and the rise in cross section seems quite
natural. For the (p,n) reaction, the 180⁰ peak for protons corresponds to 0⁰ neutrons.

The translation of the pn→pn reaction into nuclei leads to the
evaluation[5]) of potential terms of form $V_T \vec{\tau}_i \cdot \vec{\tau}_j$, $V_{\sigma T}(\vec{\sigma}_i \cdot \vec{\sigma}_j)(\vec{\tau}_i \cdot \vec{\tau}_j)$
and $V_{TT}S_{ij}$ (tensor) between nuclear shell model wave functions.
Numerical values of V_T, $V_{\sigma T}$, and V_{TT} extracted from (p,p') and
(p,n) scattering between states of well-understood structure were
presented by Austin[6]). The values of $V_{\sigma T}$ and $V_{\sigma T}$ appear to be consistent with the values expected from the one-pion exchange piece
of N-N potentials[7] However, a great deal of the past interest in

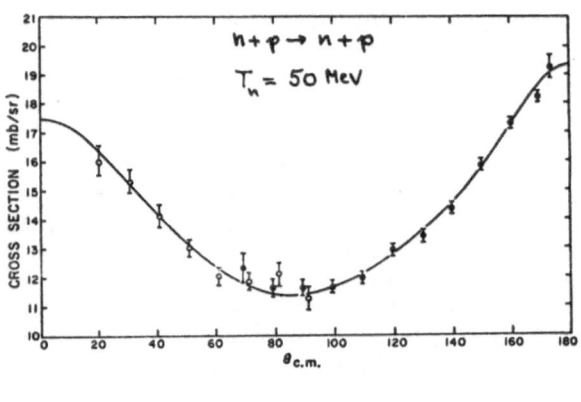

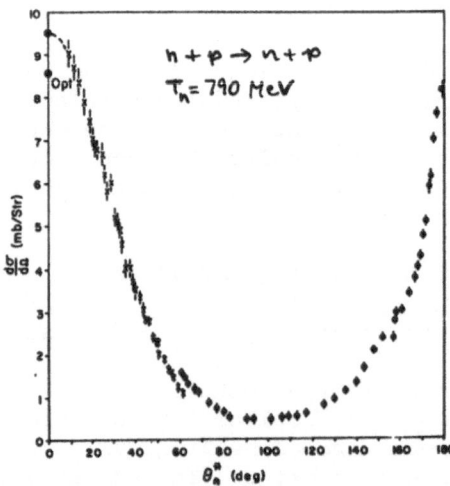

Fig. 1.   The differential cross sections for n-p scattering at
          50 MeV (Ref. 3) and at 790 MeV (Ref. 4).

the (p,n) reaction stemmed from the fact that the (p,n) reaction preferentially excited the isobaric analog state (IAS) which involves the central $V_\tau \vec{\tau}_1 \cdot \vec{\tau}_j$ term. Examination of this term in the framework of OBE potentials shows the close connection between the (p,n) and $(\pi^\pm, \pi^0)$ reactions.

The diagram for one pion exchange in the (p,n) reaction is shown in Fig. 2a. This process must be present to a large extent in (p,n) scattering at $q < 2F^{-1}$ between nuclear states connected by $\Delta S = 1$ and $\Delta T = 1$ transitions. Such transition were discussed by Goodman[8] and Debevec[9]. However, in 0^0 (p,n) scattering between IAS, the exchange of a single pseudo-scalar boson $(\pi, \eta, \omega, \rho)$ is greatly suppressed (forbidden in plane wave impulse approximation). To conserve parity at the π-N vertex, the π must have $\ell = 1$ due to the odd intrinsic parity of the π. To conserve angular momentum at the vertex, the nucleon must either flip its spin or change its direction of motion. Thus to account for 0^0 (p,n) scattering between analog states, which have identical space-spin structure, one must look to 2π exchange, or a "σ" scalar meson exchange, where the σ has T=1, and $J^\pi = 0^+$. Such a process is illustrated in Fig. 2b. Calculations by B. Bhakar[10] have shown that the inclusion of the "σ" channel can account for the energy dependence of the 180^0 np cross section. Comparing the (π^+, π^0) process as shown in Fig. 2c, with the (p,n) reaction as shown in Fig. 2b, one sees immediately the relation between the two reactions. Both at the nucleon- and at the nucleus-vertex one has pion charge exchange. It would not be unreasonable to attempt to calculate 0^0 (p,n) IAS cross sections from empirical $\pi^+ + p \to \pi^0 + n$ and $\pi^+ + A \to \pi^0 + A'$ (IAS) cross sections (with some of the usual assumptions about the off-shell behavior of the π's in Fig. 2b). So far this has not been done since the charge exchange cross sections on nuclei were not measured.

Fig. 2. Graphs which illustrate the relation between the (p,n) charge exchange reaction and the (π^+, π^0) charge exchange reaction on nuclei.

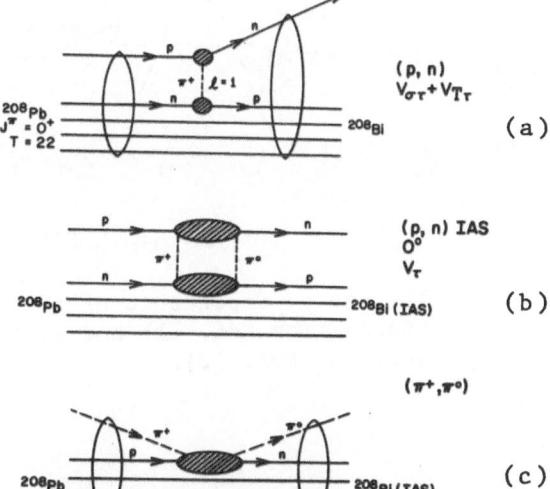

Pion charge exchange on the nucleon has of course been measured for many years. It is instructive to compare $\pi^- p \to \pi^0 n$ with $np \to np$. The energy dependence of the total cross sections is shown in Fig. 3. The curves are dramatically different. At 10 MeV the $\pi^- p \to \pi^0 n$ cross section is 150 times smaller, making pion charge exchange a much weaker process than the (p,n) reaction at Van de Graaff energies. This makes pion charge exchange on nuclei much more tractable to a many-body theory, such as the multiple scattering theory. At 150 MeV, both cross sections have a value near 50 mb. The angular distributions are also quite similar at this energy. They are approximately symmetric about 90^0 as shown in Fig. 4. We note, however, that at 50 MeV the angular distributions are very different. The 0^0 cross section for $\pi^- p \to \pi^0 n$ nearly vanishes ($\simeq .015$ mb/sr). It is this backward peaking of the $\pi^- p \to \pi^0 n$ reaction which led some theorists to predict[13] that the IAS would not stand out at 0^0 or at any other angle in (π^+, π^0) reactions on heavy nuclei.

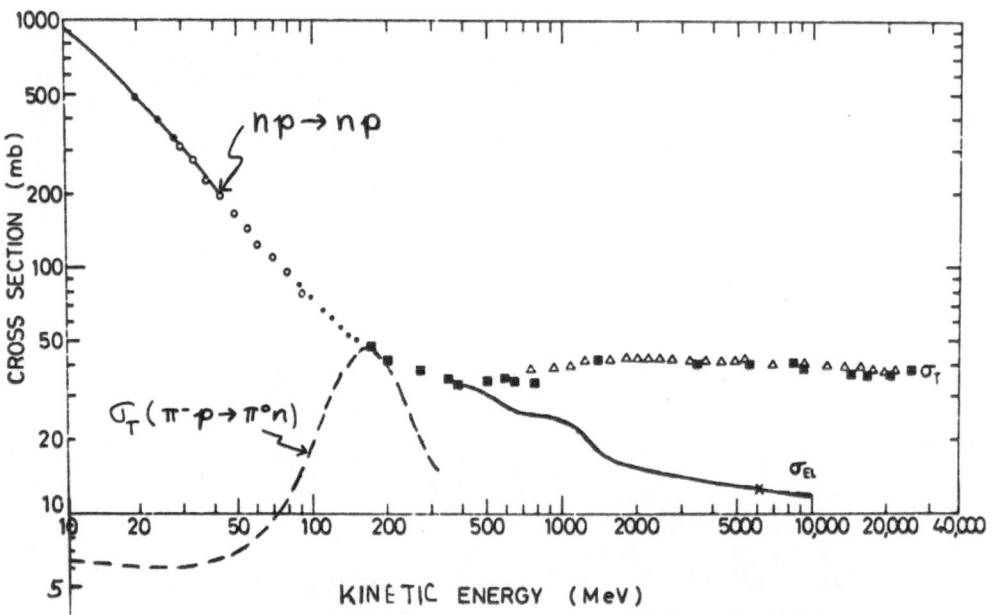

Fig. 3.   The energy dependence of the total n-p cross section
          (taken from Ref. 11) and of the total $\pi^- p \to \pi^0 n$ cross
          section (Ref. 12).

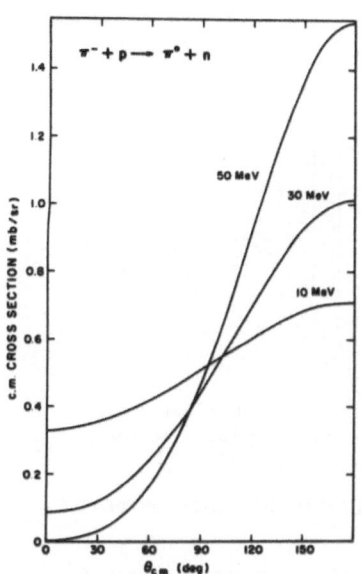
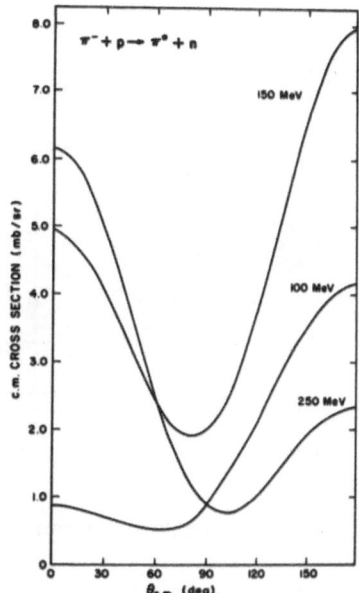

Fig. 4.   The energy dependence of the angular distribution shapes
for the $\pi^-p \to \pi^0 n$ reaction (Ref. 12).

2.   LAMPF π^0 SPECTROMETER

High resolution energy measurements on π^0's produced in (π^+,π^0)
charge exchange reactions on nuclear targets were performed at LAMPF
in November–December 1978.   The data were taken with the newly built
π^0 spectrometer which was developed by a team of scientists[14] from
Los Alamos, Tel-Aviv University and Case Western Reserve Univeristy.
The instrument was built to study pion single charge exchange and
(p,π^0) reactions connecting individual nuclear states.

The principle of operation can be understood with the help of
Fig. 5.   The π^0 has a mean life of 0.8×10^{-16} sec and decays to
two photons with a B.R.$=98.9\%$.   To measure the vector momentum $\vec{p}_{\pi^0}$,
of the π^0, one needs to measure the vectors $\vec{p}_{\gamma 1}$ and $\vec{p}_{\gamma 2}$.   To mini-
mize the propagation of uncertainties in the γ-ray measurements to
the $\vec{p}_{\pi^0}$ determination, it is advantageous to compute the π^0 energy
$W=T_{\pi^0}+m_{\pi^0}$, from the relation

$$W = m_{\pi^0} \sqrt{\frac{2}{(1-\cos\eta)(1-X^2)}} \qquad (2.1)$$

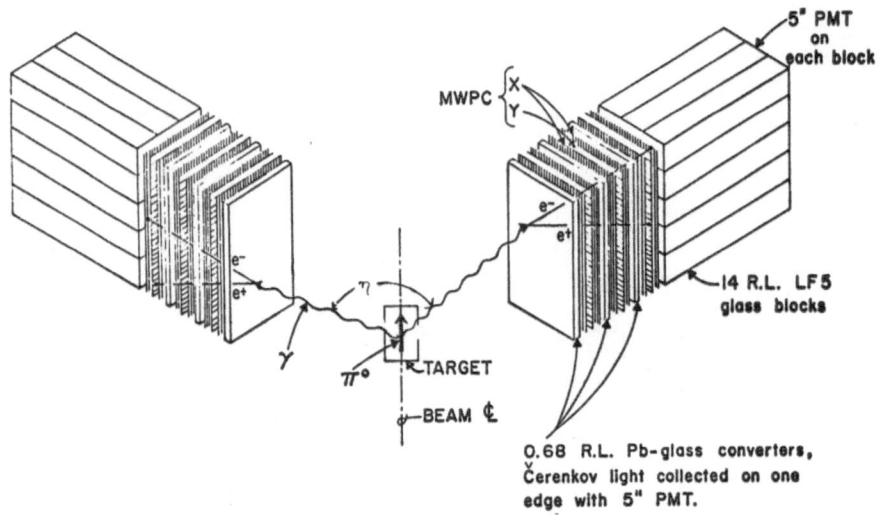

Fig. 5. The detection elements of the LAMPF π^0 spectrometer.

where η is the measured angle between the two γ-rays, and X is the
energy sharing parameter $X=(E_1-E_2)(E_1+E_2)^{-1}$ obtained from the measured γ-ray energies E_1 and E_2. The energy resolution on the π^0
thus depends on the resolution one achieves in the η measurement
and the X measurement. The η is measured by use of thin converters
and MWPC's as shown in Fig.5. The position vectors from the center
of the target to the measured conversion points for each γ-ray determine the opening angle. The conversion points can be measured
to an accuracy of $\simeq$4 mm (FWHM), which translates to $\Delta\eta=0.4\sqrt{2}/100=6mr$
for a target-to-converter distance of 100cm. The finite extent of
the beam-target intersection volume also contributes to $\Delta\eta$, but this
contribution can generally be made relatively small. The relation
between energy uncertainty $\Delta T_{\pi^0}(=\Delta W)$ and $\Delta\eta$ for π^0's of kinetic
energy 100, 200 and 300 MeV is shown in Fig. 6.

The ability to measure X depends on the resolution of the photon detectors. We decided to use Pb glass counters which totally
absorb the electromagnetic showers initiated by 70 to 500 MeV photons. Pb glass has the advantage of giving Cerenkov light pulses
with a short decay time ($\simeq 10^{-10}$ sec) and of being relatively insensitive to neutrons, protons, and heavier charged particles that one
has around accelerators and targets struck by π's. The Cerenkov
light is collected on the back face of the glass blocks with 5" photo
tubes. The light from the Pb glass converters is collected on one
edge and brought to 5" tubes via lucite light guides. The light
from the converters and blocks is added to give an analog pulse
which is linear with photon energy and which has a pulse height
resolution of $\simeq$30% at E_γ=100 MeV. At **higher** energies, the pulse-
height-resolution is dominated by photo electron statistics and it
therefore scales as $(E)^{-\frac{1}{2}}$.

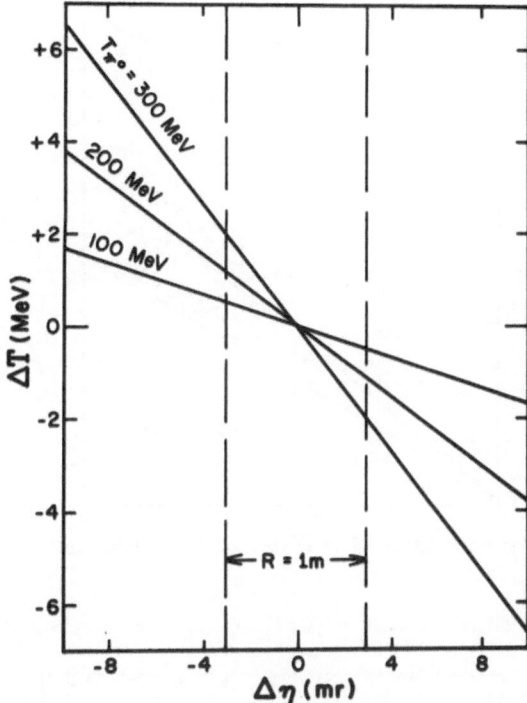

Fig. 6. Plot of π^0 energy uncertainty ΔT_{π^0} vs opening angle
uncertainty $\Delta\eta$ [$=\eta-\eta(\min)$]. The range of η expected
for a typical target-to-converter distance of 1m is
indicated.

The relation between photon resolution and X-resolution is
shown in Fig. 7a; e.g., if photons are detected with 30% resolution
(FWHM) at 100 MeV, the measured X distribution ranges from –0.1 to
+0.1 for π^0's of 100 MeV kinetic energy and decaying with X≈0 (Monte
Carlo simulation). With NaI detectors the X resolution would be
considerably improved.

In calculating W from the measured η and X, the sensitivity
to X is greatly reduced by selecting events centered on X=0 ($E_1=E_2$).
Fig. 7b shows the relation between ΔT and ΔX. One sees that for a
given ΔX interval one gets a minimum ΔT if that interval is centered
on X=0. With a photon resolution of 30%, a value of ΔT=1 MeV can
be achieved by restricting $|X| < 0.1$ for 100 MeV π^0's, and to $|X|$
≤ 0.07 for 300 MeV π^0's. Note that ΔT is independent of T_{π^0} for a
given photon resolution. We also note that for an X resolution of
0.02, which would be obtainable with NaI detectors, one could attain
a π^0 resolution of order 200 keV.

The combined effects of η resolution and X resolution on T_{π^0}
are shown in Fig. 8. The Monte Carlo calculations are for a typical

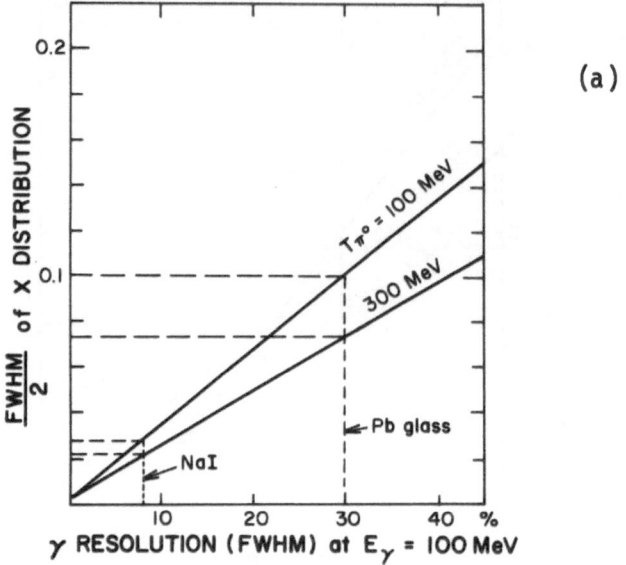

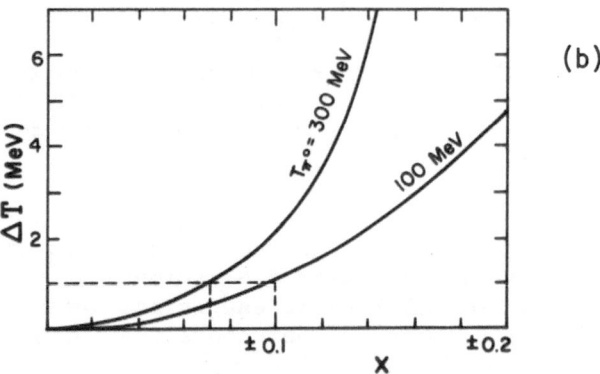

Fig. 7.   (a) The relation between photon-detection resolution
          and X-resolution for π^0's of kinetic energies of
          100 and 300 MeV.
          (b) The relation between the uncertainty in π^0 energy
          (ΔT) and the uncertainty in X.

target-to-converter distance of 1m, a photon resolution of 30%
(FWHM) and a selection on events having a measured $|X| \leq 0.1$.
The rise in $\Delta T(=\Delta E)$ with increasing T_{π^0} is due to the increase in
$\Delta\eta/\eta$ coming from the decreased η at higher T_{π^0}. The 30% photon
resolution contributes a constant $\Delta T = 1.5$ MeV, independent of energy.
In principle one can always achieve 1.5 MeV resolution by moving

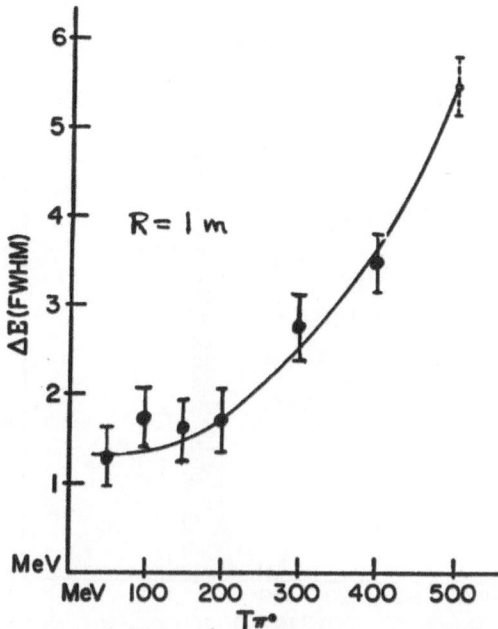

Fig. 8. Expected energy resolution of the LAMPF π^0 spectrometer
as given by Monte Carlo simulation. Note that at the
typical target-to-converter to distance of 1m, the
energy resolution is nearly constant (between 1.5 and
2.0 MeV) up to a π^0 energy of 200 MeV. It is due to
an assumed 30% photon resolution at E_γ=100 MeV.

the photon detectors away from the target until the $\Delta\eta/\eta$ contribution is negligible.

3. RESULTS

3.1 $\pi^-p\rightarrow\pi^0n$

The first goal of our run at LAMPF was to get a high resolution
π^0 spectrum from the $\pi^-p\rightarrow\pi^0n$ reaction using a CH_2 target. One such
spectrum measured at T_π=150 MeV and θ_{π^0}=30^0 is shown in Fig. 9.
The resolution is 2.5 MeV FWHM. This is larger than the 1.5 MeV
we expected from the Monte Carlo calculations. We know that this
difference is in large part due to the poor light collection efficiency we had on the converters, and we are taking steps to improve
this. A resolution of 2.5 MeV is still much better than has been
achieved before, and it is good enough to start the study of analog
state physics with pions.

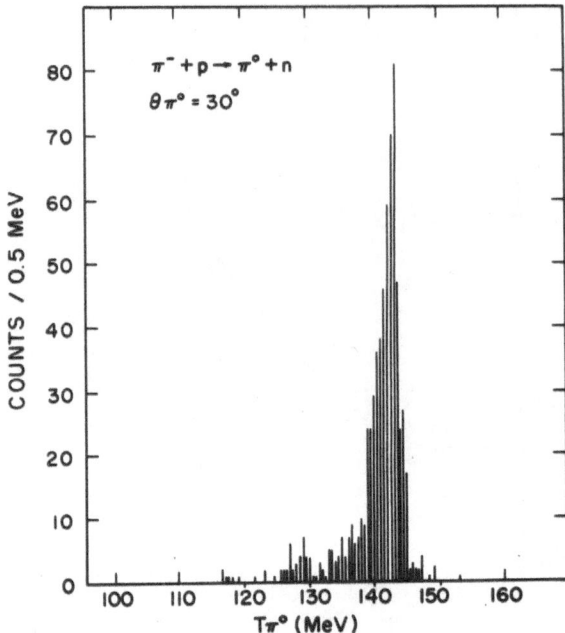

Fig. 9. Spectrum of π^0's from a CH_2 target with a beam energy
T_{π^+} = 150 MeV.

We did not measure cross sections for the $\pi^- p \to \pi^0 n$ reaction.
Rather we used the known cross sections for this reaction to determine effective solid angles of the π^0 spectrometer. These solid
angles were then used to determine cross sections for charge exchange
on the other nuclei measured.

3.2 $^7Li(\pi^+, \pi^0)^7Be$

We measured partial angular distributions for the reactions
$\pi^- + {}^3He \to {}^3H + \pi^0$, $\pi^+ + {}^7Li \to {}^7Be + \pi^0$, and $\pi^+ + {}^{13}C \to {}^{13}N + \pi^0$. The ^3He and ^{13}C
have been discussed [Ref. 15, 16] in some detail at the "Los Alamos
Workshop in Pion Single Charge Exchange" (Jan. 1979) and at the
Houston Conference on "Meson-Nuclear Physics" (March, 1979). I will
concentrate on the ^7Li results here.

A spectrum for ^7Li measured at T_π = 100 MeV is shown in Fig.
10. The range of π^0 scattering angles represented in this spectrum
is 0 to 10^0. In a 0^0 set-up of the π^0 spectrometer one gets this
broad range of scattering angles because the effective solid angle
near 0^0 follows closely the geometric value $d\Omega = 2\pi \sin\theta d\theta$. Thus the π^0
solid angle rises as $\sin\theta$ from 0 to $\sim6^0$; it then drops nearly linearly between 6 and 10^0 due to the spectrometer's geometric apertures.
The mean scattering angles in the "0^0" spectra to be shown are between 5^0 and 7^0, depending on the radius and aperture settings.

The spectrometer axis was always set at 0^0 for these spectra.

The $^7\text{Li}(\pi^+,\pi^0)^7\text{Be}$ spectrum of Fig. 10 looks quite similar to the $^7\text{Li}(p,n)^7\text{Be}$ spectrum at 0^0 and 120 MeV [Ref. 8]. Both spectra show a single peak. However, there are significant differences. On theoretical grounds one expects the (p,n) peak to be a doublet with nearly equal cross sections for the IAS ground states and the $1/2^-$ state at 0.4 MeV. The dominant pieces in the wave functions are[17]

$$|\text{IAS}; 3/2^-,\text{T}=1/2> \simeq \,| \text{ L}=1,\text{ S}=1/2,\text{ J}=3/2> \qquad (3.1)$$

$$|.4 \text{ MeV}; 1/2^-,\text{T}=1/2> \simeq \,| \text{ L}=1,\text{ S}=1/2,\text{ J}=1/2> \qquad (3.2)$$

The excited state is the spin-flip state of the IAS g.s. From (p,n) reaction measurements on ^9Be, one knows that IAS and spin-flip states have cross sections of comparable magnitude at 0^0 and at 120 MeV. Thus it is expected that the peak in the $^7\text{Li}(p,n)^7\text{Be}$ spectrum is really a doublet. Since the π has no spin, the spin-flip state is forbidden at 0^0 in $^7\text{Li}(\pi^+,\pi^0)^7\text{Be}$. The π-N spin-orbit term, $\vec{\ell}_\pi\cdot\vec{\sigma}$, does not contribute at 0^0 (see, e.g., calculations by Sparrow, Ref. 17). Thus the peak in our spectrum represents a single transition.

The energy dependence of the 0^0 cross section is shown in Fig. 11. The sharp rise with energy is in marked contrast to the behavior of the integrated CEX* cross section. This was previously measured[18] by the activation technique, and was found to be nearly constant at $\simeq 1.8$ mb from 90 to 210 MeV. The solid line shows the calculation of Sparrow.[17] It is in reasonable agreement with the data at 70 and 100 MeV, but falls below the 150 MeV and 180 MeV cross sections. The dashed curve represents the lower limit for 0^0 CEX cross section as given by the measured π^- and π^+ total cross sections and related to 0^0 CEX cross sections by use of the optical theorem. The scattering amplitude $f(\theta)$ for elastic and charge-exchange scattering involving π^+, π^0, π^- with a T=1/2 nucleus may be written in terms of an isoscalar amplitude f_0 and an isovector amplitude f_1. Assuming isospin invariance, and treating $f(\theta)$ as an operator in isospin space, we may write

$$f(\theta)=f_0 + f_1 \,\vec{t}_\pi\cdot\vec{T} \qquad (3.3)$$

For a T=1/2 nucleus, we need not retain the isotensor $(\vec{t}\cdot\vec{T})^2$ term, since only two amplitudes (T=1/2, 3/2) are allowed. The amplitudes for π^+, π^-, and (π^+,π^0) CEX are then given, respectively, by

$$f^{(+)}(\theta) = f_0(\theta) \,-\, 1/2\, f_1(\theta) \qquad (3.4)$$

$$f^{(-)}(\theta) = f_0(\theta) \,+\, 1/2\, f_1(\theta) \qquad (3.5)$$

$$f^{CEX}(\theta) = \frac{1}{\sqrt{2}}\, f_1(\theta) = \frac{1}{\sqrt{2}}\, (f^-(\theta)-f^+(\theta)) \qquad (3.6)$$

*Charge exchange (CEX).

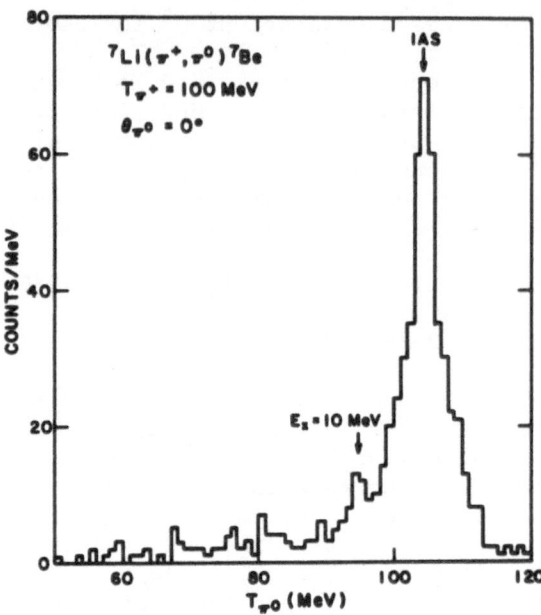

Fig. 10. Spectrum of π^0's from a ^{7}Li target.

The 0^0 CEX cross section is given by

$$\frac{d\sigma^{CEX}}{d\Omega}(0^0) = |f^{CEX}(0)|^2 \geq |Im\ f^{CEX}(0)|^2 \qquad (3.7)$$

The imaginary part of the CEX amplitude at 0 is related to the elastic amplitudes by Eq. (3.6),

$$Im\ f^{CEX}(\theta) = \frac{1}{\sqrt{2}}\ (Im\ f^-(\theta) - Im\ f^+(\theta)) \qquad (3.8)$$

We can now apply the optical theorem to obtain

$$Im\ f^{CEX}(0) = \frac{1}{\sqrt{2}}\frac{k}{4\pi}(\sigma^- - \sigma^+) = \frac{1}{\sqrt{2}}\frac{k}{4\pi}\ \Delta\sigma \qquad (3.9)$$

Inserting this in Eq. (3.7), we get the lower limit on the 0^0 CEX differential cross section

$$\frac{d\sigma^{CEX}}{d\Omega}(0) \geq \frac{k^2}{32\pi^2}\ (\Delta\sigma)^2 = 3.2 \times 10^{-4}(\frac{p\,(MeV/c)}{197})^2\,(\Delta\sigma(mb))^2$$

Using the Coulomb corrected values of $\Delta\sigma$ for ^{7}Li given by Wilkin et al[19], we obtain the dashed curve shown in Fig. 11. Our preliminary measurements fall closely to this curve which indicates that the real part of $f^{CEX}(0)$ must be relatively small.

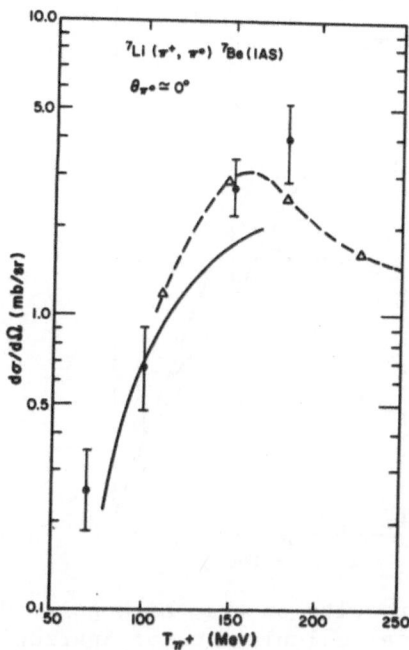

Fig. 11. Energy dependence of the 0° cross section for the
          isobaric analog state transition on ^{7}Li. The solid
          curve represents the calculation of Sparrow[17]. The
          dashed curve is the lower limit for the 0° CEX cross
          sections set by the π^- and π^+ total cross sections
          measured by Wilken et al.[19] (see text).

     The forward angular distribution is shown in Fig. 12. We are
still analyzing the data so the errors on the cross sections are
large. The solid curve is the exponential function indicated in
the figure. It integrates to a σ(CEX) = 1.64 mb. Since the exponential form does not allow for a rise in the differential cross
section at back angles, the 1.64 mb must be viewed as a lower limit.
The activation measurements of Shamai et al.[18], gave ∿1.9 mb which
is consistent with the spectrometer measurements. The calculations
of Sparrow are a factor of ∿2 low compared to the angular distribution of Fig. 12 and compared to the activation measurements near
150 MeV. The same factor of ∿2 appears in the ^{13}C. Thus the present status is that the theoretical calculations underestimate the
CEX cross sections on ^{7}Li and ^{13}C at ∿150 MeV by a factor of ∿2.

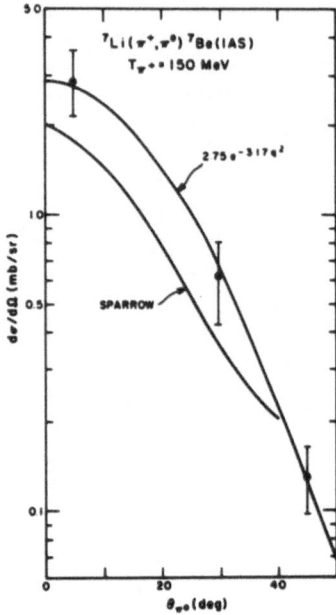

Fig. 12. Preliminary results on the ^{7}Li angular distribution.
The theoretical calculations of Sparrow[17] are shown.
The exponential curve was used to extract a lower limit
for the integrated IAS cross section.

3.3. A(π^+, π^0)A'(IAS)

The discrepancy between experiment and theory on ^{13}C, which
now has existed for a number of years, raised questions about absorption and multiple scattering of π's which led to differing
opinions on whether the IAS would stand out in a heavy nucleus.
Direct measurement of the π^0 spectrum is the only feasible means
of answering this question since the methods of activation, recoil
nucleus detection, and IAS decay proton detection are virtually
impossible for heavy nuclei. Thus it was of great interest when
the first π^0 spectra measured in December 1978 showed the IAS
clearly standing out above the background in the 0^0 spectra of
(π^+,π^0) reactions on ^{90}Zr, ^{120}Sn, and ^{208}Pb targets. The on-line
spectrum for ^{90}Zr is shown in Fig. 13. The analog state, which
occurs at an excitation of 5 MeV in ^{90}Nb is clearly visible. The
spectra for ^{90}Zr and ^{208}Pb are compared in Fig. 14. One sees peaks
at the correct excitation energies (Ex = 15.2 MeV in ^{208}Bi), but
for ^{208}Pb it does not rise above the background as clearly as for
^{90}Zr. These spectra are from the first round of the off-line analysis. We are working on improving the resolution in the hopes of
improving the signal to background ratio.

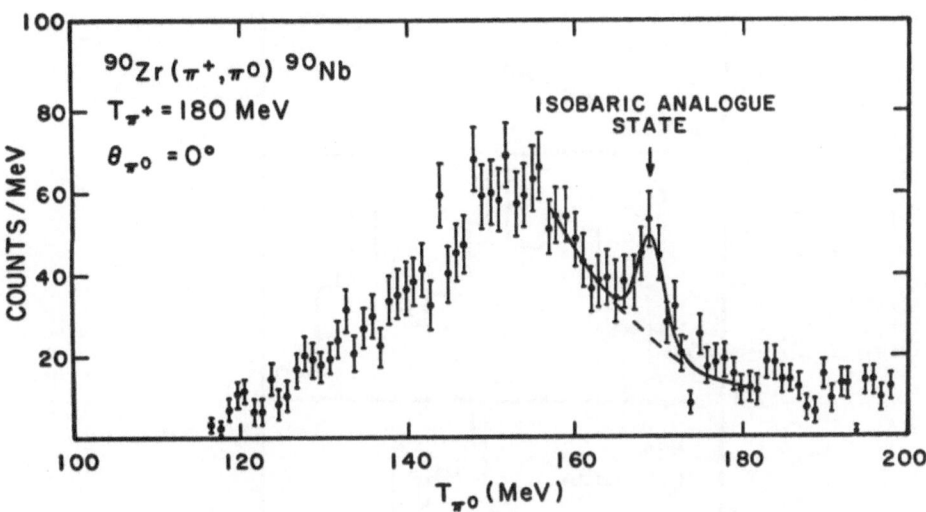

Fig. 13.   Spectrum of π^0's obtained on-line with a ^{90}Zr target.

In contrast to the ^{7}Li(π^+,π^0)^{7}Be spectrum at 100 MeV (Fig. 10), one can see that the continuum excitation is much larger at 180 MeV on heavy nuclei. Since the spectra were measured at a mean scattering angle of 6.5^0, which corresponds to a momentum transfer of $\simeq$30 MeV/c, the continuum is quite likely due to quasi free charge exchange on a loosely bound neutron, i.e., $\pi^+ + A \to \pi^0 + n + (A-1)^*$. We intend to calculate this process using the pole model[20] to see if we can account for the continuum shape and to extract a total CEX cross section. The acceptance of the spectrometer, shown in Fig. 14., cuts off the spectrum at $\simeq$130 MeV, so a model is needed to extrapolate to lower energies. The continuum spectra appear to peak at an energy corresponding to at 20-25 MeV excitation in the residual nuclei.

The mass dependence of the forward (π^+,π^0) cross sections is shown in Fig. 15. A similar curve was measured at 180 MeV. From the analysis of these two curves we intend to extract the mass and energy dependence of the $\vec{t}_\pi \cdot \vec{T}$ term of π-nucleus interactions. This will be compared to the results of first-order optical potential calculations such as those performed by Miller and Spencer[21] using Kisslinger and local Laplacian potentials.

For the present, we can extract the general features of the A-dependence using the semi-classical strong-absorption model of M. Johnson[22]. He gives the following expression for the $\pi^+ A \to \pi^0 A$ (IAS) cross section

$$\frac{d\sigma}{d\Omega}(q) = \frac{(ka)^2}{4\pi} \frac{\sigma_{tot}}{N-Z}\left(\frac{\Delta\rho(\bar{r})}{\rho(\bar{r})}\right)^2 J_0^2 \ (q\bar{r}) \qquad (3.10)$$

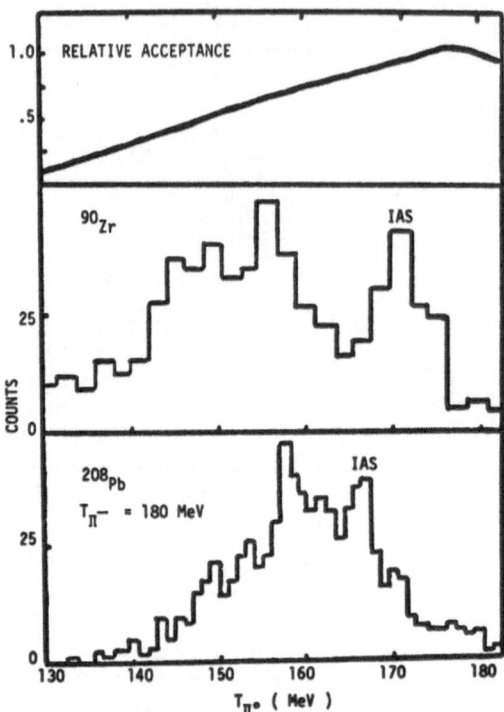

Fig. 14. Spectra measured at 0^0 for (π^+,π^0) reactions on ^{90}Zr
and ^{208}Pb at T_π=180 MeV. The relative acceptance
of the π^0 spectrometer is shown in the top panel.

where k is the pion momentum, a is the nuclear diffuseness parameter, $\sigma_{tot} = [\sigma(\pi^-)+\sigma(\pi^+)]/2$ is the average of the π^- and π^+ total cross sections on nucleus A, q is the momentum transfer, $\rho(r) = \rho_n(r) + \rho_p(r)$ is the nucleon density distribution and $\Delta\rho(r) = \rho_n(r) - \rho_p(r)$ is the difference between neutron and proton density distributions. The effective radius, $\bar{r}$, at which the densities should be evaluated is given by

$$\sigma_{tot} = 2\pi(\bar{r})^2 \qquad (3.11)$$

If we assume that the neutron and proton density functions have the same shape, i.e.,

$$\rho_n = N\rho_o, \quad \rho_p = Z\rho_o, \quad \text{and} \quad \rho = (N+Z)\rho_o$$

$$\frac{\Delta\rho}{\rho} = \frac{\rho_n - \rho_p}{\rho} = \frac{N-Z}{A} \quad \text{(independent of } \bar{r}) \qquad (3.12)$$

and also make the usual assumption that,

$$\sigma_{tot} \; \alpha \; A^{2/3} \tag{3.13}$$

the mass- and energy-dependence of the 0^0 cross section is given by

$$\frac{d\sigma}{d\Omega}^{CEX}(0^0) = (Const)(p_\pi^2)(\frac{N-Z}{A^{4/3}}) \tag{3.14}$$

The curve shown in Fig. 15 represents this formula with the constant fixed by the ^{13}C cross section. We see that the general trend of the data is described quite well. The cross section has a minimum near $A = 60$ where $(N-Z)/A^{4/3}$ has a minimum. Beyond $A = 120$, $(N-Z)/A^{4/3}$ is nearly constant. The success of the semi-classical model indicates that for $T_\pi \lesssim 100$ MeV, pion charge exchange scattering is much like black-disk scattering with differential sensitivity to neutrons.

The values of $\bar{r}$ deduced from $\sigma_{Tot}/2\pi$ are in the tails (10-20%) of the density distributions for π energies in the P_{33} resonance region. Thus within the framework of the strong absorption model, pion CEX at 0^0 is quite sensitive to the existence of a neutron halo. The data will be analyzed for such effects.

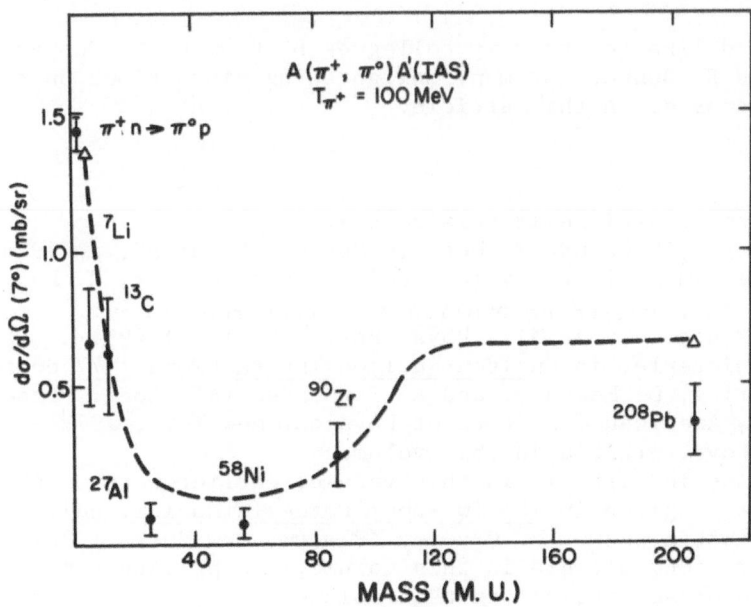

Fig. 15. The A-dependence of the forward angle (π^+,π^0) differential cross section to the isobaric analog state. The dashed curve represents the semi-classical model of Johnson[22].

4. SUMMARY

The close relation between the pion and nucleon charge exchange processes was indicated. The principle of operation and the performance parameters of the newly built π^0 spectrometer at LAMPF have been described. Some of the first results obtained with this instrument were presented. The isobaric analog state was observed in (π^+, π^0) reactions on nuclei ranging from Hydrogen to ^{208}Pb.

For light nuclei, ^7Li and ^{13}C, the new data in the 3-3 resonance region are higher than predicted by theory. Thus this long-standing puzzle persist. The measured angular distributions provide new checks on the theory, but the consequences are not fully evaluated at this time.

The mass dependence at 100- and 180 MeV of the 0^0 cross sections follow the trends predicated by strong-absorptions models. The most significant sensitivity is to $\rho_n - \rho_p$ in the tails of the nuclear densities. It appears that detailed analysis of the (π^+, π^0) cross sections will show this reaction to be a very sensitive indicater of neutron halos.

ACKNOWLEDGMENTS

I would like to thank my collegues D. Bowman, M. Cooper, M. Johnson and B. Bonner for many enlightening discussions on the matters discussed in this article.

REFERENCES

1. P. Signell, article in this volume.
2. M. Johnson, "Mesonic Effects in Nuclear Physics", in 1978 Summer Lectures in Nuclear Physics, edited by M. Cooper, R. Redwine, and R. Mischke (to be published as LASL report).
3. T. C. Montgomery et al., Phys. Rev. 16C (1977) 499.
4. B. D. Dieterle, in Nucleon-Nucleon Interactions-1977 edited by H. Fearing, D. Measday, and A. Strathdee (AIP Conf. Proceedings No. 41, American Institute of Physics, New York, 1978) p. 35.
5. W. G. Love, article in this volume.
6. S. M. Austin, article in this volume; earlier work on the same subject is given in The Two-Body Force in Nuclei, edited by S. M. Austin and G. M. Crawley (Plenum, New York, 1972) p. 285.
7. F. Petrovich, article in this volume, and private communication.
8. C. D. Goodman, article in this volume.
9. P. T. Debevec, article in this volume.
10. B. S. Bhakar, Univ. of Manitoba preprint (1979).
11. W. O. Lock and D. F. Measday, Intermediate Energy Nuclear Physics (Methuen and Co. Ltd, London, 1970) p. 183.
12. From phase-shift analysis of D. Dodder, private communication (1979).

13. W. R. Gibbs et al., Phys. Rev. Lett. 36 (1976) 85.

14. The collaboration includes: Los Alamos Scientific Laboratory;
     H. Baer, J. D. Bowman, M. Cooper, F. Cverna, R. H. Heffner,
     C. M. Hoffman, N.S.P. King: Case Western Reserve University;
     P. Bevington, E. Winkelmann, H. Baer (on leave): Tel Aviv
     University; J. Alster, A. Doran, S. Gilad, M. Moinester;
     Visitors: J. Piffaretti (Neuchatel).

15. M. D. Cooper in Proceedings of the "Los Alamos Workshop on
     Pion Single Charge Exchange", Jan. 1979, edited by H. Baer,
     D. Bowman, M. Johnson (to be published as LASL report); also
     in Proceedings of the "Second Conf. on Meson-Nuclear Physics",
     Houston, March 1979, edited by E. Hungerford (to be published
     by the American Institute of Physics, New York).

16. J. Alster in Proceedings of Los Alamos Workshop, ibid.

17. D. A. Sparrow, Nucl. Phys. A276 (1977) 365.

18. Y. Shamai et al., Phys. Rev. Lett. 36 (1976) 82; and J. Alster
     and J. Warszawski, Physics Reports (in press).

19. C. Wilkin et al., Nucl. Phys. B62 (1973) 61.

20. L. G. Dakhno and Yu. D. Prokoshkin, Sov. Jour. Phys. 7 (1968)
     351.

21. G. A. Miller and J. E. Spencer, Annals of Physics 100 (1976)
     562.

22. M. Johnson, in "Los Alamos Workshop on Pion Single Charge
     Exchange", ibid.

DISCUSSION

Madey: What factor is limiting the resolution of the pi-
zero spectrometer?

Baer: The energy resolution in the photon detection. In
our Nov.-Dec. (1978) run, we had $\sim$ 30% (FWHM) at E_γ = 100 MeV
when one converter was used, but had $\lesssim$ 35% when the full three
converters were used. We expect to get 1.5-2.0 MeV π° resolution
when we achieve 30% photon resolutions with three converters.
With NaI detectors, it looks as though it would be possible to
get π° resolutions of the order of 0.2 MeV.

K^+-NUCLEUS INELASTIC AND CHARGE EXCHANGE REACTIONS[†]

Stephen R. Cotanch

Department of Physics
North Carolina State University
Raleigh, North Carolina   27607

INTRODUCTION

The novel features of kaons, combined with state-of-the-art advances in experimental capabilities, has stimulated growing interest and research activity (1-9) in kaon-nucleus phenomena. Although "kaon factories" have not yet emerged, "kaon shops", such as the Brookhaven AGS accelerator, are producing data (9) and several new kaon facilities have been proposed. Measurements (9) have been recently performed of both elastic and inelastic (low-lying final states have been resolved) $K^{\pm}$-nucleus cross sections from ^{12}C and future experiments are planned.

The present theoretical talk anticipates the next generation of K^+-nucleus measurements: inelastic $(K^+, K^{+\prime})$ (to both low and high-lying excited states) and charge exchange (K^+, K^0). Specifically this talk focuses on the selective excitation aspects of the K^+ meson and the complementary relation between this probe and other projectiles (e, p, and π). Of special interest is to study how the nucleus responds (nuclear response) to the transfer of energy, angular momentum, and momentum and how, using both macroscopic and microscopic models, this knowledge can provide improved nuclear structure information. The arguments, ideas, and formalism developed in this discussion parallel previous studies involving electron (10), proton (11), pion (12,13), and K^- (7) inelastic excitation and these references should be consulted for additional details.

[†]Supported in part by D.O.E.

K^+ MESON PROPERTIES

The distinctive properties[†] of the K^+ meson and K^+-nucleon (KN) system previously documented (1), are briefly summarized. The +1 strangeness signature of the K^+,K^0 doublet, coupled with strangeness conservation, restricts the KN system, at energies below pion production, to single channel scattering (weak absorption). Further, within the framework of the quark model the KN system must be described by a minimum of 5 quarks and therefore, unlike the pion and $\bar{K}(K^-,\bar{K}^0)$, cannot couple to channels labeled by 3 quark quantum numbers. Resonances in the KN system are therefore classified exotic (by definition) and no resonances have yet to be convincely observed (Z^* at about 1800 MeV is still unresolved). With the aid of quark duality diagrams it is then possible to understand why the KN interaction is weaker than either $\bar{K}N$ or πN interactions. Being one of the weakest interacting hadrons the kaon is potentially a good nuclear probe. For kaon lab energies below 400 MeV ($P_{Lab} \leq$ 800 MeV/c) the dominant components of the KN interaction are the $I = 1$ (total kaon-nuclear isospin), $\ell = 0$ (s-waves), $j = 1/2$ contribution followed by the $I = 0$, $\ell = 1$ (p wave), $j = 1/2$ part (1). As discussed shortly, this spin-isospin dependence combined with weak absorption leads to selective nuclear excitation of natural parity high spin states involving $\Delta T = 0$ (T = nucleus total isospin) transitions.

K^+ NUCLEUS INELASTIC AND CHARGE EXCHANGE REACTIONS

Adopting the conventional machinery for describing meson induced reactions, the non-elastic T-matrix, T_{fi}, connecting initial and final states is represented by the first-order multiple scattering expression

$$T_{fi} = A<\chi_f\Phi_f|\tau|\phi_i\chi_i>. \tag{1}$$

Here A is the target nucleon number, τ the many-body KN transition operator, and $\chi_i(\chi_f),\Phi_i(\Phi_f)$ are the initial (final) kaon (distorted) and nuclear wavefunctions, respectively. The familiar DWIA expression is obtained by invoking the impulse approximation in which τ is replaced by the free KN transition operator t. In either coordinate or momentum space t can be decomposed into elementary partial wave contributions

$$t = \sum_{\ell j I} P(I)\ P(\ell,j)\ t_{\ell j}^I, \tag{2}$$

[†]$\bar{K}(K^+,K^0)$ properties: mass 496 MeV, spin 0, isospin 1/2, strangeness +1.

where $P(I)$ and $P(\ell,j)$ are projection operators for total KN isospin $\vec{I} = \vec{I}(K) + \vec{I}(N)$, $(I = 0$ or $1)$ and angular momentum $\vec{j} = \vec{\ell} + s$, $(j = \ell \pm 1/2)$. Specifically, for kaons $P(I)$ takes the form

$$P(I) = \begin{cases} \frac{1}{4}[1-4\vec{I}(K)\cdot\vec{I}(N)], & I = 0 \\ \frac{1}{4}[3+4\vec{I}(K)\cdot\vec{I}(N)], & I = 1 \end{cases} \tag{3}$$

while for pions $(I(\pi)=1)$ $P(I)$ becomes

$$P(I) = \begin{cases} \frac{1}{3}[1-2\vec{I}(\pi)\cdot\vec{I}(N)], & I = 1/2 \\ \frac{1}{3}[2+2\vec{I}(\pi)\cdot\vec{I}(N)], & I = 3/2 \end{cases} \tag{4}$$

For both kaons and pions $P(\ell,j)$ is

$$P(\ell,j) = \begin{cases} \dfrac{1 - 2\vec{s}\cdot\vec{\ell}}{\ell + 1}, & j = \ell - 1/2 \\ \dfrac{\ell + 1 + 2\vec{s}\cdot\vec{\ell}}{\ell + 1}, & j = \ell + 1/2 \end{cases} \tag{5}$$

The T matrix, eq. (1), is now simplified through further approximation and interpreted to provide qualitative understanding of the selective aspects present in K induced processes. Because t is a one-body operator in the space of the A nucleons, the many body wave functions Φ_i, Φ_f can only differ by quantum numbers for a single nucleon. Let α denote this set of quantum numbers which include the nucleon spin and isospin projections. Further, define the initial and final kaon isospin projections by m and m´ respectively and let $\gamma = (\alpha,m)$ denote the combined set of quantum number. Using spin-isospin completeness relations the T-matrix for K-nucleus inelastic scattering (m´ = m) or charge exchange (m´ ≠ m) becomes

$$T_{fi} = T_{m´m} = A \sum_{\alpha\,\alpha´} \langle \chi_{m´}\Phi_{\alpha´}|t_{\gamma´\gamma}|\Phi_\alpha\chi_m\rangle \tag{6}$$

where the matrix elements of the KN transition operator are defined by

$$t_{\gamma'\gamma} = <\alpha'm'|t|\alpha m> \tag{7}$$

If plane waves are used for $\chi_{m'}, \chi_m$ and the factorization approximation for the t matrix, eq. (6) reduces to

$$T_{m'm} = \sum_{\alpha \alpha'} t_{\gamma'\gamma} F_{\alpha'\alpha}(\vec{q}). \tag{8}$$

$\vec{q}$ being the momentum transferred to the nucleus. The quantity $F_{\alpha'\alpha}(\vec{q})$ is the transition form factor governing the nuclear response

$$F_{\alpha'\alpha}(\vec{q}) = A <\Phi_{\alpha'}|e^{-i\vec{q}\cdot\vec{r}}|\Phi_{\alpha}>, \tag{9}$$

and is identical to the longitudinal form factor which appears in electron scattering. Integrating over all possible nuclear coordinates produces the simple expression

$$F_{\alpha'\alpha}(\vec{q}) = \int e^{-i\vec{q}\cdot\vec{r}} \rho_{\alpha'\alpha}(\vec{r}) \, d\vec{r}, \tag{10}$$

where $\rho_{\alpha'\alpha}(\vec{r})$ is the nuclear transition density containing all structure information.

Microscopic Model

To proceed further, specific nuclear structure models are necessary. In the simplest microscopic model, the excited state is viewed as a coherent superposition of 1 particle-1 hole (1p-1h) states. Consider K^+ excitation of closed shell J,S,T = 0 nuclei to final states J, S = 0 or 1, T = 0 or 1. Here S,T, and J are the spin, isospin and total angular momentum of the particle-hole pair respectively. The transition density is then given in terms of single particle particle-hole wavefunctions $\phi_{a\alpha}(\vec{r})$

$$\rho_{\alpha'\alpha}(\vec{r}) = \sum_{a,a'} c_{aa'} \phi_{a'\alpha'}(\vec{r})\phi_{a\alpha}(\vec{r}). \tag{11}$$

The mixing coefficients, $c_{aa'}$, are obtained from diagonalizing the residual interaction in a truncated model space and the sum is over all particle-hole basis vectors spanning this space.

Within the framework of the above approximations qualitative predictions can now be made concerning the selectivity of K-nucleus reactions. Recalling that the dominant partial wave KN amplitudes are the S_{11} ($\ell=0, I=1, 2j=1$) and then P_{11} ($\ell=1, I=0, 2j=1$), eqs. (2), (3), and (5) yield the approximate transition operator

$$t \cong \frac{1}{4} S_{11}[3 + 4\vec{I}(K)\cdot\vec{I}(N)] +$$

$$\frac{P_{11}}{3} [1-4\vec{I}(K)\cdot\vec{I}(N)][1-2\vec{s}\cdot\vec{\ell}] \tag{12}$$

or to lowest order

$$t \cong \frac{S_{11}}{4} [3 + 4\vec{I}(K)\cdot\vec{I}(N)]. \tag{13}$$

Hence the isoscalar component of the interaction is effectively three times larger than the isovector term ($<4\vec{I}(K)\cdot\vec{I}(N)> \leq 1$). Apart from other considerations, kaons should therefore preferrentially excite final states involving $\Delta T = 0$ transitions by almost an order of magnitude over $\Delta T = 1$ transitions (14). Of course the other amplitudes will effect this prediction as well as the degree of collectivety which is contained in the structure wave functions. For pions, which are dominated by the (3,3) resonance, the same argument using eq. (4) would also favor $\Delta T = 0$ transition, but now by only 4 to 1 in the cross section. Calculations (12) for pion inelastic scattering are qualitatively consistent with these predictions. Further, from eq. (5) and the p wave nature of the pion (3,3) resonance, the possibility for spin-flip, $\Delta S = 1$, transitions to unnatural parity states exist and has been numerically predicted to strongly compete. In contrast, the dominate s wave nature of the KN interaction suggest only $\Delta S = 0$, (no spin-flip) natural parity states should be selectively excited in kaon inelastic and charge exchange reactions at kaon lab energies below 250 MeV (1).

The selective aspects of kaon non-elastic scattering are also different from electron and proton excitation. In electron scattering the transverse form factor dominates at large angles (large $\vec{q}$ transfer). For this term the electromagnetic excitation interaction predominantly involves the nuclear isovector ($\Delta T = 1$) magnetic moment (spin currents, $\Delta S = 1$) (10). Accordingly, transitions involving $\Delta T = 1$, $\Delta S = 1$ (unnatural parity states for closed shell nuclei) are favored. The proton, because of both tensor and two-body spin-orbit interaction, has a more complicated excitation mechanism which depends upon momentum transfer: at low q, $\Delta S = 0$

(natural parity), $\Delta T = 1$ states are excited while at high q, $\Delta S = 1$ (unnatural parity), $\Delta T = 1$ states are preferentially populated. Finally, if plane waves do provide a reasonable description of K-nucleus relative motion then high spin states should also dominate the excitation spectrum for large momentum transfer. This follows from eq. (9) which, when decomposed into partial waves, involves the spherical bessel function j_J (qr) that peaks for $\tilde{J} = qr$.

Macroscopic Model

Because $\Delta T = 0$, $\Delta S = 0$ (natural parity) states are expected to be selectively excited by kaons it is appropriate to consider a collective macroscopic model, especially for low-lying excited states. Following the conventional treatment the transition density is deformed ($\delta\rho$) and collective wave functions are used for initial and final states. In the simplest case (13) the transition density is diagonal in spin-isospin variables

$$\rho_{\alpha'\alpha}(\vec{r}) = \delta\rho = \delta_{\alpha'\alpha} \frac{\beta_J}{\sqrt{2J+1}} \frac{\partial\rho_\alpha(r)}{\partial r} Y_J^M(\hat{r}),$$ (14)

$\rho_\alpha(r)$ being the undeformed (proton or neutron) density and β_J the deformation parameter for state J. From eqs. (8), (14) it is possible to express both the elastic U (optical) and inelastic $\hat{U}$ K-nucleus interactions ($U = t\rho$ and $\hat{U} = t\delta\rho$) for $Z \neq \mathcal{N}$ nuclei ($T \neq 0$)

$$U = [t_{Kp} \frac{Z}{A} + t_{Kn} \frac{\mathcal{N}}{A}] \rho(r),$$ (15)

$$\hat{U} = [t_{Kp} \frac{Z}{A} + t_{Kn} \frac{\mathcal{N}}{A}] \delta\rho(r).$$ (16)

Here t_{KN} is the appropriate spin-isospin averaged KN t-matrix and for convenience the neutron and proton densities are assumed proportional. Eq. (15) can be expressed in terms of central U_0 and symmetry U_1 potentials

$$U = U_0 - \frac{(\mathcal{N}-Z)}{A} U_1,$$ (17)

$$U_0 = \frac{(t_{Kp} + t_{Kn})}{2} \rho, \quad U_1 = \frac{(t_{Kp} - t_{Kn})}{2} \rho,$$ (18)

or more succinctly in Lane-model form

$$U = U_0 + 4 \frac{\vec{I}(K) \cdot \vec{T}}{A} U_1. \tag{19}$$

Eq. (19) also follows rigorously from eq. (1) using the projection theorem for vector operators and restricting the target and excited state to the same T. Consequently, the simple Lane model can be extended to describe analogue transitions induced by kaons. The relative real V_0, V_1 and imaginary W_0, W_1 strengths can be crudely estimated using numerical values for the KN amplitudes[1,15]. This yields

$$\frac{V_1}{V_0} \simeq \frac{1}{4} \, , \, \frac{W_1}{W_0} \simeq 1/3$$

for kaon lab energies up to 200 MeV. At larger energies the ratios become smaller. Comparing to the real and imaginary ratios, -1/2 and 1/2 respectively for proton projectiles suggest that analogue states should not be as strongly dominating in kaon charge exchange as they are in the (p,n) reaction. Also notice that because V_0 is repulsive for kaons the sign of the symmetry potential is the same for both kaons and nucleons.

Returning to kaon inelastic scattering, Figs. 1 and 2 illustrate macroscopic, DWIA calculations for exciting T = 0 states in ^{16}O. (see ref. 8 for details of the calculation). Figure 1 compares plane wave (PWIA) with distorted wave calculations to assess the importance of distortions for two different sets of KN amplitudes (Martin (15) and BGRT (16)). Distortion effects are approximately constant as a function of energy and are less than 50%. Interestingly, the cross section uncertainties introduced by different KN amplitudes (different distortions) is of the same order as the effect of no distortions at all. In Fig. 2, differential cross sections for exciting high spin natural parity states $(2^+, 3^-, 4^+, 5^-)$ are presented as a function of energy and momentum transfer. As predicted by the above plane wave argument based on weak absorption, high spin states are selectively excited at high momentum transfer, low spin states at low momentum transfer.

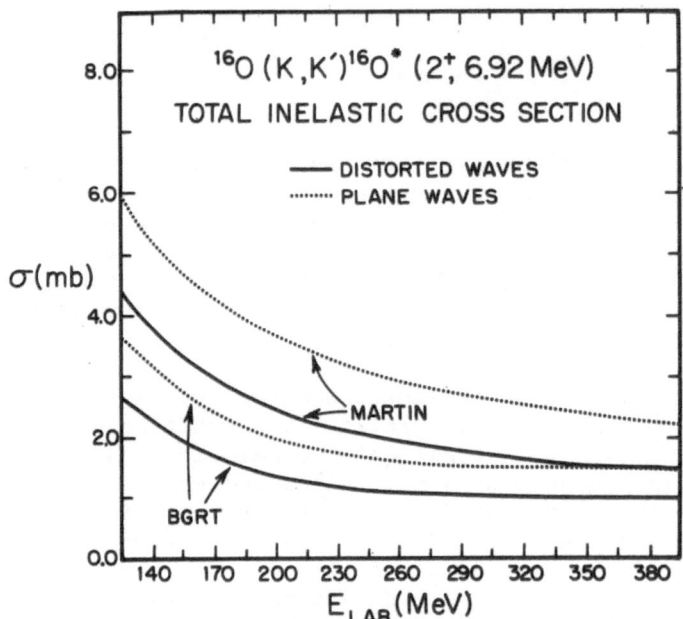

Fig. 1. Kaon-nucleus distortion effects. Comparison of DWIA
         (solid curve) and PWIA (dotted curve) total inelastic
         cross sections using Martin and BGRT kaon-nucleon
         amplitudes. See ref. 6 for further details.

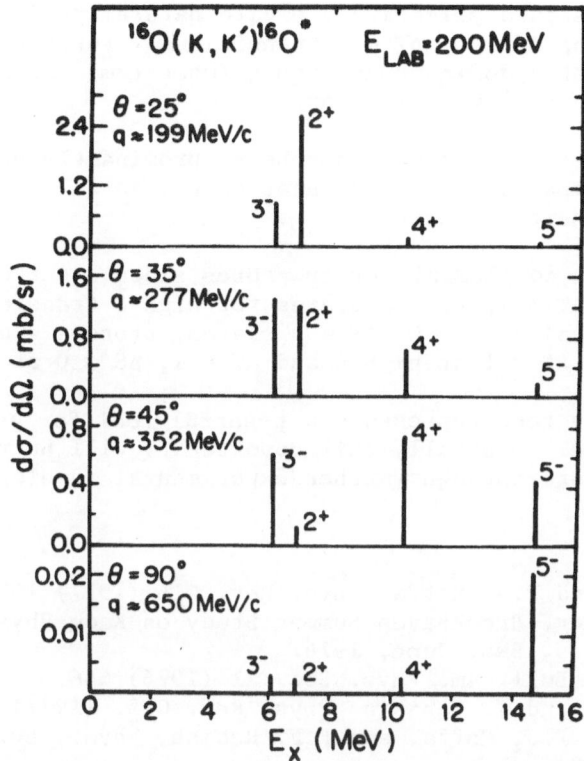

Fig. 2.   Selective excitation of high spin states. Inelastic
          differential cross section as a function of energy,
          momentum, and angular momentum transfer.

CONCLUSION

    This talk has focused upon K⁺-nucleus inelastic and charge
exchange reactions with emphasis upon selective excitation and
comparison to other nuclear probes. Summarizing, the main points
are:

    1. The weakly interacting nature of the kaon facilitates
       the decoupling between reaction mechanisms and nuclear
       structure.

2. The K^+ should selectively excite natural parity states involving $\Delta S = 0$, $\Delta T = 0$ transitions. Further, high spin states dominate at high $\vec{q}$ (momentum) transfer, low spin at low $\vec{q}$ transfer.

3. Analogue states should not be as prominantly excited, when compared to elastic scattering, in (K^+, K^0) as in (p,n) charge exchange.

4. The kaon complements other probes since electrons excite $\Delta T = 1$, $\Delta S = 1$ states for high $\vec{q}$ transfer, pions excite $\Delta T = 0$, $\Delta S = 1$ states, protons excite $\Delta T = 1$, $\Delta S = 1$ at high $\vec{q}$ and $\Delta T = 1$, $\Delta S = 0$ at low $\vec{q}$.

Clearly, the kaon represents a powerful tool for studying the nuclear response. Hopefully this opportunity will be realized in the near future through further experimental developments.

REFERENCES

1. C.B. Dover and P.J. Moffa, Phys. Rev. C16 (1977) 1087; see also C.B. Dover, Brookhaven Summer Study on Kaon Physics and Facilities, BNL, June, 1976.
2. G.E. Walker, Bull. Am. Phys. Soc. 21 (1976) 646.
3. S.R. Cotanch and F. Tabakin, Phys. Rev. C15, (1977) 1379.
4. R.D. Koshel, P.J. Moffa, and E.F. Redish, Phys. Rev. Lett. 39 (1977) 1319.
5. Y. Alexander and P.J. Moffa, Phys. Rev. C17 (1978) 676.
6. S.R. Cotanch, Phys. Rev. C18 (1978) 1941.
7. C.B. Dover and G.E. Walker, Bull. Am. Phys. Soc. 23 (1978) 554 and Phys. Rev. C10 (in press).
8. S.R. Cotanch, Nucl. Phys. A308 (1978) 253.
9. E.V. Hungerford, private communication.
10. T. de Forest and J.D. Walecka, Adv. in Phys. 15 (1966) 1.
11. P.J. Moffa and G.E. Walker, Nucl. Phys. A222 (1974) 140.
12. M.K. Gupta and G.E. Walker, Nucl. Phys. A256 (1976) 444.
13. T.S.H. Lee and F. Tabakin, Nucl. Phys. A226 (1974) 253.
14. G.E. Walker, private communication.
15. B.R. Martin, Nucl. Phys. B94 (1975) 413.
16. G. Giacomelli, et al., Nucl. Phys. B71 (1974) 138.

EFFECTIVE INTERACTIONS AND REACTION DYNAMICS[*]

J. R. Comfort

Department of Physics and Astronomy
University of Pittsburgh
Pittsburgh, Pennsylvania 15260

ABSTRACT

Consideration is given to the effects of strong coupling of the nucleon elastic-scattering channel with other reaction channels. Pickup channels are found to be much more important than inelastic couplings in a coupled-channels environment. For high energies, however, the conventional optical-model potential simulates the coupling effects rather well. At lower energies ($E \lesssim 50$ MeV), coupled-channels effects can be quite significant and the problem is made worse by difficulties regarding the treatment of intermediate channels.

INTRODUCTION

The testing of theoretically constructed effective interactions in nuclear reactions, or their extraction from cross-section data, requires a knowledge of the reaction dynamics. Calculations of nuclear-reaction cross sections rely heavily on the nuclear optical-model description of elastic scattering for the wave functions in the incident and outgoing channels and the use of the distorted-wave Born approximation (DWBA). Although it is possible to construct optical potentials from the effective interactions themselves,[1,2] phenomenological potentials still provide the best fits to the data. The effective interaction then enters the calculation of inelastic-scattering or charge-exchange reactions only in the construction of the form factor.

---

[*]Work supported in part by the National Science Foundation.

Numerous developments in recent years, both experimental and theoretical, have clearly shown the need for higher-order mechanisms in the reaction dynamics. Two-step processes can contribute in significant ways to both the cross sections and the shapes of the angular distributions of states of interest. Furthermore, in the common approach of coupled reaction channels (CRC), the amplitudes resulting from the back couplings to the elastic channel may destroy the fit to the elastic-scattering data given by the phenomenological optical potentials. Compensations for this must then be devised. Even though these considerations may not have much influence on the interpretation of particular data, each needs to be examined before firm conclusions are reached.

I wish to focus some attention especially on the latter issue and to look at some of its consequences. Mention has already been made of recent $^{12}C(p,p')^{12}C$ data at 122 MeV and the general success of an analysis made in terms of a realistic effective interaction.[2] Unfortunately ^{12}C is often considered to be a somewhat intractable nucleus for realistic studies, since its large deformation and strong coupling to the 2^+ state at 4.44 MeV are difficult to handle in reaction calculations. Thus the reaction dynamics must be given special attention. We shall also examine some nucleon elastic-scattering data from the Mo isotopes, for which the Lane model provides a description of a macroscopic effective interaction.

ELASTIC SCATTERING AND COUPLED CHANNELS

Proton interactions with nuclei may lead to both (p,p') and (p,d) reactions and these often have comparable cross sections. Since couplings to collective excitations in (p,p') reactions are known to affect the results of both elastic-scattering and reaction calculations, there is good reason to suppose that couplings to pickup channels could also produce significant effects. Indeed, phenomenological studies by Mackintosh[3] have demonstrated some effects quite clearly. The analysis of Coulter and Satchler[4] has elucidated many of the features of such couplings.

To be specific, we consider proton interactions with ^{12}C. In second order DWBA, it is possible to compute cross sections in the elastic channel that result from the back-coupling amplitudes for various two-step paths. Such cross sections are shown in Fig. 1 for (p,p',p) and (p,d,p) two-step processes proceeding through, respectively, the first 2^+ state of ^{12}C and the ground state of ^{11}C. The (p,p',p) cross section had to be multiplied by 100 for display purposes. Thus it is clear that the pickup-stripping process may be far more important than the inelastic couplings implying, perhaps, that ^{12}C is even more intractable than was pre-

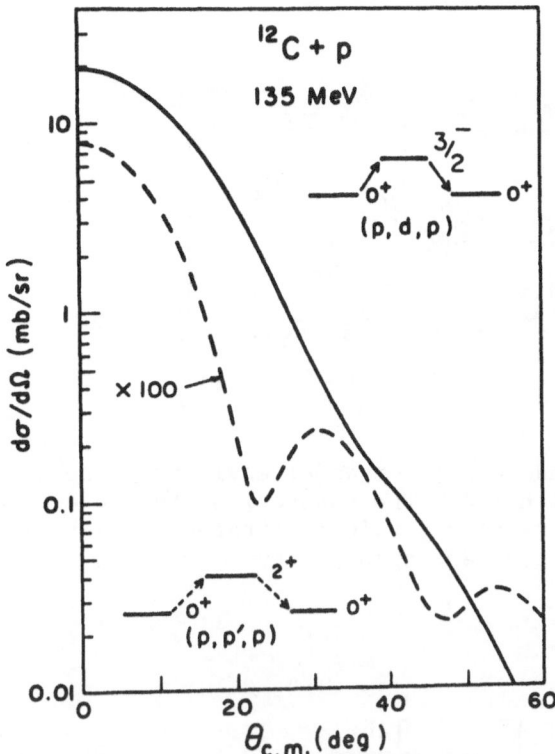

Fig. 1. Cross sections corresponding to two-step amplitudes back
to the elastic channel for 135-MeV proton scattering from
^{12}C. The solid curve corresponds to the pickup-stripping
process.

viously supposed. The energy dependence of this ratio is not large.

The two-step amplitudes are found to be destructive to the
direct amplitudes so that the full elastic cross section is reduced considerably from that computed with a conventional optical-
model potential U. To correct for this, the optical-model potential must be revised. In order to fit the data best, one would
like to use an optical-model search code with a coupled-channels
subroutine. Such a subroutine should be able to handle, for example, the full CRC environment for proton interactions with ^{12}C
that is illustrated in Fig. 2. Needless to say, such a direct
approach would be prohibitively time consuming and expensive.

Fortunately, a solution to this problem has been developed.[5]
The method is shown schematically in Fig. 3. The CRC program
CHUCK2[6] is used to calculate amplitudes in the elastic channel
that result from conventional DWBA or optical-model calculations
(β_1) and coupled-channel calculations (β_2).   The difference

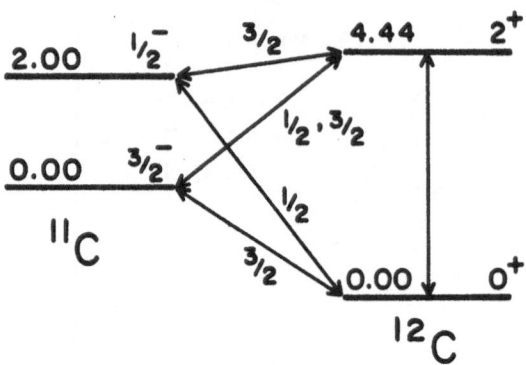

Fig. 2.   The coupled-reaction-channel environment for proton inter-
          actions with ^{12}C. The connections to ^{11}C are labelled
          with the transfer j values. States are labelled with
          their energies and spin-parity.

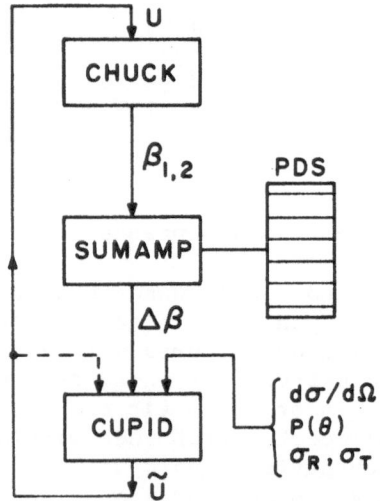

Fig. 3.   A sketch of the coupled-channel fitting procedure.
          Amplitudes β_n produced by CHUCK are stored in a
          "partitioned data set" (PDS). Differences $\Delta\beta = \beta_2 - \beta_1$ are
          supplied to the optical-model search program CUPID.

$\Delta\beta=\beta_2-\beta_1$ then approximates the back-coupling amplitudes in the elastic channel. These are supplied to an optical-model search code CUPID[5] where they are added back to the direct amplitudes computed with a modified optical potential $\breve{U}$ prior to comparisons with the data. For consistency, the procedure must be iterated, but converges rapidly with only a few occasional runs with code CHUCK2. The user has considerable flexibility in his search procedures, without entailing much expense.

RESULTS FOR PROTONS ON ^{12}C

The procedures just discussed have been applied to the analysis of proton scattering from ^{12}C. In order to understand more fully the effects of the strong couplings, data were obtained from the literature over a broad range of energies from about 60-183 MeV. Some new data were also obtained at 122 MeV at the Indiana University Cyclotron Facility in conjunction with the inelastic-scattering project. Full details of the analysis will be presented elsewhere.[7] A summary of the principle results may be given here.

The proton optical-model potential consisted of a complex volume Woods-Saxon potential and a conventional real spin-orbit potential. Fits to the elastic-scattering data were obtained with both the conventional optical model as well as in the CRC environment of Fig. 2. In each case, the potential strengths and geometrical parameters were found to vary rather smoothly with energy with very reasonable and realistic values. The real well depths of the $\breve{U}$ potential were found to be larger than for the normal optical potential U, due to the repulsive nature of the back-coupling amplitudes.[4] As expected, the imaginary potentials were smaller in the $\breve{U}$ potential since direct account is now being made of some of the absorption processes. The fits to the elastic-scattering data were about the same for each method, with some hints of improvement in the CRC approach. Some are shown elsewhere.[7,8]

There was also a very important result with regard to calculations of reaction cross sections. As was the case for the elastic scattering, the transition differential cross sections were reduced substantially from the DWBA values if the original optical potential U was used in the CRC calculations. However, the DWBA results were almost fully restored when the $\breve{U}$ potential was used instead. The validity of this generalization increases with the bombarding energy. Above 100 MeV there was little distinction between the two approaches, while at 60 MeV some distinctions could be discerned.

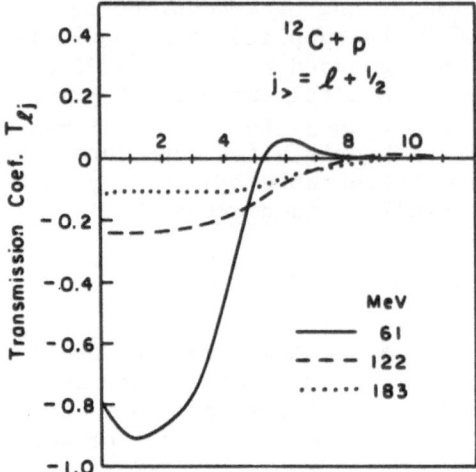

Fig. 4.   Transmission coefficients corresponding to the back-
          coupling amplitudes in the elastic channel for the
          12C+p system at three energies. These are plotted vs
          partial wave number ℓ.

     A measure of the energy dependence is shown in Fig. 4. Trans-
mission coefficients are computed with conventional formulas from
the back-coupling amplitudes at three energies and plotted against
ℓ. We see that the effects are associated primarily with the lowest partial waves, and also that they decrease substantially with
energy.

     The significance of this is that the optical potential does
appear to be able to simulate many of the general features resulting from explicit couplings of the elastic channel with other
reaction channels, provided that it is adjusted so that elastic-
scattering data are reproduced and the bombarding energy is sufficiently high. At energies below 50 MeV the characteristic of
simulation is still rather high, but details of the couplings to
specific channels may have a singular influence on certain observables such as cross sections and polarizations.[3]

     The heartening aspect of this is that it appears to be quite
legitimate to use the DWBA method to compute the $^{12}C(p,p')^{12}C$
cross sections for energies greater than 100 MeV. The back-
coupling amplitudes resulting from the strong couplings of CRC calculations are represented rather well by a conventional optical-
model potential. ^{12}C is indeed a tractable nucleus.

Does this imply that two-step contributions are negligible in the ^{12}C(p,p')^{12}C reaction? Not necessarily. But it suggests that the full CRC apparatus may not be necessary. Explicit calculations of two-step (p,d,p') contributions to the excited states of ^{12}C indicate that they are small, except possibly for the 1$^+$ T=0 state at 12.71 MeV, which is poorly fit by the one-step calculations.

There is one other difficulty of the CRC approach that needs to be mentioned and that is the problem of what optical potential should be used for the intermediate channels. Reaction theory indicates[9] that folded potentials are necessary. Explicit phenomenological support for this has been found[10] for the (p,d,p) contributions to 60-MeV proton elastic scattering from ^{12}C. However, there remains some ambiguity regarding the folding procedures. The results of two prescriptions[11,12] are shown in Fig. 5 for ^{12}C(p,d)^{11}C reactions at 61 and 183 MeV. The ambiguity is much more serious at the lower energies, where the CRC influences are also larger. These folded potentials attempt to simulate some of the contributions of deuteron breakup (including the continuum),[13] a difficult problem that needs further study. (See also the paper by P. D. Kunz in these proceedings.)

NUCLEON SCATTERING FROM MOLYBDENUM ISOTOPES

A second application of the methods considered here concerns proton and neutron elastic scattering from ^{92,94,96,98,100}Mo. Since many of the details have already been published,[14] the discussion here shall be limited to a brief mention of the principle results.

The Lane model of nucleon scattering splits the optical-model potential into an isoscalar and an isovector term, the latter being proportional to the symmetry parameter $\varepsilon=(N-Z)/A$. The isovector term has opposite signs for proton and neutron scattering. Assuming that the isoscalar term is approximately independent of mass, the strengths of the neutron and proton optical-model potentials should have equal but opposite slopes when plotted vs. A for a set of isotopes. However, in a conventional analysis of the elastic scattering of 11-MeV neutrons and 15-MeV protons from Mo isotopes, significant discrepancies from expectations were found.

Even though the deformation is increasing slowly between ^{92}Mo and ^{100}Mo, it is to be noted that neutrons are filling several different shells as well. As was the case with ^{12}C, two-step cross sections in the elastic channel were considerably larger for the pickup-stripping process than for the inelastic process. Shell effects would then imply that significant discrepancies would occur especially for proton scattering. This was indeed the situation observed in the optical-model analysis.

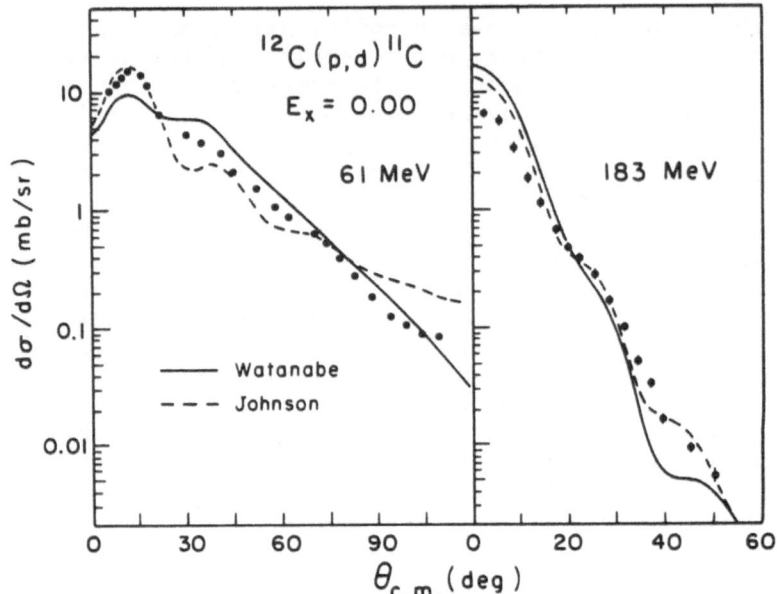

Fig. 5. Cross sections for $^{12}C(p,d)^{11}C$ reactions computed in the
         CRC environment for two energies. Two separate procedures
         are used for constructing a folded deuteron optical-model
         potential.

A reanalysis of the same data in the CRC approach gave some
improvements. The slopes were more in agreement with expectations
but, unfortunately, the shell effects could not be removed entirely.
However, at these energies, the problem of the treatment of the
intermediate deuteron channels (namely breakup and continuum p-n
propagation) is very important. The use of an optical-model deuteron potential was positively excluded. Even though a folded potential was used, the results are still in need of improvement.

SUMMARY

The strong couplings of elastic-scattering channels with reaction channels, particularly pickup channels for nucleon projectiles, opens up numerous technological problems. Some of the problems have been considered here. The ability to search for best-fit
optical potentials in a CRC environment is a considerable advance.
Although there is enhanced understanding of the optical potentials
for intermediate channels, numerous difficulties and ambiguities
remain.

For this conference, the most important result of this work
is the indication that a conventional DWBA treatment of the

$^{12}C(p,p')^{12}C$ reaction at 122 MeV gives a very satisfactory description of the reaction dynamics. Two-step effects are not excluded, but they appear to be small except possibly for the 12.71-MeV state.

It may also be suggested that the best comparisons of data with theory may be at intermediate energies (E>100 MeV, say) where many of the difficulties of reaction dynamics are no longer especially relevant. At low energies (E<50 MeV) one must remain on guard for the singular effects of the structure of individual nuclei on such dynamics.

REFERENCES

1. F. A. Brieva and J. R. Rook, Nucl. Phys. A291:299, 317 (1977).
2. W. G. Love, these proceedings.
3. R. S. Mackintosh, Nucl. Phys. A209:91 (1973) and A230:195 (1974).
4. P. W. Coulter and G. R. Satchler, Nucl. Phys. A293:269 (1977).
5. J. R. Comfort, Computer Phys. Commun. 16:35 (1978).
6. P. D. Kunz, unpublished.
7. J. R. Comfort and B. C. Karp, paper in preparation.
8. J. R. Comfort in "Proc. International Symposium on Nuclear Direct Reaction Mechanisms," Fukuoka, Japan, Oct. 25-28 (1978), to be published.
9. S. R. Cotanch and C. M. Vincent, Phys. Rev. C14:1739 (1976).
10. J. R. Comfort and B. C. Karp, Phys. Letters 80B:1 (1978).
11. S. Watanabe, Nucl. Phys. 8:484 (1958).
12. R. C. Johnson and P. J. R. Soper, Phys. Rev. C1:976 (1970).
13. J. P. Farrell, C. M. Vincent, and N. Austern, Ann. Phys. (N.Y.) 96:333 (1976) and 114:93 (1978).
14. J. R. Comfort, Phys. Rev. Letters 42:30 (1979).

EXPERIMENTAL CAPABILITIES - PRESENT AND FUTURE

C. D. Zafiratos

Nuclear Physics Laboratory
University of Colorado
Boulder, CO 80309

INTRODUCTION

The first purpose of this talk is to describe existing experimental capabilities for the study of (p,n) reactions at energies which are clearly in the direct-reaction regime. Due to the unification of (p,p), (p,n), (n,p) and (n,n) reactions under an isospin formalism some slight attention also will be given to (n,n) and (n,p) capabilities at appropriate energies.

The second purpose of this talk is to point out new developments which will enhance our experimental capabilities in the future. Some of these new developments involve improved accelerator technology and some involve improved neutron detection techniques.

PRESENT FACILITIES

The standard experimental method for (p,n) studies at these energies uses a pulsed beam with a neutron time-of-flight spectrometer. The pulsed beam is provided by cyclotrons, by tandem Van de Graaffs with bunchers, or by a combination of these two. Overall beam burst lengths typically are in the nanosecond range. Detectors are hydrocarbon scintillators with a recoil proton from n,p collisions providing the light signal. Standard fast photomultipliers and timing discriminators lead to detector time resolutions in the nanosecond range.

For work below 60 MeV or so n-γ discrimination is essential to remove unwanted γ-ray and cosmic ray muon backgrounds. The commonly used pulse-shape discimination method for n-γ discrimination

requires that the scintillator have slightly different light decay
properties for the recoil proton events that signal arrival of a
neutron than for the Compton electron events that signal detection
of a gamma ray.

The two most widely used scintillators for n-γ discrimination
are the liquid scintillators NE-213 and NE-224.* The latter has
the advantage that it does not attack lucite so that arbitrary
shapes can be chosen and machined in a lucite vessel. An example
of such a detector is shown in Fig. 1 taken from Ref. 1. The
ability of such a detector to discriminate between neutrons and
gamma-rays for neutrons above 5 MeV is shown in Fig. 2.

Angular distributions are taken in a variety of ways as indicated schematically in Figs. 3 and 4. The moveable detector shown
in the top of Fig. 3 is typically shielded and is thus rather
massive. The facility at TUNL is an example of this approach.[2]
The approach shown in the bottom of Fig. 3 is followed by the
Lawrence Livermore Cyclo-Graf facility.[3] One drawback is the discrete step-size in angular distributions set by the design. This
can be alleviated by providing a small range of steering of the
incident beam to interpolate between the fixed angles--a rudi-
mentary form of beam swinger.

True beam swingers are shown schematically in Fig. 4. The top
of the figure shows a mechanically rotated set of magnets such as
those in use at the University of Colorado[1] and Michigan State
University.[4] In this approach a fixed, well-shielded neutron
detector is located out of the plane of the page, toward the reader

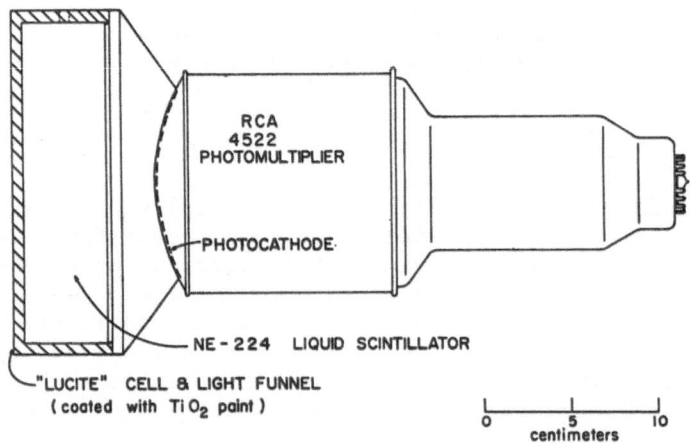

Fig. 1. Neutron detector in use at the University of Colorado.

*Supplied by Nuclear Enterprises Ltd., San Carlos, CA.

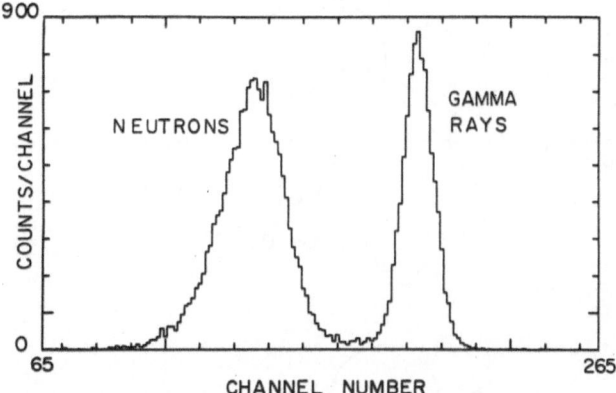

Fig. 2. Clean separation of neutrons and gamma rays is shown in this pulse-shape discriminator spectrum obtained with the detector shown in Fig. 1.

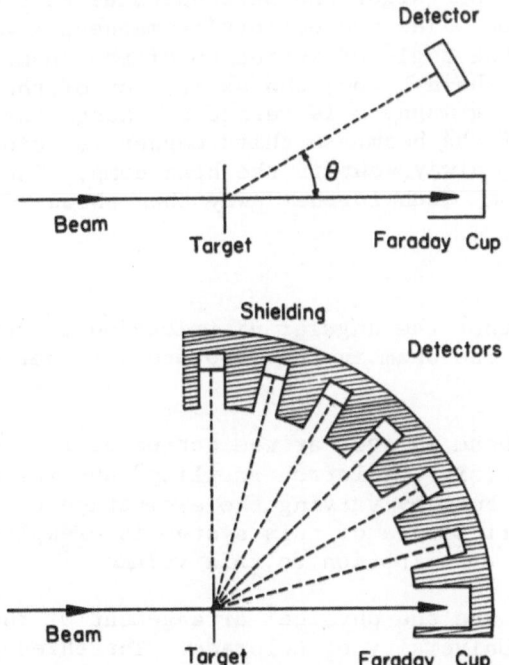

Fig. 3. Angular distributions can be obtained by moving a well-shielded detector, as at TUNL, or by a fixed array as at Lawrence Livermore Laboratory.

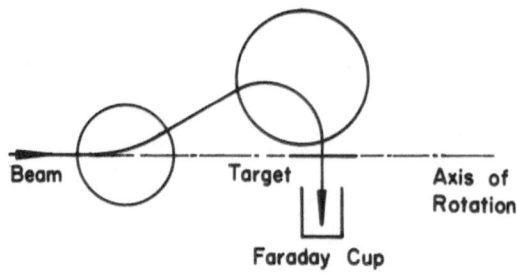

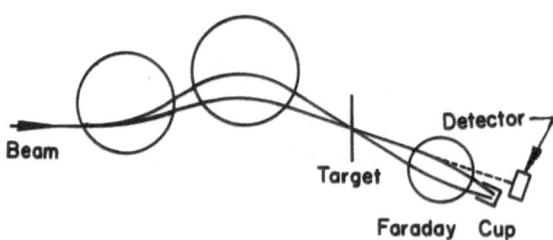

Fig. 4.   In the beam swinger at the top of the illustration a
          fixed well-shielded detector is located in a plane passing
          through the target and perpendicular to the initial beam
          direction. The two deflecting magnets are rotated to
          change the angle of incidence of the beam. In the beam
          swinger shown below, the excitation of the two beam
          deflecting magnets is varied to change the angle of inci-
          dence of the beam. A third magnet is adjusted so that
          the beam always enters the beam dump. The detector is,
          of course, much further away than shown in this schematic.

for example, so that the angular distribution is obtained by varying the angle of the beam in the laboratory rather than by motion
of the detector.

     The type of beam swinger at the bottom of Fig. 4 is in use at
the Indiana University Cyclotron Facility[5] and varies the angle of
incidence of the beam by varying the excitation of a set of steering magnets. Performance of this system is exemplified by results
shown in another contribution to this volume.[15]

     In Fig. 5 we see the physical arrangement of the beam swinger
facility at the University of Colorado. The three detectors in
the shield wall are at 9 meters flight path while the exterior
bunker is at 30 meters. Fig. 6 shows a spectrum taken with a
mixed-isotope tin target. The small shift in Coulomb-displacement
of the analog state allows simultaneous measurement of the quasi-

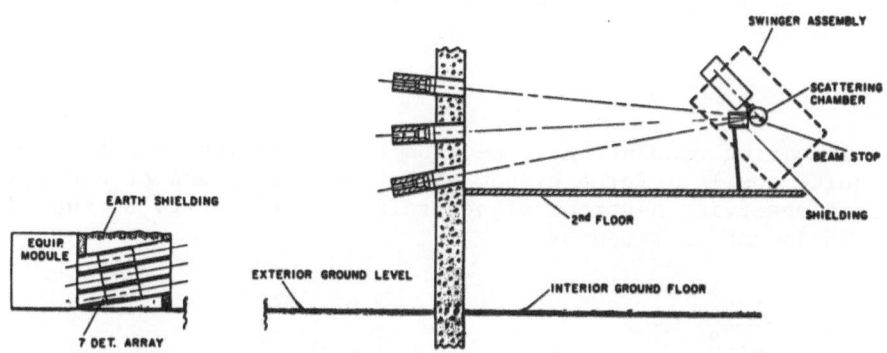

Fig. 5. The beam swinger at the University of Colorado. The detectors in the shield wall are at 9 meters flight path while the exterior bunker is at 30 meters.

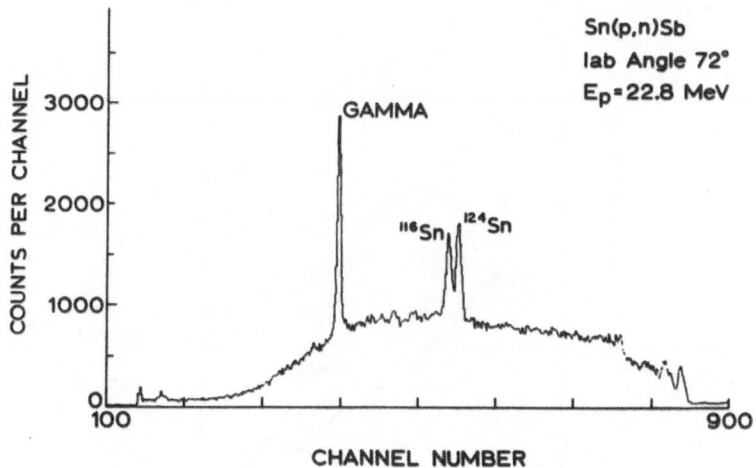

Fig. 6. The isobaric analog states for ^{116}Sn and ^{124}Sn are easily resolved in this spectrum taken at 9 meters with a mixed isotope target.

elastic (p,n) reaction for both isotopes thus greatly reducing
relative uncertainties. The angular distributions shown in Fig. 7
demonstrate the small errors possible in (p,n) studies and show
a systematic trend along the isotope sequence which is attributed
to a neutron matter radius variation.[6]

IMPROVED RESOLUTION

Improving resolution in neutron time-of-flight spectroscopy
is quite costly. For a system with time resolution Δt and flight
path L observing neutrons of non-relativistic energy E, the energy
resolution ΔE is given by

$$\Delta E = \frac{2^{3/2}}{m^{1/2}} \frac{\Delta t}{L} E^{3/2}.$$

We see that resolution deteriorates rapidly with increasing energy;
going as $E^{3/2}$. At relativistic velocities resolution deteriorates
even more rapidly with increasing energy, as shown in a graph

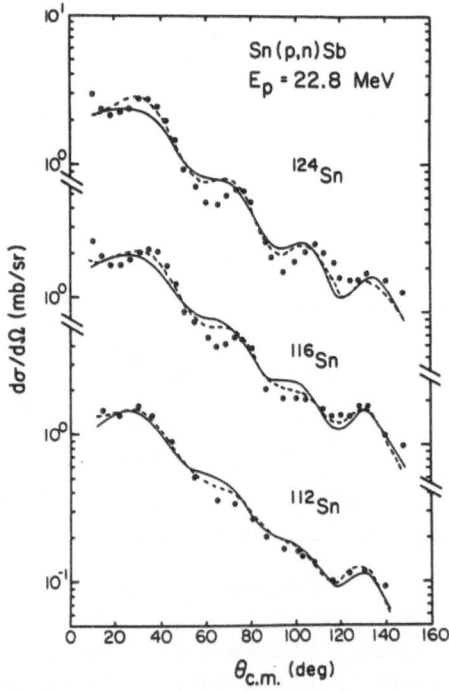

Fig. 7. The small relative errors in the quasi-elastic (p,n) data
for these three tin isotopes allows study of the systematic variation seen along the isotope sequence.

contained in C. D. Goodman's contribution to this volume.[15] We also see that for a given energy and for a given time resolution Δt, the only way to improve energy resolution is to increase L. However, this improvement is extremely costly in count rate. First, for a detector of fixed size the count rate falls as L^{-2}. Second, as L is increased the beam pulse repetition rate must be reduced proportionately to prevent overlap of fast neutrons in one beam burst with slow neutrons from the preceding beam burst. Most schemes for reducing beam repetition rates reduce the time average beam proportionately. Finally, to take advantage of the improved energy resolution capability the target thickness must be reduced. Thus, improving resolution by increasing L comes at the price of an overall count rate decrease given by L^{-4}. The decrease due to the need for reduced target thickness is common to improved resolution by any type of spectroscopy. The remaining factor of L^{-3}, however, is quite unique to time-of-flight spectroscopy.

As an example, at the University of Colorado the 500 keV resolution of the 9-meter facility for 30 MeV neutrons can be reduced to 160 keV at 30 meters but at a cost of a factor of $(9/30)^4 = 8 \times 10^{-3}$ in count rate.

LARGE DETECTORS

Some of the decrease in count rate at large flight paths can be offset by using large volume neutron detectors. However, since the speed of light propagating in a typical scintillator is 20 cm/ns, the size of the detector cannot be much larger than 20 cm without seriously degrading the time resolution. The effect of time un-certainties due to the thickness of the detector and the even lower speed of the incident neutrons limits the thickness of the detector to between 3 cm and 15 cm over the range 10 MeV to 200 MeV. This limit, in turn, keeps neutron detection efficiencies in the region of 5-10%. Thus, detectors of either large area or thickness require some means of compensation of the effects of light and neutron transit times.

One successful approach has been that of mean-timing with the arrangement shown in Fig. 8-c. The sum of the times measured at the two ends of such a detector is independent of position. This approach has been used by the Michigan State University Cyclotron Laboratory[4] and by the Kent State group.[7] The latter group has developed detectors 10" x 80" x 4" with intrinsic time resolution (detected particle at velocity c) of 350 ps. A drawback with this type of detector is the loss of recoil protons out of the back of the detector at higher energies. At 200 MeV, for example, the range of a recoil proton is 25 cm. At this energy, a 15 cm thick detector would have, at best, 80 MeV deposited before the recoil escaped. To achieve reasonable efficiency the threshold would

have to be set at 40 MeV or lower. However, 40 MeV neutrons are
overtaken by 200 MeV protons in only 30 meters at the lowest repetition rates presently available at IUCF. Multiple detectors, one
behind the other, with appropriate logic could overcome this difficulty.

Another large detector is sketched in Fig. 8-b. This type of
detector, developed at Ohio University,[8] is very deep and compen-
sates for the large effect of neutron transit time by using the
information given by the differing light arrival times at the two
ends. It has the advantage that very long range recoil protons
can be accommodated.

The detector type shown in Fig. 8-a was developed for use at
the Indiana University Cyclotron Facility and employs a cancellation of the effects of light transit time by tilting the far end
of the scintillator toward the neutron source.[9] The degree of tilt
depends on the neutron velocity of interest and causes the neutrons
to strike more distant portions of the scintillator earlier by an
amount equal to the scintillation light transit time. Partial cancellation of neutron and light transit times also occurs. Further,
the effective thickness for long-range recoil protons is greater
than the actual thickness. Sub-nanosecond timing has been achieved
with this detector. An example of a spectrum taken with this
detector is shown in another contribution to this volume.[15]

Future increases in the area of neutron detectors may involve
a method similar to that sketched in Fig. 8-d. Good timing with
this detector could be achieved by mean timing or by on-line computer correction utilizing differences in light arrival time or
differences in pulse heights. These differences could be employed
to determine the position of the scintillation and time correction
accomplished by a look-up table. At higher energies the effects of
recoil proton escape would have to be handled as discussed above.

The large volume detectors of the future, and to a lesser
extent the largest detectors available today, are subject to
extremely large backgrounds produced by cosmic rays. The bulk of
these background events are due to energetic muons passing through
the scintillator. These events are largely eliminated either by
conventional n-γ discrimination for liquid scintillators or by
the use of anti-coincidence paddle-shaped scintillators surrounding the detector.

Experience with the large detectors of the type described in
ref. 8 at the 30-meter station of the University of Colorado time-
of-flight facility has shown that rejection of cosmic ray muons
alone is insufficient for extremely low count rate experiments.
It was found that neutrons produced by muons in the shielding
around the detector contributed substantially to background events.

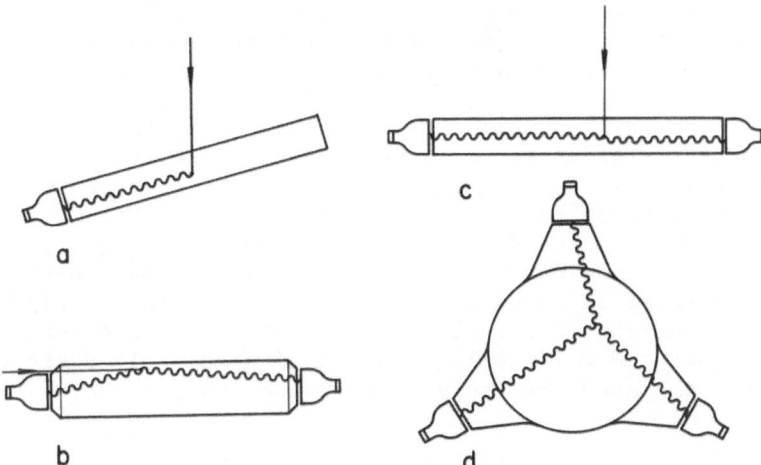

Fig. 8. a) This detector cancels the effects of scintillation
            light transit time against variations in neutron
            flight time.
         b) In this detector neutrons are incident through one of
            the photomultipliers and variation of the time of
            detection with depth is corrected by use of the infor-
            mation produced by the differing light arrival times
            at the ends.
         c) This mean-timing detector utilizes the mean of light
            arrival times at each end of the detector. This mean
            is independent of the position of the event.
         d) Possible extremely large detectors of the future may
            utilize several photomultipliers to provide informa-
            tion on position and hence allow compensation of
            light transit times.

These neutrons can not be rejected by an anti-coincidence paddle
nor by the use of n-γ discrimination. Current plans are to use
veto counters placed <u>outside</u> the local shielding surrounding the
detector so that even neutrons produced within the local shielding
will have a veto signal associated with them.

NEUTRON INDUCED REACTIONS

    Studies of neutron elastic and inelastic scattering with a
precision comparable to (p,n) studies are being made at Ohio University,[10] at Duke University,[11] and Michigan State University[12]
in the energy range of interest for this conference. The detection techniques are similar to those for (p,n) studies. Neutron

sources in use are the reactions $^2H(d,n)^3He$, $^3H(d,n)^4He$, $^3H(p,n)^3He$, and $^7Li(p,n)^7Be$. Energy resolution is comparable to that obtained in (p,n) measurements.

As reported elsewhere in this volume[16] the (n,p) reaction has been studied extensively at the Crocker Laboratory of the University of California at Davis. Most of these studies have been made with 60 MeV neutron beams with approximately one MeV resolution. The source reaction used is $^7Li(p,n)^7Be$ and the geometry places the target a little over 3 meters from the neutron source. This type of geometry is shown schematically in the top of Fig. 9. Though the neutron flux is low at this distance, the neutron beam is very clean and well collimated allowing multiple detector telescopes to be placed extremely close to the target.

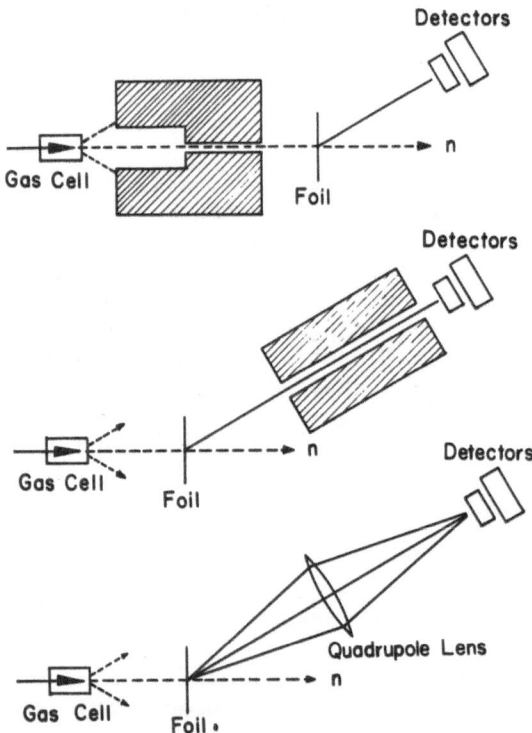

Fig. 9. Three types of geometry for (n,p) studies. At the top the source is well-collimated so the target foil is far from the source but the detectors can be placed close to the foil. In the center the situation is reversed with the foil close to the source necessitating heavy shielding for the detector which places the detector far from the target foil. At the bottom the geometry is similar to that in the center except that a quadrupole lens preserves a large solid angle for the distant detector.

The opposite type of geometry, shown at the bottom in Fig. 9, is utilized in the quadrupole spectrometer recently completed at Ohio University, also reported elsewhere in this volume.[17] There, the target is placed close to the source and a quadrupole lens transports the reaction protons to a distant well-shielded detector telescope. Solid angles of 10 msr can be achieved with such an arrangement.

NEW ACCELERATORS AND TIME-OF-FLIGHT SPECTROMETERS

Future improvements in experimental techniques will be directed toward improved energy resolution, lower cross section measurements and polarized beams. This review will not discuss polarized beams but work in the energy range of interest for this conference is underway at TUNL, Texas A & M, and at IUCF.

As mentioned earlier, long flight paths bring about a count rate decrease proportional to the fourth power of the energy resolution for a given detector and given time resolution. We have discussed large volume detectors which can offset the L^{-2} decrease caused by decreased solid angle. If average beam currents could be maintained while reducing the repetition rate for long flight path work another factor of L^{-1} could be offset.

A recent new accelerator proposal from the University of Colorado[13] includes provision for a 200-meter flight path and extremely high intensity beam bursts. In that proposal the high intensity beam is achieved through the use of a high luminosity ion source, possibly using pulsed extraction. Investigation of the effects of pulsed extraction upon instantaneous current are now beginning at the University of Colorado.

Another approach toward maintaining high average beam currents while reducing repetition rates is the use of a storage ring.[14] As shown in Fig. 10 a change of charge state (such as H^- stripping to H^+ or He^+ stripping to He^{++}) is needed to allow injected beam pulses to enter a stable orbit. Design of such a storage ring is underway at the University of Colorado and may have some application to the IUCF time-of-flight facility.

Further improvements in energy resolution may be achieved by a technique analogous to the energy-loss spectrometer for charged particles. The technique to be described could well be called time-loss spectroscopy. As indicated at the top of Fig. 11, energy spread causes beam bursts to disperse in time as they propagate. It is possible to design beam transport systems so that this spread is compensated. Ideally, however, the highest energy beam particles should be made to strike the target last, as shown at the bottom of Fig. 11. The degree of tipping of the phase space

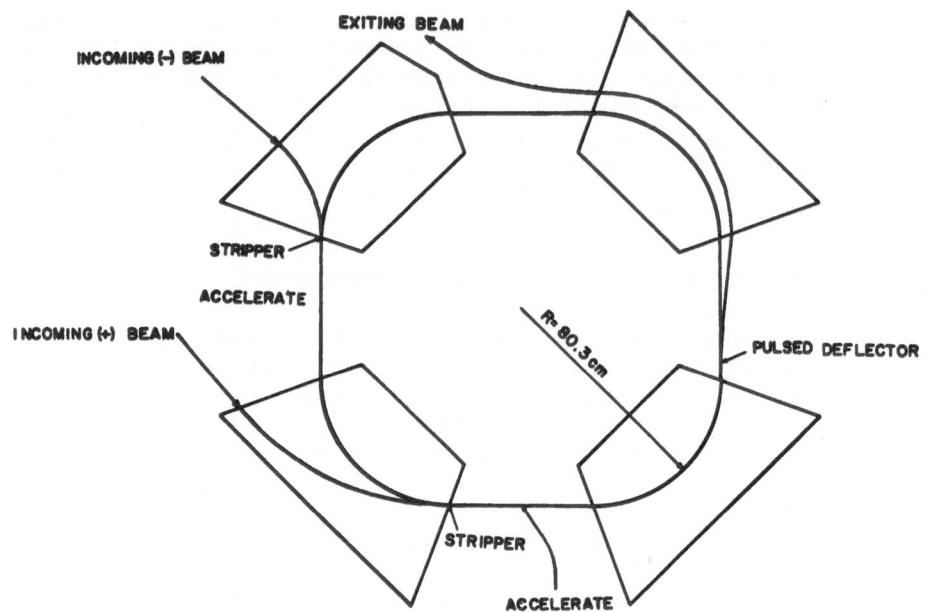

Fig. 10.   In this storage ring the incident beam is brought into
           a stored orbit by changing its charge state in a stripper
           foil. After N beam bursts had been stored, a pulsed
           deflector extracts a single N-fold intensified beam burst.

ellipse would be chosen to produce a minimum time spread in the
group of reaction product neutrons as they strike the detector.[14]

CONCLUSION

     Through the use of new detectors of large area, high intensity
accelerators with low repetition rates, and complete compensation
of time spreading effects in the time-of-flight spectrometer,
energy resolution can be improved by one or two orders of magnitude
over present figures.

REFERENCES

1.   D. A. Lind, R. F. Bentley, J. D. Carlson, S. D. Schery, and
     C. D. Zafiratos, Nucl. Instr. and Methods 130, 93 (1976).
2.   G. W. Glasgow, F. O. Purser, J. C. Clement, G. Mack, K. Stelzer,
     J. R. Boyce, D. H. Epperson, H. H. Hogue, E. G. Bilpuch, H. W.

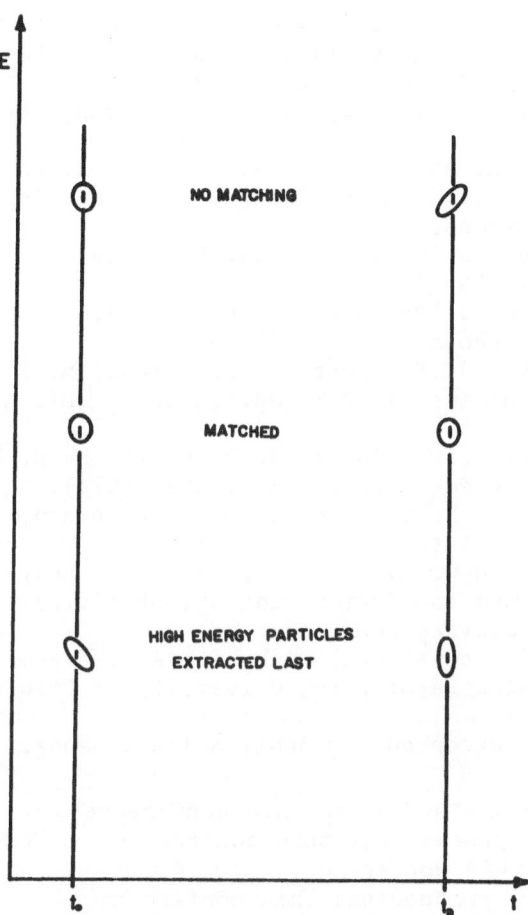

Fig. 11.   In this E vs. t phase-space plot the growth of time
          spread due to energy spread is shown at the top.
          Isochronous beam transport systems can compensate this
          effect up to the point of striking the target. Optimal-
          ly, the transport system should cause high energy parti-
          cles to strike the target late, as at the bottom, so
          that all neutrons reach the <u>detector</u> with a minimum
          time spread.

Newson and C. R. Gould, Proc. of Neutron Cross Sections and
Technology Conference, March, 1975, NBS 425 Vols. 142, edited
by R. A. Schrack and C. D. Bowman.

3.   J. C. Davis, J. D. Anderson, E. K. Freytag, and D. R. Rawles,
     IEEE Trans. Nucl. Science <u>20</u>, 213 (1973).

4.   Ranjan K. Bhomik, Robert R. Doering, Lawrence E. Young, Sam M.
     Austin, Aaron Galonsky, and Steve D. Schery, Nucl. Instr. and
     Methods <u>143</u>, 63 (1977).

5. Charles D. Goodman, Charles C. Foster, Mark B. Greenfield, Charles A. Goulding, David A. Lind, and Jacobo Rapaport, IEEE Trans. Nucl. Science 26, 2248 (1979).

6. S. D. Schery, D. A. Lind, and H. H. Wieman, Phys. Rev. C 14, 1800 (1976).

7. R. Madey, B. D. Anderson, A. R. Baldwin, J. Knudson, J. W. Watson, T. Witten, and C. Foster, to be published in Nucl. Instr. and Methods.

8. J. D. Carlson, R. W. Finlay, and D. E. Bainum, Nucl. Inst. and Methods 147, 353 (1977).

9. C. D. Goodman, J. Rapaport, D. E. Bainum, and E. Bryant, Nucl. Instr. and Methods 151, 125 (1978).
    C. D. Goodman, J. Rapaport, D. E. Bainum, M. B. Greenfield, and C. A. Goulding, IEEE Trans. in Nucl. Sci. NS-25 (1978) 577.

10. D. E. Bainum, R. W. Finlay, J. Rapaport, M. H. Hadizadeh, and J. D. Carlson, Nucl. Phys. A311, 492 (1978).

11. H. H. Hogue, P. L. van Behren, D. H. Epperson, S. G. Glendenning, P. W. Lisowski, C. E. Nelson, H. W. Newson, F. O. Purser, W. Tornow, C. R. Gould, and L. W. Seagondolor, Nuclear Science and Engineering 68, 38 (1978).

12. Sam Austin, private communication.

13. A Proposal for a National Light Ion Accelerator Facility, Nuclear Physics Laboratory, University of Colorado, 1978 (unpublished).

14. P. A. Smith, accepted for publication in Nucl. Instr. and Methods.

15. C. D. Goodman, proceedings this conference.

16. F. P. Brady, proceedings this conference and N.S.F. King, proceedings this conference.

17. R. N. Finlay, proceedings this conference.

APPENDIX

Responses to facility questionnaire.

Facility: Indiana University Cyclotron Facility
Respondent: Charles C. Foster
Accelerator Type: Separated sector cyclotron

Beam Properties	Protons	Deuterons
Energy Range (MeV):	20–225	20–100
Maximum Average Intensity (μA):	1.25	1.25
Polarized Average Intensity (nA):	100	100
Repetition Rate (MHz):		

   Time of Flight Spectrometer
Flight Path: 20–70 m   Angular Range: 0–26°
Time Resolution: < 1 ns   Detector: 1 meter x 15 cm x 15 cm tilted
                           and 1 m x 10 cm x 20 cm mean timing detector.

```
      Neutron Sources
Source Reaction:        $^3$H(p,n)      $^2$H(d,n)      $^3$H(d,n)      $^7$Li(p,n)


Unpolarized Intensity:    --            --            --            --
Polarized Intensity:      --            --            --            --
```

Special Features: Future plans for 20–100 m flight path and 0–90° angular range.

Areas of Research Related to Conference Topics: (p,n) at 60, 120, and 135 MeV on ^{7}Li, ^{9}Be, ^{12}C, ^{13}C, ^{16,17,18}O, ^{24,25,26}Mg, ^{90}Zr, ^{112,116,124}Sn.

Facility: JAERI, Tokai-mura, Ibaraki-Ken 319-11, Japan
Respondent: Shigeya Tanaka
Accelerator Type: 5.5 MV Van de Graaff (CN)

Beam Properties	Protons	Deuterons
Energy Range (MeV):	1.8-5.5	1.8-5.5
Maximum Average Intensity (μA):	5	5
Polarized Average Intensity (nA): --		--
Repetition Rate (MHz): 1.0		

Time of Flight Spectrometer
Flight Path 8 m   Angular Range: -30° to 150°
Time Resolution: 3 ns Detector: 5" x 4" NE213

```
      Neutron Sources
Source Reaction:        $^3$H(p,n)      $^2$H(d,n)      $^3$H(d,n)      $^7$Li(p,n)

Unpolarized Intensity: 2.5x10$^9$n/s  5x10$^8$n/s  5x10$^8$n/s      --
Polarized Intensity:      --            --            --            --
```

Special Features: Multi-angle T.O.F. spectrometer for fast neutron scattering experiments. 20 UR pelletron tandem accelerator under construction.

Areas of Research Related to Conference Topics: Neutron scattering from ^{94,96}Mo at 5-8 MeV and from ^{12}C, ^{32}S, and ^{28}Si at 21 MeV.

Facility: Lawrence Livermore Laboratory
Respondent: Carl Poppe
Accelerator Type: EN Tandem Van de Graaff and 80 cm AVF cyclotron

Beam Properties	Protons	Deuterons
Energy Range (MeV):	0-12; 15-27	0-19.5
Maximum Average Intensity (μA):	2 μA	2 μA
Polarized Average Intensity (nA):	--	--
Repetition Rate (MHz): 0.625 - 5		

Time of Flight Spectrometer
Flight Path: 11 m   Angular Range: 3° to 160° (16 fixed detectors)

Time Resolution: 1 ns   Detector: $4\frac{1}{2}$" x 2" NE213

          Neutron Sources
Source Reaction: ^{3}H(p,n) ^{2}H(d,n) ^{3}H(d,n) ^{7}Li(p,n)

Unpolarized Intensity: --   --   --   --
Polarized Intensity: --   --   --   --

Special Features: An extremely intense (10^{12}n/s·cm^2) 14 MeV
neutron source is also available. This facility has a quadrupole
spectrometer for (n,z) reaction studies.

Areas of Research Related to Conference Topics: (p,n) to analog
and excited analog states. (p,p') studies including spin-flip,
(p,nγ), (n,p) studies.

Facility: Michigan State University
Respondent: Sam M. Austin
Accelerator Type: Sector-focussed cyclotron
Beam Properties   Protons   Deuterons
Energy Range (MeV):   5.5-12;25-48   2.6-5.7;13-25
Maximum Average Intensity (μA):   3   3
Polarized Average Intensity (nA):   --   --
Repetition Rate (MHz):   14-22 MHz (can be reduced up to 11-fold)
      Time of Flight Spectrometer
Flight Path: 4-34 m   Angular Range: -5° to 165°;180°
Time Resolution: 0.6 ns   Detector: 3 liter mean timing, several
                                       1 liter

          Neutron Sources
Source Reaction: ^{3}H(p,n) ^{2}H(d,n) ^{3}H(d,n) ^{7}Li(p,n)

Unpolarized Intensity: --   --   -- 2×10^{7}n/cm^2·s*
Polarized Intensity: --   --   --   --

   *For ΔE=1 MeV, E$_p$=45 MeV, 3 μA at 15 cm.

Special Features: Beam swinger (57 Q^2/A MeV) with 34 meter
                   flight path.

Areas of Research Related to Conference Topics: (p,n)--search for
giant M1 strength in nuclides near ^{90}Zr; the energy dependence of
$V_{\sigma\tau}$; the strength of the tensor interaction, (n,n)--elastic
scattering from ^{12}C, ^{28}Si, ^{32}S, ^{40}Ca, and ^{208}Pb at 30, 40 MeV.

Facility: Ohio University
Respondent: J. Rapaport
Accelerator Type: 5 MV tandem

Beam Properties	Protons	Deuterons
Energy Range (MeV):	1-10	1-10
Maximum Average Intensity (μA):	100 DC, 10 pulsed	80DC, 5 pulsed
Polarized Average Intensity (nA):	--	--

Repetition Rate (MHz): $5/2^n$ $0 < n < 6$
      Time of Flight Spectrometer:
Flight Path: $\leq$ 12 m   Angular Range: 0-160° (flight path $\leq$ 8 m
0-80°)

Time Resolution: 1 ns   Detector: Ne-224 2" x 8" diam. and
12" x 8" diam.

Neutron Sources				
Source Reaction:	^{3}H(p,n)	^{2}H(d,n)	^{3}H(d,n)	^{7}Li(p,n)
Unpolarized Intensity:	2×10^8*	3×10^8*	8×10^8*	7×10^8*
Polarized Intensity:	--	--	--	--

      * n/sr·s for 100 keV energy loss

Special Features: Large intensities of monoenergetic neutrons
            from 0-12 MeV and 20-26 MeV

Areas of Research Related to Conference Topics: Scattering of
monoenergetic neutrons with energies up to 26 MeV. Optical model
analysis of neutron elastic scattering. Isospin effects in neutron
inelastic scattering. Empirical determination of the isovector
component of the optical model potential.

Facility: Osaka University, Japan
Respondent: M. Kondo
Accelerator Type: Isochronous Cyclotron

Beam Properties	Protons	Deuterons
Energy Range (MeV):	7.5-75	6-60
Maximum Average Intensity (A):	20	20
Polarized Average Intensity (nA):	150	150

Repetition Rate (MHz): 6-18
      Time of Flight Spectrometer
Flight Path: 1 meter   Angular Range ---
Time Resolution: 2 ns   Detector: 3" x 5" and 5" x 5" NE213

Neutron Sources				
Source Reaction:	^{3}H(p,n)	^{2}H(d,n)	^{3}H(d,n)	^{7}Li(p,n)
Unpolarized Intensity:	--	--	--	--
Polarized Intensity:	--	--	--	--

Special Features: Studies of neutrons emitted in pre-equilibrium
            processes.

Areas of Research Related to Conference Topics: p-p, p-d, p-α
scattering.

Facility: Texas A & M University
Respondent: L. C. Northcliffe
Accelerator Type: 88" cyclotron
Beam Properties   Protons   Deuterons
Energy Range (MeV):   10–45   10–50
Maximum Average Intensity (μA):   ?   ?
Polarized Average Intensity (nA):   100–1500 100–1000
Repetition Rate (MHz): Cyclotron RF
        Time of Flight Spectrometer
Flight Path: --   Angular Range: --
Time Resolution: 1 ns   Detector: --

         Neutron Sources
Source Reaction ^{3}H(p,n) ^{2}H(d,n) ^{3}H(d,n) ^{7}Li(p,n)

Unpolarized Intensity: --   --   --   --
Polarized Intensity: -- 8×10^4n/s·cm^2*   --   --
     * at 4.5 meters
Special Features: Polarized neutron source

Areas of Research Related to Conference Topics:

Facility: Tohoku University, Japan
Respondent: H. Orihara
Accelerator Type: AVF cyclotron
Beam Properties   Protons   Deuterons
Energy Range (MeV):   2.5 – 40 5 – 26
Maximum Average Intensity (μA):   40   40
Polarized Average Intenstiy (nA):   --   --
Repetition Rate (MHz): 20–40
        Time of Flight Spectrometer
Flight Path: 45 m   Angular Range: -4° to 165°
Time Resolution: 1 ns   Detector: NE213,NE224 80(H)x1000(W)
                                                   x25(D) mm^3

         Neutron Sources
Source Reaction: ^{3}H(p,n) ^{2}H(d,n) ^{3}H(d,n) ^{7}Li(p,n)

Unpolarized Intensity: --   --   --   --
Polarized Intensity: --   --   --   --

Special Features: High resolution neutron spectroscopy

Areas of Research Related to Conference Topics:

Facility: Triangle Universities Nuclear Laboratory
Respondent: R. Walter
Accelerator Type: FN tandem, 15 MeV cyclotron injector

Beam Properties	Protons	Deuterons
Energy Range (MeV):	2-17	2-17
Maximum Average Intensity (µA):	3	3
Polarized Average Intensity (nA):	100	100
Repetition Rate (MHz):	0.03-2	0.03-2

Time of Flight Spectrometer

Flight Path: 4 m, 6 m Angular Range: $-155°$ to $+160°$

Time Resolution: 1.6 ns   Detector: NE218   8 cm x 5 cm and 12 cm x 5 cm

Neutron Sources

Source Reaction:	^{3}H(p,n)	^{2}H(d,n)	^{3}H(d,n)	^{7}Li(p,n)
Unpolarized Intensity:	10^7*	2.5×10^7*	0.6×10^7*	--
Polarized Intensity:	See below			

*neutrons/$cm^2 \cdot sec \cdot µA \cdot$ atmosphere target gas, sample 8 cm from source

Special Features: Polarized neutrons to be generated via polarization transfer. Anticipate $p \approx 0.7$. Pre-
buncher and harmonic addition buncher capture
80% of DC beam into 2 ns beam burst at 2 MHz
repetition rate.

Areas of Research Related to Conference Topics: (p,n) to discrete
states for $E_p \leq 17$ MeV. Neutron polarization and analyzing power
measurements for (p,n) reactions. Neutron optical model studies;
(n,n') to discrete states.

Facility: University of California, Davis

Respondent: F. P. Brady

Accelerator Type: Cyclotron

Beam Properties	Protons	Deuterons
Energy Range (MeV):	67	44
Maximum Average Intensity (µA):	20-40	40-80
Polarized Average Intensity (nA):	--	--

Repetition Rate (MHz): 20 at 50 MeV protons

Time of Flight Spectrometer

Flight Path: 10 m   Angular Range: $0-40°$ (5 m beyond $40°$)

Time Resolution: Detector: 8"x3" NE224

Neutron Sources

Source Reaction;	^{3}H(p,n)	^{2}H(d,n)	^{3}H(d,n)	^{7}Li(d,n)
Unpolarized Intensity:	--	--	--	10^6n/s·MeV
Polarized Intensity:	--	--	10^5n/s·MeV	--

Special Features: Improving deuteron beam intensity to 120 µA at 40 MeV.

Areas of Research Related to Conference Topics: Nucleon-nucleon
scattering using polarized and unpolarized beams. Studying (n,p)
reactions to analog giant resonances. Beginning (n,n') studies.

Facility: University of Colorado
Respondent: C. D. Zafiratos
Accelerator Type: AVF cyclotron

Beam Properties	Protons	Deuterons
Energy Range (MeV):	0.2-28	0.2-15
Maximum Average Intensity (μA):	10	10
Polarized Average Intensity (nA):	--	--

Repetition Rate (MHz): RF (7-20 MHz) ÷ n where n is any even
                        integer
        Time of Flight Spectrometer
Flight Path: 9 m and 30 m   Angular Range: $0^{o}-155^{o}$
Time Resolution: 1 ns   Detector: NE224 1"x8", 2"x8" and 14"x8"

Neutron Sources Source Reaction:	^{3}H(p,n)	^{2}H(d,n)	^{3}H(d,n)	^{7}Li(p,n)
Unpolarized Intensity:	--	--	--	--
Polarized Intensity:	--	--	--	--

Special Features: Beam swinger for angular distributions.

Areas of Research Related to Conference Topics: (p,n) to analog
and non-analog states.

Facility: WNR, Los Alamos Scientific Laboratory
Respondent: N. S. P. King
Accelerator Type: Linac

Beam Properties	Protons	Deuterons
Energy Range (MeV):	800	--
Maximum Average Intensity (μA):	10	--
Polarized Average Intensity (nA):	--	--
Repetition Rate (MHz):	0.1-0.5	

        Time of Flight Spectrometer
Flight Path: 35 m   Angular Range: $30^{o},45^{o},60^{o},110^{o}$
Time Resolution: 700 ps   Detector: Conversion telescope and
                                       NE213 Scintillator

Neutron Sources Source Reaction:	^{3}H(p,n)	^{2}H(d,n)	^{3}H(d,n)	^{7}Li(p,n)
Unpolarized Intensity:	--	--	--	--
Polarized Intensity:	--	--	--	--

Special Features: Thick Ta target produces intense white-neutron
                   spectrum from 100 keV to 800 MeV.
Areas of Research Related to Conference Topics: (p,n) on targets
of ^{6,7}Li, ^{27}Al, Cu, In, Pb, and U.

CALCULATIONS OF NEUTRON DETECTOR EFFICIENCIES

Bryon D. Anderson, Robert A. Cecil, and Richard Madey

Department of Physics
Kent State University
Kent, Ohio   44242

1. INTRODUCTION

Several improvements have been made during the last few
years to computer calculations of efficiencies for detecting
neutrons with hydrocarbon scintillators. The most recent work of
Cecil et al. (1979) claims an accuracy of a few percent for a
substantial range of neutron energies, detector thresholds, and
counter sizes. These improvements include the adoption of new
inelastic cross sections and energy and angular distributions
for neutron-induced reactions on carbon, adoption of new light-
response functions, use of relativistic kinematics, and proper
determination of light deposited by escaping charged particles.
Although the number of possible nuclear reactions for medium-
energy neutrons in a scintillator is large, good agreement of
calculations with many different measurements of detector
efficiencies up to neutron energies of 300 MeV indicates that
the efficiencies may be described adequately by grouping the
different reactions into a few channels in the efficiency code.

Although earlier workers developed various methods of
calculating neutron detector efficiencies, most of the recent
work has been done on computer codes which use the Monte-Carlo
method to simulate the interaction of a neutron in a hydrocarbon
scintillator. Stanton (1971) developed a popular Monte-Carlo
code based on the original computer code of Kurz (1964). The
Monte-Carlo method allows one to include neutron rescattering
explicitly. Edelstein et al. (1972) and McNaughton et al. (1975)
modified the Stanton code to obtain good agreement with their
own efficiency measurements. Recently, Del Guerra (1976)

developed a Monte-Carlo code which included new cross sections for the inelastic reactions on carbon based on his extensive compilation of those cross sections. Since measurements do not exist to define unambiguously the separate inelastic cross sections for neutrons on carbon, we adjusted inelastic cross sections in energy regions where experimental points do not exist to provide good agreement of calculated efficiencies with measurements. Furthermore, since any one set of measurements of neutron detector efficiencies contains unknown systematic errors, the most reliable tuning is to try to fit many different sets of efficiency measurements simultaneously.

Specific improvements to the Monte-Carlo code of Stanton which use and extend the work of McNaughton and Del Guerra are described below.

2. IMPROVEMENTS TO THE MONTE-CARLO CODE

The important recent improvements made to the Monte-Carlo type of calculation of neutron detector efficiencies include: (a) use of relativistic kinematics, (b) adoption of new light-response functions, (c) proper determination of light deposited by escaping charged particles, and (d) adjustment of inelastic cross sections and kinematics for neutron-induced reactions on carbon. Relativistic kinematics are necessary for neutron energies above 100 MeV and constitute a straightforward improvement.

Light-response functions are used in the Monte-Carlo code to convert proton-recoil energy into equivalent electron-energy. The light-response measurements of Madey et al. (1978) and Czirr et al. (1964) for the NE-102 plastic scintillator may be represented by the semi-empirical expression shown in Fig. 1. Similar semi-empirical expressions are available also for the measurements on NE-224, NE-228, and NE-228A scintillators. For the liquid scintillator NE-213, which is popular for its pulse-shape discrimination capability, the light response measurements of Verbinski et al. (1968) and Czirr et al. (1964) have been fit by Cecil et al. (1979). Note that one needs to have reliable light response functions only over the range of energies where one may set pulse-height thresholds. Fitted expressions for the relative light response from the less important deuteron and alpha-particle recoils are also incorporated into the Monte-Carlo code.

The proper determination of light deposited by escaping charged particles is important for calculating efficiencies accurately at medium energies. The range of a 100 MeV proton recoil in hydrocarbon scintillator such as NE-102 is about 3 in.,

$$T_E = -8.0[1.0-EXP(-0.10T_P^{0.90})]+0.95T_P$$

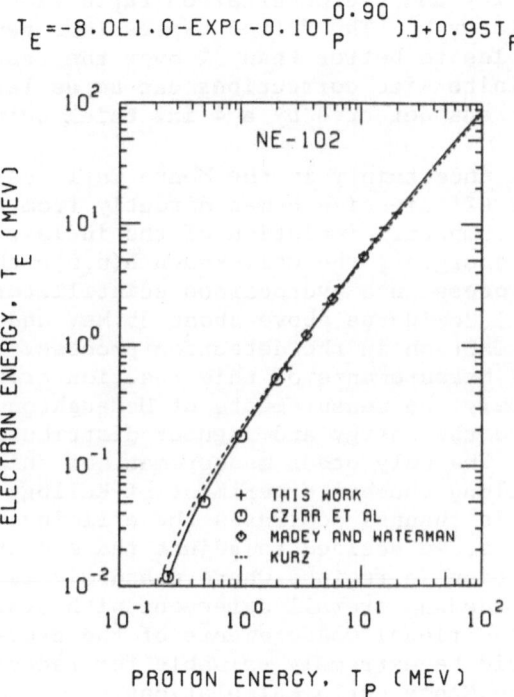

Fig. 1. Response of NE-102 plastic scintillator to protons from
the work of Madey et al. (1978). The dashed curve is
the response function used in the earlier computer
codes of Kurz (1964) and Stanton (1971).

which is a significant fraction of the thickness of a typical
detector. The range of each charged-particle recoil is
calculated in the Kent State version (Cecil et al., 1979) of
the Stanton Monte-Carlo code. If this range is longer than the
distance to the boundary of the scintillator, the code calculates
the energy deposited in the scintillator and the amount of light
produced. The particle ranges and energy losses are calculated
from logarithmic-polynomial fits to the range-energy tables of
Janni (1966) for Pilot B scintillator. (Most plastic scintillators have a density and hydrogen/carbon ratio close to that of
Pilot B scintillator.) The fitted expressions reproduce the
range-energy tables to better than 3% over the range from 0.1 to
500 MeV. The finite size corrections can be as large as 30%
for 100 MeV neutrons detected by a 4 in. thick scintillator.

The largest uncertainty in the Monte Carlo calculations of
neutron detector efficiencies comes directly from the uncertainties
involved in the computer simulation of the inelastic reactions on
carbon. At low energies, the well-known H(n,n)H channel dominates
the detection process in a hydrocarbon scintillator; however, the
^{12}C(n,np) channel dominates above about 35 MeV and becomes the
most important reaction in the detection process. Unfortunately,
few experimental measurements of this reaction cross section have
been made, and only the measurements of McNaughton et al. (1974)
at 56 MeV studied the energy and angular distributions of the
emitted proton. The only other measurement of this important
channel is the cloud-chamber experiment of Kellogg (1953) at 90
MeV. Because this channel dominates the efficiency calculation
at medium energies, we decided to adjust the energy dependence
of this cross section in regions where measurements do not exist
in order to provide best overall agreement with available efficiency
measurements. Additional measurements of the cross section for
this channel would be extremely valuable for reducing the uncertainties of these Monte Carlo calculations.

Shown in Fig. 2 are the inelastic cross sections for neutrons
on carbon used in the Kent State code. Available experimental
measurements of the separate inelastic cross sections are shown
as individual points and the cross sections used in the Monte-Carlo
code as solid lines. These cross sections are generally similar
to those adopted by McNaughton et al. (1975) and by Del Guerra
(1976). We added an explicit (n,2n) channel which was formerly
included in the strength for the (n,p) channel. Although the
(n,2n) channel does not provide a charged-particle recoil, it
does rescatter the neutron and lower its energy. As shown,
the (n,2n) channel is larger than all the other inelastic

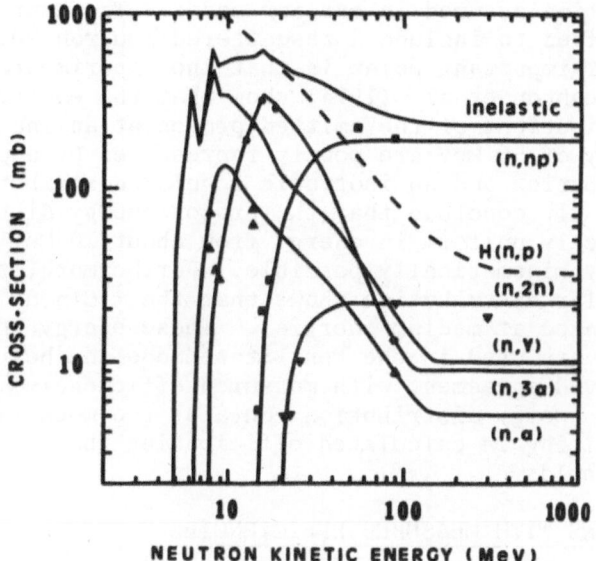

Fig. 2. Neutron-carbon inelastic cross-sections. The solid
lines represent the cross-sections used in the Monte-
Carlo code of Cecil et al. (1979). The symbols denote
cross-section measurements compiled by Del Guerra
(1976).

reactions except for the (n,np) channel above about 100 MeV.

The energy dependence of the important (n,np) channel is basically the same as that adopted by McNaughton et al. and Del Guerra except that it rises from threshold somewhat slower in the Kent State version. This slower rise was found to be necessary to fit efficiency measurements in the 20 to 30 MeV energy range. Because this one channel is so important at medium energies, two things should be noted regarding the computer simulation of this reaction. First, at 90 MeV, Kellogg notes that a rescattered neutron is present 90% of the time. Thus, the channel is more correctly the ^{12}C(n,np) channel than the ^{12}C(n,p) reaction assumed in earlier codes. The Kent State code has been modified to include a rescattered neutron 90% of the time. Another important point is that the experimental measurements of McNaughton et al. (1974) show that the energy and angular distributions of the emitted proton at an incident neutron energy of 56 MeV are poorly represented by a phase-space energy distribution and an isotropic angular distribution. McNaughton et al. conclude that the proton energy distribution is more correctly uniform in energy from about 10 MeV up to the maximum energy kinematically possible. Furthermore, their measured angular distribution shows that the ^{12}C(n,np) reaction is forward-peaked at medium energies. These energy and angular distributions are used in the Kent State code and help to provide improved agreement with measured efficiencies. Note that a phase-space energy distribution contains too much yield at low energies resulting in calculated efficiencies that are too small at high thresholds.

3. COMPARISONS WITH MEASURED EFFICIENCIES

Since the purpose of the Monte-Carlo code is to provide reliable calculations of neutron counter efficiencies for different detector geometries, neutron energies, and detector thresholds, the best test of the code is comparison with different efficiency measurements. Cecil et al. (1979) tested the Kent State code by comparison with twelve different sets of efficiency measurements for both plastic and liquid hydrocarbon scintillator detectors. In these measurements, neutron energies vary from about 1 MeV to 340 MeV and counter thresholds from 0.12 MeV to 22 MeV equivalent-electron energy. The calculations agree with the measurements to better than 10 percent, and usually much better. Note that when calculating the efficiencies, an attempt was made to reproduce the conditions relevant to each experiment, such as the correct geometry of the counter, the threshold setting, and the energy spread of the incident neutrons; however, sometimes it was difficult to determine these parameters reliably from the literature. Note also that

efficiencies near threshold are especially sensitive to the
various experimental parameters and are expected to be more
difficult to reproduce accurately. No parameters were adjusted
differently for individual measurements in order to obtain
improved agreement.

In Fig. 3 we compare Monte-Carlo calculations with the Kent
State code with the measurements of Riddle et al. (1974) for a
7.6 cm thick by 17.8 cm diameter NE-102 counter and with the
measurements of Betti et al. (1976) for a 15.3 cm diameter by
27.0 cm thick NE-110 scintillator. Both experiments report data
at four thresholds which vary from 1 MeV to 22 MeV equivalent-
electron energies. The calculations reproduce the measurements
well at all thresholds. The excellent agreement of the
calculations with the highest threshold data shown here is not
seen with the earlier computer calculations of neutron detector
efficiencies.

4. CONCLUSIONS

Monte-Carlo calculations of neutron counter efficiencies now
provide good agreement with experimental measurements for neutron
energies up to 300 MeV and detector thresholds up to 22 MeV
electron-equivalent energy. Since the calculations agree with
many different experiments to better than 10%, and usually much
better, we estimate that the calculations are accurate to a few
percent (except near threshold) for the range of experimental
parameters tested.

The recent improvements in Monte-Carlo calculations of
neutron detector efficiencies include a new adjustment of the
cross sections and the kinematics for the carbon inelastic
reaction channels, new measurements of light-response functions
for various scintillators, and a more refined determination of
the light deposited by charged-particle recoils which escape
the counter.

REFERENCES

Betti, G.
    1976: and A. Del Guerra, A. Giazotto, M.A. Giorgi,
           A. Stefanini, D.R. Botterill, D.W. Braben, D. Clarke,
           and P.R. Norton, Nucl. Instr. and Meth. 135, 319.

Cecil, R.A.
    1979: and B.D. Anderson and R. Madey, accepted for
           publication in Nucl. Instr. and Meth.

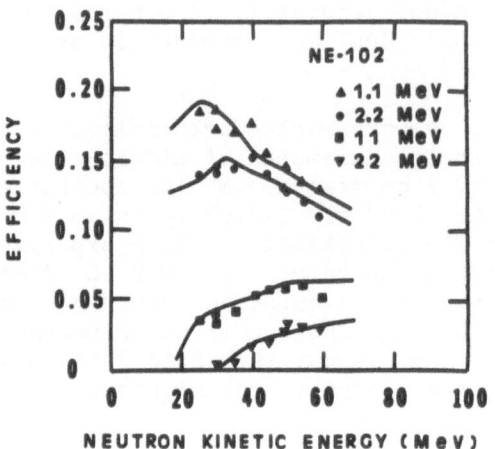

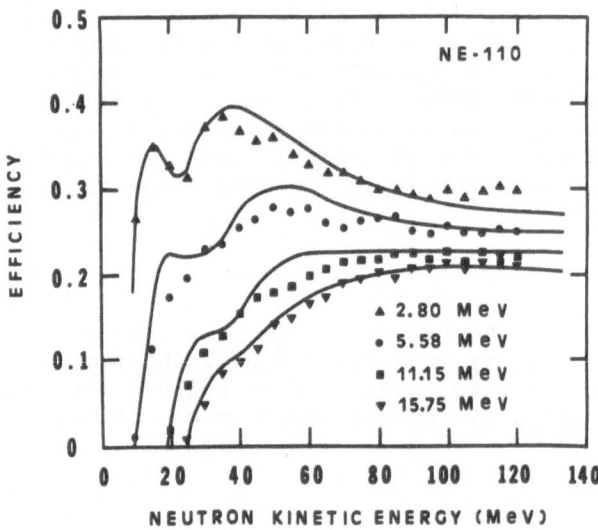

Fig. 3. Comparison of efficiency measurements with calculations
         of the Monte-Carlo computer code of Cecil et al. (1979)
         for plastic scintillators with thresholds set from 1.1
         to 22.2 MeV equivalent-electron energies. The upper
         figure are the measurements of Riddle et al. (1974) and
         the lower figure are the measurements of Betti et al.
         (1976).

Czirr, J.
    1964:   and D.R. Nygren, and C.D. Zafiratos, Nucl. Instr.
            and Meth. 31, 226.

Del Guerra, A.
    1976:   Nucl Instr. and Meth. 135, 337.

Edelstein, R.M.
    1972:   and J.S. Russ, R.C. Thatcher, M. Elfield, E.L. Miller,
            N.W. Reay, N.R. Stanton, M.A. Abolins, M.T. Liu,
            K.W. Edwards, and D.R. Gill, Nucl. Instr. and Meth.
            100, 355.

Janni, J.F.
    1966:   AFWL-TR-65-150, September.

Kellogg, D.A.
    1953:   Phys. Rev. 90, 224.

Kurz, R.J.
    1964:   UCRL-1139, March.

Madey, R.
    1978:   and F.M. Waterman, A.R. Baldwin, J. Knudson,
            J.D. Carlson, and J. Rapaport, Nucl. Instr. and
            Meth. 151, 445.

McNaughton, M.W.
    1974:   and F.P. Brady, W.B. Broste, A.L. Sagle, and
            S.W. Johnson, Nucl. Instr. and Meth. 116, 25.
    1975:   and N.S.P. King, F.P. Brady, and J.L. Ullman,
            Nucl. Instr. and Meth. 129, 241.

Riddle, R.A.J.
    1974:   and G.H. Harrison, P.G. Roos, and M.J. Saltmarsh,
            Nucl. Instr. and Meth. 121, 445.

Stanton, N.R.
    1971:   COO-1545-92, February.

Verbinski, V.V.
    1968:   and W.R. Burrus, T.A. Love, W. Zobel, N.W. Hill, and
            R. Testor, Nucl. Instr. and Meth. 65, 8.

DISCUSSION

    Rapaport: What is the uncertainty you think you have in the calculated neutron efficiencies?

    Anderson: Because we agree with the various experimental measurements of neutron detector efficiencies always within 10%, and usually much better, and because each measurement probably contains some systematic error, we believe the calculations are accurate to about 5% over the range of parameters tested, viz., for neutron energies from about 1 to 300 MeV and for counter thresholds from 1 to 22 MeV equivalent-electron energy.

EXPERIMENTAL DETERMINATIONS OF NEUTRON DETECTOR EFFICIENCIES

Charles A. Goulding

Florida A & M University
Tallahassee, Florida 32307

The most difficult part of obtaining neutron differential cross sections is the determination of the efficiency of the neutron detection system. One method is the use of sophisticated computer codes to calculate the absolute efficiency, as previously described by B. Anderson,[1] and the other is to measure the efficiency experimentally, which will be discussed here.

The most popular method of experimental efficiency determination is that of the tagged neutron beam. Tagged neutrons can be made, for example, by scattering neutrons from a proton target, either liquid H_2 or CH_2. Detection of the recoil protons then tags the presence of a kinematically well defined neutron beam.[2-4] Among other reactions amenable to tagging are the $\pi^- + p \rightarrow \pi^\circ + n$ reaction[5] and $\gamma + p \rightarrow \pi^+ + n$.[6] At IUCF, Debevec[7] used the $^7Li(p,n)$ reaction. By first using proton elastic scattering from ^7Li, he was able to map the flux of protons corresponding to detection of recoils by the recoil counter. Then, taking into account a small kinematic shift, he was able to define a neutron beam of known flux and spatial extent by detecting the recoil ^7Be from the $^7Li(p,n)^7Be$ reaction.

Another procedure that has been used by some workers is to calibrate a reference detector and then compare the unknown detector and reference detector at a facility that provides neutron fluxes over the required energy range.[8] Of course, with this procedure one must be careful to reproduce the same threshold setting between calibration and actual use.

The efficiency determination techniques we have used are quite different from those of the previously described methods.

Since the tilted geometry of our detection system[9] complicates
the accurate calculation of detector efficiencies, a method of
experimental determination was needed. The efficiency of a
neutron detector depends strongly on the pulse height discrimination level, and we felt a technique that measures efficiency
under actual operating conditions should be developed. Then, the
efficiency measurements would become an integral part of the data
runs. We have chosen to determine unknown (p,n) cross sections
by direct comparison with known or easily measured "calibration"
cross sections. This directly accomplishes the desired measurement and implicitly determines the detector efficiency. Our
calibration reactions are $^{12}C(p,n)^{12}N$ and $^7Li(p,n)^7Be$. The
$^{12}C(p,n)^{12}N$ cross section is determined by measuring $^{12}C(p,p')^{12}C$
at the same energy and using isospin arguments to relate the (p,n)
to the (p,p') cross sections. The $^7Li(p,n)^7Be$ cross section is
determined by counting the 53 day residual radioactivity.

$^{12}C(p,n)$ vs. $^{12}C(p,p')$

     If isospin is conserved the $^{12}C(p,n)^{12}N(g.s.)$ and
$^{12}C(p,p')^{12}C$ (15.11) should be related[10] by the equation

$$\frac{d\sigma}{d\Omega}(p,p') = \frac{d\sigma}{d\Omega}(p,n) \tag{1}$$

and the same relation should hold between the cross sections to
the 16.11 MeV state in ^{12}C and the first excited state in ^{12}N.
The effect of isospin mixing in the 15.1 MeV state is not
expected[11] to change the cross section by more than 5%, a value
within the experimental uncertainty of the results. Figure 1
shows the results obtained at 61.8 MeV. The curve is a DWBA
calculation for the $^{12}C(p,p')$ reaction and using Eq. (1) to apply
the normalization ot the (p,n) results. One should note that,
using the same normalization as for the $^{12}N(g.s.)$, the cross section to the first excited state in ^{12}N is well reproduced, as one
would expect. Also, the absolute cross section agrees to within
5% of the cross sections calculated by the KSU group,[1] well
within the 10% uncertainty of this method of normalization.

     We have also used this method at 120 MeV. Here, the first
excited state in ^{12}N is not resolved in the (p,n) measurement.
Thus, the (p,p') cross sections measured by Comfort[11] to the 15.1
and 16.1 MeV states must be summed to provide a valid comparison.

^7Li ACTIVATION

     A second means of normalization, is the use of activation
techniques to measure the $^7Li(p,n)^7Be$ (0.0 + 0.4 MeV) total

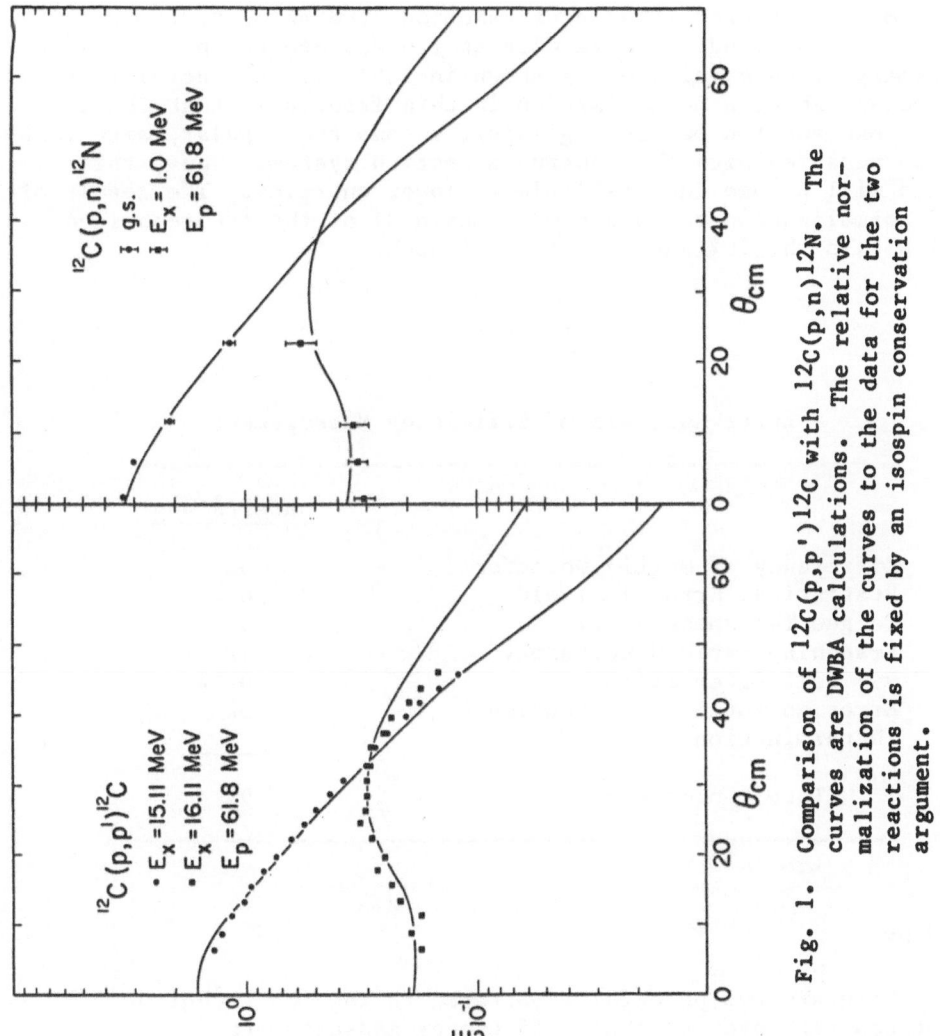

Fig. 1. Comparison of $^{12}C(p,p')^{12}C$ with $^{12}C(p,n)^{12}N$. The curves are DWBA calculations. The relative normalization of the curves to the data for the two reactions is fixed by an isospin conservation argument.

cross section. In a method similar to that used by Schery,[12] the beta decay of the residual ^{7}Be nuclei is measured to obtain the total cross section. The 53.4 day half life is long enough so that the targets can be counted at a convenient time after activation. A relative angular distribution is obtained for the ^{7}Li(p,n)^{7}Be (0.0 + 0.4 MeV) reaction. The angular distribution is then fit with a DWBA calculation. By normalizing the calculation to the measured total cross section, the absolute cross section is obtained. The results at 120 MeV are shown in fig. 2. A summary of uncertainties is shown in table 1. A crucial feature for establishing a normalization in this fashion is that the relative cross section become negligible beyond the angular range that can be measured with the neutron detection system. Thus, this method will become less reliable at lower energies. The result of this normalization technique was within 3% of the ^{12}C technique, well within the 10% uncertainties of each.

Table 1

Error Analysis of Efficiency Measurement

	Uncertainty
Efficiency of Ge(Li) Detector	5%
Statistical Error in Yield	4%
Target Thickness Error	5%
Branching Ratio Uncertainty	1%
γ – Self Attenuation	0.5%
Error in Total Cross Section Determination	3%
Total Uncertainty	9%

SUMMARY

There are two powerful advantages to the two techniques discussed here over separate efficiency measurements.

First, these measurements are an integral part of the experiment. The efficiency measurement is made with the detectors in their normal locations under normal operating conditions. Thus, the discrimination levels are certain to be those used during the experiment itself.

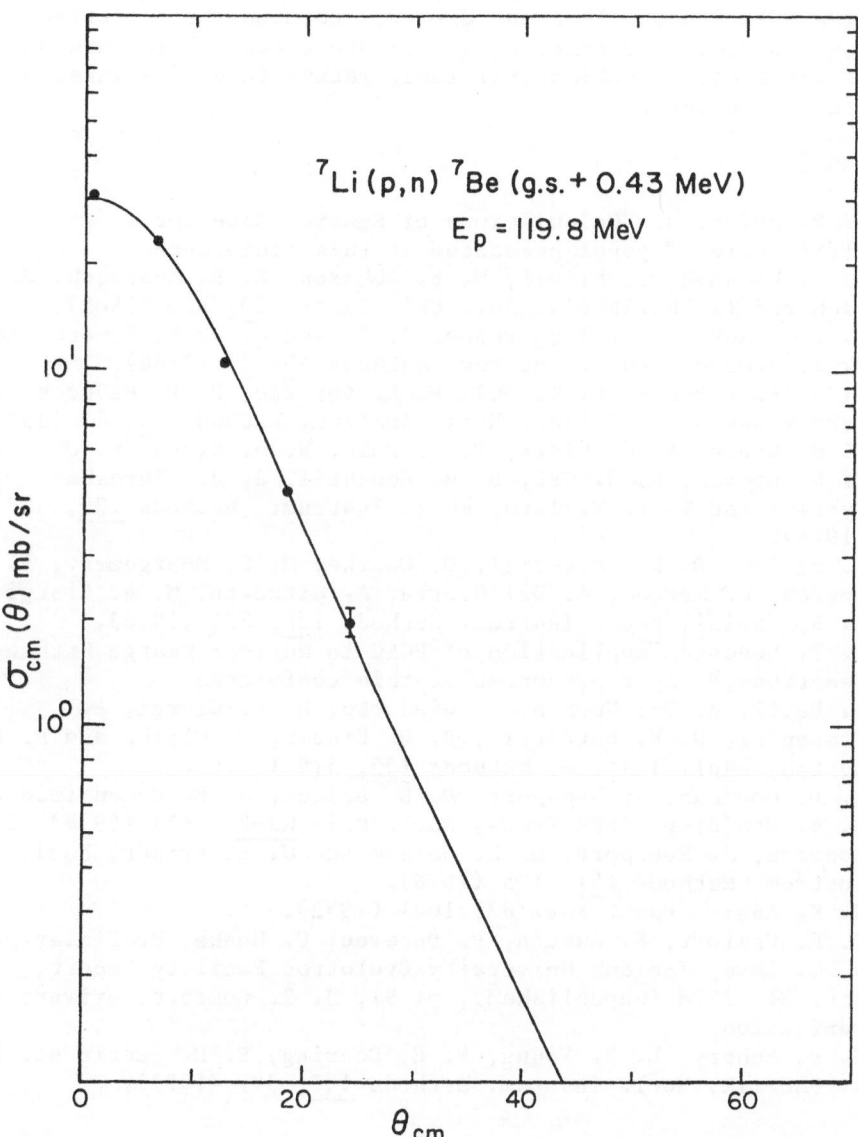

Fig. 2. Differential cross section for ^{7}Li(p,n)^{7}Be. The absolute normalization is obtained by setting the total cross section represented by the DWBA calculation equal to that measured from counting the ^{7}Be radioactivity.

Secondly, these techniques do not directly measure the
intrinsic efficiency but rather measure the product of solid
angle, attenuation of neutrons in the air in the flight path, and
detection efficiency. Thus one directly obtains the effective
efficiency needed to determine an absolute cross section and derives the intrinsic detection efficiency rather than vice versa as
with other techniques.

REFERENCES

1. B. D. Anderson, "Calculations of Neutron Detector
   Efficiencies," paper presented at this conference.
2. C. E. Wiegand, T. Elioff, W. B. Johnson, L. B. Auerbach, J.
   Lach and T. Ypsilantis, Rev. Sci. Instr. $\underline{33}$, 526 (1962).
3. F. P. Brady, J. A. Jungermann, J. C. Young, J. L. Ronero and
   P. J. Symonds, Nucl. Instrum. Methods $\underline{58}$, 57 (1968).
4. A. S. L. Parsons, P. Truoel, P. A. Berardo, R. P. Haddock, L.
   Verhey and M. E. Zeller, Nucl. Instrum. Methods $\underline{79}$, 43 (1970).
5. R. M. Brown, A. G. Clark, P. J. Duke, W. M. Evans, R. J. Gray,
   E. S. Groves, R. J. Ott, H. R. Renshall, J. J. Thresher, M. W.
   Tyrrell and T. B. Willard, Nucl. Instrum. Methods $\underline{136}$, 307
   (1976).
6. J. Bailey, D. R. Botterill, D. Clarke, H. E. Montgomery, P. R
   Norton, G. Matone, A. Del Guerra, A. Giazotto, M. A. Giorgi and
   A. Stefanini, Nucl. Instrum. Methods $\underline{135}$, 331 (1976).
7. P. T. Debevec "Application of PCAC to Nuclear Charge Exchange
   Reactions," paper presented at this conference.
8. G. Betti, A. Del Guerra, A. Giazotto, M. A. Giorgi, A.
   Stefanini, D. R. Botterill, D. W. Braben, D. Clarke and P. R
   Norton, Nucl. Instrum. Methods $\underline{135}$, 319 (1976).
9. C. D. Goodman, J. Rapaport, D. E. Bainum, M. B. Greenfield and
   C. A. Goulding, IEEE Trans. Nucl. Sci. $\underline{NS-25}$, 577 (1978); C. D.
   Goodman, J. Rapaport, D. E. Bainum and C. E. Briant, Nucl.
   Instrum. Methods $\underline{151}$, 125 (1978).
10. R. K. Adair, Phys. Rev. $\underline{87}$, 1041 (1952).
11. J. R. Comfort, S. Austin, P. Debevec, G. Moake, R. Finlay and
    W. G. Love, Indiana University Cyclotron Facility Report,
    Jan. 31, 1978 (unpublished), p. 91; J. R. Comfort, private com-
    munication.
12. S. D. Schery, L. E. Young, R. R. Doering, S. M. Austin and R.
    K. Bhowmik, Nucl. Instrum. Methods $\underline{147}$, 399 (1977).

A SPECTROMETER FOR NEUTRON-IN, CHARGED-PARTICLE-OUT REACTIONS*

Roger W. Finlay

Ohio University

Athens, Ohio 45701

INTRODUCTION

Recent appearance in the nuclear physics literature of high quality measurements of neutron-induced charged-particle reactions has helped to rekindle interest in this largely-ignored branch of nuclear reaction studies. Other contributors to this conference[1,2] describe results obtained with the techniques developed at the University of California at Davis. This contribution will focus on our efforts to develop and extend an alternative technique-- using magnetic quadrupole lenses--which was pioneered at the Lawrence Livermore Laboratory. Since the Livermore methods have been previously described[3,4], comparison of the design features of the Ohio University spectrometer with the Livermore system should provide an efficient approach to the subject.

The first and most significant difference is that the Livermore spectrometer has been in service collecting data for nearly four years. The Ohio spectrometer was first assembled early in 1979 and has not yet passed the acid test of successful use in nuclear reaction studies. Nonetheless, enough progress has been achieved on the design and testing of the new instrument that a discussion at this time of its major features should be informative for many readers.

THE SPECTROMETER

The principal components of any detection system for charged particles produced by neutrons are: the neutron source, the target (radiator) chamber, the transport system, and the charged-particle

detector. A schematic drawing of the spectrometer system is shown
in Fig. 1.

The Neutron Source

Neutron beams at the Ohio University Tandem Van de Graaff
Laboratory are produced by means of the $T(p,n)^3He$, $D(d,n)^3He$, and
$T(d,n)^4He$ reactions. The gas cell which contains the deuterium or
tritium gas has been previously described[5] and has been used
extensively for studies of elastic and inelastic neutron scattering.
At the available incident beam energies, these reactions can be used
to produce neutrons in the energy regions from 0-12 MeV and 20-26 MeV
MeV. Neutron fluxes at 0° of about 2×10^9 n/sr sec (at 11 MeV) and
5×10^8 n/sr sec (at 26 MeV) are available in nanosecond bunches
with a 5 MHz repetition rate at the Ohio University facility. These
fluxes are two orders of magnitude smaller than those available at
the Rotating Target Neutron Source (RTNS) at Livermore, so it is
clear that a considerable price must be paid in terms of counting
rate if (n,z) reactions are to be studied at energies other than
14 MeV.

The neutron yield of the RTNS source reaction is isotropic;
the tandem Van de Graaff source reactions are strongly forward
peaked. This difference suggests that lower background rates and
improved signal-to-noise ratios should be available at outgoing-
particle angles other than 0° with the present experimental arrangement.

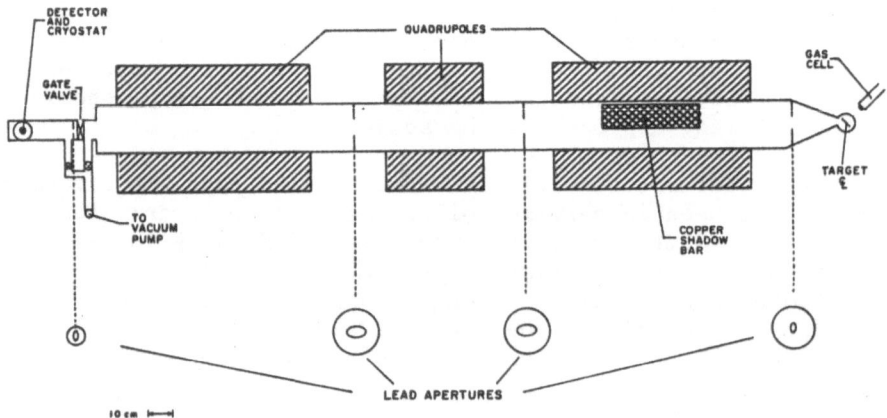

Fig. 1. Schematic view of magnetic quadrupole triplet spectrometer
         as viewed from above. An additional shadow bar may
         placed in the symmetric location in the exit lens at no
         loss in solid angle.

The T(d,n)^{4}He reaction at low (~200 keV) energy is a very clean
source reaction, i.e. very few neutrons are produced at energies
other than 14 MeV. At Van de Graaff energies, unwanted neutrons
from deuteron breakup and from reactions in the gas cell entrance
foil and beam stop constitute a significant fraction of the total
neutron flux. These lower-energy neutrons could produce charged
particles which are indistinguishable from the particles of interest
but are associated with different regions of excitation in the final
nucleus. This is a serious complication with the present design if
the aim of the program is to study regions of high excitation in
the residual nucleus but should not present a major difficulty in
the study of reactions of interest to this conference. In any case,
we are attempting to deal with the problem by developing an
appropriate sample-in, sample-out technique in which we attempt to
simulate the effects of the unwanted neutrons.

The Target Chamber

Targets are mounted on a conventional motor-driven target
ladder which moves in the vertical direction inside a 7.5 cm diameter, thin-walled aluminum cylinder which intersects the entrance
cone to the spectrometer. All surfaces of the chamber, the ladder,
and the entrance cone which could be a source of charged particles
are lined with lead of sufficient thickness to stop the most
penetrating charged particles of interest. Experience at Livermore
has shown that, at 14 MeV, very few charged particles are produced
by a clean surface of lead. The targets are 3.2 cm diameter foils
with typical thickness of a few mg/cm^2.

The Transport System

The central feature of the spectrometer is a magnetic quadrupole
triplet lens system of imposing proportions. The aperture of the
lens system is slightly greater than 20 cm. The first and third
elements are 82 cm in length while the central lens is 41 cm long.
The lenses have a venerable history. They were constructed for the
old Princeton-Penn accelerator, resided for a time at Cornell, and
reached Ohio as excess property. Although mass is not necessarily
a relevant quantity in the description of a magnetic transport
system, the 10,000 kg mass of Ohio system presented qualitatively
different engineering problems from the 400 kg system at Livermore.

Of course, the reason for such a dramatic upscaling of the
physical dimensions of the quadrupole system is to compensate in
part for the equally dramatic loss in neutron flux described in an
earlier section. Since the magnets were originally intended to serve
in the transport system of a few-GeV accelerator, their magnetic
characteristics are fully adequate for the present purposes.

Maximum field gradients of 970 G/cm are available at currents of
1000 amperes, and this is well in excess of the requirements for
focusing 30 MeV protons at the adopted radiator-to-detector distance
of 3.4 m.

A magnetic quadrupole triplet spectrometer has one important
advantage over the more familiar magnetic quadrupole doublet: it
can be operated with unit magnification in each plane. Charged
particles emerging from a circular radiator are focussed on a
circular detector.

The magnetic performance of the quadrupole lens system can be
characterized in terms of two parameters: the effective solid angle
and the bandpass. Although both of these quantities are best determined empirically, it is useful to be able to estimate their
values from magnet-optics calculations. The computer codes TURTLE[6]
and TRANSPORT[7] were used at many stages of the system design.
For a given particle type and energy, radiator-to-detector distance,
and a specified detector size, TRANSPORT can be used to search for
those gradients which provide the desired focal condition, i.e.,
unit magnification in each plane. The gradients are then entered
into TURTLE which calculates the effective solid angle by determining
the histories of particles of energy E emitted at various angles
from various points on the radiator surface. The calculation may
then be repeated for particle energies E ± δE at the same gradients.
A plot of the effective solid angle vs. energy--at fixed gradient--
yields the bandpass which we define as ΔE (FWHM) divided by the
energy at which the effective solid angle is a maximum. The bandpass
is a sensitive function of the areas of the radiator and the detector. The larger the bandpass, the fewer magnet settings are
required to cover a wide spectrum of outgoing particle energies.
Various constraints such as the inside diameter of the spectrometer
tube and collimator and detector dimensions can be entered into
TURTLE. Alternatively, the envelope of the extreme rays which reach
the detector may be determined in order to choose appropriate
dimensions for collimators and additional neutron shielding.

The essential correctness of the magnet optics calculation
was verified for the Ohio University spectrometer by measuring the
counting rate of a large-area ^{244}Cm alpha source with and without
magnetic field. The anticipated solid angle (10 msr) was obtained
at the predicted gradient settings. The eventual calibration
of both the effective solid angle and the bandpass will be determined by measuring the recoil-proton and -deuteron spectra from
thick targets of CH_2 and CD_2.

Fig. 1 also shows the tentative positions of a thick copper
shadow bar and a series of thin lead collimators. The collimators
intercept charged particles which originate at points other than

the radiator. The shadow bar serves two vital functions: it pre-
vents charged particles from travelling down the zero-magnetic-
field region along the central axis thus increasing the momentum
selectivity of the spectrometer, and it blocks the direct path of
source neutrons toward the detector. In the present application,
the source-to-radiator distance is less than the radius of the
spectrometer tube. Thus, the path from source to detector lies
inside the spectrometer tube. No shadowing of direct neutrons is
obtained from the 2 meters of magnet iron; in fact, the inner surfaces of this enormous mass are a formidable source of in-scattered
neutrons.

Detectors

     Assuming that the problems of neutron and gamma ray shielding
can be adequately mastered, there is much to gain by increasing the
area of the charged-particle detector. Recent development of the
large-area intrinsic Germanium detector make it an obvious choice
for this application since it is readily available in the appropriate
sensitive thickness (~4 mm) and, more importantly, since neutron
damage can be annealled out by modest heating in situ. Since the
energy spread of the neutron source reaction is usually greater
than 100 keV, there are no stringent requirements on detector energy
resolution.

     At a given gradient setting, all particles with the same value
of ME/Z^2 will reach the detector, so it is necessary to adopt some
means of particle identification. At Livermore this is accomplished
by means of a standard E-ΔE particle telescope. Since it is often
of interest to measure outgoing charged particles of very low energy,
very thin ΔE counters are required. Two problems associated with
thin Si ΔE counters are small area and small pulse height for the
high energy particles. Moreover, the thin Si counters deteriorate
fairly rapidly in the neutron flux. Recent development of a thin-
window gas proportional counter at Livermore[3,8] shows great
promise for alleviating this situation.

     The Ohio University spectrometer was designed to use the
charged-particle time-of-flight technique for particle identification. At a flight path of 3.4 m, protons, deuterons, tritons, etc.
of the same magnetic rigidity are well separated in flight time.
Deuterons and alphas of the same rigidity will have the same time
of flight but will differ by a factor of two in energy. With our
present gas cell technology, no loss in neutron flux results from
beam pulsing since the average current in the pulsed beam is near
the upper limit of the tolerance of the gas cell. The method seems
very promising provided that the counting rate for neutron-induced
reactions in and around the detector can be reduced to an acceptable

level. In the long run, it may prove advantageous to invest some time and effort into the development of more intense neutron sources which can utilize the full 100 μamp (D.C.) capability of the Ohio University Tandem. Even if the pulsed-beam method proves satisfactory, the development of a large-area ΔE counter (serving more as a background-reducing coincidence requirement than as a mass-identification device) may be very beneficial.

Two practical features of the new spectrometer deserve brief mention. 1) Since hydrocarbon oils from the vacuum pump could be a prolific source of recoil protons, an oil-free vacuum system was chosen. It consists of two Varian Vacsorb[9] cryopumps which are capable of evacuating the 100 liter volume of the system to ~10 μm of Hg in a few minutes. 2) Angular distributions will be measured by rotating the spectrometer about a vertical axis through the fixed radiator position. In the present design, the entire spectrometer can be moved by one person with the help of commercially available air pads[10] which require no fixed track. The laboratory space remains available for other uses such as neutron scattering experiments. Both of these features are clearly visible in Fig. 2.

Fig. 2. The Spectrometer. The target chamber is the vertical cylinder at the extreme right. The motor-driven target ladder is not shown. The pivot post (lower right) attaches rigidly to the floor close to the neutron source. The vertical cylinder in the upper left is the liquid nitrogen reservoir for the intrinsic Germanium detector.

SUMMARY

A large-aperture, magnetic quadrupole triplet spectrometer has been constructed and has undergone preliminary testing. The spectrometer utilizes--but is not limited to--time-of-flight as a means of mass identification. When used with the Ohio University pulsed neutron source under optimum conditions (E_n = 11 MeV, detector area = 800 mm^2, target mass number = 27, thickness = 4 x $10^{-3}g/cm^2$) a counting rate of 0.4 cts/sec/mb is expected. Other circumstances may lower this rate by one or more orders of magnitude.

The author wishes to acknowledge the tireless efforts of S.M. Grimes and J. Rapaport in bringing the project to its present state. The cooperation of Boyce McDaniel, Director of the Laboratory of Nuclear Studies at Cornell is also gratefully acknowledged.

REFERENCES

1. F.P. Brady, Contribution to this Conference.
2. N.S.P. King, Contribution to this Conference.
3. S.M. Grimes, R.C. Haight, J.D. Anderson, K.R. Alvar and R.R. Borchers, Symposium on Neutron Cross Sections from 10 to 40 MeV, Upton, N.Y. (1977), BNL-NCS-50681, M.R. Bhat and S. Pearlstein, ed., July 1977, pp. 297-304.
4. K.R. Alvar, H.H. Barschall, R.R. Borchers, S.M. Grimes and R.C. Haight, Nucl. Inst. and Meth. 148:303 (1978).
5. J.D. Carlson, Nucl. Instr. and Meth. 113:541 (1973).
6. D.C. Carey, NAL-64, 2041.000 (1978).
7. K.L. Brown, F. Rothacker, D.C. Carey and C.H. Iselin, SLAC-91, Rev. 2, UC-28 (1977).
8. R.C. Haight, private communication.
9. Vacuum Division, Varian Corp., Palo Alto, California.
10. Rolair Systems, Inc., Santa Barbara, California.
*   Supported in part by the National Science Foundation and by the Department of Energy.

THE (n,p) REACTION ON LIGHT N=Z NUCLEI AT 60 MeV

F. Paul Brady and G.A. Needham[*]

Crocker Nuclear Laboratory and Department of Physics
University of California
Davis, CA 95616

INTRODUCTION

The excitation of highly collective giant resonance states via inelastic electron and hadron scattering has received considerable attention recently.[1,2] These measurements supplement earlier photo-nuclear[3] and nucleon-capture[4] data. Population of the analogs of giant resonances by radiative pion[5] and muon[6] capture and by charge exchange reactions has extended the scope of such studies. Pion capture and charge-exchange reactions with $\Delta T_3 = +1$ [e.g. (n,p) and (t,h)] offer the advantage of populating only the analogs of isovector excitations. Charge exchange reactions also allow the study of cross section variation with momentum transfer via the measurements of angular distributions, and hence determination of resonance multipolarities.

The isospin selection rules of (n,p) are easily understood by noting that in this reaction T_3 increases by one unit. Except for ^{3}He, stable nuclei have $T_3 \geq 0$; therefore $\Delta T_3 = +1$ implies that $\Delta T = +1$. Figure 1 (left) shows the isospin of the states resulting from isovector excitations with various reactions. The relative strengths, given by the isospin factor of $(T_0 1 \ T_0 q / T_f \ T_0 + q)^2$ are tabulated in Fig. 1. Also shown is the other isospin factor from the light ions which gives a factor of two for σ_{np}/σ_{pp}. The total isovector strength of the (n,p) reaction is concentrated in the $T = T_0 + 1$ state (or states) while that for (p,p') and (p,n) is split. For large T_0 (other factors being equal) the (p,n) tends to populate $T_f = T_0 - 1$ while the (p,p') populates $T_f = T_0$. The (n,p) always populates $T_f = T_0 + 1$.

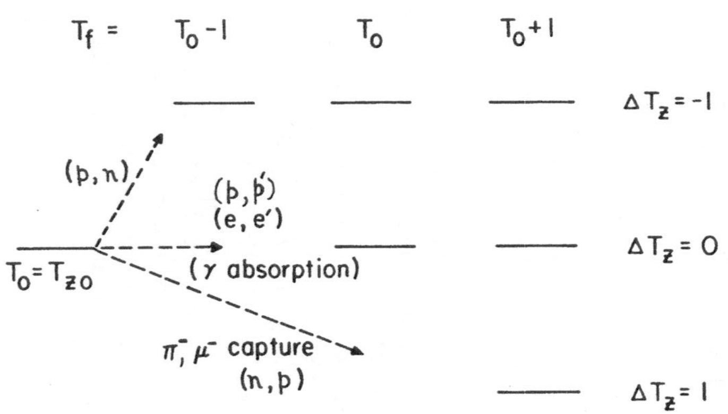

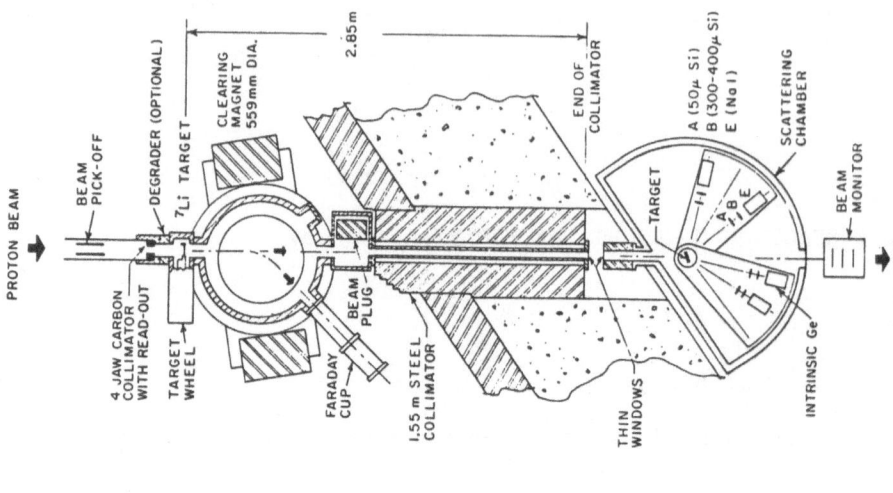

$$T_f = \quad T_0-1 \qquad T_0 \qquad T_0+1$$

$$\Delta T_z = -1$$

$$\Delta T_z = 0$$

$$\Delta T_z = 1$$

(p,n) (p,p') (e,e')

$(\gamma \text{ absorption})$

π^-, μ^- capture (n,p)

$T_0 = T_{z0}$

Fig. 1. Isovector excitations and strength factors.

q \\ T_0	T_0-1	T_0	T_0+1	Geometrical Strength Factors $\left[C^{T_0\ 1\ T}_{T_0\ q\ T_0+q}\right]^2$ for Isovector Excitations	$\left[C^{t_f\ 1\ t_i}_{n_f\ q\ n_i}\right]^2$
-1	$\left(\dfrac{2T_0-1}{2T_0+1}\right)$		$\dfrac{1}{(T_0+1)(2T_0+1)}$		$(p,n):\ \dfrac{2}{3}$
0		$\dfrac{1}{(T_0+1)}$	$\left(\dfrac{1}{T_0+1}\right)$		$(p,p'):\ \dfrac{1}{3}$
1		$\left(\dfrac{T_0}{T_0+1}\right)$	1		$(n,p):\ \dfrac{2}{3}$

Fig. 2. Experimental facilities for (n,p) studies.

Besides isospin there are dynamical factors which play an important role.[7] For example, in heavy nuclei the $T_>$ dipole transition is almost completely blocked because neutrons have nearly filled the orbitals of the next major shell.

In this work we report on the measurements of (n,p) cross sections for N=Z targets. Much of the work is exploratory. Comparisons are made with other reactions such as (p,n), (d,^{2}He), (γ,N) and with DWBA calculations. The following paper (by N.S.P. King) deals with N>Z targets.

EXPERIMENTAL METHOD

The facilities used for these studies are shown in Fig. 2, and described more fully elsewhere.[8] The unpolarized neutron beam is obtained from the ^{7}Li(p,n) reaction at 0°, collimated over a 3m flight path into a scattering chamber. The neutron flux on target is 10^6 n/sec (over a 1.8 cm wide by 3.5 cm high beam spot) in a peak of $\Delta E \simeq 850$ keV for proton beam currents of 14 µA. Scattered charged particles are detected in six ΔE–E telescopes with ΔE being measured in Si surface barrier detectors (300–400µ thick) and E in Ge (intrinsic), NE 102, and four NaI detectors. Three parameters, E, ΔE, and time-of-flight (TOF) for each telescope are stored event-by-event on magnetic tape. Overall energy resolution is typically $\simeq 1$ MeV FWHM resulting from neutron beam width, energy straggling in the target, intrinsic resolution of the E detectors, and kinematic effects. TOF is used to isolate events due to the full-energy neutron beam peak, and particle cuts on ΔE vs. E (or E + ΔE) are used to select the protons, deuterons, etc., and to construct particle energy spectra.

Natural targets were used in all cases. The gases were cooled to near liquid nitrogen temperature and pressurized to $\simeq 50$ psi in a 1.25 mil thick nickel bag of 5 cm diameter. Solid CH$_2$ and H$_2$ gas targets were used to normalize to the absolute n-p cross sections. The (n,d) spectra peaks were valuable for energy calibration. The uncertainty in the location of peak energies is $\simeq 0.3$ MeV. The finite dimensions of the beam and detectors were such that 68% of the angular acceptance is within 2° lab of the central detector angle. The resulting corrections to the cross sections were only about 4%. The corrections for nuclear interactions ranged up to about 4.5%.

EXPERIMENTAL RESULTS AND ANALYSIS

The ^{4}He(n,p)^{4}H spectrum, at $E_n = 59.6$ MeV (fig. 3), shows evidence for enhancements at Q = −23.2 MeV, and 27.2 MeV. Conven-

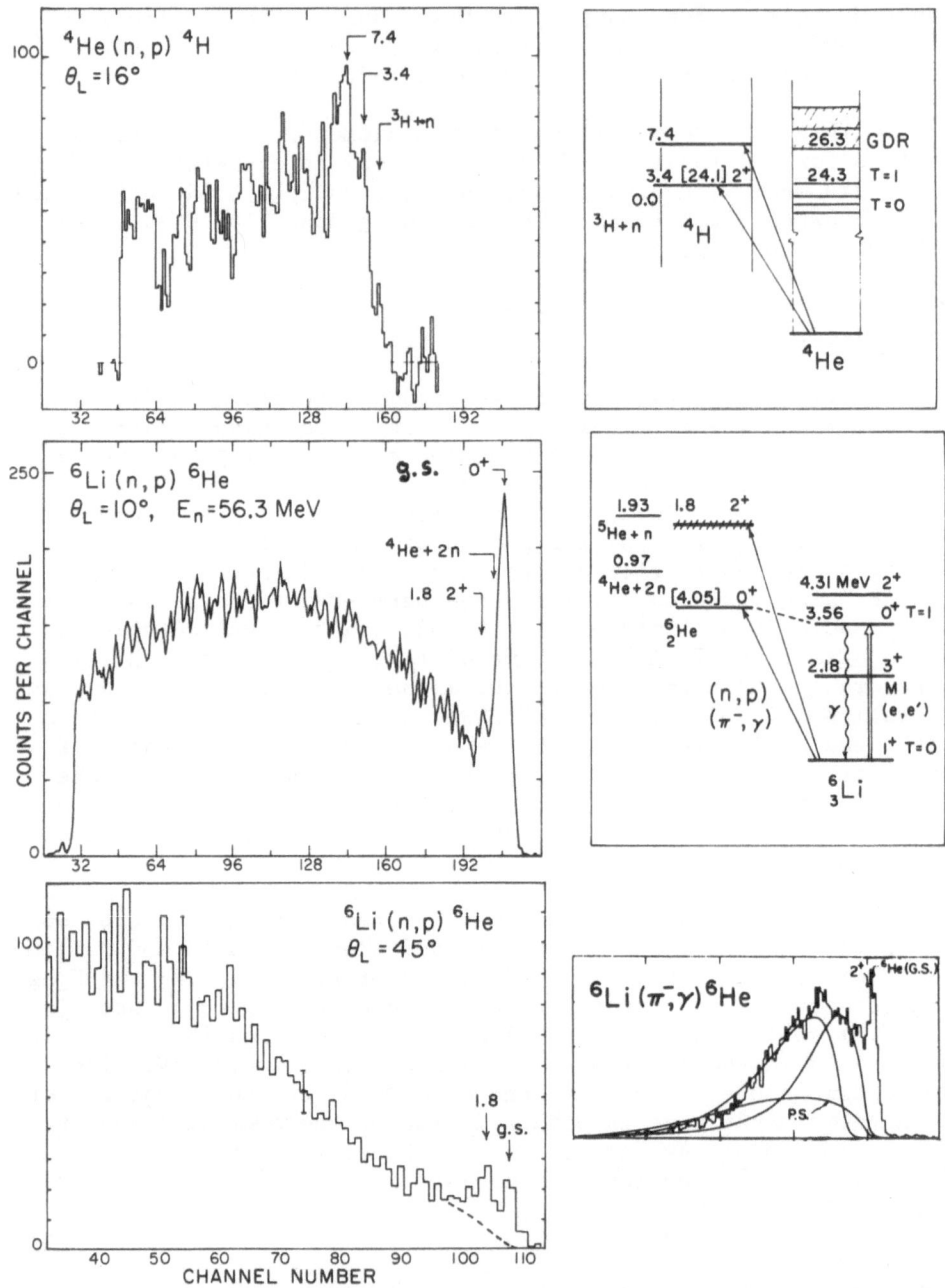

Fig. 3.   ⁴He(n,p)⁴H, ⁶Li(n,p)⁶He energy spectra and
⁶Li(π⁻,γ)⁶He spectrum (ref. 5).

tionally[9] the "ground state" of ^{4}H is taken to be 3.4 MeV above the ^{3}H + N mass, 24.1 MeV above the ^{4}He mass.[9] The second enhancement (Q = -27.2) is thus at an excitation of 7.4 MeV above the ^{3}H + n mass, or 4 MeV above the ^{4}H ground state.

The ^{6}Li(n,p)^{6}He spectrum at 10° (fig. 3) shows the prominent transition to the analog of the M1 state, which is very strongly populated in backward (e,e') reactions.[10] Such strength, as predicted,[11] is seen in all N=Z and other p-shell nuclei.[10] In the 45° spectrum one can also see the 2$^+$ first excited state of ^{6}He. The GDR is very broad and we are not able to separate it clearly from the continuum due to (n,np) and other reactions. The angular distribution (g.s.) is almost pure L = 0, S = 1 transfer,[12] with little or no L = 2. Allowing for kinematic effects and isospin factors, agreement with (p,p')[13] and (p,n)[14] cross sections is quite good.

The ^{12}C spectra (fig. 4) show three prominent structures[15]: the g.s. (1$^+$), which is M1 analog, and 0.95 MeV (2$^+$) doublet, the 4.37 (2$^-$) and 4.54 MeV (4$^-$) doublet,[16] and the 7.7 MeV, GDR analog.[17] The dashed line in the phase space (PS) subtracted spectrum at 10° gives the shape of the total photonuclear cross section[18] which, along with the (γ,xn) cross section,[17] agrees well with the (n,p) excitation of the GDR analog. This PS continuum is calculated assuming three-body phase space with the end-point usually specified by the (n,np) Q value, and the normalization determined by fitting to the high excitation region of the (n,p) spectrum. The three structures stand out so clearly in the (n,p) spectra as compared to (p,p') spectra because the strong isoscalar excitations are absent or very small in the (n,p) case. The ^{12}C (π^-,γ)^{12}B photon spectrum[5] is remarkably like an angle integrated (n,p) spectrum. As expected (p,n)[19] and (n,p) reactions on N=Z nuclei are very similar. Another charge exchange reaction of interest is (d,He^2)[20], in which the projectile neutron undergoes both spin and isospin flip. Comparison between ^{12}C (n,p)^{12}B and ^{12}C (d,^{2}He)^{12}B (fig. 4) shows strong similarity in the excitation of the magnetic spin-flip transitions to the (1$^+$) and (2$^-$,4$^-$) states. At forward angles the (n,p) excites strongly the GDR analog (1$^-$) at 7.7 MeV while the (d,^{2}He) does so only weakly, thus verifying that the GDR excitation is largely non-spin-flip.

The ^{12}C(n,p)^{12}B angular distributions[15] are shown in fig. 5 with DWBA fits[21] using optical parameters of the 1p shell[22] and macroscopic form factors.[23] An incoherent sum of L = 0 and L = 2 describes the summed ^{12}B g.s. (1$^+$) and 0.95 MeV (2$^+$) angular distribution. The angular distribution of the 4.3 MeV doublet, 4.37 (2$^-$) and 4.54 (4$^-$), in ^{12}B is well fit by a superposition of L = 1 and L = 3 transfer (and possibly some L = 5). These are also strongly excited in the spin-isospin-flip (d, ^{2}He) reaction (fig. 4).

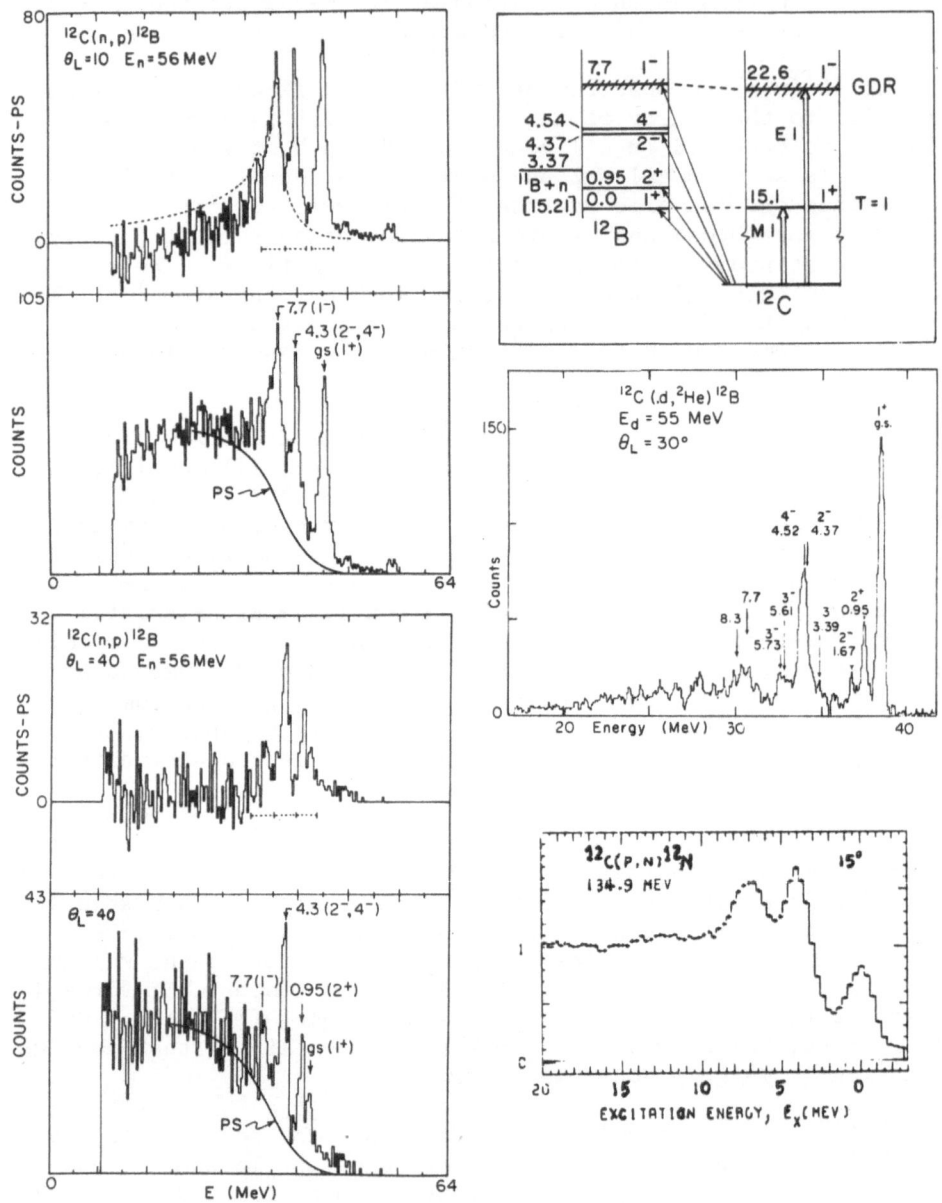

Fig. 4. $^{12}C(n,p)$, $(d,^2He)$ (ref. 19) and (p,n) (ref. 20) spectra.

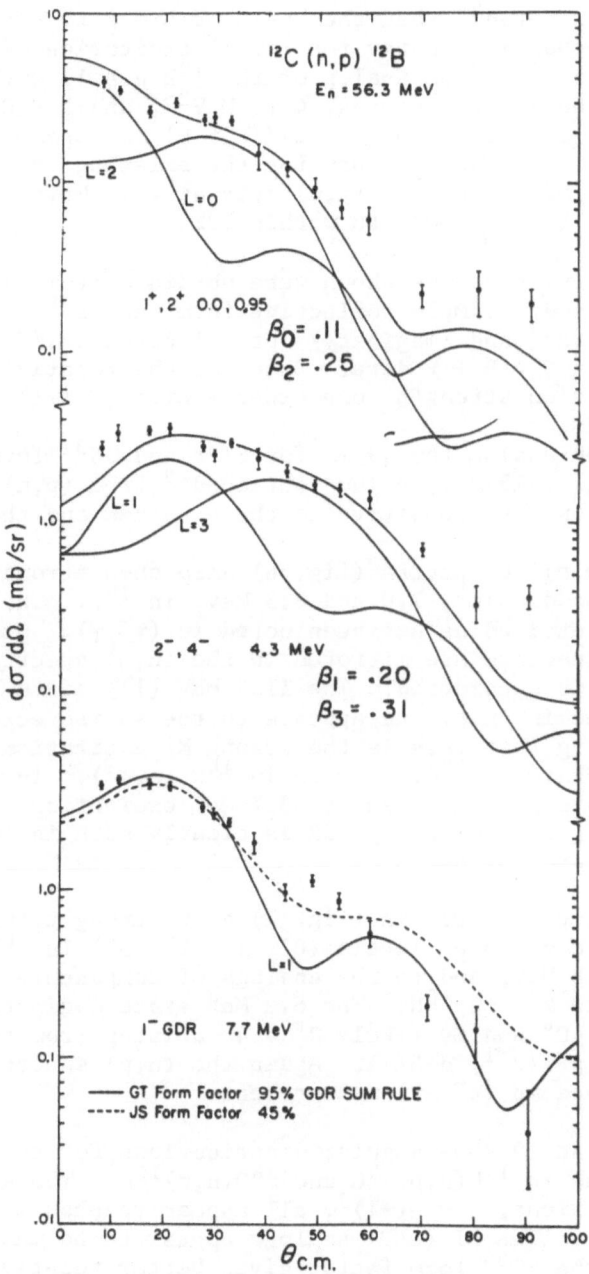

Fig. 5. ^{12}C(n,p)^{12}B angular distributions, with $\beta_L = \beta_L$ (eff).
Recent analysis indicate that all the data should be
renormalized by a factor of 0.81.

It has been noted[24] that the ratio of L = 2 to L = 0 appears
to vary with bombarding energy for (p,p') excitation of the
15.1 MeV state in ^{12}C (the analog of the ^{12}B g.s.), with L = 2
dominating at the lower energies, 45.5 MeV[25], and L = 0 at high
energy (186 MeV). The ratio (L = 2)/(L = 0) is apparently lower
for the present (n,p) data[15] than for the existing (p,p') data at
45.5[25] and 61[26] MeV. The ^{12}C (p,n) data at 49.5 MeV[27] show agreement with (n,p) (out to 50°) to within 10%.

The effective β values shown were obtained from fits made by
Satchler, who used a simple collective form factor[28] obtained by
deforming the (real and imaginary) optical potential[29] extrapolated
from ^{12}C, N^{14}, and ^{11}B + P data. Based on the relative nucleon-
nucleon interaction strengths one expects β(S=1, T=1) ≃ 3 β(eff).

For the GDR analog the fits for GT[30] and JS[31] form factors,
use V_1 = 10, W_{D1} = 15 MeV, values obtained[23] from (p,n) data. The
sum rule fractions are sensitive to these values and their ratio.

The ^{14}N (n,p)^{14}C spectra (fig. 6) also show strong analog M1
excitation: The doublet, 7.0 and 8.3 MeV, in ^{14}C. The ^{14}C g.s.
is very weakly excited as has been noted in (π^-,γ). Other features
of the (π^-,γ) spectrum are mirrored in the (n,p) spectra depending
on which angle is considered. The 11.3 MeV (1$^+$) state in ^{14}C is
strongly excited in (n,p) and appears to the an isovector E2 analog.
It is interesting that this is the strong M1 excitation relative
to the ground state of ^{14}C, as seen in ^{14}C (e,e'). In ^{14}N (e,e')
no particular strength is seen at 13.7 MeV excitation, the analog
excitation energy. The analog GDR is clearly seen in (n,p) particularly at forward angles.

The ^{16}O (n,p)^{16}N spectra (fig. 7) show strong transitions to
the ground state group of levels (0$^-$, 1$^-$, 2$^-$, 3$^-$) in ^{16}N, to states
near 6.2 and 7.8 MeV, and to the analogs of components of the GDR
centered near 10 MeV in ^{16}N. The 6.2 MeV state dominates the
spectra beyond 40° and is likely J^π = 4$^-$ arising from the stretched
configuration (p 3/2^{-1}, d 5/2). Again the (n,p) spectra are similar
to those obtained in (π^-, γ) measurements.

Figures 8 and 9 show angular distributions for the most prominent excitations in ^{14}N(n,p)^{14}C and ^{16}O(n,p)^{16}N. The analog
magnetic transitions, π = -(-1)^J, all appear to show mixing of L
values, L = J±1. The E1 (GDR) analogs appear to be mainly L = 1.
In both cases the GT[30] form factor gives better relative and
absolute fits than does the JS[31].

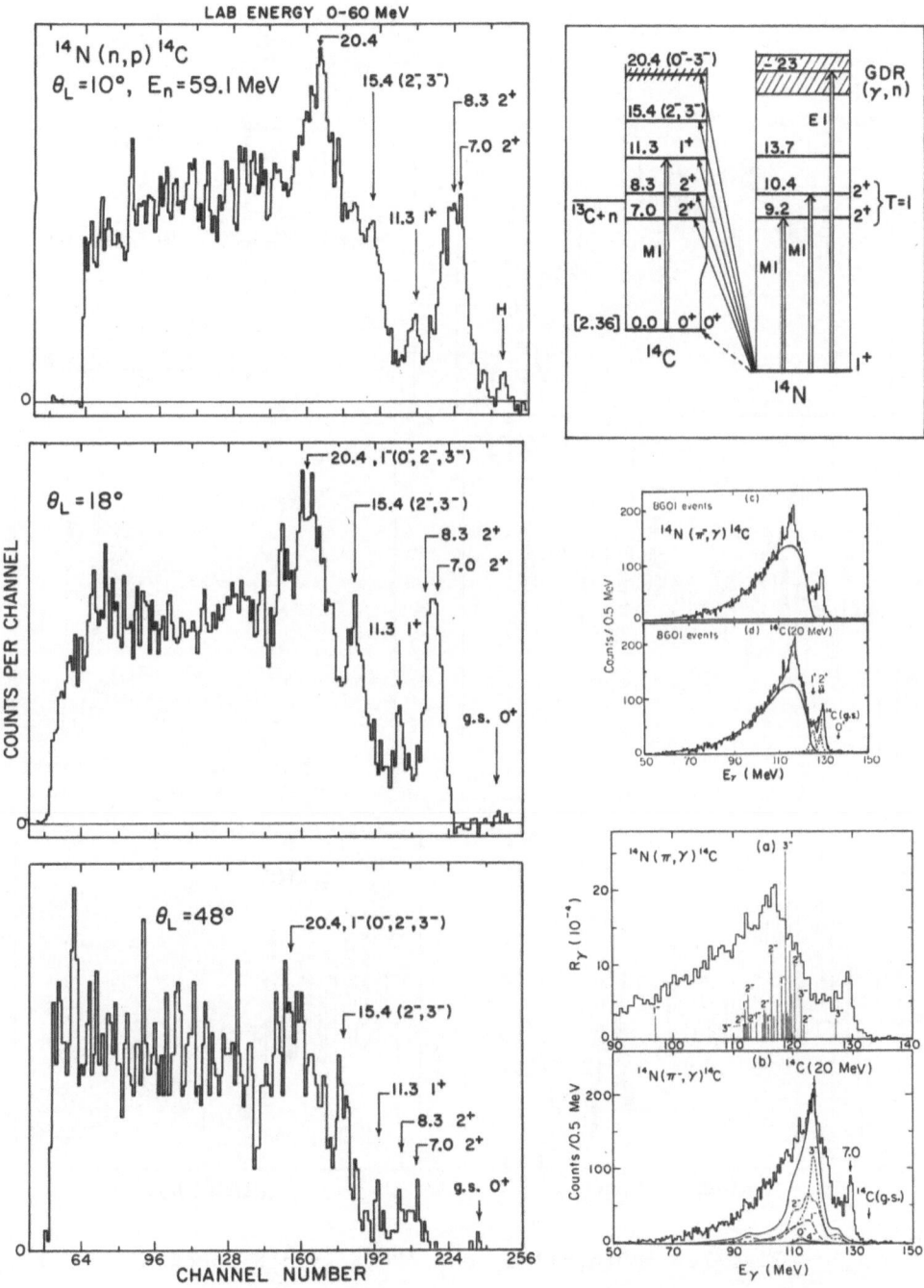

Fig. 6. ^{14}N(n,p)^{14}C spectra and ^{14}N(π^-, γ)^{14}C spectrum.

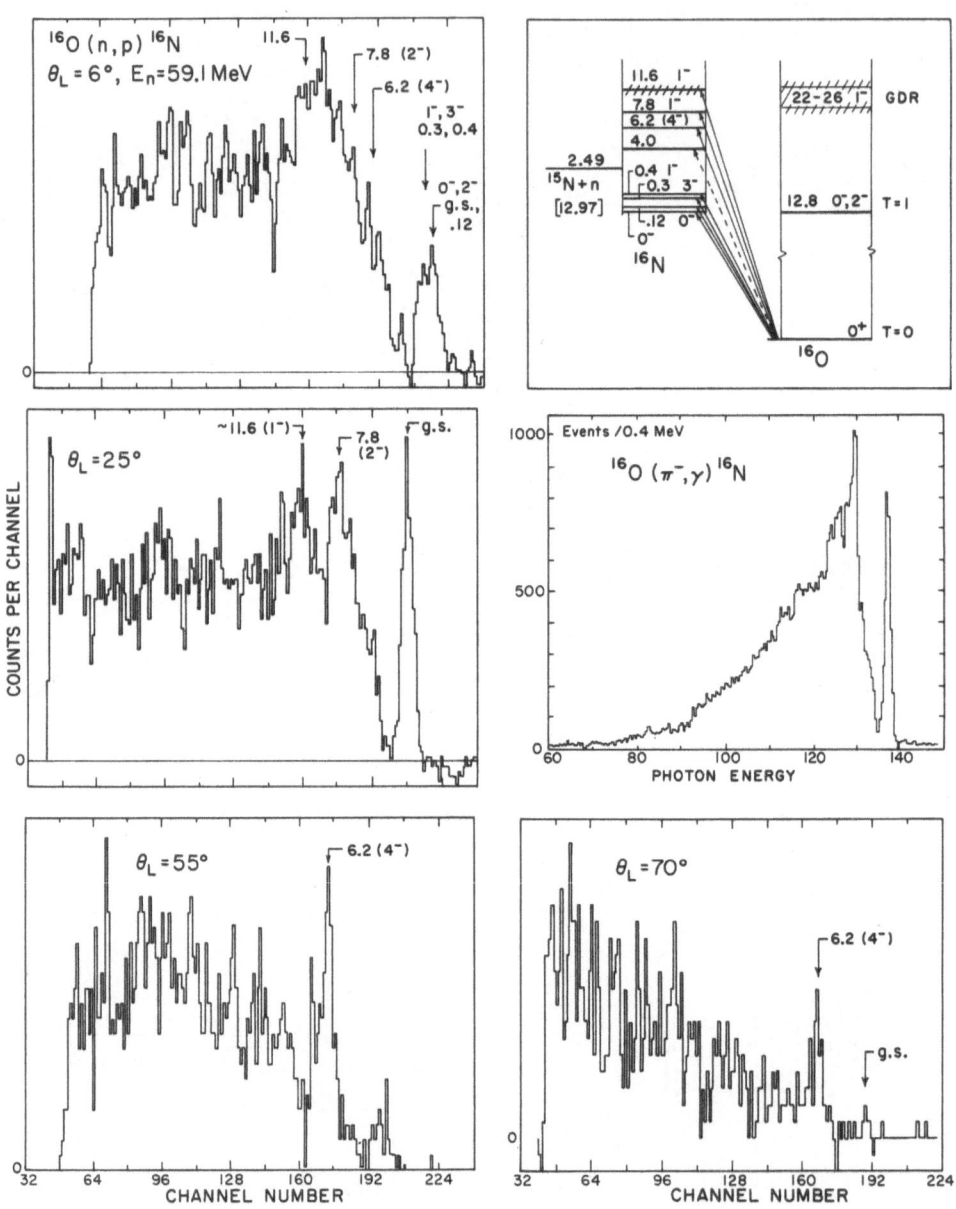

Fig. 7. $^{16}O(n,p)^{16}N$, and $^{16}O(\pi^-,\gamma)^{16}N$ spectra.

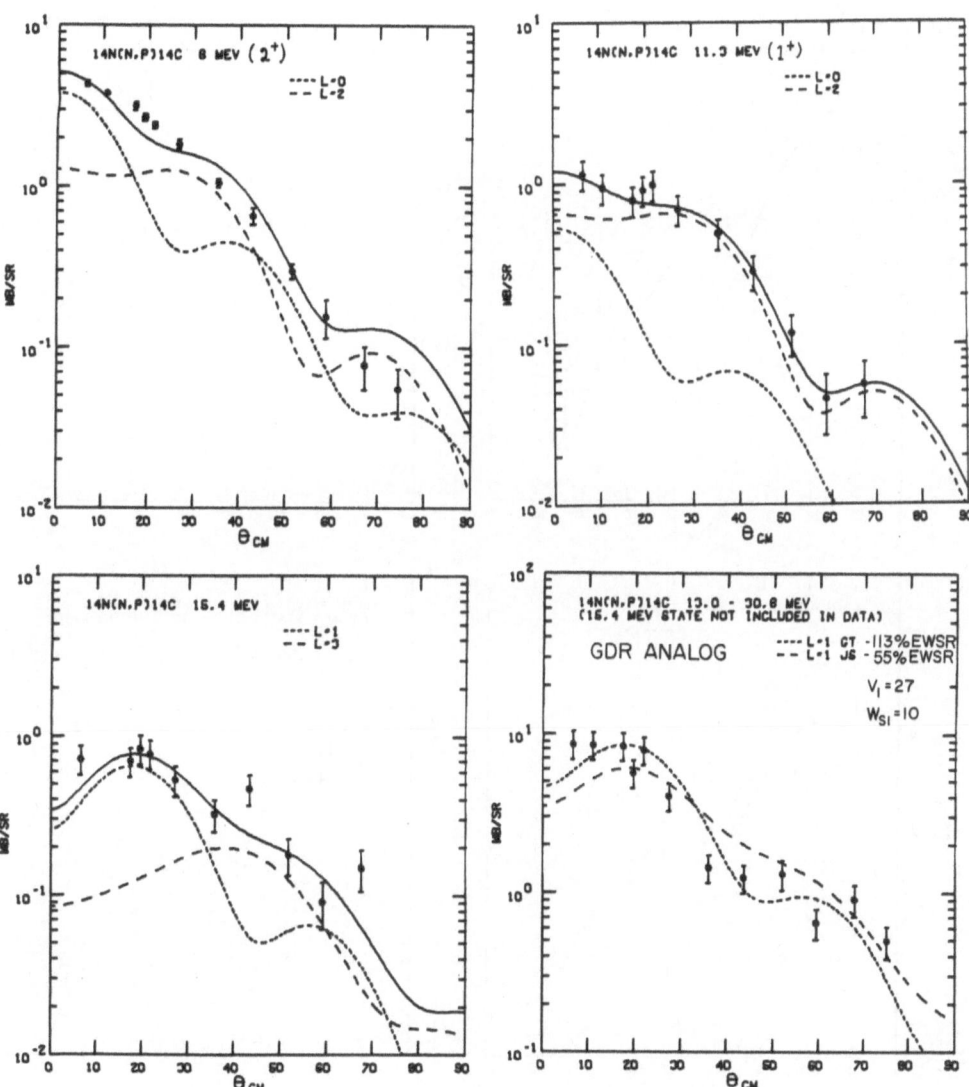

Fig. 8. ^{14}N(n,p)^{14}C angular distributions.

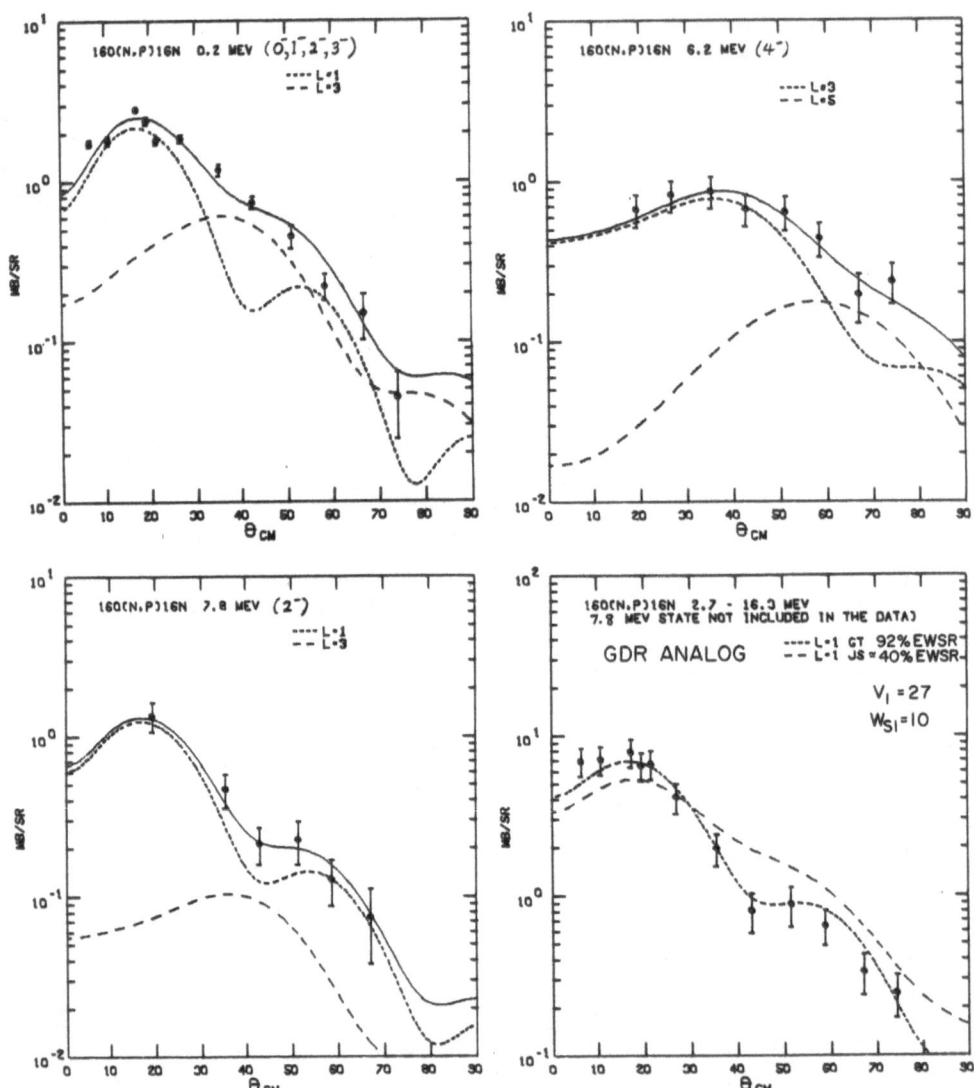

Fig. 9. $^{16}O(n,p)^{16}N$ angular distributions.

Fig. 10 shows spectra from ^{20}Ne(n,p)^{20}F and ^{28}Si(n,p)^{28}Al.
Very little analysis has been carried out. In both cases the GDR
analog is present. In ^{20}Ne the M1 is seen at 11.2 MeV[10], so the
analog is expected to be excited via (n,p) at ≃1.0 MeV in ^{20}F.
Some strength appears there in fig. 9 (top). In ^{28}Si(n,p) the
excited state at 2.1 MeV is assumed to be the analog of the M1
resonance at 11.4 MeV in ^{28}Si.[10] The state near 5.1 MeV in ^{28}Al
dominates the larger angle spectra indicating a large angular
momentum transfer. Comparison with (p,p')[32] and (e,e')[33] suggests
that this may be the 6⁻ seen in these reactions.

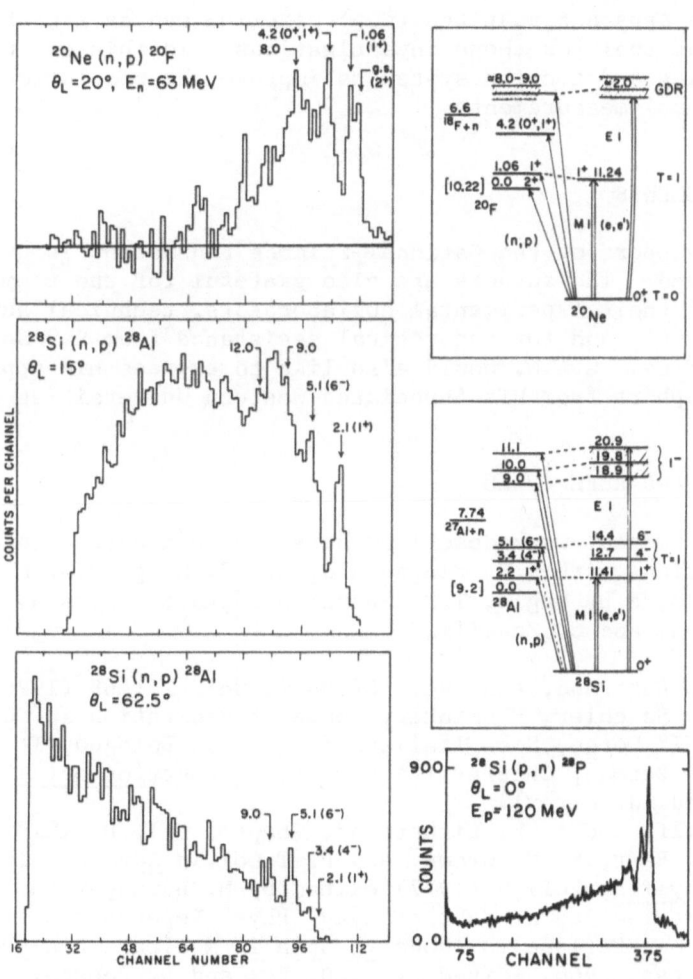

Fig. 10. ^{20}Ne(n,p)^{20}F (63 MeV) ^{28}Si(n,p)^{28}Al (59 MeV) and
^{28}Si(p,n)^{28}P spectra.

CONCLUSIONS

The (n,p) reaction appears to produce isovector collective excitations with a wide range of multipolarities, including those due to stretched particle-hole configurations. The magnetic analog angular distributions, except for that in $^6Li(n,p)^6He$, show mixing of L values, L = J±1. In terms of a microscopic one-step description, this indicates the need for spin-flip as can be provided by the spin-spin and/or tensor force components of the nucleon-nucleon interaction. The angular distributions of the analog electric dipole excitations appear to be better fit (in magnitude and shape) by the GT than by the JS form factor.

It is apparent that the (n,p) reaction can be a useful and even unique tool for these investigations. To this end we are constructing a detection system to improve the resolution and count rate in (n,p) measurements.

ACKNOWLEDGMENTS

The support of the National Science Foundation is gratefully acknowledged. The authors are also grateful for the tremendous efforts of their experimental collaborators, technical support of CNL personnel, and for theoretical assistance from Ray Satchler and Bill True. G.A.N. would also like to express his appreciation for the support from his Associated Western Universities Fellowship.

FOOTNOTES AND REFERENCES

*The experimental work described here has been carried out in collaborations with D.H. Fitzgerald, N.S.P. King, M.W. McNaughton, J.L. Romero, A.L. Sagle, T.S. Subramanian, J.L. Ullmann, P.P. Urone, J.W. Watson, and C. Zanelli.

1. F. E. Bertrand, Ann. Rev. of Nucl. Sci. 26:468 (1976).
2. G. R. Satchler, Elementary Modes of Excitation in Nuclei (1977) LXIX Corso, Soc. Italiana de Fisica, Bologna, Italy.
3. B. L. Berman, Chapter IXA of Nuclear Spectroscopy II (ed. J. Cerny).
4. C. Rolfs and A. E. Litherland, Chapter VII. D, ibid.
5. H. W. Baer, K. M. Crowe, and P. Truöl in Advances in Nuclear Physics, Vol. 9 (1977) edited by M. Baranger and E. Vogt.
6. For review see N. Mukhopadhyay, Phys. Reports 30C:1 (1977).
7. M. H. MacFarlane in Isobaric Spin in Nuclear Physics, Academic Press, 1966; edited by J. D. Fox and D. Robson.
8. J. L. Romero, T. S. Subramanian, F. P. Brady, N. S. P. King, and J. F. Harrison, Symposium on Neutron Cross-Sections

From <u>10</u> to <u>40</u> <u>MeV</u>, BNL-NCS-50681; (1977) ed. M. R. Bhat and S. Pearlestein.

9. S. Fiarman and W..E. Meyerhof, Nuclear Physics A206 (1973) 1.
10. See L. W. Fagg, R.M.P. 47:683 (1975), and references therein.
11. D. Kurath, Phys. Rev. 130:1525 (1963).
12. F. P. Brady, Progress Report UCD-CNL 196 and to be published.
13. G. S. Mani, D. Jacques, and A. B. Dix, Nucl. Phys. A165:45 (1971).
14. C. J. Batty, B. E. Bonner, E. Friedman, C. Tschalar, C. E. Williams, A. S. Clough, and J. B. Hunt, Nucl. Phys. A120:297 (1968).
15. N. S. P. King, F. P. Brady, M. W. McNaughton, and A. L. Sagle, UCD-CNL 190 and to be published.
16. R. O. Lane, C. E. Nelson, J. L. Adams, J. E. Monohan, A. J. Elwyn, F. P. Mooring, and A. Langsdorf, Jr., Phys. Rev. C2:2097 (1970).
17. Atlas of Photoneutron Cross Sections, B. L. Berman, UCRL-74622.
18. J. Ahrens, H. Borchert, H. B. Eppler, H. Gimm, H. Gundrum, P. Riehn, G. Sita Ram, A. Ziegler, M. Kroning, and B. B. Ziegler, Nuclear Structure Studies Using Electron Scattering and Photoreaction (Eds.: K. Shoda, H. Ui) Tokoku University Report 5 (1976).
19. B. Anderson, R. Madey, G. L. Moake, IUCF Report (1977).
20. R. Jahn, D. P. Stahel, G. J. Wozniak, and J. Cerny, LBL-8636.
21. DWUCK, written by P. D. Kuns, U. of Colorado, Boulder.
22. B. A. Watson, P. P. Singh, and R. E. Segal, Phys. Rev. 182:977 (1969).
23. G. R. Satchler, Nuclear Physics A195:1 (1972).
24. G. R. Satchler, Nucl. Phys. A100:497 (1967).
25. E. L. Peterson, I. Slaus, J. W. Verba, R. F. Carlson, and J. R. Richardson, Nucl. Phys. A102:145 (1967).
26. F. E. Bertrand and R. W. Peele, ORNL-4799, UC-34-Physics (1973).
27. A. S. Clough, C. J. Batty, B. E. Bonner, and L. E. Williams, Nucl. Phys. A143:385 (1970).
28. G. R. Satchler and W. G. Love, Nucl. Phys. A172:449 (1971).
29. G. R. Satchler, Particles and Nuclei 1:412 (1971).
30. M. Goldhaber and E. Teller, Phys, Rev. 74:1046 (1948).
31. H. Steinwedel and J. N. D. Jensen, Z. Naturforsch A5:413 (1950).
32. G. S. Adams, A. D. Bacher, G. T. Emery, W. P. Jones, R. T. Kouzes, D. W. Miller, A. Picklesimer, and G. E. Walker, Phys. Rev. Letters 38:1387 (1977).
33. T. W. Donnelly, Jr., J. D. Walecka, G. E. Walker, and I. Sick Physics Letters 32B:545 (1970).

THE (n,p) REACTION AT 60 MeV ON N>Z TARGETS

N.S.P. King

Los Alamos Scientific Laboratory
Los Alamos, New Mexico

J. L. Ullmann

Crocker Nuclear Laboratory and Department of Physics
University of California, Davis, California

INTRODUCTION

Historically, the work of Obu & Terasawa[1] and Measday[2] provided evidence that the (n,p) reaction showed some selectivity for population of collective states. However, until the recent experiments at UC Davis, no systematic studies with resolution $\leq$ 1 MeV were undertaken. A comparison of the (n,p) reaction on $T_0 = 0$ targets with photonuclear experiments, inelastic electron scattering (particularly at $\theta = 180°$), radiative pion capture and hadronic probes has revealed considerable sensitivity to known collective M1 and E1 states.[3] These properties of the (n,p) reaction are identical to those of the (p,n) reaction on the same targets since only $\Delta T = 1$ is allowed for both. The extension of the (n,p) reaction to $T_0 \neq 0$ targets leads to the same isovector ($\Delta T = 1$) selection rule. This is not the case for the (p,n) reaction which can populate $\Delta T = 0, \pm 1$ states favoring those with $\Delta T = -1$ and $\Delta T = 0$. This paper presents results of the (n,p) reaction at 60 MeV on ^{7}Li, ^{9}Be, ^{27}Al, ^{58,60,62,64}Ni, ^{90}Zr, and ^{209}Bi targets.

The emphasis will be on discussing qualitative features of the data through a comparison with existing results from other nuclear probes as well as observed properties of the isovector transitions anticipated from known isospin selection rules.

The usefulness of the (n,p) reaction for studying isovector excitations is readily made apparent by examining the isospin dependence in (p,p'), (n,n'),(p,n), and (n,p) scattering. The transition matrix for the (n,p) reaction contains an interaction potential of the form

$$V_{eff} = \sum_{i=1}^{A} V\vec{t}_1 \cdot \vec{t}_i (1 + \alpha\vec{\sigma}_1 \cdot \vec{\sigma}_i)$$

where $\vec{t}_1 \cdot \vec{t}_i$ can be expanded in terms of isospin raising and lowering operators between the incident projectile and a target nucleon. This interaction can lead to both isospin and isospin-spin flip transitions. Additional isospin independent terms also contribute to the (p,p') and (n,n') transitions. The ratio of isospin dependent to independent interaction strength for potentials commonly utilized in shell model calculations is ~1/3 so that isoscalar transitions in (p,p') and (n,n') reactions dominate by ~9:1. In regions where states of different isospin occur the isovector states may therefore be difficult to observe.

The isospin Clebsch-Gordon coefficients obtained by projecting out the dependence associated with the entrance and exit channels for a $\Delta T = 1$ transition leads to $\sigma(n,p) = 2\sigma(p,p'$ or n,n') for members of an isospin multiplet.

A more important geometric isospin effect for large T_0 is that due to the splitting of a particular multipolarity transition strength into isospin components for a $T_0 \geq 1/2$ target nucleus. Different allowed transitions are indicated in Fig. 1 as well as their relative strengths.[4] This shows that only the (n,p) reaction has no reduction in strength of the reduced matrix element to the $T_> = T_0 + 1$ states. If $M_{T>} = M_{T<}$ the entire, unsplit strength is available to the (n,p) reaction.

Additional factors involving available configurations for $T_>$ vs $T_<$ transition matrix elements will modify the above ratios since the interactions are in general isospin dependent.

EXPERIMENTAL RESULTS

The present data were obtained at the Crocker Nuclear Laboratory unpolarized neutron facility.[5] Time of flight restrictions on the incident neutrons allows (n,p) reaction data to be obtained with bombarding neutron energy resolution of ≤ 1 MeV FWHM for energies ~ 60 MeV. Recoil proton energy resolution is also ≤ 1 MeV. All of the targets used in the present work were ≤ 50 mg/cm^2 in thickness and isotopically pure to greater than 99%.

$^7Li(n,p)^7He$ and $^9Be(n,p)^9Li$

The isovector M1 selectivity found for low momentum transfer in $T_0 = 0$ targets should be modified due to the $T_0 = 1/2$ ground states in ^7Li and ^9Be. In this case a splitting of the $\Delta T = 1$ M1 strength to the g.s. is allowed between $T_> = 3/2$ and $T_< = 1/2$ components. Theoretical calculations [8] indicate most of the M1 strength should be concentrated in the $T_<$ states for both ^7Li and ^9Be due to less spatial symmetry in the $T = 3/2$ and $T = 1/2$ wavefunctions. This result is nicely confirmed by 180° (e,e') data.[6] More recently, Baer[7] pointed out the weak $\Delta T = 1$ M1 strength to the $T = 3/2$ parent analogs in the $^7Li(\pi^-,\gamma)^7He$ and ^9Be $(\pi^-, \gamma)^9Li$ data from SIN. Further confirmation of this effect is shown in Figures 2 and 3 for the (n,p) reaction on ^7Li and ^9Be. The ground state of ^7He ($J^\pi = 3/2$, $T = 3/2$) is the parent analog to the 11.25 MeV state in ^7Li. In ^9Li the $(3/2^-, 3/2)$ g.s. and the 2.69 MeV $(1/2^-, 3/2)$ state form parent analogs to the M1 states at 14.4 and 17.0 MeV in ^9Be seen in the inelastic electron scattering. The angular distributions for these three states are consistent with an $L = 2$ transfer based on DWBA calculations utilizing a macroscopic form factor.

Microscopic form factor DWBA calculations including exchange terms will be required before a detailed comparison can be made with measured M1 matrix elements. The strength of the ^9Li g.s. transition is however, ~.2mb/sr for $\theta_{c.m.} \leq 40°$ which is at least an order of magnitude less than observed cross sections for collective M1 strength in the (n,p) reaction data for parent analog M1 transitions for A = 6 and 12. Inelastic electron results also indicate a ratio of roughly 10 to 1 for even A vs odd A $T_>$ M1 strengths.

Evidence can be seen in Fig. 3 for two peaks above a smooth phase space or pre-equilibrium energy distribution centered at 7.5 and ~18 MeV excitation energy. Similar structure has been found in (π^-,γ) data on 1p shell nuclei and interpreted as configurational splitting of the $T_>$ giant dipole resonance (GDR). This effect has been predicted by Dogotar et al[8] as having its origins in the fact that $1\hbar\omega$ transitions have different energies for core and valence nucleons. The relevant major oscillator shell 1^- transitions for A = 9 involve the $1s(1/2) \to 1p(3/2)$ spacing of ~28 MeV and $1p(3/2) \to 1d(5/2)$ spacing of ~16 MeV. The difference of 12 MeV is consistent with the observed splitting of the $T_>$ GDR region in the $^9Be(n,p)^9Li$ reaction. The corresponding GDR region in ^9Be at 22 and 32 MeV may well be obscured in $^9Be(p,p')$ since both $T_<$ and $T_>$ components are populated. This is also the case for photonuclear reaction data.

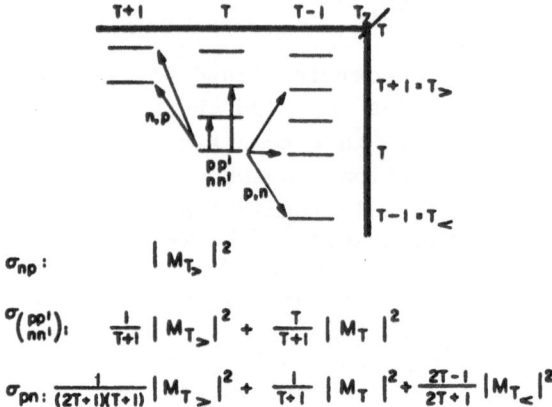

$$\sigma_{np}: \qquad |M_{T_>}|^2$$

$$\sigma\binom{pp'}{nn'}: \qquad \frac{1}{T+1} |M_{T_>}|^2 + \frac{T}{T+1} |M_T|^2$$

$$\sigma_{pn}: \frac{1}{(2T+1)(T+1)} |M_{T_>}|^2 + \frac{1}{T+1} |M_T|^2 + \frac{2T-1}{2T+1} |M_{T_<}|^2$$

Fig. 1. Allowed charge exchange transititions with
isospin geometric factors.

^{27}Al(n,p)^{27}Mg

The ^{27}Al(n,p) reaction provides an example of isovector
transitions from a (1/2$^-$,1/2) g.s. in the s-d shell.[3] The results are well summarized in Fig. 4 showing an energy spectrum
at θ_L = 15.5°. This is compared to an ^{27}Al(p,p') spectrum[9] at
θ_L = 15° as well as ^{27}Al(γ,xn) and ^{27}Al(γ,tot) photonuclear
spectra.[10] The angular distribution for the peak at 14.4 MeV
is consistent with an L = 1 transfer and exhausts about 20% of
the energy weighted GDR sum rule. A macroscopic form factor is
used in a DWBA calculation in a model proposed by Satchler[11] for
a Goldhaber-Teller GDR excitation.

Total exhaustion of the GDR sum rule is related to the calculated cross section by

$$\left(\frac{d\sigma}{d\Omega}\right)_{np} = 2 \ \beta_{GT}^2 \left(\frac{d\sigma}{d\Omega}\right)_{p,p'}^{DWBA}$$

The unobserved strength is quite likely distributed outside the
14.4 MeV peak region similiar to the photonuclear results. The
14.4 MeV peak in ^{27}Mg is the parent analog for a 21.3 MeV excitation in ^{27}Al which is very close to the observed peak in the
photonuclear distribution. It is worth noting that the momentum
transfer dependence in a photonuclear reaction is different than
for the (n,p) reaction at a fixed angle so that exact correspondence between the two should not be expected.

Although some evidence for M1 strength (L = 0 angular distribution) was found below 8 Mev in ^{27}Mg, its lack of concentration in excitation energy makes it difficult to obtain quantitative information. The prominent peak in the ^{27}Al(p,p') data is

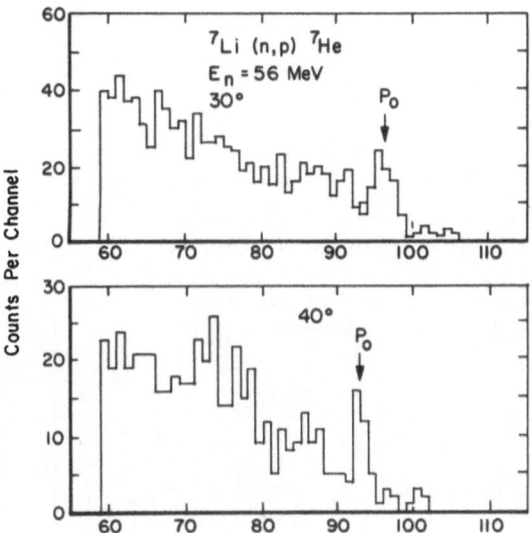

Fig. 2. ^{7}Li(n,p)^{7}He energy spectra at θ = 30° and 40°.

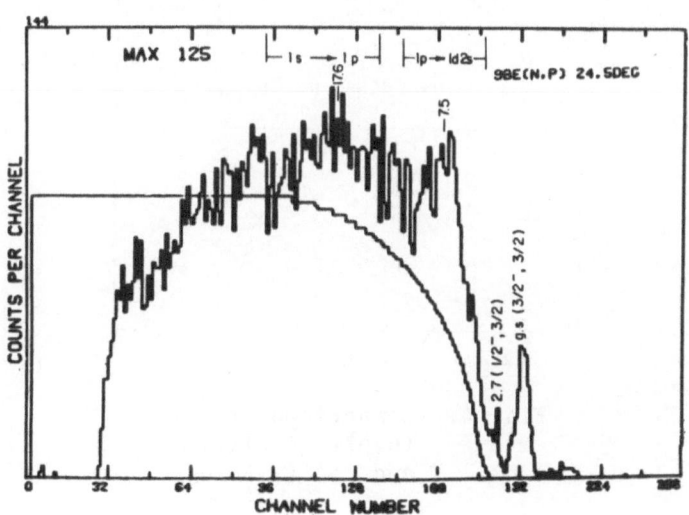

Fig. 3. ^{9}Be(n,p)^{9}Li energy spectrum.
           E_n = 59.4 MeV.

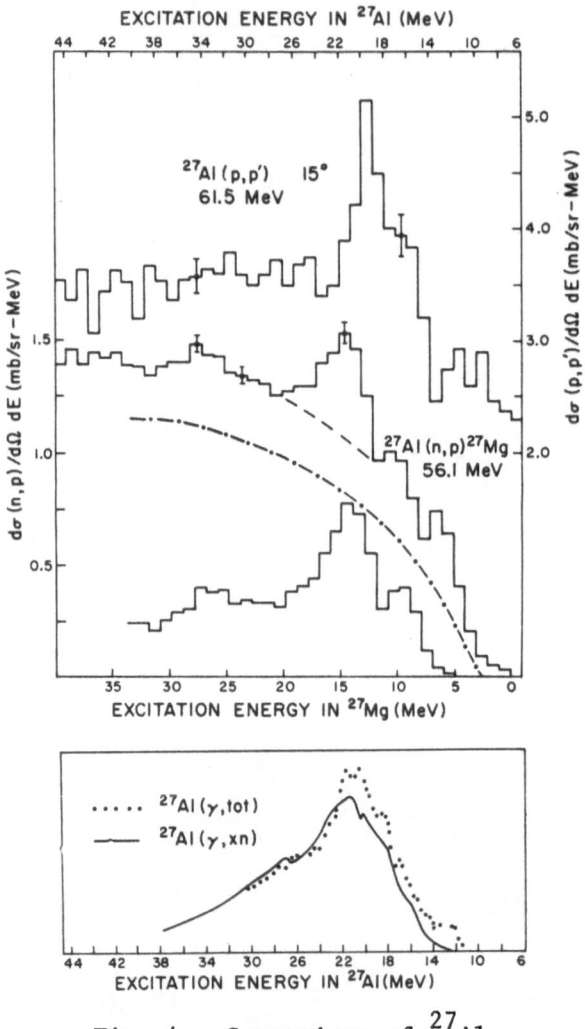

Fig. 4.   Comparison of ^{27}Al
(n,p), ^{27}Al(p,p'),
and ^{27}Al(γ,tot).

predominantly isoscaler quadrupole strength which obscures the GDR region. This shows the advantage of having a probe selective to only isovector transitions in helping to sort out collective excitations.

^{58,60,62,64}Ni(n,p)^{58,60,62,64}Co

This series of targets was selected as a means of systematically investigating the N-Z dependence of both the relative $T_>$ GDR strength as well as the $T_>$ vs $T_<$ energy splitting in the target. Ngo-Trong and Rowe[12] have carried out a RPA calculation giving the dipole strength distribution for both $T_>$ and $T_<$ components.

Comparison of (γ,n) and (γ,p) have been the primary source of information for the GDR distribution. The (γ,p) reaction is presumed to be selective to $T_>$ states due to an isospin selection rule inhibiting neutron decay. However, neutron decay of $T_>$ states to the IAS of the daughter can exceed the proton decay.[12] The (n,p) reaction should in principle be much more selective to $T_>$ components.

An energy spectrum from the ^{62}Ni(n,p)^{62}Co reaction at $\theta_L = 16°$ for 59.4 MeV neutrons is shown in Fig. 5. The continuum background is assumed to arise from a 3-body channel involving the particle from the decay of a ^{62}Co excited state and the usual ejectile proton. The continuum therefore has the indicated threshold. A pre-equilibrium model would tend to predict a lower contribution above the background by including available unresolved states in ^{62}Co up to the maximum allowed proton energy. The picture for

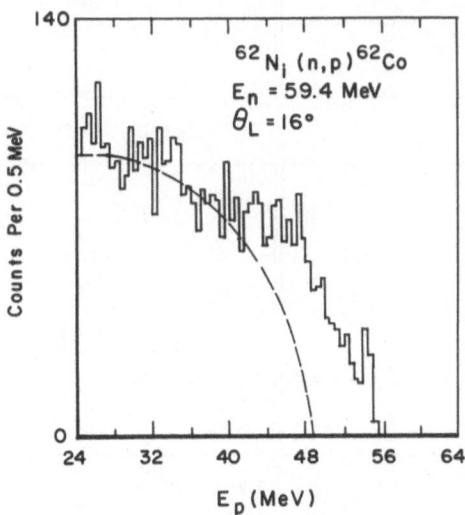

Fig. 5. ^{62}Ni(n,p)^{62}Co energy spectrum.

collective states discussed below will not be altered by this
except in the absolute overall strength. The result of removing
the background in a consistent manner from all the target data at
16° is shown in Fig. 6. The energy scale has been adjusted to represent analog excitation in the target nucleus by taking the coulomb energy shift and mass differences into account. The calculated Co-Ni excitation shifts are 8.8 MeV, 11.1 MeV, 13.4 MeV, and
15.1 MeV for A = 58, 60, 62, and 64, respectively. The vertical
bars are the results from Ngo-Trong and Rowe for $T_>$ GDR strength.[12]
Remarkably good agreement is found between the data and calculations for the GDR strength distributions and L = 1 angular distributions. The fractions of the GDR EWSR exhausted for a JS DWBA form
factor[11] are 61, 60, 63, and 68% for A = 58, 60, 62, and 64, respectively. The location of the $T_>$ vs $T_<$ E1 strength is predicted to
be $\Delta E = U(T_0+1)/T_0$ MeV by Goulard and Fallieros[4] where the scale
factor is related to the isospin symmetry potential. The equation
$\Delta E = 76(T_0+1)/A$ reproduces the approximate location of the weighted
average $T_>$ compared to the $T_<$ GDR energy.

Another obvious feature of the data is the rapid decrease in
strength for the region below the 1⁻ strength as A increases from

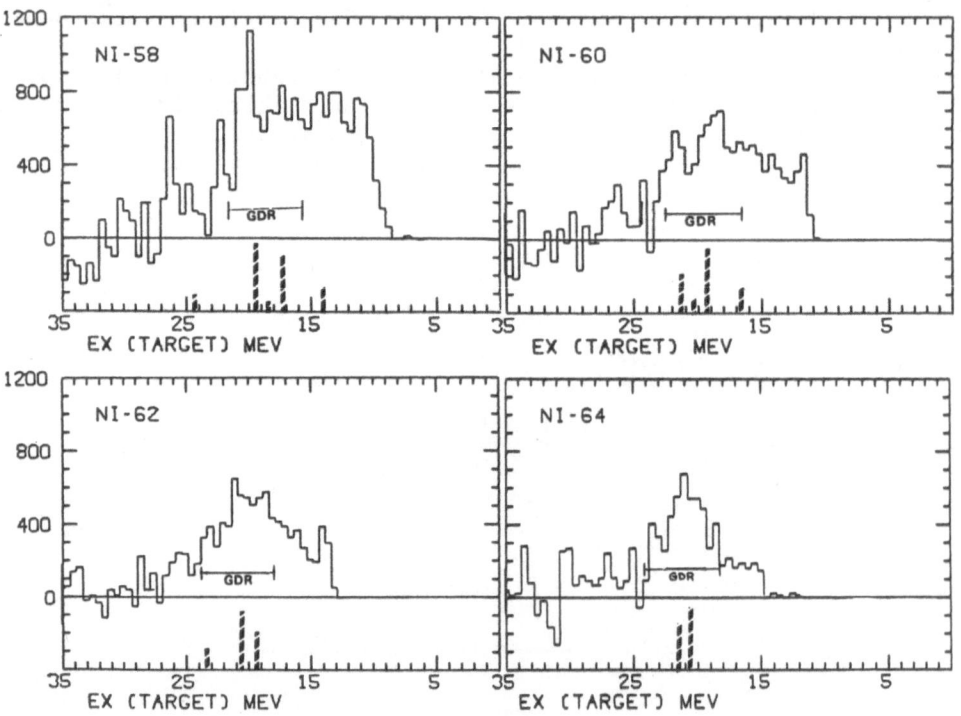

Fig. 6 Background subtracted Ni(n,p) spectra. Dotted bars are
theoretical E1 strength.

58 to 64. An obvious candidate is M1 strength since the predomi-
nant contributions are from $1f(7/2) \rightarrow (5/2)$ and $2p(3/2) \rightarrow 2p(1/2)$ transitions which are essentially completely blocked for ^{64}Ni and un-
blocked in ^{58}Ni for the (n,p) reaction. This is since the $\Delta T = 1$
transition removes a proton and fills a neutron orbital. Recent
(e,e') experiments by Lindgren et al[13] have found $T_>$ M1 strength
in ^{58}Ni at ~10.5 MeV which exhausts close to 50% of the M1 closure sum rule for ^{58}Ni. Some M1 strength in ^{60}Ni was also located
at ~12.1 MeV. This upward shifting of available $T_>$ M1 strength by
a few MeV appears to continue up to ^{64}Ni based on the present data.
It should be pointed out that (e,e') $T_>$ M1 strength is reduced by
the isospin geometric factor so that higher weak components are
difficult to observe.

The angular distributions for this "M1" region do peak in
cross section for low momentum transfer. However, they do not
fall off rapidly enough to be pure M1. Since evidence for, $T_>$ 8-
stretched configuration strengths which will not be blocked, have
been found overlapping the M1 region,[13] one might expect contributions to back angle cross sections.

^{90}Zr(n,p)^{90}Y

The A = 90 system is a good testing ground for studying $T_>$
giant multipoles due to the (p,γ) work of Hasinoff et al.[14]
locating $T_>$ E1 strength at 14.4, 16.3 19.4, and 21.0 MeV in ^{90}Zr
and the inelastic electron scattering experiments of Fukuda and
Torizuka[15] providing evidence for isovector E2 strength at 17 and
26 MeV and E3 strength from 20 to 30 MeV. A number of theoretical
calculations are also available for $T_>$ E1, E2, and E3 energy
distributions.[16]

A ^{90}Zr(n,p)^{90}Y spectrum at 16° is shown in Fig. 7. The smooth
curve is the (n,np) phase space assumed for a background. The
subtracted data is given in Fig. 8 with the observed $T_>$ E1
strength from (p,γ) and E2 strength from (e,e') shown as solid
and dotted lines, respectively. The corresponding energies in
^{90}Y were obtained by subtracting 13.3 MeV from the ^{90}Zr energies.
The overall E1 energy distribution is not too different from that
obtained from the photonuclear data,[14] however some contribution
from higher multipoles is evident in the angular distribution for
the proton energy region above 40 MeV. This should be expected
based on the E2 strength from the (e,e') data. In addition, the
observed cross section of 6.5 mb/sr at 16° is 54% larger than that
required to exhaust the GDR sum rule. The angular distribution
for Ep = 40 to 30 MeV (~12 MeV excitation in ^{90}Y) is consistent
with an L transfer of greater than one. This region has an analog
in ^{90}Zr at 25.3 MeV where Fukuda and Torizuka found considerable
isovector E2 strength. The possibility for isovector E3 strength

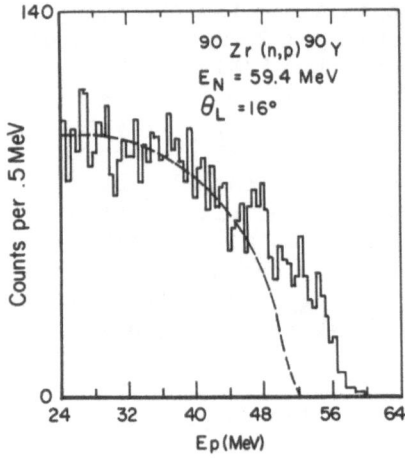

Fig. 7. ^{90}Zr(n,p)^{90}Y
spectrum.

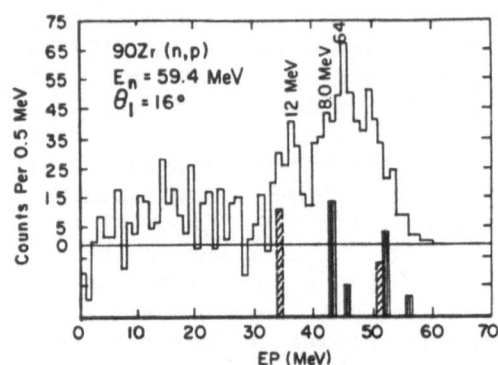

Fig. 8.   Background subtracted spectrum.
Cross hatched and solid bars
are experimental E2 and E1
strengths from ^{90}Zr(e,e')
and ^{89}Y(γ,p), respectively.

cannot be ruled out based on energy systematics or the angular
distribution.

^{209}Bi(n,p)^{209}PB

Three experiments prompted the investigation of the ^{209}Bi(n,p)
reaction; (1) the indication of T> E2 strength in ^{209}Bi from the
^{208}Pb(p,γ)^{209}Bi work of Snover et al.[17] and (2) the subsequent
location of a peak in ^{209}Pb at ~7.9 MeV close to the excitation
energy of the parent analog of the same resonance via ^{209}Bi(π^-,γ)
^{209}Pb,[18] and (3) the collective E2 strength found in ^{208}Pb(e,e')
data between Ex ~18 to 27 MeV.[19] The energy spectrum in Fig. 9
shows a peak at 7.5 MeV which would have an analog at 26.3 MeV.

Since most of the neutron orbitals are full for simple 1hω
transitions in A = 209 only very weak parent analogs to the GDR of
^{209}Bi are possible. Few 2hω M1 transitions are available so that
little M1 strength is expected. The excitation energy is consistent with that for parent analogs to collective isovector E2
excitations at $120/A^{1/3}+\Delta E(T_>-T_<) \approx 24$ MeV. The angular distribution is given in Fig. 10 with a DWBA calculation utilizing a JS
form factor exhausting 100% of the isovector E2 sum rule as calculated by Brown and Madsen.[20] The calculated strength is based
on the known isoscaler E2 strength from ^{209}Bi (p,p') isoscaler E2
measurements [21] and assumes $|V_1/V_0|$~1/2 for isovector to iso-
scaler potential strengths.

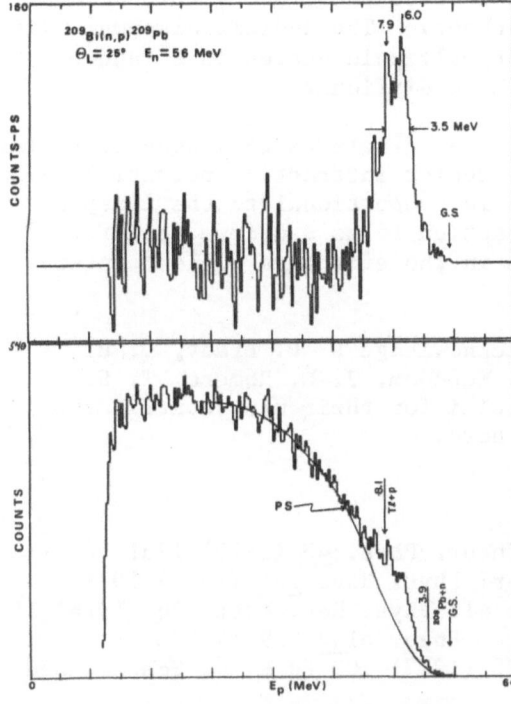

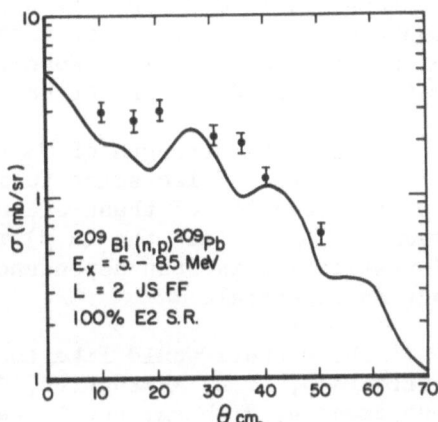

Fig. 10. Angular distribution
for E_p = 5 to 8.5
MeV in ^{209}Pb. L =
2 DWBA calculation
is normalized to
100% of isovector
$T_>$ E2 sum rule.

Fig. 9. ^{209}Bi(n,p)^{209}Pb energy
spectrum before and after
background subtraction.

CONCLUSIONS

The ^{7}Li(n,p)^{7}He and ^{9}Be(n,p)^{9}Li reactions both show evidence
for population of parent analogs to well known M1 transitions in
the target nuclei. Configuration splitting of the GDR for A = 9
was also found in agreement with (π^-,γ) data.

For A $\geq$ 27, enhancements over a continuum background are consistent with parent analog E1 and E2 states. The results are
much less dramatic than for the T_0 = 0 targets and angular distributions are not uniquely fitted by one L transfer. The geometric
isospin factors favoring ΔT = 1 (n,p) transitions do provide an
advantage over inelastic (p,p'),(n,n'), or (e,e') scattering and
(p,n) reactions for investigating $T_>$ isovector giant multipole
states. The Ni isotope data provide perhaps the best example to

be found of T> GDR population as a function of N-Z within an isotopic chain for comparison with theory. The additional complexity
of possible overlapping isovector multipole states in Zr and Bi
make interpretation of the data more difficult.

Since the strength of T> states is related to isoscaler states
via the ratio of isovector to isoscaler interaction potentials[20]
and the location of these states is proportional to the isospin
symmetry potential[4] the (n,p) reaction forms a unique tool for
investigating isospin dependence in the effective nuclear interaction potential.

The authors would like to acknowledge F. P. Brady, D. H.
Fitzgerald, M. W. McNaughton, G. Needham, J. L. Romero, T. S.
Subramanian, J. Wang, and C. Zanelli for their many contributions
in obtaining the data presented here.

REFERENCES

1. M. Obu & T. Terasawa Prog. Theor. Phys. $\underline{43}$ (1970) 1231
2. D. F. Measday & J. N. Palmieri Phys. Rev. $\underline{161}$ (1967) 1071
3. UCD Progs. Reports, Brady et al. Phys. Rev. Lett. $\underline{36}$ (1976) 15
4. S. Fallieros & B. Goulard Nuc. Phys. $\underline{A147}$ (1971) 593
5. F. P. Brady et al., NBS SP425 (1975) 103 Ed R. A. Schrack and
    C. D. Bowman
6. L. W. Fagg, Rev. Mod. Phys. $\underline{47}$ (1975) 683
7. H. W. Baer, 7th Int. Cont. on High Energy Physics and Nuc.
    Structure El. M. Locker, Birkhauser Verlag, Basel (1978)
8. G. B. Dogotar et al., Nuc. Phys. $\underline{A282}$ (1977) 474
9. M. B. Lewis and F. E. Bertrand Nuc. Phys. $\underline{A196}$ (1972) 337
10. Atlas of Photoneutron cross sections obtained with Mono-
    energetic Photons B. L. Berman UCRL - 94622
11. G. R. Satchler, Nuc. Phys. $\underline{A195}$ (1972) 1
12. C. Ngo-Trong and D. J. Rowe, Phys. Lett. $\underline{36B}$ (1971) 553
13. R. A. Lindgren et al., Phys. Rev. C. $\underline{14}$ (1976) 1789
14. M. Hasinoff et al. Nuc. Phys. $\underline{A216}$ (1973) 221
15. S. Fukuda and Y. Torizuka, Phys. Lett. $\underline{62B}$ (1976) 146
16. J. D. Vergados and T. T. S. Kuo. Nuc. Phys. $\underline{A168}$ (1971) 225;
    T. A. Hughes & S. Fallieros, in Nuc. Isospin, ed. J. D.
    Anderson, S. P. Bloom, J. Cerny, and W. W. True (Acad. Press,
    N. Y. 1969) p109
17. K. Snover et al., Phys. Rev. Lett. $\underline{32}$ (1974) 317
18. H. Baer et al., Phys. Rev. C. $\underline{10}$ (1974) 267
19. M. Sasao and Y. Torizuka Phys. Rev. C. $\underline{15}$ (1977) 217
20. V. Brown and V. Madsen Phys. Rev. C. $\underline{17}$ (1978) 1943
21. F. Bertrand Ann. Rev. Nuc. Sci. $\underline{26}$ (1976) 457

DISCUSSION

    <u>Wong</u>: Does the ^{58}Ni(n,p) measurement agree with the ^{58}Ni(e,e') measurements for populating the M1 1$^+$ analog states? In addition, do they agree with the ^{58}Ni(t,^{3}He) LASL measurements?

    <u>King</u>: We see strength at the correct parent analog energies for both ^{58}Co and ^{60}Co in agreement with the (e,e') data. Results for ^{62}Co and ^{64}Co show a decrease in strength as one would expect from blocking effects for $f_{7/2} \rightarrow f_{5/2}$ M1 transitions. The (e,e') and (t,^{3}He) experiments are both consistent with the ^{58}Ni(n,p)^{58}Co results. However, our resolution is not adequate to show individual M1 states.

EXCITATION OF SPIN-FLIP, ISOSPIN-FLIP STATES IN (p,n), (e,e')

AND (γ,γ): A COMPARATIVE STUDY ON ^{24,25,26}Mg

U.E.P. Berg

Cyclotron Laboratory, Michigan State University,
East Lansing, Michigan 48824 and
Justus-Liebig-Universität, Strahlenzentrum, Institut
für Kernphysik, D-6300 Giessen, West Germany

INTRODUCTION

The giant magnetic dipole resonance can be described in a simple picture as a single-particle spin-flip transition from a $\ell + 1/2$ into a $\ell - 1/2$ shell with an isospin-flip although nucleon-nucleon recoupling within a subshell can contribute a considerable amount of M1 strength. The giant M1 resonances have been extensively studied in recent years by inelastic electron scattering, elastic photon scattering, photoneutron processes and radiative capture reactions. References regarding these experiments can be found in Fagg's review article on "Electroexcitation of Nuclear Magnetic Dipole Transitions"[1] and lectures of Richter, Hayward, and Hanna at the International School on Electro- and Photonuclear Reactions[2].

Spin-flip, isospin-flip transitions can also be induced by (p,n) reactions. Isobaric analog state (IAS) transitions no longer dominate the (p,n) spectra at proton energies above 100 MeV range[3] and spin-flip transitions seem to become more important with increasing proton energy. These transitions lead to isospin analogs of the giant M1 resonance as will be discussed in this contribution. Closely related to the spin-flip (p,n) reaction is the Gamow-Teller β[+] decay, which connects the same states. For convenience we refer to a spin-flip (p,n) transition as a Gamow-Teller (p,n) transition. A Fermi (p,n) transition involves only a flip in the isospin space and links the ground state with its isobaric analog state.

We chose the three stable Mg isotopes for an investigation of the (p,n) reaction to the analogs of the giant M1 resonance.

Magnetic dipole transitions in ^{24,25,26}Mg are well explored by
resonance fluorescence experiments[4] and inelastic electron scattering work.[5] The level density of the states involved was expected
to be low enough that individual states could be resolved in (p,n),
and the availability of transition matrix elements from full sd-
space shell model calculations[6] provides the basis for a microscopic DWBA treatment of the (p,n) cross section and angular distributions. The transition probabilities from shell model calculations can be checked by a comparison of the theoretical data with
values from the less model dependent photonuclear studies. If
these Ml matrix elements reproduce the experimental facts, we can
have some more confidence in the nucleon-nucleon force derived
with the help of the theoretical transition densities and microscopic DWBA calculations.

SOME GENERAL REMARKS ON THE RELATIONSHIP BETWEEN Ml, GAMOW-TELLER,

(p,n) and β^+ TRANSITIONS

The relationship between Ml, Gamow-Teller (p,n), and β^+ transitions will be discussed using ^{26}Mg as an example. Fig. 1 shows
the different 1^+ states which can be reached by (p,n), (e,e'), and
(γ,γ). The ground state spin and parity of ^{26}Mg is $J^\pi = 0^+$; its
isospin is T = 1 and the T_3 component, defined here as $T_3 = (Z-N)/2$,
is $T_3 = -1$. Absorption of magnetic dipole radiation leads to 1^+
states with T = 1 and $T_3 = -1$, which, as will be shown below, are
concentrated at about 10 MeV. Their analogs with $J^\pi = 1^+$, T = 1
and $T_3 = 0$ can be reached by a (p,n) reaction. These states should
be located at nearly the same excitation energies above the 0^+;
T = 1 ($T_3 = 0$) isobaric analog state of the ^{26}Mg ground state.
This excitation region is hatched in Fig. 1. The antianalog states
of the giant Ml resonance in ^{26}Mg differ only in the isospin quantum numbers (T = 0, $T_3 = 0$) and are expected to be in the low
energy part of the ^{26}Al level scheme.

The inverse process of a Gamow-Teller (p,n) transition, namely
the Gamow-Teller β^+ decay, can experimentally be observed from the
$T_3 = +1$ member of the T = 1 isospin triplet composed of the ground
state of ^{26}Mg, the 0.228 MeV level in ^{26}Al, and the ground state
of ^{26}Si. The corresponding Gamow-Teller transition probability
can be extracted from the measured ft values.

Theoretically, the matrix elements for a collective transition as a Ml giant resonance transition or a giant Gamow-Teller
transition can be treated in a single-particle representation:[7]

$$\mathcal{M} \propto \overline{\sum_{jj'\Delta J\Delta T}} \{SPME_{jj'\Delta J\Delta T}\} <\Psi^{A''J''T''} | [a_j^+ a_{j'}]_{\Delta J\Delta T} |\Psi^{A'J'T'}> \qquad (1)$$

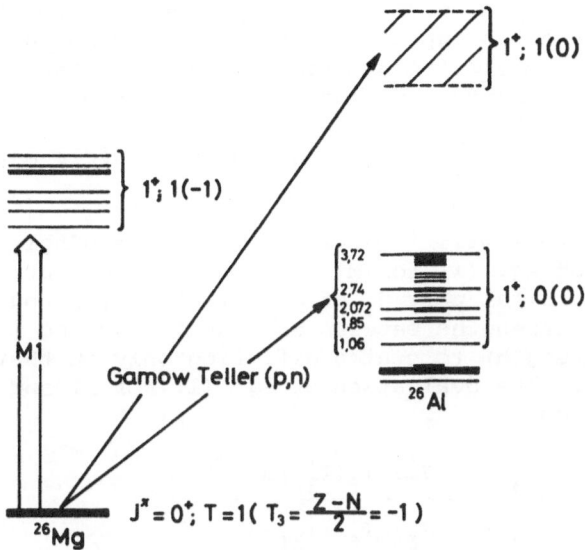

Fig. 1.   Excitation of spin-flip, isospin-flip transitions to 1^+ states, who differ only in isospin space. The notation is J^π; $T(T_3)$. The 1^+; 1(-1) states in ^{26}Mg are at about 10 MeV excitation energy.

where the first factor in the sum is the Hamiltonian independent 'single particle matrix element' and the second factor is the Hamiltonian dependent, 'transition density matrix element'. The single particle matrix elements for our operators in question are the M1 operator and the Gamow–Teller operator. If we neglect the isoscalar part of the M1 operator, which is an order of magnitude smaller than the iosvector part, we obtain for the isovector M1 operator[8]

$$\mathcal{O}(M1, \begin{matrix} T = 1 \\ J = 1 \end{matrix}) \propto \sum_i \vec{\tau}^{(i)} \{\nabla r_i Y_1\}\{(\mu_- + 1/2)\vec{\sigma}_i - \vec{j}_i\} \tag{2}$$

for the Gamow-Teller operator

$$\mathcal{O}(GT, \Delta J = 1) \propto \sum_i \vec{\tau}^{(i)} \{\nabla r_i Y_1\}\vec{\sigma}_i \tag{3}$$

and for the (p,n) operator[9]

$$\mathcal{O}(GT, pn) \propto V_{\sigma\tau} \, (\vec{\sigma}_i \cdot \vec{\sigma}_p)(\vec{\tau}_i \cdot \vec{\tau}_p)$$

All operators are identical in the spin isospin space of the target and contain $\vec{\sigma}$ and $\vec{\tau}$, which flip the spin and isospin. The addition of the $\vec{j}$ term in the M1 operator enhances the isovector single particle matrix element for a rearrangement of nucleons in the $0d_{5/2}$ shell and quenches the $0d_{3/2}$ matrix elements.   As a consequence transitions with large $d_{5/2} - d_{5/2}$ particle-hole transition densities can be up to a factor of two stronger in experiments which use the electromagnetic interaction than in Gamow-Teller type measurements. On the other hand strong transitions between the $d_{5/2}$ and $d_{3/2}$ shell will be nearly the same in all experiments since the corresponding single particle matrix elements are almost equal.

We neglect the small isoscalar part in the single particle matrix element of eq. (1) for M1 transitions and, furthermore, we exclude the orbital part j_i of eq. (2) to get a rough estimate of the ratio of strengths between M1 transitions and (p,n) strength (or β^+ decay strength) to states differing only in their isospin quantum numbers. The evaluation of eq. (1) for M1 and Gamow-Teller transitions yields:

$$\frac{B(M1)[J_i; T_i(T_{3i}) \rightarrow J_f; T_f^\gamma(T_{3f}^\gamma)]}{B(GT)[J_i; T_i(T_{3i}) \rightarrow J_f; T_f^{GT}(T_{3f}^{GT})]}$$

$$= \frac{2}{2T_f^\gamma + 1} \frac{(T_i T_{3i} 10 | T_f T_{3f})^2}{(T_f^{GT} T_{3f}^{GT} 1 - 1 | T_i T_{3i})^2} \tag{4}$$

This ratio is 2, 2, 2/3 if we compare M1 and Gamow Teller (p,n) transitions from the ground states of ^{24,25,26}Mg to the 1^+; 1(1) in ^{24}Al, to the $3/2^+$, $5/2^+$, and $7/2^+$; 3/2 (1/2) states in ^{25}Al, and to the 1^+; 1(0) states in ^{26}Al, respectively.

EXPERIMENTAL METHODS

The experimental methods are discussed only briefly.

The (p,n) Experiment

The neutron time of flight spectra (see Fig. 2) have been measured at E_p = 35 MeV and a flight path of 32 m at the beam swinger facility of the Michigan State University Cyclotron Laboratory. The total time resolution obtained during the (p,n) runs of Fig. 2 was 0.8 ns, corresponding to an energy resolution of about 200 keV for neutrons at 30 MeV and about 90 keV for neutrons at 15 MeV.

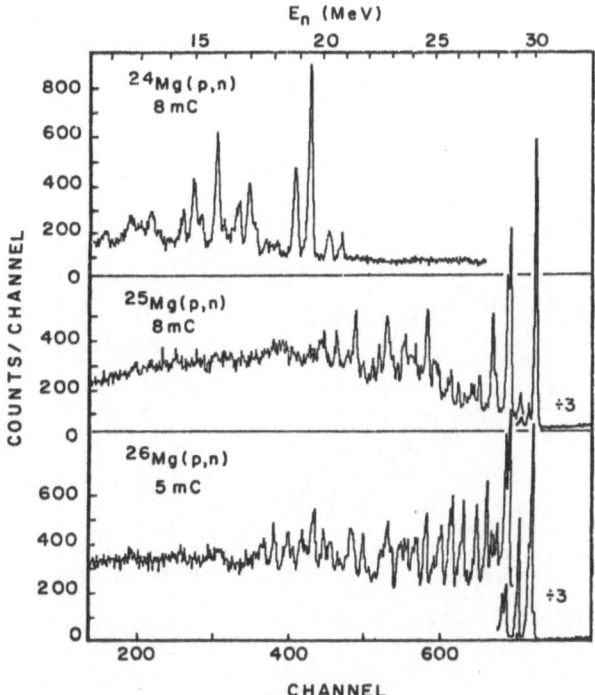

Fig. 2. Neutron time of flight spectra from the (p,n) reaction on the three stable Mg isotopes at E_p = 35 MeV and θ_{lab} = 30°.

The (e,e') Experiment

Inelastic electron scattering experiments at 180° and low momentum transfer are well suited to observe magnetic multipole transitions.[1] The energy resolution of the (e,e') data is comparable with the (p,n) data at high neutron energies. Multipolarities of the levels observed are determined in a model dependent way though not always without ambiguities. An inelastic electron scattering spectrum from ^{25}Mg at 180° is compared with a ^{25}Mg(p,n)^{25}Al spectrum in Fig. 3.

The (γ,γ) Experiment

Resonance fluorescence experiments are sensitive to dipole and electric quadrupole transitions. These elastic photon scattering experiments using bremsstrahlung photons which are scattered into Ge(Li) detectors are currently being carried out at Giessen, at Urbana-Champaign (Illinois), and Sendai. The energy resolution achieved is less than 10 keV FWHM at 10 MeV excitation energy. A ^{24}Mg(γ,γ) spectrum is plotted in Fig. 4. The multipolarity of

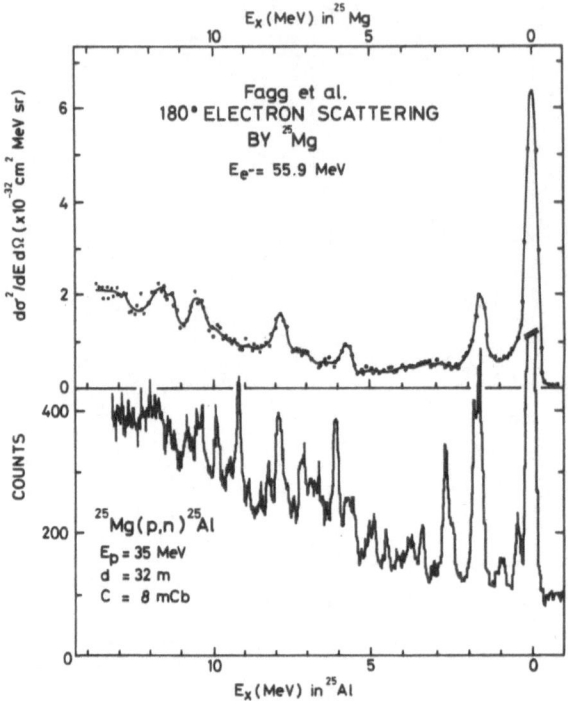

Fig. 3. Comparison of an inelastic electron scattering spectrum from
^{25}Mg with a ^{25}Mg(p,n)^{25}Al measurement. The (e,e') spectrum
was measured by Fagg et al.,[10] the (p,n) spectrum at MSU.

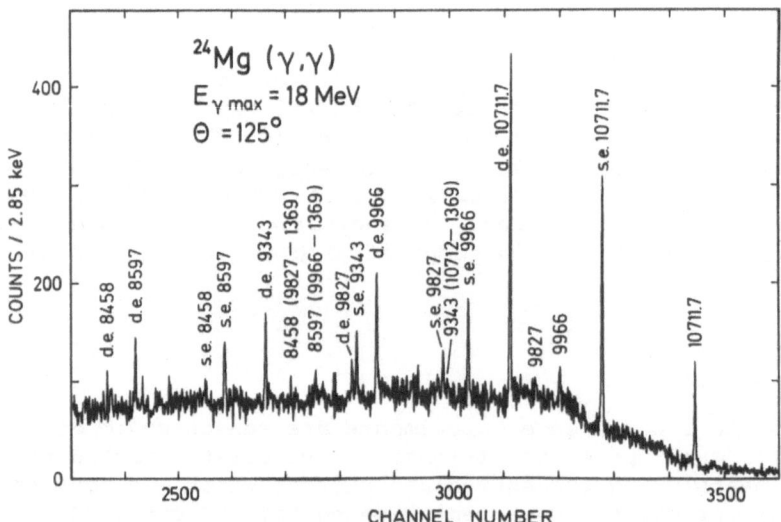

Fig. 4. Nuclear resonance fluorescence spectrum from ^{24}Mg recorded
with Ge(Li) detectors at the University of Giessen electron
accelerator bremsstrahlung beam.

the states excited can be determined model independently from a measurement of the angular distribution. To determine the parity of the states involved a measurement with polarized photons is necessary. First experiments with polarized bremsstrahlung photons have been recently performed at Giessen and Urbana-Champaign.

EXPERIMENTAL RESULTS

^{24}Mg

The two strongest Ml transitions in ^{24}Mg occur to the 1^+ states at 9.966 and 10.712 MeV. The analog states of these Ml states are at 0.439 and 1.12 MeV in ^{24}Al. Unfortunately, the 1^+ states in ^{24}Al are not resolved in the (p,n) experiment from close lying 2^+ states, which are located at 0.514 and 1.14 MeV. We compared the measured (p,n) angular distribution for the 1.12 MeV doublet with the angular distributions of the 10.712 MeV 1^+; 1(0) and 10.731 MeV 1^+; 1(0) states in ^{24}Mg from inelastic proton scattering.[11] The sum of differential cross sections from the (p,p') measurement is plotted in Fig. 5 together with the angular distribution of the 1.12 MeV doublet. Since the (p,p') experiment has been performed at 40 MeV proton energy, we have corrected the scattering angles for the different momentum transfers. A change of the absolute cross sections at 35 MeV and 40 MeV is assumed to be small. As can be seen from Fig. 5 the shape of the angular distributions is quite similar. The ratio of $\sigma(p,n)/\sigma(p,p')$ is expected to be 2. This requirement is obviously not always fulfilled; deviations may be due to the different incident proton energies and to experimental errors. We quote an error of ± 20% for the absolute (p,n) cross section.

A comparison of B(Ml) values from inelastic electron scattering and elastic photon scattering with the differential (p,n) cross section at $\theta_{cm} = 30^o$ is given in Table 1. The (p,n) cross section is multiplied by a factor of 3 to account for a change of cross sections due to different isospins if we compare $\sigma(p,n)$ for ^{24,25,26}Mg (see eq. (4)). The (p,n) cross sections in this case are upper limits for transition strengths to 1^+ states because they were not resolved from the close lying 2^+ states; the numbers in brackets are estimates from a comparison of our $\sigma(p,n)$ with the (p,p') data.

No (γ,γ) experiments exist for ^{25}Mg. Therefore, we can only compare our (p,n) data to (e,e') measurements. A comparison of a ^{25}Mg(e,e') with a ^{25}Mg(p,n) spectrum (Fig. 3) reveals correspondence between Ml transitions and Gamow-Teller (p,n) transitions. The analog states in ^{25}Al of the 1.60, 5.77, 7.03, 7.81, 10.43, 11.43, and 11.76 levels in ^{25}Mg are populated. But on the other

Table 1. Comparison of B(M1) values from (e,e') and (γ,γ) with $d\sigma(p,n)/d\Omega$ at 30°_{cm}, E_p = 35 MeV, for ^{24}Mg.

(e,e)[a]		(γ,γ)[c]		(p,n)	
E_x (MeV)	B(M1) (μ_N^2)	E_x (MeV)	B(M1) (μ_N^2)	E_x in ^{24}Al (MeV)	$3(\frac{d\sigma}{d\Omega})30^\circ$ (mb/sr)
9.97	0.39[b]	9.966	0.54	0.44	$\leq$ 0.24 (0.12)
10.70	1.24	10.712	1.25	1.12	$\leq$ 1.5 (1.0)

[a]Ref. 1   [b]Ref. 5   [c]Ref. 4

hand, strong transitions to states at 2.67, 6.11, 9.12, and 9.82 MeV are induced in addition.

The B(M1) values for the first four M1 transitions in ^{25}Mg from electron scattering are compared to the measured (p,n) cross sections in Table 2. The numbers for the transitions to the 5.78 and 7.12 MeV levels in ^{25}Al are in brackets because the angular distributions show deviations from that of a typical spin-flip isospin-flip transition. Nevertheless, the energies are in good agreement and the (p,n) cross section fits well into the systematical comparison with B(M1) values. The ^{25}Mg(p,n)^{25}Al cross sections in Table 2 have been multiplied by a factor of 3 again because of the ratio of Clebsch-Gordans in eq. (4).

Table 2. Comparison of B(M1) values from (e,e') with $d\sigma(p,n)/d\Omega$ at 30° for ^{25}Mg.

(e,e')[a]		(p,n)	
E_x (MeV)	B(M1) (μ_N^2)	E_x in ^{25}Al (MeV)	$3(\frac{d\sigma}{d\Omega})30^\circ$ (mb/sr)
1.60	0.87	1.61	0.69
5.77	0.42	(5.78	0.29)
7.03	0.55	(7.12	0.33)
7.81	0.85/g	7.90	0.54

g = $(2J_o + 1)/(2J_f + 1)$
[a]Ref. 1.

^{26}Mg

The region at about 10 MeV excitation energy in ^{26}Al was expected to be near where the isobaric analog states of the giant M1 resonance should occur. Fig. 6a shows the differential cross sections of four strong states having the angular distribution of a 0^+ to 1^+ transition. This can be clarified by a comparison with Fig. 6b picturing the angular distribution of the well separated, low lying 1^+ state at 1.058 MeV in ^{26}Al. In addition to the 9.44, 9.89, 10.47, and 10.83 MeV levels a state at 11.21 MeV has a similar angular distribution.

Table 3 quotes the B(M1) values from (e,e') and (γ,γ) together with the (p,n) cross sections at $30°$ in the center of mass system. The energy of the IAS has been subtracted for the excited states in ^{26}Al to get the equivalent energies of the giant M1 resonance in ^{26}Mg.

The resonance fluorescence experiment,[15] which has been carried out at MUSL II as a Giessen Urbana-Champaign collaboration, resolves three states not being resolved in the (e,e') or (p,n)

Table 3. Comparison of B(M1) values from (e,e') and (γ,γ) with $d\sigma(p,n)/d\Omega$ at $\theta = 30°_{cm}$, $E_p = 35$ MeV, for ^{26}Mg.

(e,e')[a]		(γ,γ)[c]			(p,n)	
E_x (MeV)	B(M1) (μ_N^2)	E_x (MeV)	B(M1) (μ_o^2)	Sum (μ_N^2)	^{26}Al E_x -0.23 MeV (MeV)	$(\frac{d\sigma}{d\Omega})30°$ (mb/sr)
9.24	0.36	8.960 9.239	0.072 0.090 }	0.171	9.21	0.10
9.67	0.16	9.563 9.769	0.049 0.024 }	0.073	9.66	0.078
10.20	0.39[b]	10.103 10.148	0.176 0.223 }	0.39	10.24	0.15
10.65	0.46[b]	10.647		0.45	10.60	0.26
11.20	0.24	above particle threshold			10.98	0.12
13.33	0.53	above particle threshold			no strong peaks observed	

[a]Ref. 1   [b]Ref. 15   [c]Ref. 16

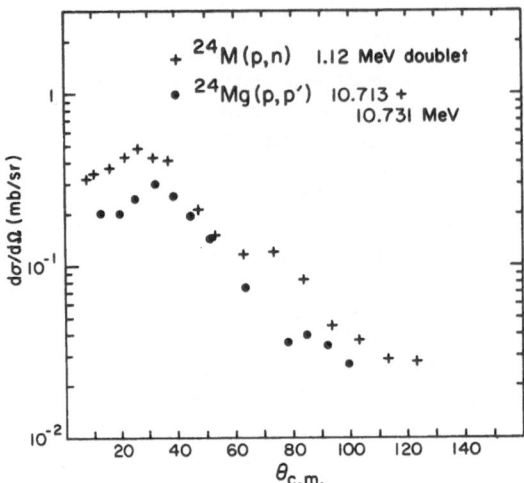

Fig. 5.   Comparison of the ^{24}Mg(p,n) cross section to the 1.12 MeV
          doublet with the sum of ^{24}Mg(p,p') cross sections to the
          1$^+$ state at 10.713 MeV and 2$^+$ state at 10.731 MeV from
          Ref. 11.   The (p,p') spectrum has been corrected for the
          different q transfers.

experiment.   Therefore, the sum of certain states from resonance
fluorescence is given in Table 4.   It should be pointed out that
the (γ,γ) data are preliminary and may change slightly.
Within the experimental uncertainties which are in general in the
order of ± 25% for each experiment, the agreement of the derived
B(M1) values and (p,n) cross sections is reasonable.   The excitation energies are in good agreement.

        The results from Tables 1, 2, and 3 are summarized in Table 4,
indicating that there is a close relationship between B(M1) values
and (p,n) cross sections for isospin-flip spin-flip transitions.
The ratio between B(M1) and σ(p,n) increases if we go from ^{24}Mg
to ^{26}Mg.   It looks hopeful that the (p,n) reaction at higher proton
energies can be used to extract matrix elements for spin-flip
isospin-flip transitions.

ACKNOWLEDGMENTS

        This material is based upon work supported by the National
Science Foundation under Grant No. PHY-7822696 and by the Deutsche
Forschungsgemeinschaft.   I thank Professor S.M. Austin for his

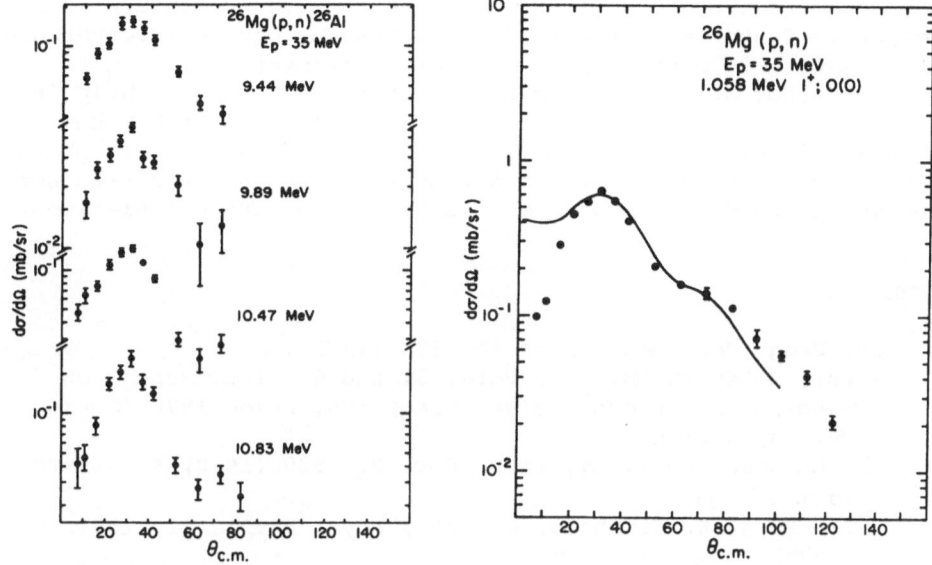

Fig. 6a. Angular distribution of four highly excited states in ^{26}Al from the ^{26}Mg(p,n) reaction at E_p = 35 MeV.

Fig. 6b. Angular distribution for the well resolved 1.058 MeV 1^+ state. Solid curve is from a microscopic DWBA calculation with the recommended parameter set from Ref. 13 and "best fit" optical parameters from Ref. 14. Normalization factor is 0.55. The transition matrix elements are from Ref. 6.

Table 4. Ratio between B(M1) values from the electromagnetic inter-action and $d\sigma/d\Omega$ at 30°_{cm} from Tables 1, 2, and 3.

		(e,e')/(p,n)	(γ,γ)/(p,n)
^{24}Mg	9.97	3.3	4.5
	10.70	1.24	1.25
^{25}Mg			
	1.60	1.8	--
	5.77	1.4	--
	7.03	1.7	--
	7.81	1.6	--
^{26}Mg			
	9.47	2.1	1.1
	9.90	2.1	0.9
	10.43	2.7	2.7
	10.83	1.8	1.7
	11.21	2.0	--

always encouraging discussions. I appreciate the extensive assistance of Professor A. Galonsky, Mr. R. DeVito, Drs. W. Sterrenburg,
and Y. Iwasaki in performing the (p,n) experiments. I wish to
thank Professor M. Brussel and Dr. T. Chapuran for their help in
acquisition and analysis of the (γ,γ) data obtained at Urbana-
Champaign. Last but not least I am indebted to the Giessen photonuclear group for their support during the resonance fluorescence
experiments and Mr. W. Naatz for his help with DWBA calculations.

REFERENCES

1.   L.W. Fagg, Rev. Mod. Phys. <u>47</u>, 683 (1975).
2.   Lecture Notes in Physics, Vols. 61 and 62, Int. School on
        Electro- and Photo-nuclear Reactions, Erice 1976 (Springer
        Verlag, Berlin).
3.   C.D. Goodman, Bull. Am. Phys. Soc. <u>23</u>, 939 (1978) and these
        proceedings.
4.   U.E.P. Berg, K. Wienhard, H. Wolf, Phys. Rev. C <u>11</u>, 1851
        (1975).
5.   O. Titze, Z. Phys. <u>220</u>, 66 (1969), L.W. Fagg et al., Phys.
        Rev. C <u>1</u>, 1137 (1970), and A. Johnston and T.E. Drake,
        J. Phys. <u>A7</u>, 898 (1974).
6.   B.H. Wildenthal and W. Chung, private communication.
7.   B.H. Wildenthal in "Elementary Modes of Excitation in Nuclei,"
        1977, LXIX Corso, Soc. Italiana de Fisica-Bologna-Italy.
8.   E.K. Warburton and J. Weneser in "Isospin in Nuclear Physics,"
        ed. D.H. Wilkinson, North-Holland Publishing Co., Amsterdam
        (1969).
9.   Sam M. Austin in "The Two-Body Force in Nuclei," Plenum Pub-
        lishing Corp, New York (1972).
10. L.W. Fagg, W.L. Bendel, E.C. Jones, Jr., H.F. Kaiser, and T.F.
        Godlove, Phys. Rev. <u>187</u>, 1384 (1969).
11. B. Zwieglinski, G.M. Crawley, H. Nann, and J.A. Nolen, Jr.,
        Phys. Rev. C <u>17</u>, 872 (1978).
12. P.M. Endt and C. van der Leun, Nucl. Phys. <u>A214</u>, 1 (1973).
13. G. Bertsch, J. Borysowicz, H. McManus, and W.G. Love, Nucl.
        Phys. <u>A284</u>, 399 (1977).
14. F.D. Becchetti, Jr. and G.W. Greenlees, Phys. Rev. <u>182</u>, 1190
        (1969).
15. E.W. Lees, A. Johnston, S.W. Brain, C.S. Curran, W.A.
        Gillespie, and R.P. Singhal, J. Phys. <u>A7</u>, 936 (1974).
16. U.E.P. Berg, M. Brussel, T. Chapuran, and K. Wienhard, to
        be published.

DISCUSSION

Austin: This comment is a general response to a question of
H. Baer which I paraphrase: What do you learn from such (p,n)
studies? I think that at present charge exchange work is in a
preliminary stage where we are essentially calibrating a new
tool. Transitions to discrete states yield information which in
many cases can be better obtained from (γ,γ), (e,e'), etc. But
studies of this sort do tell one the extent to which the charge
exchange probe is understood and can be trusted. In the future
one hopes that the field will move toward studies of information
otherwise unavailable. For example, location of spin-flip
strength in heavy nuclei (see contributions by Galonsky and King)
or studies of subtle details of wavefunctions by probes with
slightly different properties [e.g. Gamow-Teller vs. (p,n)].

Wong: Can you resolve the 2^+, T=1 analog state from the
close-lying T=0 states in ^{26}Mg(p,n)? These states occur at $\sim$2
MeV excitation energy in ^{26}Al.

Berg: We can not resolve the first 2^+, T=1 analog state at
2.07 MeV from a 1^+ and a 4^+ state which are only 2 keV apart.
However, we can estimate the 2^+ strength from the angular distribution and from a comparison of our (p,n) cross sections with ft
values from β decay of ^{26}Si. The second 2^+, T=1 state at 3.16
MeV is completely resolved.

Comparisons of (p,p') and (p,n) Reaction Measurements

C. C. Foster

Indiana University Cyclotron Facility
Physics Department, Indiana University
Bloomington, Indiana 47405

INTRODUCTION

Comparison of absolute differential cross-section values for the ^{12}C(p,p') reaction to the T=1, 15.11 MeV (1+) and 16.11 MeV (2+) states with absolute values for their respective analog states at 0.0 and 1.0 MeV excited by the ^{12}C(p,n)^{12}N reaction may be used to determine neutron detection efficiencies for large volume neutron detectors at intermediate energies as reported by C. A. Goulding at this conference. In doing this, use is made of the relation[1]

$$2\frac{d\sigma}{d\Omega} (p,p') = \frac{d\sigma}{d\Omega} (p,n)$$

which holds for such analog states when isospin conservation is assumed. In the case of ^{12}C(p,p') and ^{12}C(p,n) reactions careful independent determinations of neutron detection efficiencies by using the ^{7}Li(p,n)^{7}Be reaction[2] or by Monte Carlo calculation[3] have established the validity of the relation to an accuracy of ± 10%. It is now interesting to consider how well comparisons using this relation, together with the facts that the (p,n) reaction T=0 targets excite T=1 states while the (p,p') reaction excites both T=0 and T=1 states, may be employed to study other nuclei. In order to investigate this question, ^{28}Si(p,p') and ^{28}Si(p,n)^{28}P reaction data are compared at 62 and about 120 MeV proton energies in this paper.

DATA

Time-of-flight spectra have been taken for ^{28}Si(p,n)^{28}P in the

angular range $0° \le \theta_{lab} \le 26°$ at E_p = 62 and 120 MeV at the Indiana
University Cyclotron Facility. Large volume plastic scintillator
(NE 102) neutron detectors were used at a flight path of about 26
meters. A dump magnet directly after the target swept the proton
beam away from the neutron flight path. Low-lying 1+, T=1 states[4]
were observed to be strongly excited in this reaction.

Spectra of inelastically scattered protons from ^{28}Si were taken
for E_p = 62 MeV with the IUCF QDDM spectrograph.[5] The limited,
3.5 MeV at 62 MeV, energy acceptance range of the spectrograph was
placed to include the analogs of the M1 states seen in the ^{28}Si(p,n).
For study of broad structures at higher excitation energies in
^{28}Si(p,p'), the limited energy acceptance of the spectrograph is too
restrictive. Therefore, a 3 cm thick intrinsic germanium detector
telescope was used. Proton inelastic scattering data for ^{28}Si over
the excitation energy range 0-30 MeV with 150-180 keV resolution in
the angular range $8° \le \theta_{lab} \le 34°$ has been obtained with the IUCF
64-inch scattering chamber.[6]

COMPARISONS

Figure 1 shows a ^{28}Si(p,p') spectrograph spectrum at θ_{lab} = 14°
and E_p = 62 MeV (solid line) and a ^{28}Si(p,n) time-of-flight spectrum
(dotted line) at θ_{lab} = 20° on the same energy scale. The peak at
2.1 MeV excitation in the (p,n) spectrum has been matched in energy
and normalized by peak height to its analog 11.45 MeV peak in the
(p,p') spectrum. Shaded peaks in the (p,p') spectrum, which has a
resolution of 44 keV (FWHM), indicate T=1 peaks reported in the literature.[8] Energy resolution of the (p,n) data is 800 keV (FWHM). From
this figure, it is clear that it is not possible in the (p,n) case
to resolve single T=1 states and that, if one is to calculate the
measured (p,n) cross-section angular distributions theoretically, contributions from all unresolved states should be considered. It is
difficult, as well, to cleanly resolve T=1 states from T=0 states in
the (p,p') spectrum due to the high density of states. Nevertheless,
a comparison has been made of the angular distributions of the peaks
centered at about 11.445 MeV in the (p,p') and 2.1 MeV in the (p,n).
Figure 2 shows this comparison. Two sets of measurements made at the
same time are shown for the (p,n). Those made by Madey et al. used
six fixed detectors and calculated efficiencies with a Monte Carlo
code. Those made by Goodman et al. employed two moved detectors and
determined dfficiencies by measurement with the ^{7}Li(p,n)^{7}Be reaction.[2]
The ^{28}Si(p,p') data was taken with the QDDM spectrograph. Errors
shown are statistical only. Normalization uncertainties are ± 16%
for the (p,n) data due mostly to a 14% target non-uniformity and
± 10% for the (p,p') data due to a 7% uncertainty in target thickness.
Within these uncertainties, 2 times the ^{28}Si(p,p')^{28}P cross-section
equals the ^{28}Si(p,n) cross-section. Attempts to make similar compar-

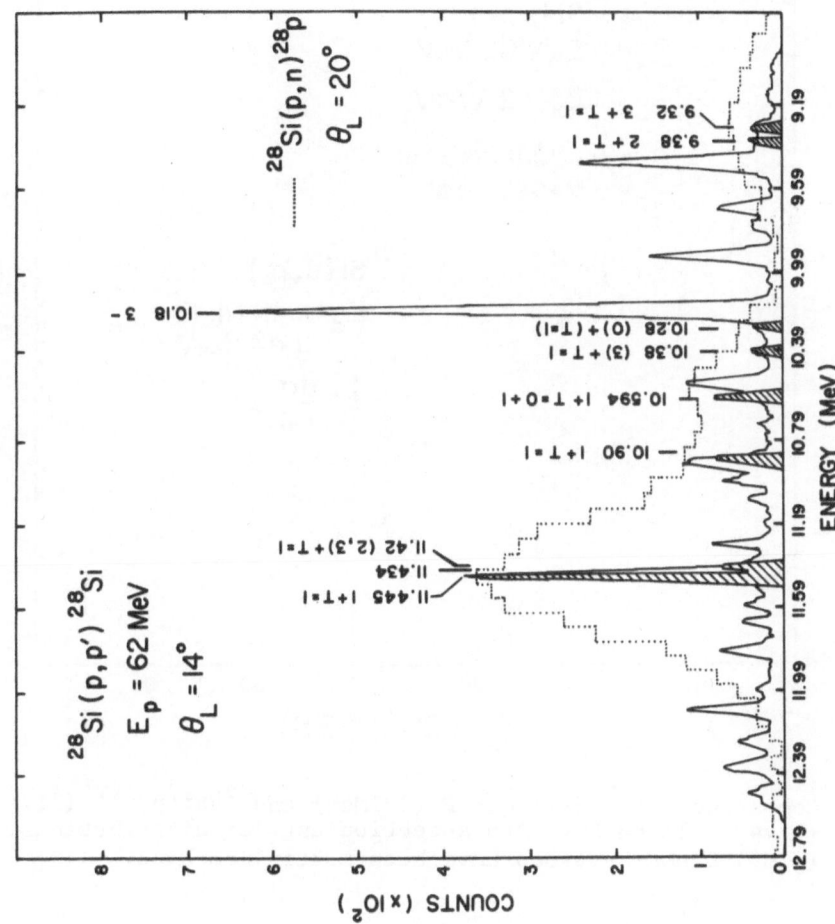

Fig. 1. ^{28}Si(p,n)^{28}p time-of-flight spectrum (dotted line) compared to ^{28}Si(p,p') spectrograph spectrum (solid line). Shaded peaks indicate location of T=1 states reported in the literature.[8]

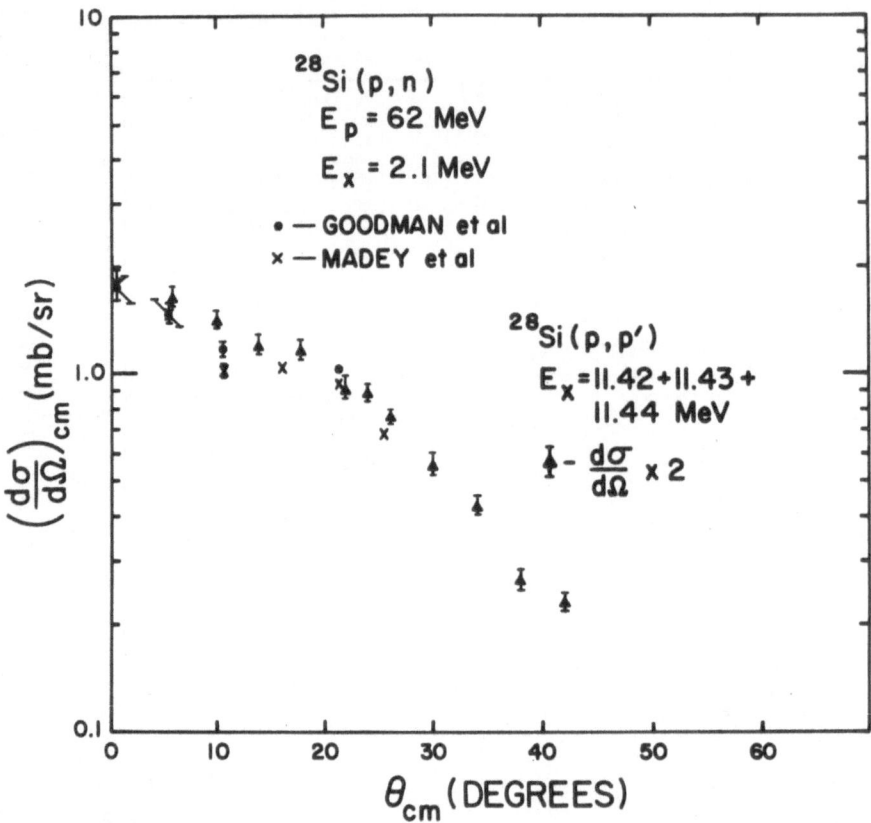

Fig. 2.   Comparison of $^{28}Si(p,n)^{28}P$ (2.1MeV) and $^{28}Si(p,p')$ (11.42
+11.43 + 11.44 MeV) cross-section angular distributions.
(p,p') cross-sections have been multiplied by 2.

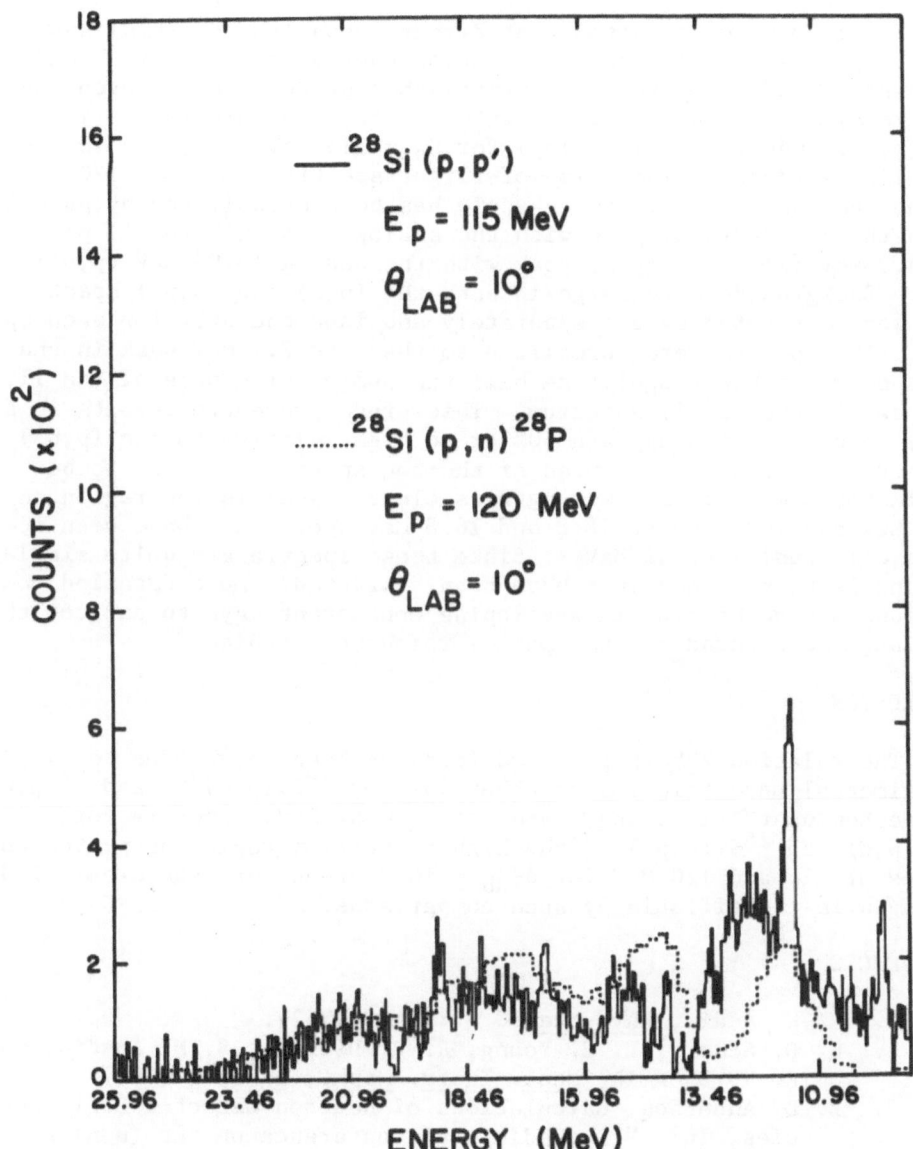

Fig. 3. ^{28}Si(p,n)^{28}P time-of-flight spectrum (dotted line) at E_p = 120 MeV compared to ^{28}(p,p') spectrum (solid line) at E_p = 115 MeV for θ_{lab} = 10° and in the high excitation region. Energy scale is for excitation energy in ^{28}Si.

isons for the other T=1 peaks seen in Fig. 1 have not been as successful because of difficulties resolving the peaks in the (p,p') spectra.

Figure 3 shows an attempt at direct comparison of (p,n) and
(p,p') spectra in the higher excitation energy region where broad
resonances and/or clusters of unresolvable states are expected and
good resolution is not as important. Plotted in this figure is a
^{28}Si(p,p') spectrum (solid line) for E_p = 115 MeV at θ_{lab} = 10° and
a ^{28}Si(p,n) (dotted line) time-of-flight spectrum for E_p = 120 MeV
at the same angle. The energy scale has been established by associ-
ating the 2.1 MeV(p,n) peak with the analog 11.45 MeV (p,p') peak and
the 4.7 MeV (1+, T=1) (p,n) peak with the analog 13.95 MeV (p,p')
peak. Backgrounds were large in both the (p,n) and (p,p') spectra.
They were extracted by eye separately and independently for each spectrum. The spectra were normalized so that the 2.1 MeV peak in the
(p,n) spectrum has roughly one half the number of counts of the 11.45
MeV peak in the (p,p') spectrum. This crude procedure results in a
(p,n) spectrum which appears 50% or so high relative to the (p,p')
spectrum. The general features of the two spectra are seen to be
remarkably similar. T=0 strength is clearly seen in the region near
12.7 MeV and in peaks at 18.8 and 16.8 MeV where they have been seen
in (α,α') studies at 62 MeV.[7] Since these spectra are quite similar
reasonable backgrounds must have been extracted. More detailed comparisons may be helpful in developing consistent ways to extract the
continuum background which troubles resonance studies.

CONCLUSION

The relation $2\frac{d\sigma}{d\Omega}$ (p,p') = $\frac{d\sigma}{d\Omega}$ (p,n)has been shown true to within
experimental uncertainties of about 25% for ^{28}Si(p,n)^{28}P and ^{28}(p,p')
to the dominant T=1 analog states at E_p = 62 MeV. Spectra for
^{28}Si(p,n) and ^{28}Si(p,p') in the high excitation region at proton energies of 115 and 120 MeV for θ_{lab} = 10 degrees were compared. T=0
strength is identifiable by such comparisons.

REFERENCES

1. R. K. Adair, Phys. Rev., 87:1041 (1952).
2. S. D. Schery, L. E. Young, R. R. Doering, S. M. Austin, and
    R. K. Bhowmik, Nucl. Instr. Meth., 147:399 (1977).
3. B. D. Anderson, Calculations of Neutron Detector Efficien-
    cies, in: "Proceedings of Conference on the (p,n) Re-
    action and the Nucleon-Nucleon Force" (1979).
4. C. D. Goodman, F. E. Bertrand, R. Madey, B. Anderson, A.
    Baldwin, J. Knudson, T. Whitten, J. Rapaport, D. Bainum,
    M. B. Greenfield, and C. C. Foster, Bull. Am. Phys. Soc.,
    22:544 (1977).

5. C. C. Foster, J. Rapaport, D. Bainum, C. D. Goodman, M. B. Greenfield, C. A. Goulding, R. Madey, and B. Anderson, unpublished (1978).

6. P. P. Singh, S. Kailas, P. Schwandt, A. D. Bacher, J. Wiggins, D. L. Friesel, C. C. Foster, A. Van der Woude, and A. G. Drentje, Bull. Am. Phys. Soc., 24:647 (1979).

7. K. Van der Borg, M. N. Harakeh, S. Y. Van der Werf, A. Van der Woude, and F. E. Bertrand, Phys. Lett., 67B (1977).

8. P. M. Endt and C. van der Leun, Nucl. Phys., A310 (1978) 238.

A FOLDING MODEL ANALYSIS OF

THE (p,n) QUASIELASTIC REACTION*

S. D. Schery

Moody College
Texas A&M University System
Galveston, TX 77553

I. INTRODUCTION

Since the discovery in 1961 of the prominence of the isobaric analogue of the target ground state in (p,n) spectra[1], theoretical study of this reaction, often called the (p,n) quasielastic reaction, has evolved along several lines. One approach, sometimes called the microscopic model,[2,3,4] represents the target and residual nuclei by shell model wave functions. A nucleon-nucleon interaction is assumed and the reaction is described in terms of a sum of two-body forces between the projectile and target nucleons. This fundamental approach can include such refinements as multistep processes[5,6] and the effect of antisymmetrization of the projectile and target nucleons.[7] An alternate approach describes the quasielastic reaction by means of a nucleon-nucleus optical potential that possesses an isospin dependence. This approach originated with A. M. Lane[8,9] and the term macroscopic model is used when a phenomenological isospin dependent optical potential is used to describe the quasielastic reaction.[10,11,12] These two models are not necessarily inconsistent and both have met with success in predicting quasielastic data.

Recently another approach which has fundamental basis intermediate between that of the microscopic and macroscopic models has been applied with success to (p,n) quasielastic reactions. This approach can be termed the folding model since its potential is obtained by folding components of the nucleon-nucleon interaction with nuclear matter densities. This model resembles the microscopic model

* Research supported in part by the National Science Foundation

in that a nucleon-nucleon interaction is used; however, unlike
the microscopic model, the target nucleus is described by proton
and neutron matter densities rather than shell model wave functions.
A convenient method of applying this model is to retain an isospin
dependent optical potential and to generate its isovector component
by the folding procedure. Such a formulation of the full optical
potential has met with success in describing proton elastic scattering.[13,14] Those studies have laid much of the groundwork for the
present study of the closely related (p,n) quasielastic reaction.

In this paper the folding model approach will be used to analyze
a wide body of recent precision (p,n) quasielastic data in the energy range 22.8 to 45 MeV. Such a wide ranging study, though tedious
and time consuming, provides a uniquely rigorous test of (p,n) reaction models. Refinements in the model such as density dependent
interactions and exchange corrections will be discussed. Uncertainties in the analysis due to distorted wave potentials and the
strength and range of the nucleon-nucleon interaction will be addressed. Attention will be given to isotopic, isotonic, and isobaric sequences since calculations involving a sequence of nuclides
can sometimes reduce uncertainties inherent in absolute calculations
of a single nuclide. A major motivation for folding model analysis
of precision (p,n) experiments is a desire to extract information
about the neutron matter distributions in nuclei. The status and
prospects of this application of the folding model will be discussed.

Section II will describe the formalism of the folding model and
its use through distorted wave Born approximation (DWBA) calculations
to predict the quasielastic reaction. Section III will give results
of folding model calculations for various assumptions. Section IV
will discuss prospects of extracting nuclear matter radii from the
(p,n) quasielastic reaction and Section V will summarize the present
status of the folding model and make recommendations for future study.

Some of the results reported in this paper have not been previously published and the contribution of collaborators should be recognized. Special acknowledgement is given to A. Galonsky and S. M.
Austin at the neutron time-of-flight facilities of Michigan State
University and D. A. Lind and C. Zafiratos at the time-of-flight
facilities of the University of Colorado.

II. THE FOLDING MODEL

The folding model formalism of the (p,n) quasielastic reaction
can be developed conveniently by retaining Lane's isospin description of the nucleon-nucleus optical potential and then generating
the isospin component of the optical potential by folding matter
densities with appropriate components of the nucleon-nucleon interaction. An alternate development which will not be discussed here

would start directly with the microscopic expression for the cross
section and then relate the wave function terms to the nucleon densities. Theoretical background related to the present discussion
can be found in references 4, 15, and 16.

Lane's model[8],[9] of the nucleon-nucleus optical potential represents the optical potential in the form

$$V = V_0 + \frac{4}{A} \vec{t} \cdot \vec{T} V_1 \tag{1}$$

where $\vec{t}$ and $\vec{T}$ are the nucleon and nucleus isospin operators respectively, V_0 is the isoscalar potential, and V_1 is the isovector
potential.

If we let χ_{pT} represent the relative wave function of the proton-
target system and χ_{nA} represent the relative wave function of the
neutron-analogue system, the (p,n) quasielastic reaction is described
by the coupled equations

$$(K + V_0 - 2T(V_1/A) + V_c - E_{pT})\chi_{pT} = -2(2T)^{\frac{1}{2}}(V_1/A)\chi_{nA}$$

$$\tag{2}$$

$$(K + V_0 + 2(T - 1)(V_1/A) + \Delta_c - E_{pT})\chi_{nA} = -2(2T)^{\frac{1}{2}}(V_1/A)\chi_{pT}$$

where $T = \frac{1}{2}(N-Z)$ for the target nucleus, K is the kinetic-energy
operator, V_c is the Coulomb potential, Δ_c is the Coulomb displacement energy, and E_{pT} is the relative energy of the proton-target
system. The ratio of the coupling term to the main potential
$2(2T)^{\frac{1}{2}}V_1/(AV_0)$ is usually sufficiently small that these equations
can be solved with good approximation in the distorted wave Born
approximation to give[4]

$$\frac{d\sigma}{d\Omega} = \left(\frac{\mu}{2\pi\hbar^2}\right)^2 \frac{k_n}{k_p}\left|T_{pn}\right|^2 \tag{3}$$

where μ is the reduced nucleon mass and k_p and k_n are the wave numbers of the incident proton and outgoing neutron respectively. If
the relative wave functions for the proton-target and neutron-analogue systems of equation (2) are approximated by distorted waves
determined from elastic scattering, ξ_p and ξ_n, the transition
amplitude T_{pn} is given by

$$T_{pn} = \langle\xi_n|(8T)^{\frac{1}{2}} V_1/A|\xi_p\rangle. \tag{4}$$

So far no mention has been made of the folding model and our
theory is the same as that used for the phenomenological analysis
of the (p,n) quasielastic reaction where a phenomenological V_1 is
used in equation (4) to calculate the differential cross section.

Major assumptions involved in using equations (3) and (4) are that
the reaction can be represented by an optical potential which can
be expressed in the form of equation (1), that the coupling in equation (2) is weak, and that potentials determined (in practice) from
elastic scattering on target and residual nuclei ground states can
be used for the distorted waves. For computational convenience it
is also customary to use "local equivalent" potentials to approximately correct for the presence of non-locality.[17] These assumptions
are expected to be reasonable on theoretical grounds and the
success of the phenomenological analysis of the quasielastic reaction
provides further substantiation of this expectation.

The folding model approach involves representing the optical potential equation (1) in terms of matter distributions folded with appropriate components of the nucleon-nucleon interaction. The derivation of the full optical potential from the many body point of view
has been reviewed by Sinha[14] and only those results for the isovector
potential V_1 that will be used for analysis of the (p,n) reactions
will be summarized here. The notation $V_1 = U_1 + iW_1$ will be used
where U_1 and W_1 are respectively the real and imaginary components
of the isovector potential. The nucleon-nucleon interaction will be
assumed expressable in the following form

$$v = v_d + 4\vec{t}_1 \cdot \vec{t}_2 v_t + \ldots \tag{5}$$

where no restriction has been placed on the functional form of v_t
or on whether it represents a free interaction or bound state effective interaction. Neglecting exchange, the first order approximation for V_1 is given by $W_1 = 0$ and

$$U_1 = \frac{A}{(N-Z)} \int v_t \ (|\vec{r} - \vec{r}'|)(\rho_n(r') - \rho_p(r'))d^3r' \tag{6}$$

where ρ_n and ρ_p are the distributions for the neutron and proton
mass centers. In the lowest order approximation v_t is the free particle nucleon-nucleon interaction. Next order corrections involve
effects of exchange, and virtual and real excitations of target nucleons by the incident proton. Calculation of these corrections from
first principles is not straightforward and their exact form remains
an open question. Estimates of these effects to be used in the present analysis will be given here, but the reader should be cautioned
that the theoretical foundation is not firm and that alternate, sometimes contradictory, formulations exist. [18,19,20]

Correction for virtual excitations of nucleons in the target
nucleus can be made by replacing the free body interaction with an
"effective" bound state interaction. This approach has been successful both in analysis of proton elastic scattering and in traditional
microscopic analysis of the (p,n) reaction. One common form used
for the isospin component of the effective interaction is the

Gaussian function

$$v_t = -v_o e^{-r^2/b^2} \tag{7}$$

where b is a parameter that determines the range of the interaction and is typically between 1 and 2fm.   Other common forms include the Yukawa function or sums of Yukawa functions. There is some indication that for studies such as those directed toward neutron matter radii that the rms range of the interaction rather than its functional form is the crucial question.[21]

Some evidence exists for a density dependence in the nucleon-nucleon potential and it is possible that there is specifically a density dependence in the isospin component of this interaction. The principal effect of density dependence is to make the interaction stronger at the nuclear surface where more scattering states are available to the unsaturated target nucleons. For example, the weak and strong forces due to Green[22] based on predictions of binding energies in nuclear matter implies a density dependence in the isospin component of the nucleon-nucleon interaction. A calculationally convenient form[23] for density dependence based on the assumptions of a linear density dependence and a two-parameter Fermi form for nuclear matter is

$$v_{t,D} = v_{to} \left[1 + \alpha_t \frac{e^{(r-R_m)/a_m}}{(e^{(r-R_m)/a_m} + 1)} \right] \tag{8}$$

where v_{to} is the interaction at the nuclear interior, α_t is a parameter determined by the strength of the density dependence, and R_m and a_m are the radius and diffuseness for nuclear matter.

Real excitation of the nucleons of the target nucleus leads in part to the imaginary isovector potential. This effect is difficult to calculate, and traditionally both the macroscopic and microscopic approaches have used a phenomenological potential. However, recently a promising estimate[24] of this effect has been made from the forward scattering amplitude approximation (FSAA):

$$W_1 = \frac{-W_{10} \hbar V_p(r) A}{4(N-Z)} (\rho_n'(r) - \rho_p'(r)) \times (\bar{\sigma}_{pn} - \bar{\sigma}_{pp}), \tag{9}$$

where V_p is the velocity of the proton at nuclear radius r, ρ_n' and ρ_p' are effective densities obtained by folding ρ_n and ρ_p with a finite range interaction, and σ_{pn} and σ_{pp} are total cross sections averaged in the local Fermi gas model approximation to exclude scattering to forbidden states. W_{10} is a normalization constant that has a value of 1 in first approximation. Equation (9) is basically a high energy approximation and is not expected to be particularly valid at lower energies. Nevertheless both equation (9) and the corresponding expression for the full imaginary potential have proven useful in analysis of the (p,n) quasielastic reaction and proton elastic scattering. A comparison of predictions of equation (9) to more fundamental Brueckner-Hartree-Fock calculations shows surprising similarity,[24] although a value for W_{10} greater than one is indicated. The need for increased strength is not surprising since the derivation of equation (9) excludes collective excitations of the target nucleus[25] that in reality contribute to the imaginary potential.

Given an appropriate nucleon-nucleon interaction the calculation of exchange effects in principle is straightforward in the microscopic model. For example, both Doering[7] and Love[26] have included exchange in microscopic calculations. An appropriate estimate of exchange in terms of nuclear densities in the folding model is much less clear. A local density approximation[27,14] has been used to estimate exchange effects for proton elastic scattering. The corresponding result for the real isospin potential is

$$
U_1 = \frac{A}{(N-Z)} \int v_t^d \left(|\vec{r}-\vec{r}\,'|\right) \left[\rho_n(r') - \rho_p(r')\right] d^3 r' \tag{10}
$$

$$
+ \frac{A}{(N-Z)} \int v_t^e \left(|\vec{r}-\vec{r}\,'|\right) \left[\rho_n(\vec{r},\vec{r}\,') - \rho_p(\vec{r},\vec{r}\,')\right] \times j_0 \left(k\left(\frac{\vec{r}+\vec{r}\,'}{2}\right)|\vec{r}-\vec{r}\,'|\right) d^3 r'
$$

where the second term gives the local equivalent form for the exchange correction. Here k is the wave number of the incident proton evaluated at the midpoint $\frac{1}{2}(\vec{r}+\vec{r}\,')$, $\rho_n(r,r')$ and $\rho_p(\vec{r},\vec{r}\,')$ are mixed densities, and j_0 is the spherical Bessel function of order zero. The densities are evaluated by a modification of the Slater approximation[28] due to Pandharipande[29] and are expressed in terms of a mean Fermi momentum determined from local densities. The terms v_t^d and v_t^e are respectively the direct and exchange parts of the effective nucleon-nucleon interaction. Although Sinha[14] found the local density approximation approach useful for elastic scattering, Georgiev[20] made comparisons to more exact microscopic calculations and found the approximation worked well at energies above 60MeV but was less suitable at lower energies. The present analysis contains the first report of this approach to the exchange correction for the isovector component of the optical potential in calculations of the (p,n) quasielastic reaction.

III. ANALYSIS

This section will discuss analysis which starts with cross sections calculated with equations (3) and (4) using equation (6) with an

effective interaction. The various estimates for the higher order
processes such as exchange, density dependence, and absorption are
then included in separate calculations to see if improved agreement
with data results. Since proton densities are comparatively well
known from electron scattering experiments and the study of muonic
x-rays, the proton densities are fixed at values obtained from the
literature.[30,31,32] In order to reduce time in rather lengthy computer calculations, only two-parameter Fermi functions are used and
the unfolding of mass centers from charge distributions is done by
the approximate description of Uberall.[30] In some cases two-parameter Fermi functions have not been listed directly in the literature.
In these cases, a two-parameter Fermi function is chosen that gives
the same rms radius as the listed value, with the ambiguity of pro-
portion of radius and diffuseness resolved by reproducing the 10-90%
skin thickness or comparison with results for nuclides of similar
mass numbers. The neutron densities are also parameterized by two-
parameter Fermi functions. However, these parameters are usually
treated as free variables since their value is much more poorly known.

Experimantal data has been selected to cover proton energies in
the range 22.8 to 45 MeV, a wide range of N-Z, and isotopic, isobaric,
and isotonic sequences.[7,33,34,35,24,36] Within these guidelines preference in selection is given to data that is the highest precision
available since it is found that the combination of a wide range of
nuclides and precise data provides quite demanding tests of (p,n)
reaction models. Figure (1) shows angular distributions of the data
selected for study.

The computational procedure followed allows the neutron density
radius R_n and diffuseness a_n and the nucleon-nucleon isospin strength
to be free parameters. The optical potentials used to generate the
distorted waves are conventional phenomenological potentials taken
from the literature. No attempt is made to generate these potentials
from the folding model since the computational time required would be
prohibitive. In principle, a well-determined phenomenological potential should give a result quite similar to that obtained from a folding model. For a given nuclide and energy the potentials were chosen
that provided the best prediction of the elastic scattering. This
criterion means that different potentials could be used for different
nuclides or for proton or neutron potentials. It is hence possible
that in a given DWBA calculation the proton and neutron potentials
were not necessarily consistent with the Lane model. The most fre-
quently used global potentials were those due to Becchetti and Greenlees,[37] Wilmore,[38,39,40] and Patterson.[41] In cases where two different potentials were applicable and provided comparable predictions of
elastic scattering the one providing the best prediction of the (p,n)
data was used subsequently.

The procedure for determining the free parameters that best predicted the data used a least squares routine with the DWBA code

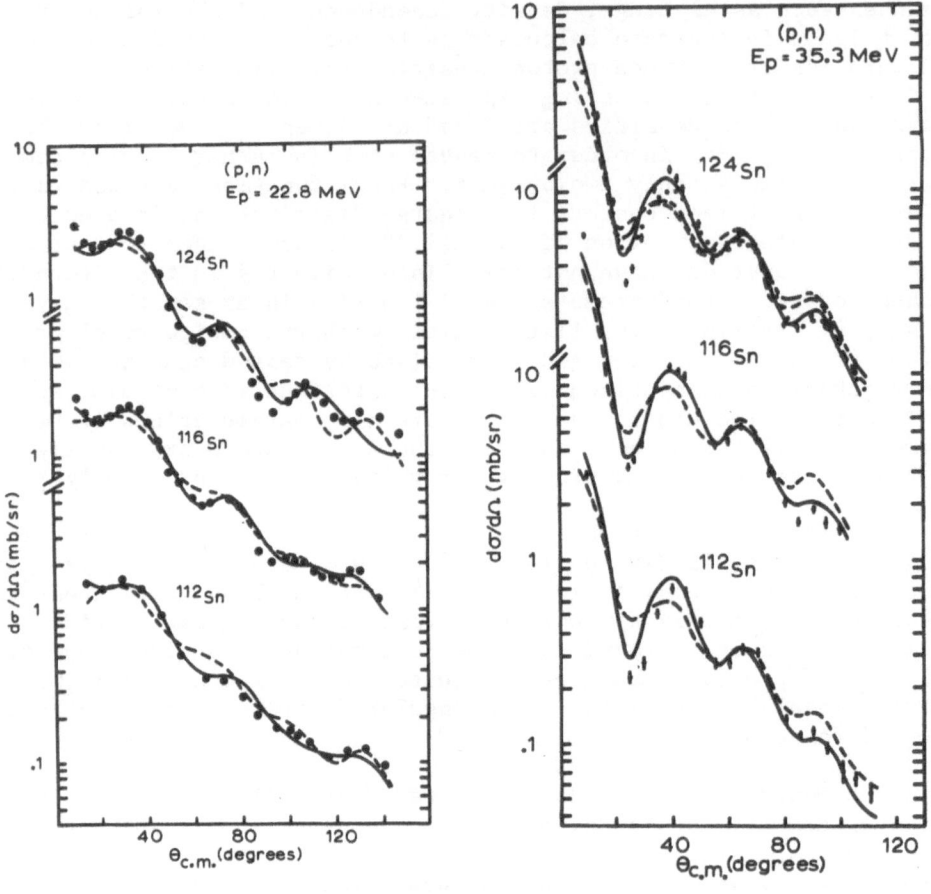

Figure 1a.

Figure 1a,1b,1c.   Angular distributions for the quasielastic
                   (p,n) reaction. Experimental data are from
                   references 7, 33, 34, 35, 24, 36. The short
                   dashes are DWBA calculations using a pure
                   real form factor. The solid curves are cal-
                   culations with a complex form factor with the
                   imaginary component determined from the for-
                   ward scattering amplitude approximation in-
                   cluding a finite range correction. The short-
                   long dashes curve for ^{124}Sn at 35.3MeV uses
                   the forward scattering amplitude approxima-
                   tion with no finite range correction. The
                   dotted curve for ^{124}Sn at 35.3MeV uses a
                   phenomenological potential for W_1.

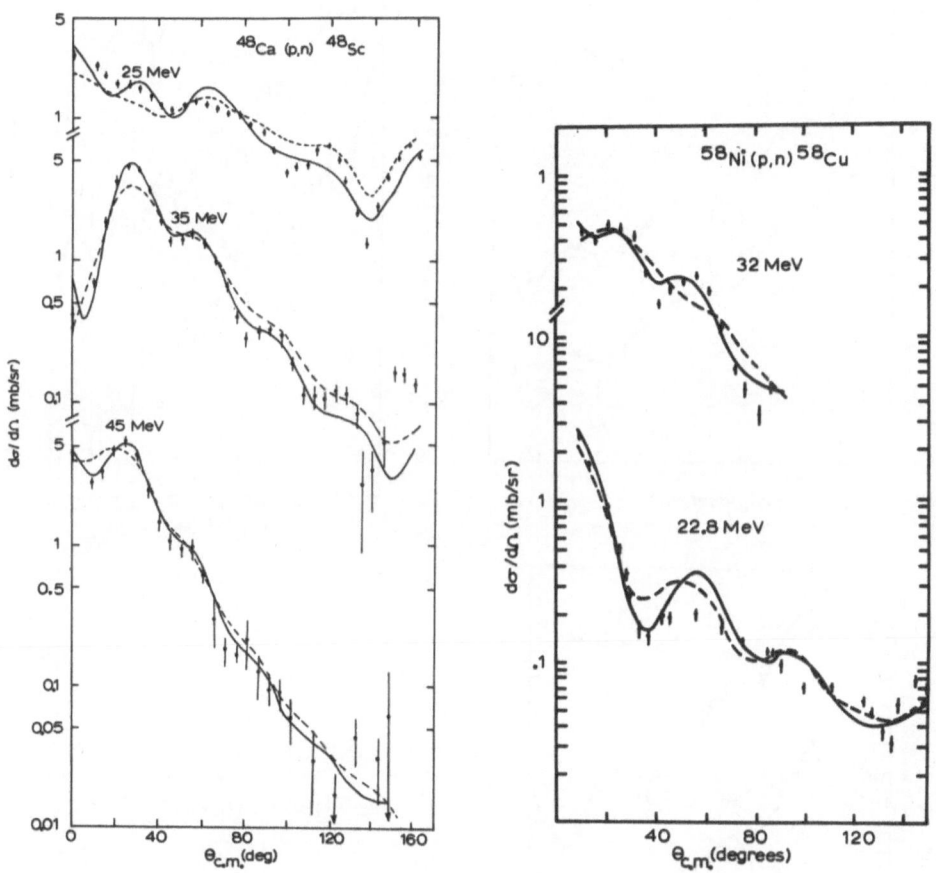

Figure 1b.

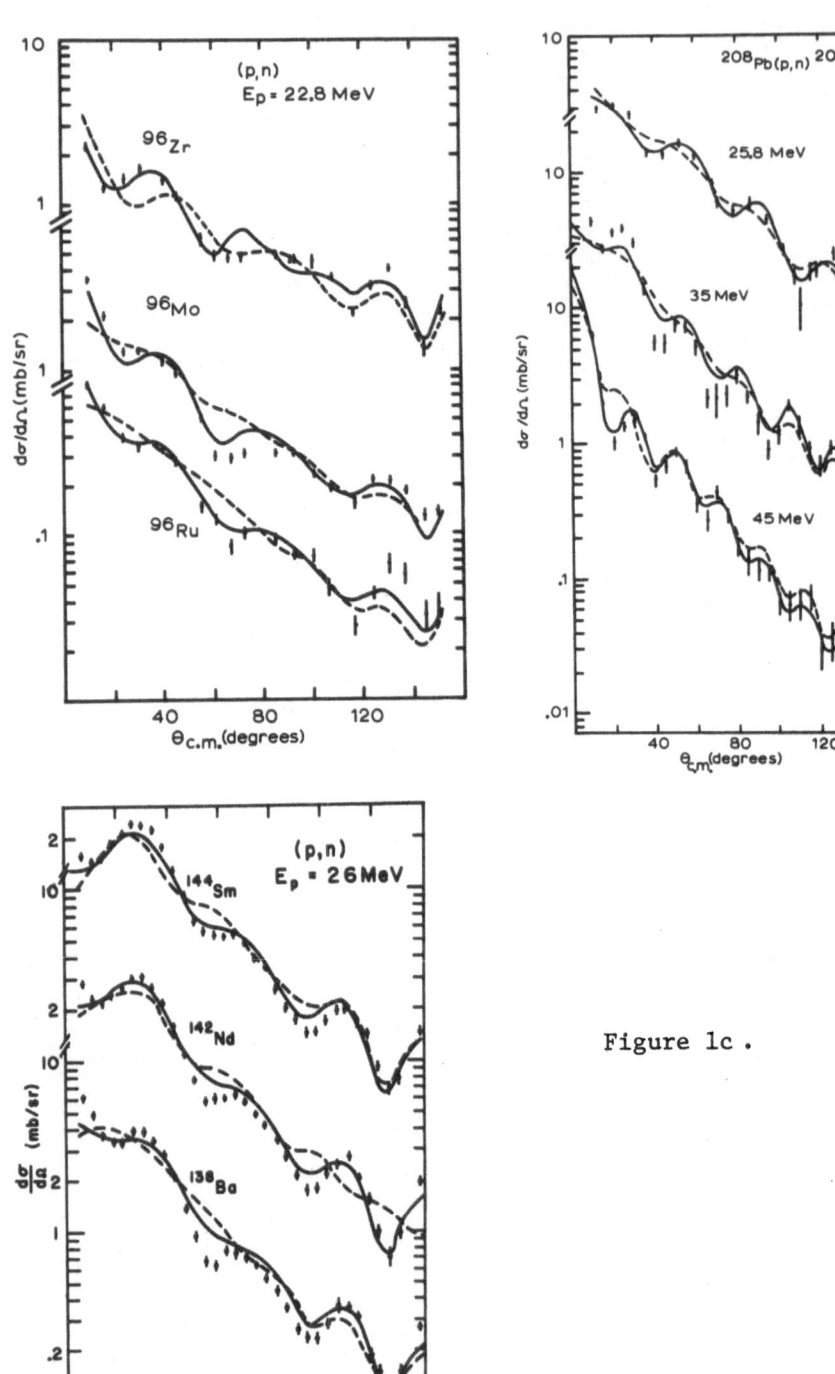

Figure 1c .

DWUCK[42] as a subroutine. For a given set of parameters, V_1 and the form factor were generated numerically in a subroutine for DWUCK. The least squares code was written for this research and used the Broyden-Fletcher-Goldfarb-Shanno algorithm for unconstrained optimization. Due to the length of these calculations it was usually prohibitive to have more than three parameters free, one of which was the strength.

A non-standard χ^2 test was used to determine the agreement between calculations and data. This step was taken because of the desire to fit the overall shape of the angular distributions and the well-known tendency of the standard χ^2 test to give added weight to the forward angles. This problem arises in part because often not all errors, especially systematic, are included in the analysis. Inclusion of just the experimental statistical counting error will often result in a disproportionally small error for regions of smallest statistical error, which are usually the forward angles for the (p,n) reaction. This problem was treated by the physically reasonable but statistically non-rigorous procedure of Carlson[33] that uses logarithms to reduce the sensitivity to large variations in experimental error:

$$\chi^2 = \frac{1}{N} \sum_{i=1}^{N} \frac{(\ln\sigma(\theta_i) - \ln\sigma_0(\theta_i))^2}{(\Delta\sigma_0(\theta_i)/\sigma_0(\theta_i))^2} \tag{11}$$

where $\sigma_0(\theta_i)$ = experimental cross section at θ_i, $\sigma(\theta_i)$ = calculated cross section at θ_i, $\Delta\sigma_0(\theta_i)$ = experimental error in θ_i, and N = the number of points in the angular distribution. A Gaussian function with a root mean square range 2.1fm was used for the effective interaction in the initial analysis with equation (6). This function and range had proven successful in previous (p,p) and (p,n) analysis.[43,34,35] Calculations in which this range was varied will be discussed later.

The short dashes in figure (1) shows calculated angular distributions for a three parameter optimization of v_0, R_n, and a_n using the pure real form factor generated from equation (6). Table 1 lists the integrated strengths for U_1 for a few of these nuclides. Although most angular distributions are approximately reproduced, there is a clear trend of the calculated angular distributions underestimating the structure of the diffraction minima and maxima.

The angular distributions can be improved by adding an imaginary isovector potential to the calculations. Following Schery[24] we use equation (9) from the forward scattering amplitude approximation with $W_{10}=1$ and a rms finite range correction for ρ_n and ρ_p of 2.8 fm. The value of v_0 was fixed at 6.0 MeV and then R_n and a_n of this complex form factor and overall normalization were adjusted for best agreement with data. The number of free parameters remained three.

Table 1.*

Nuclide, energy	Real	Complex	
	$\int U_1 d^3 r / A$	$\int U_1 d^3 r / A$	$\int W_1 d^3 r / A$
^{58}Ni, 22.8MeV	215MeVfm^3	164MeVfm^3	124MeVfm^3
^{124}Sn, 22.8MeV	178	174	99
^{208}Pb, 25.8MeV	218	185	95
^{208}Pb, 45MeV	202	188	102

*Integrated strengths of the isovector potential
for a few of the calculations shown in figure (1).
Column 2 gives the integrated strength per nucleon
for the pure real calculation and columns 3 and 4
give the integrated strengths per nucleon of the
real and imaginary parts for the complex calculation with finite range correction. A major source
of error in these numbers is the experimental contribution which is of the order of 10%.

The results are shown as the solid lines in figure (1), and table 1
lists the integrated strength for U_1 and W_1 for a few nuclides. The
improvement in the quality of the predictions is obvious, although
there is still a slight tendency to underestimate the structure of
the minima and maxima. For comparison, the dotted line in figure (1)
for ^{124}Sn shows the result of a calculation with the phenomenological form for W_1 from the optical potential of Becchetti and Green-
less.[37] For this calculation R_n, a_n and v_o of the real isovector
potential were varied and thus the number of free parameters was
again three. A comparison of the forward scattering amplitude
approximation calculation with the phenomenological calculation
indicates comparable agreement. Although limitations on computer
time prohibited systematic comparisons of phenomenological imaginary
potentials to the forward scattering amplitude approximation, selected calculations have supported this finding. Discussion of optimum
strength W_{10} and finite range correction will be discussed later.

The effects of a density dependence and exchange on the isovector
potential can be seen in figure (2). Here calculations of the real
isovector potential for ^{124}Sn using the strong Green density depen-

dent form in equation (10) are compared to equation (6) using the density independent Gaussian force without exchange. Each calculation uses the same value for ρ_n and ρ_p. The direct part of equation (10) indicates that the strong surface peaking in this calculation is due to the density dependence of the Green force. The exchange correction is roughly similar in shape though much weaker. The essential result of the direct plus exchange calculations is a potential with a shape more surface peaked because of the density dependence and with a strength renormalized due to exchange.

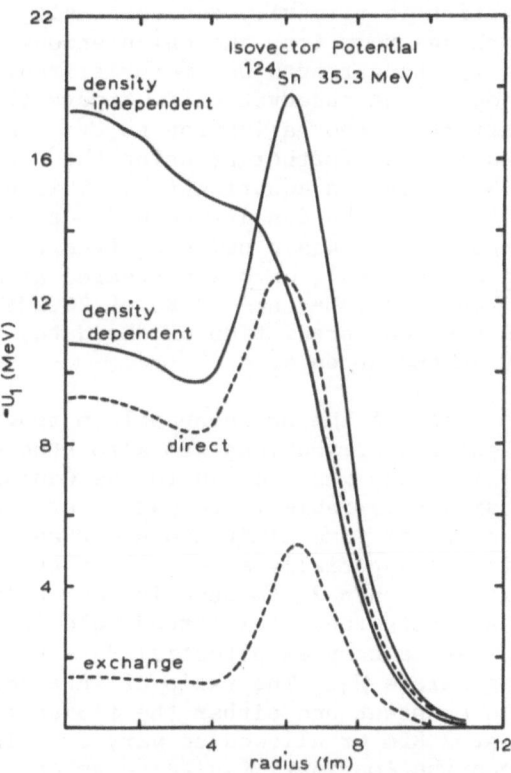

Fig. 2.   Calculation of the real isovector potential using a density independent Gaussian interaction of rms range 2.1fm and strength 8.0 MeV in equation (6) and a density dependent calculation with exchange using the strong Green force in equation (10). The direct and exchange contributions to the density dependent calculation are shown separately (dashes) which combine to give the total potential (solid line). Due to the different parameters of the Gaussian and Green forces, the integrated strength and rms radii of these potentials are not necessarily the same.

Calculations of angular distributions using equation (10) were limited since the evaluation of equation (10) requires considerable computer time. However, sufficient calculations (about a half dozen nuclides) were done to establish an apparent trend. The FSAA was used for W_1 and the free parameters were R_n and a_n plus overall normalization. Except in a few isolated cases the results were not promising and there was a systematic trend of slight deterioration of predictions when equation (10) was used to calculate U_1. Figure (3) shows a calculation using the weak Green force for ^{138}Ba (dashes). The agreement with data is obviously inferior to the density independent exchange-free calculation with equation (6) (solid line). This deterioration is primarily due to density dependence since removing the exchange term from the calculations yielded essentially the same χ^2 if the strength was renormalized. This result is not surprising based on the observation in figure (2) that exchange primarily contributes renormalization to U_1. The question of density dependence was pursued further by using the density dependence from equation (8) for v_t in equation (6). This procedure provided a different form for the density dependence with a variable density dependent strength α_t. Again using W_1 from the FSAA (and omitting the exchange correction), $\alpha_t \geq 0$ was treated as a fourth free parameter for both ^{124}Sn at 35.3MeV and ^{138}Ba at 26.0MeV. In both cases $\alpha_t \cong 0$ provided the best prediction of the data, although the χ^2 was fairly weakly dependent on α_t.

The effect of the range of the nucleon-nucleon reaction on the predictions of the angular distributions was also studied on a more qualitative basis. This study was limited to the Gaussian interaction equation (7) using a variable range parameter b and was not comprehensive due to computer time limitations. Predictions of angular distributions were not especially sensitive to this parameter but in general a rms range less than 2.1fm used in the analysis shown in figures (1) and (2) was preferred. This trend held for both a pure real potential $V_1 = U_1$ and a complex potential $V_1 = U_1 + iW_1$ where the FSAA was used to generate W_1. The ratio of integrated strengths for U_1 and W_1 was held constant and either the finite range correction of W_1 was fixed at 2.8fm or allowed to vary to maintain the same difference between the imaginary finite range correction and the real interaction. For example, for ^{208}Pb at 45MeV with $\int U_1 d^3r / \int W_1 d^3r = 2.0$ and an imaginary finite range correction of 2.8fm, an rms value of 1.6fm provided the best agreement with data. The experimental data was precise enough and the selection of nuclides broad enough that it appeared feasible to make a more definitive study of the range of U_1 if sufficient computer time were available.

On a qualitative basis we have also studied effects of variation of W_{10} and the finite range correction in the FSAA calculation of W_1. The most definitive result is that zero range provides significantly poorer predictions of data typified by the short-long dashes

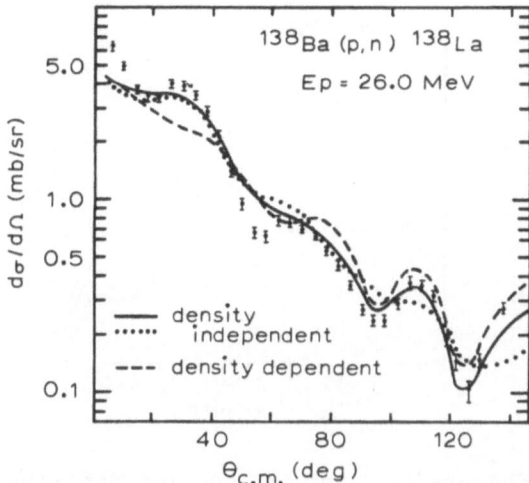

Fig. 3.   A Comparison of calculations for ^{138}Ba using a
density independent isovector potential (solid
line) and a density dependent potential with
exchange correction (dashed line). The effect
of varying the distorted wave parameters is also
shown. The dotted line is a calculation identical
to that of the solid line except the neutron parameters of Wilmore[38] were used in place of the
neutron parameters of Patterson.[41]

in figure (1) for ^{124}Sn. It appears important that the finite range
correction for the imaginary potential exceed the range of the real
interaction used in U_1. Typically for a real rms range of 2.1fm and
W_{10} = 1, an rms finite range correction of 2.8fm seemed satisfactory,
although neither an optimum value nor possible energy or target
nuclide mass dependence was determined. Studies of the strength
parameter W_1 were not detailed enough to reveal a mass or energy
dependence but there did seem to be a systematic trend favoring a
stronger value for W_{10} than the unit value used for the FSAA calculations of figure (1) and table 1. The χ^2 dependence is not particularly strong but, for example, a value of W_{10} = 2 gave consistently
better predictions for the few cases studied.

IV.   NEUTRON MATTER DENSITIES AND THE (p,n) QUASIELASTIC REACTION

An important feature of the folding model approach to the (p,n)
quasielastic reaction is that it potentially could be used to obtain
information about neutron densities if uncertainties in the other
parameters of the analysis were understood. The sensitivity of the

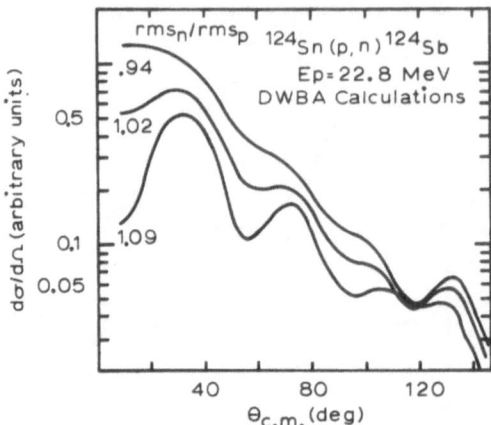

Fig. 4. DWBA predictions for angular distributions for
^{124}Sn using a complex isovector potential in
the folding model illustrating the sensitivity
of the reaction to the neutron density ρ_n.
The radius R_n of ρ_n was varied to give the
different ratios $\langle r^2_n \rangle^{\frac{1}{2}}/\langle r^2_p \rangle^{\frac{1}{2}}$. In all other
respects the calculations were the same.

(p,n) quasielastic reaction to the neutron distribution is well
known[4],[16] and can easily be seen in the folding model formalism
where the reaction is a function of the difference between the
neutron and proton densities. This sensitivity is illustrated in
figure (4) where a small change in the ratio $\langle r^2_n \rangle^{\frac{1}{2}}/\langle r^2_p \rangle^{\frac{1}{2}}$ makes
large changes in the folding model calculations of the angular dis-
tribution. Although in principle the microscopic approach can provide similar insight into neutron densities, the use of densities
in folding model calculations, rather than wave functions generated
from additional assumptions (such as the shell model), may be a more
direct approach to the problem. For example, one popular microscopic
approach[16] uses

$$U_1 = \frac{A}{(N-Z)} \int v_t \left(|\vec{r}-\vec{r}'| \right) \rho_{ex} (r') d^3r'$$

where ρ_{ex} refers to the excess neutron density generated by shell
model wavefunction calculations. This approach involves not only
the added uncertainty of shell model calculations but also assumes
the "core" neutrons do not participate in the reaction.

Additional details are reported here for the real and complex
model calculations shown in figure (1) relevant to a discussion of
neutron matter densities. As discussed in Section III the calculations of figure (1) do not quite make use of optimum parameters

for b, W_{10}, and the finite range correction, but details reported here should be indicative of the type of results that can be obtained from this approach to matter distribution analysis. In any case, due to uncertainties in analysis that will be discussed, any literal interpretation of the parameters R_n and a_n of ρ_n is premature.

Table 2 lists extracted values for $<r^2_n>^{\frac{1}{2}}/<r^2_p>^{\frac{1}{2}}$ along with χ^2 for best fits to data for several different versions of the folding model. Only the ratio $<r^2_n>^{\frac{1}{2}}/<r^2_p>^{\frac{1}{2}}$ is listed since this is believed to be the result least sensitive to the parameterization chosen for ρ_p and ρ_n. Columns 2 and 3 contain the results of the real and complex with finite range correction calculations of figure (1) and table 1. In addition, a few results using the weak Green density dependent force in equation (10) for U_1 combined with the FSAA calculation for W_1 are shown (column 4). Due to the use of equation (11) for χ^2 and different bases for reporting experimental errors, the value of χ^2 can be used for comparisons of calculations for the same nuclide but not for comparisons between different nuclides. Examination of the χ^2 values indicate that the complex model using equation (6) for U_1 clearly provides the best overall predictions of data.

If interpreted literally as the ratio of the neutron to proton rms radius the quantity $<r^2_n>^{\frac{1}{2}}/<r^2_p>^{\frac{1}{2}}$ should not change with projectile energy. Also, in the spirit of the optical model and the bulk properties of nuclear matter we should probably anticipate, on the average, smooth trends of extracted parameters as a function of the mass number and energy. With these comments in mind a study of the details of the basic complex model indicates some promising trends and some difficulties. The consistency of the results for $<r^2_n>^{\frac{1}{2}}/<r^2_p>^{\frac{1}{2}}$ for ^{208}Pb at different energies is encouraging. Also, the consistency of an increased $<r^2_n>^{\frac{1}{2}}/<r^2_p>^{\frac{1}{2}}$ value with increased neutron excess for the tin isotopes at both 22.8MeV and 35.3MeV is encouraging, although some discrepancies appear in the absolute value of this ratio for a given isotope. Another obvious discrepancy exists for ^{48}Ca at 25MeV, a result which seems clearly anomalous compared to results for 35 and 45MeV.

Additional calculations have been carried out to gain insight into the significance of these $<r^2_n>^{\frac{1}{2}}/<r^2_p>^{\frac{1}{2}}$ values. One point is that table 2 illustrates that the model providing smaller χ^2 for a given nucleus in general provides more consistent results. Based on the discussion in Section III we anticipate that a shorter range in v_t and a stronger W_{10} would improve the complex model of table 2. A second point is that it is well known that (p,n) calculations are moderately sensitive to the optical potentials for the distorted waves, particularly the neutron potential[10]. This effect was investigated for several nuclei and ^{138}Ba illustrates a typical result. Figure (3) shows the calculations for ^{138}Ba where separately

Table 2.*

| Nuclide, energy | $\langle r^2_n\rangle^{\frac{1}{2}}/\langle r^2_p\rangle^{\frac{1}{2}}, \chi^2$ | | |
	Real	Complex	Density dependent
^{48}Ca, 25MeV	1.102, 38	1.073, 29	1.055, 39
^{48}Ca, 35MeV	1.056, 15	1.042, 17	
^{48}Ca, 45MeV	1.012, 6	1.040, 2	
^{58}Ni, 22.8MeV	1.030, 12	1.017, 11	
^{58}Ni, 32.0MeV	0.998, 13	1.010, 6	
^{96}Zr, 22.8MeV	1.104, 9	1.047, 41	
^{96}Mo, 22.8MeV	1.025, 52	1.046, 13	
^{96}Ru, 22.8MeV	0.992, 13	1.020, 3	
^{112}Sn, 22.8MeV	1.024, 30	1.004, 16	
^{112}Sn, 35.3MeV	1.052, 19	1.034, 10	1.025, 13
^{116}Sn, 22.8MeV	1.026, 26	1.017, 10	
^{116}Sn, 35.3MeV	1.072, 62	1.042, 14	
^{124}Sn, 22.8MeV	1.053, 77	1.024, 31	
^{124}Sn, 35.3MeV	1.090, 67	1.055, 13	1.036, 17
^{138}Ba, 26.0MeV	1.007, 23	1.002, 13	0.996, 21
^{142}Nd, 26.0MeV	1.017, 34	1.012, 15	
^{144}Sm, 26.0MeV	1.053, 28	1.016, 13	1.000, 21
^{208}Pb, 25.8MeV	1.075, 5	1.058, 2	1.050, 3
^{208}Pb, 35.0MeV	1.044, 16	1.072, 10	
^{208}Pb, 45.0MeV	1.090, 8	1.055, 2	1.028, 3.5

*The ratio $\langle r^2_n\rangle^{\frac{1}{2}}/\langle r^2_p\rangle^{\frac{1}{2}}$ extracted from (p,n) data for various models and the accompanying value of χ^2. The column labeled real corresponds to the real calculations (short dashes) of figure (1) while the column labeled complex corresponds to the complex calculation (solid curve) of figure (1). The column labeled density dependent uses equation (10) with the weak Green force and includes an exchange correction. Uncertainties in the values for the ratios due to experimental error and mathematical determination of the χ^2 minimum are typically $\pm.005$. These uncertainties are probably small compared to systematic uncertainties due to choice of the proper (p,n) model and parameters for the DWBA calculations.

the Patterson[41] and Wilmore[38,39] neutron potentials have been used for the neutron distorted waves (dotted and solid lines). Both these global potentials are presumably suitable for this nuclide and out-going neutron energy. There is a measurable difference in the agreement with data and the extracted ratio of rms radii changed from 1.015 using Wilmore's parameters to 1.002 using Patterson's parameters. Generally we have found that different global potentials suitable for the same nucleus and energy can give up to a 2 per cent difference in the ratio of rms radii. For a few nuclides such as ^{208}Pb at 25.8MeV there exists both proton and neutron potentials from individual fits to elastic data, but most often it has been necessary to rely on global potentials for calculations and accept the accompanying error.

The study of isotopic, isobaric, and isotonic sequences is particularly valuable since some of the uncertainties involved in an absolute calculation can be removed in a comparison of nuclides. The tin isotopes are an important example. Even the relatively poor real model follows the trend of increased ratio of rms radii with increased neutron excess. This sequence at 22.8MeV has previously been studied by Schery[35] who found the same trend although his absolute numbers are different since his calculations used different distorted wave sets and did not include some of the present refinements of the folding model. Proton distribution parameters are well-known for the tin isotopes so a major remaining question is whether this isotopic trend could be explained by an incorrect isotopic dependence in the optical potentials for the distorted waves. To investigate this dependence we have used Becchetti and Greenlees's "best fit" proton and neutron potentials[37], the proton potential of Menet[44], and the neutron potential of Wilmore[38,39] in separate calculations. In all cases this trend remained. Unfortunately, there is no published proton or neutron elastic data at the appropriate energies in this reaction to specifically check the elastic scattering predictions of these potentials. However, we have checked the Becchetti "best fit" potentials against the precise proton data of Hardacre[45] at 30.4MeV. We found that the isotopic variation was fairly well reproduced and concluded that the Becchetti and Greenlees potential gave a reasonable description of the isotopic effect in this energy range. This proton potential was then used for the 35.3MeV (p,n) data with the neutron potential generated in a Lane model consistent fashion using the same V_1 as used in the (p,n) form factor. Again the trend of increased ratio of rms radii with neutron excess was found, giving additional evidence that the observed effect may be real.

V. CONCLUSIONS

It is apparent that there is now sufficient high precision (p,n) quasielastic data in the range 22.8MeV to 45MeV to make exacting tests of (p,n) folding models. The results typified by the calculations with the complex isovector potential shown in figure (1) seem promising. Predictions of data are comparable to that with phenomenological calculations while providing a more fundamental interpretation of the isovector potential parameterization. The evidence for an imaginary isovector component W_1 in the nucleon-nucleus optical potential seems clear. At present, the best overall prediction of data is obtained when a complex isospin potential calculated with equations (6) and (9) with a density independent form for the isospin component of the nucleon-nucleon interaction v_t is used. Our results indicate an rms range of v_t slightly smaller than that used in earlier calculations might be preferable, say a Gaussian interaction with rms $\simeq$ 1.2 to 1.6fm. An imaginary strength somewhat greater than that used for figure (1) also seems in order. A value of $W_{10} \simeq 2$ with rms finite range correction of 2.8fm improves the predictions somewhat and would be more consistent with phenomenological integrated strengths for W_1[46] and the Brueckner-Hartree-Fock calculations of Jeukenne.[47,48] An energy and mass dependence in W_{10} and the nucleon-nucleon range have not been investigated and remain topics for further study.

The finding that exchange effects estimated by equation (10) give approximately the same results as renormalized calculations without exchange is not surprising and is in agreement with the findings of Doering for microscopic calculations.[7] However, the finding of no evidence for a strong density dependence in the isospin component of the real effective nucleon-nucleon interaction is somewhat surprising. There is some evidence for density dependence in effective nucleon-nucleon interactions and many formulations specifically suggest it should appear in the isospin piece. Perhaps the density dependence is weaker than supposed or is contained predominantly in the imaginary part of the isospin interaction. On the other hand, the present (p,n) calculations may involve poor models for density dependence and improved approximations may be required. These points require further study.

The topic of whether the folding model is precise enough to provide information about nuclear matter densities is a more speculative subject. The present analysis gives evidence for a ratio $\langle r_n^2 \rangle^{1/2} / \langle r_p^2 \rangle^{1/2}$ greater than one in ^{208}Pb and an increase in this ratio with neutron excess for the tin isotopes. The results for the complex calculation of Table 2 for ^{48}Ca and ^{208}Pb are in essential agreement with some recent analysis of 0.8 GeV proton scattering and Hartree-Fock calculations.[49,50] However, in general the need for further study is clear. Elastic scattering data and optical potentials need to be obtained specific to the energies and

nuclides of the (p,n) reaction data, especially for isotopic, isobaric and isotonic sequences. Secondly, parameterization of the
neutron and proton densities more sophisticated than the two-
parameter Fermi forms assumed in the present analysis needs to be
used. Although quite lengthy, model independent analysis similar
to that done for electron and proton scattering should be carried
out for nuclides such as ^{208}Pb and the tin isotopes.

The uncertainties due to distorted wave optical potentials and
the parameterization of the nuclear densities are quantities presently known to be limiting analysis of neutron densities and alone
could account for inconsistencies discussed in Section IV. Their
reduction followed by careful evaluation of remaining uncertainties
should be the next step toward determining the value of the (p,n)
quasielastic reaction for studying nuclear matter densities. The
recent body of precise (p,n) data over a range of energies and
nuclides permits quite exacting tests of proposed models. A model
whose nuclear density parameterization is to be taken literally
should provide high quality predictions of the data and consistent
results for the same target nuclide at different projectile
energies. Given the sensitivity of the (p,n) quasielastic reaction
to the difference of the neutron and proton densities, should such
a model be found it might provide an important tool for the study
of nuclear matter densities.

REFERENCES

1. J. D. Anderson and C. Wong, Phys. Rev. Lett. 7,250 (1961).
2. G. R. Satchler, Nucl. Phys. A95, 1 (1967).
3. V. A. Madsen, Nucl. Phys. 80, 177 (1966).
4. N. Austern, Direct Nuclear Reaction Theories (Wiley-Inter-
    science, New York, 1970) p. 148.
5. L. D. Rickertsen and P. D. Kunz, Phys. Lett. 47B, 11 (1973).
6. V. A. Madsen, V. R. Brown, S. M. Grimes, C. H. Poppe, J. D.
    Anderson, J. C. Davis, and C. Wong, Phys. Rev. C 13, 548 (1976).
7. R. R. Doering, D. M. Patterson, and A. Galonsky, Phys. Rev.
    C 12, 378 (1975).
8. A. M. Lane, Phys. Rev. Lett. 8, 171 (1962).
9. A. M. Lane, Nucl. Phys. 35, 676 (1962).
10. G. R. Satchler, R. M. Drisko, and R. H. Bassel, Phys. Rev.
    136, 637 (1964).
11. G. R. Satchler, Isospin in Nuclear Physics (North-Holland,
    Amsterdam, 1969) p. 389.
12. C. S. Batty, B. E. Bonner, E. Friedman, C. Tschalär, L. E.
    Williams, A. S. Clough, and J. B. Hunt, Nucl. Phys. A116,
    643 (1968).
13. G. W. Greenlees, G. J. Pyle, and Y. C. Tang, Phys. Rev. 171,
    1115 (1968).

14. B. Sinha, Phys. Reports 20, 1 (1975).
15. V. A. Madsen, Nuclear Isospin (Academic Press, New York, 1969), p. 149.
16. C. J. Batty, E. Friedman, and G. W. Greenlees, Nucl. Phys. A127, 368 (1969).
17. G. R. Satchler, Phys. Lett 44B, 13 (1973).
18. W. H. Bassichis and M. R. Strayer, Phys. Rev.C 18, 632(1978).
19. W. G. Love and L. W. Owen, Nucl. Phys. A239, 74(1975).
20. B. Z. Georgiev and R. S. Mackintosh, Phys. Lett. 73B, 250(1978).
21. E. Friedman, Phys. Rev.C 10,2089 (1974).
22. A. M. Green, Phys. Lett. 24B, 384 (1967).
23. E. Friedman, Phys. Lett. 29B, 213 (1969).
24. S. D. Schery, S. M. Austin, A. Galonsky, L. E. Young, and U. E. P. Berg, Phys. Lett. 79B, 30 (1978).
25. E. Gadioli, E. Gadioli Erba, and G. Tagliaferri, Phys. Rev. C17, 1294 (1978).
26. W. G. Love, Phys. Rev.C 15, 1261 (1977).
27. G. L. Thomas, B. C. Sinha, and F. Duggan, Nucl. Phys. A203, 305 (1973).
28. J. C. Slater, Phys. Rev. 81, 385 (1951).
29. V. R. Pandharipande, Nucl. Phys. A166, 317 (1971).
30. H. Überall, Electron Scattering from Complex Nuclei (Academic Press, New York, 1971).
31. C. W. de Jager, H. de Vries, and C. de Vries, At. Data Nucl. Data Tables 14, 479 (1974).
32. I. Angeli and M. Csatlos, ATOMKI Közlemenyek 20, 1 (1978).
33. J. D. Carlson, C. D. Zafiratos, and D. A. Lind, Nucl. Phys. A249, 29 (1975).
34. S. D. Schery, D. A. Lind, H. W. Fielding, and C. D. Zafiratos, Nucl. Phys. A234, 109 (1974).
35. S. D. Schery, D. A. Lind, and H. Wieman, Phys. Rev.C 14, 1800 (1976).
36. S. D. Schery, D. A. Lind, and C. D. Zafiratos, Phys. Rev. C 9, 416 (1974).
37. F. D. Becchetti, Jr. and G. W. Greenlees, Phys. Rev. 182, 1190 (1969).
38. D. Wilmore and P. E. Hodgson, Nucl. Phys. 55, 673 (1964).
39. C. M. Perey and F. G. Perey, At. Data Nucl. Data Tables 13, 293 (1974).
40. C. M. Perey and F. G. Perey, At. Data Nucl. Data Tables 17, 1 (1976).
41. D. M. Patterson, R. R. Doering, and A. Galonsky, Nucl. Phys. A263, 261 (1976).
42. P. D. Kunz, University of Colorado, unpublished.
43. G. W. Greenlees, W. Makofske, and G. J. Pyle, Phys. Rev. C 1, 1145 (1970).
44. J. J. H. Menet, E. E. Gross, J. J. Malinify, and A. Zucker, Phys. Rev. C 4, 1114 (1971).
45. A. G. Hardacre, J. F. Turner, J. C. Kerr, G. A. Gard, P. E. Cavanagh, and C. F. Coleman, Nucl. Phys. A173, 436 (1971).

46. S. Kailas and S. K. Gupta, Phys. Lett. 71B, 271 (1977).
47. J.-P. Jeukenne, A. Lejeune, and C. Mahaux, Phys. Rev. C 15, 10 (1977).
48. J.-P. Jeukenne, A. Lejeune, and C. Mahaux, Phys. Rev. C 16, 80 (1977).
49. L. Ray, W. Coker, G. W. Hoffman, Phys. Rev. C 18, 2641 (1978).
50. L. Ray, Los Alamos Report LA-UR-79-93, 1979.

DISCUSSION

King: I have a two-part question: 1) Given the various radii in the problem -- entrance, exit channels, form factor and real vs. imaginary -- is there a correlation between these in defining an "rms radius" for protons vs. neutrons? And, 2) given the uncertainties in fitting the data and in the theory, can you give an uncertainty in the 4 significant figures quoted in your neutron vs. proton rms radius ratio?

Schery: Question 1: The rms radii I have quoted refer only to that of the neutron and proton matter radii. The real isovector potential, the imaginary isovector potential, etc. are formulated in terms of these distributions and in general will have different rms values. The rms radius of a potential can be obtained through the folding equations I have quoted. Question 2: The 4 digit ratios I quoted are meant to reflect the uncertainty in the mathematical determination of the minimum chi squared and the experimental error only. Other errors due to choice of model or potential parameters are probably much larger. For example one presently known source of these larger errors is uncertainty in the distorted wave parameters which has affected the ratio of rms radii by as much as two percent in some cases.

PROPERTIES OF MULTISTEP AMPLITUDES IN CHARGE-EXCHANGE REACTIONS*

V.A. Madsen and V.R. Brown

Oregon State University, Corvallis, Oregon and

Lawrence Livermore Laboratory, Livermore, California

INTRODUCTION

In this talk we shall first review historically what is known about inelastic multistep processes in analog charge-exchange transitions at lower energies, in the few tens of MeV, particularly the possibility of measuring directly with the (p,n) reaction the isovector deformation parameter β_1. We shall then show the expected energy dependence of multistep relative to one-step processes and attempt to explain the behavior.

(p,n) ANALOG TRANSITIONS AT LOW ENERGY

At low energies, the (p,n) spectrum is dominated by transitions to the analog of the target ground state. These transitions, discovered[1] at Livermore by Anderson and Wong, are the Fermi transitions of charge exchange. As such the nuclear structure factors are approximately model independent except for the assumption of isospin conservation, and the cross section is proportional to N-Z. The charge-exchange mechanism is fairly well described in terms of the isospin matrix elements of the Lane potential

$$V_{Lane} = \frac{V_1}{A} T \cdot t \qquad (1)$$

---

*Part of this work was performed under the auspices of the U.S. Department of Energy for Lawrence Livermore Laboratory under contract NO. W-7405-Eng-48.

In recent years there has been considerable interest in meas- uring differences in neutron and proton behavior in nuclear in- elastic transitions. The simplest case for discussion is that of collective vibrational or rotational states of the nucleus. In the usual collective model of the nucleus there is a deformation param- eter β describing the non-spherical shape, and both neutron and proton distributions are characterized by this parameter. All iso- vector effects in the model come simply from the neutron excess. The inelastic-scattering interaction is the difference between the optical potential for the deformed nucleus and its spherical equil- ibrium or average. If we also want to use the model for charge- exchange reactions, we can use the Lane form of the optical potential to get

$$\Delta V = \begin{cases} -\beta R \dfrac{\partial}{\partial r}[V_0 + \dfrac{T \cdot t}{A} V_1] & \text{general} \\[2ex] -\beta R \dfrac{\partial}{\partial r}[V_0 + \dfrac{N-Z}{4A} V_1] & \text{inelastic} \end{cases} \tag{2}$$

On the other hand, from the point of view of the nuclear shell model there is no reason to believe that the isospin effects, differences between behavior of nuclear neutrons and protons, should be due to the neutron excess. As an example consider the case of a neutron- closed-shell nucleus such as ^{90}Zr. If the closed shell were truly inert, the low-lying collective states would consist entirely of proton vibrations, for which the isovector term has a negative sign. Generalizing Eq. (2) slightly, such a case could still be represented in terms of the collective model by

$$\Delta V = -R \frac{\partial}{\partial r}[\beta_0 V_0 + \beta_1 \frac{T \cdot t}{A} V_1] \tag{3}$$

in which the isoscalar and isovector strengths have their own defor- mation parameters β_0 and β_1. The quadrupole β_1 would then have to be negative for ^{90}Zr and other nuclei in which the protons contrib- ute more strength than the neutrons to a particular inelastic tran- sition.

In any case the parameter β_1 contains important nuclear struc- ture information on differences in neutron and proton contributions to inelastic scattering. It can be related to the isovector effec- tive charge from a shell-model description of an electromagnetic transition. It is therefore not surprising that attempts are being made to measure β_1 in various ways.

Because the Lane model Eq. (1) has been so successful in ex- plaining $0^+ \rightarrow 0^+$ analog transitions, an extension[2] to $0^+ \rightarrow 2^+$ analog transitions[3] using Eq. (3) would seem appropriate, and the charge- exchange field would give a direct measurement of β_1. Unfortunately

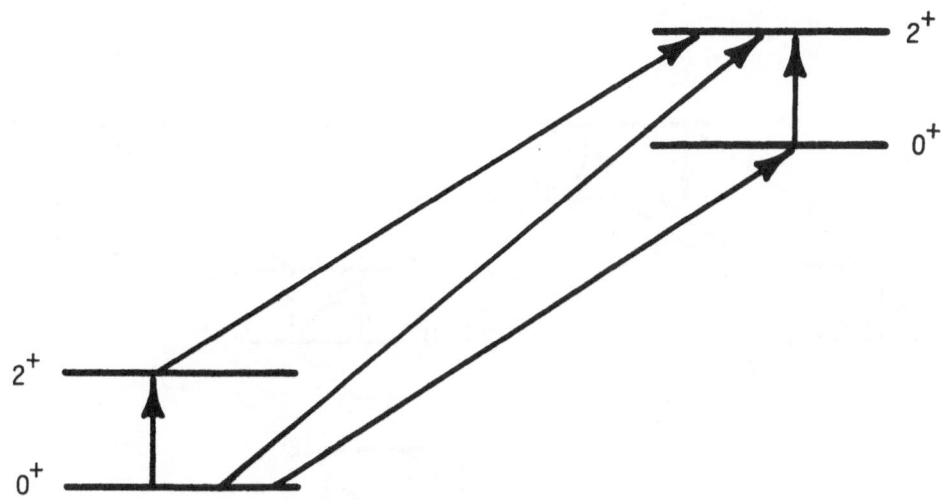

Figure 1--The one- and two-step mechanisms involved in the $(p,n)2^+$ analog transition. The reaction is dominated by the two 2-step mechanisms.

it turns out in practice[3] that the two-step mechanisms[4] indicated in Fig. 1 completely dominate[5] this transition and thus mask the desired direct amplitude. Not only is the latter small, but it is also about $90°$ out of phase with the two-step term, which means that these amplitudes are almost completely incoherent. They are otherwise very similar as a function of scattering angle. Furthermore, what coherence the amplitudes have is usually destructive and nearly nullifies the small incoherent one-step contribution to the cross section. This means that a determination of β_1 on the basis of interference is difficult or impossible.

It has been stated that the Lane model has been very successful in explaining $0^+ \rightarrow 0^+$ analog transitions. These are perhaps the simplest of all nuclear reactions, having identical nuclear structure wave functions initially and finally except that a proton replaces each neutron. Surprisingly they are also strongly affected[6] by the multistep amplitudes shown in Fig. 2a. In this case the three three-step amplitudes are not incoherent with the one step; they are all approximately in phase with each other and about $180°$ out of phase with the one-step amplitude.

This is a very systematic effect, which also occurs, for

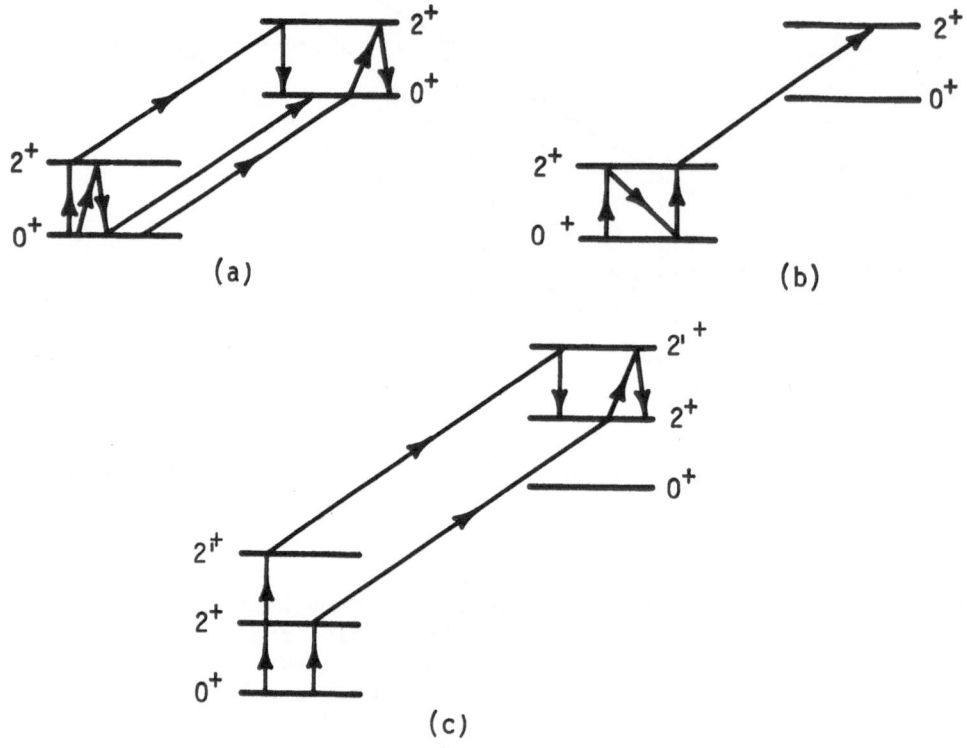

Figure 2--Multistep mechanisms. (a) Three 3-step routes whose amplitudes interfere destructively with the dominant 1-step amplitude. (b) A 4-step route contributing destructively to the dominant 2-step mechanism for excited analogs. (c) A 4-step route involving a 2-phonon state.

inelastic scattering and can be interpreted physically as follows: It is well known that, if a single specific inelastic channel such as the 2^+ state of Fig. 1 is coupled to the ground state, the transition strength is weaker than in DWBA. In order to get agreement

with DWBA one must reduce the imaginary potential to compensate for
the loss of probability current due to the explicit inclusion of
coupling to the ground state. In considering the effect of coupling
of the 0^+ ground state and the 2^+ first excited inelastic state from
the point of view of multistep perturbation theory, the next leading
order after first which contributes to inelastic scattering is third
order through the sequence $0^+ \to 2^+ \to 0^+ \to 2^+$. In order to reproduce
the reduction which channel coupling gives, this third-order
amplitude should be primarily destructive compared to the first-order
amplitude.

In the $0^+ \to 0^+$ analog transition there are, however, three such
routes (See Fig. 2a) when the 0^+, 2^+ states and their analogs are
included. We therefore expect to get for the excited analog 3 times
as much reduction as for the inelastic state. An example is shown
in Table I of the $^{100}Mo(p,p')2^+$ and $^{100}Mo(p,n)2^+$ analog transitions
calculated with and without coupling but with no changes in optical
parameters. With a fairly big collective nucleus such as this,
the channel coupling effects are very large. The 3-step inelastic
amplitude required for the 27% reduction of Table I is .15 times
the one-step, whereas for the charge exchange, assuming all three
3-step amplitudes are equal, each one is .11 times the one-step
charge exchange amplitude.

Table I. Comparison of DWBA and coupled-channel
(C.C.) runs for the 16 MeV $^{100}Mo(p,p')$ and (p,n)
0^+ transition. The 0^+ ground state, the 2^+ and
3^- excited states, and their analogs are coupled.

Case	$\sigma(p,p')$ mb	$\sigma(p,n)0^+$ mb
DWBA	41.4	5.15
CC	30.6	2.21
% reduction	27%	57%

Fig. 3 shows that for a set of even isotopes of Mo, in which
the collective deformation parameter steadily increases from the
single-closed-shell nucleus ^{92}Mo to the highly collective ^{100}Mo,
the ratio $\sigma/(N-Z)$ decreases by about a factor of 2 over this range.
A DWBA calculation using the Lane model would give approximately a
constant for this ratio, so one would have to allow V_1 to have a
different value for each isotope. Fig. 4 shows the differential
cross sections for analogs of the ground state, the 2^+ quadrupole
vibration and the 3^- octupole vibration in ^{92}Mo. There are no
free parameters in the calculations of the latter two transitions,

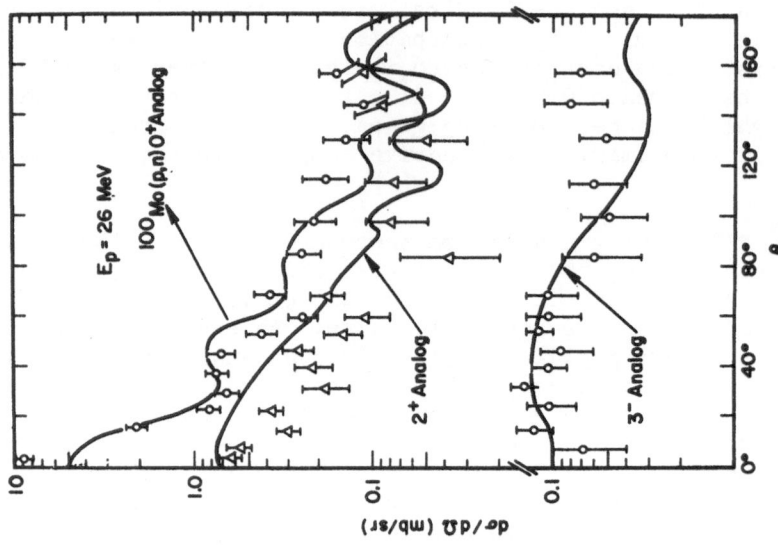

Figure 4—The ^{92}Mo(p,n)0^+, 2^+, 3^- analog differential cross sections for 26 MeV incident protons.

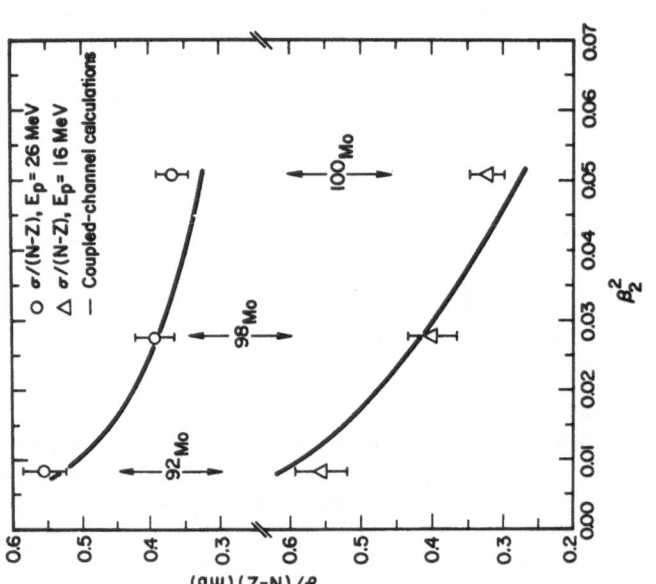

Figure 3—The dependence of the Mo 0^+(p,n)0^+ analog ratio $\sigma/(N-Z)$ on the deformation parameter.

the deformation β_+ having been determined from inelastic scattering and the V_1 from 0^+-analog charge-exchange transition for all the isotopes of Mo.

EXPLANATION OF THE MULTISTEP PHASES

The multistep amplitudes each involve one or more factors of the optical Green's function representing the propagation of the projectile in the intermediate state. For example in the calculation of the 2^+ excited-analog transition by one of the two-step processes, second order perturbation theory yields an amplitude

$$A^{(2)} = \int X_f^{(-)}(\vec{r}\,')V_{43}(r')g_3^+(r',r)V_{31}(r)X_i^{(+)}(r)d^3r\,d^3r' \qquad (4)$$

where the 4 states 0^+, 0^+ analog, 2^+, 2^+ analog are represented, respectively, by 1, 2, 3, 4 and

$$V_{43} = \langle \phi_4 | \Delta V | \phi_3 \rangle = \sum_\ell v_\ell^{4,3}(r)Y_\ell^m(\hat{r}) \qquad (5)$$

is the nuclear matrix element of the interaction responsible for the transition. Of course, if exchange were taken into account explicitly, this interaction would be non local.

The optical Green's function has outgoing boundary conditions and satisfies the equation

$$(E - E_3 - h)g_3^+(r,r') = \delta^3(r - r'), \qquad (6)$$

where E_3 is the energy of nuclear state 3 and

$$h = t + V + iW \qquad (7)$$

gives the optical Hamiltonian in terms of the potentials. The Green's function solution to Eq. (6) can be written as a partial wave expansion (ignoring spin complications)

$$g^+(\vec{r},\vec{r}\,') = \sum_{\ell m} g_\ell^+(r,r')Y_{\ell m}(\hat{r})Y_{\ell m}^*(\hat{r}\,') \qquad (8)$$

with

$$g_\ell^+(r,r') = \frac{a}{i} R_\ell(r_<)R_\ell^+(r_>) \qquad (9)$$

where R_ℓ is the solution of the optical-model equation regular at the origin, R_ℓ^+ is a solution of the same equation with outgoing asymptotic boundary conditions,

$$R_\ell^{(+)} \to H_\ell^{(1)}(kr) \to (kr)^{-1} \exp i[\, kr-(\ell+1)\pi/2 - \eta \ell n 2kr + \sigma_\ell] \qquad (10)$$

and $a = \dfrac{2mk}{h^2}$. Generally we may also write

$$R_\ell = \tfrac{1}{2}(e^{2i\delta_\ell} H_\ell^{(1)} + H_\ell^{(2)}) \qquad . \qquad (12)$$

Thus the Green's function can be written

$$g_\ell(r,r') = \frac{a}{2i}[\, e^{2i\delta_\ell} R_\ell^{(+)}(r_<) + R_\ell^{(-)}(r_<)]\, R_\ell^+(r_>) \qquad (13)$$

The first term of Eq. 13 is somewhat erratic in phase, but it is also reduced compared to the second by the complex phase shift factor. The second term is very steady in phase; in fact beyond the nuclear force it is

$$g_\ell^+(r,r') \to \frac{a}{2i}[\, e^{2i\delta_\ell} H_\ell^{(1)}(kr_>) H_\ell^{(1)}(kr_<) + H_\ell^{(1)}(kr_>) H_\ell^{(1)^*}(kr_<)]$$

$$(14)$$

At locality, that is $r = r'$, the second term is in fact an absolute square. Thus the second term dominates the partial-wave optical Green's function, and its sign is negative imaginary at locality, which tends to dominate in multistep processes.

If we applied Eq. (4) to elastic scattering, the imaginary (non local) optical potential

$$w(r,r') = Im \sum_n V_{1n}(r) g_n^+(r,r') V_{n1}(r') \qquad , \qquad (15)$$

would be expected on the basis of the above argument to have a well-defined negative sign near locality. Away from locality $w(r,r')$ falls off[7] in $r-r'$ in 1 to 2 Fermis and oscillates in sign beyond that distance. In a local approximation the imaginary optical potential would then be negative definite giving always absorption, as expected both from empirical optical potentials and from intuitive arguments.

Continuing with Eq. (4) for the two-step amplitude $0^+ \to \lambda^+ \to \lambda^+$

analog and ignoring the small contributions from $\Delta J \neq 0$ for the charge exchange step, we may rewrite Eq. (4) by expanding in multipoles and integrating over solid angle to give

$$A^{(2)} = (4\pi)^{3/2} \sum_{\ell_3 \ell_1} i^{\ell_1 + \ell_4} (-1)^\lambda \frac{\langle Y_{\ell_4} || Y_\lambda || \ell_1 \rangle}{\hat{\lambda}} [Y_{\ell_4}(\hat{k}_f) Y_{\ell_1}(\hat{k}_i)]_{\lambda \mu}$$

$$\cdot \iint R_{\ell_4}^{(-)*}(r') V_0^{4,3}(r') g_{\ell_4}^+(r',r) V_\lambda^{3,1}(r) R_{\ell_1}^{(+)}(r) r^2 r'^2 \, dr \, dr'$$

$$(16)$$

By contrast, the direct excitation $0^+ \rightarrow 1^+$ analog is

$$A^{(1)} = (4\pi)^2 \sum i^{\ell_1 + \ell_4} (-1)^\lambda \frac{\langle \ell_4 || Y_\lambda || \ell_1 \rangle}{\hat{\lambda}} [Y_{\ell_4 m_4}(\hat{k}_f) Y_{\ell_1 m_1}(\hat{k}_i)]_{\lambda \mu}$$

$$\int R_{\ell_4}^{(-)*} V_\lambda^{4,1}(r) R_{\ell_1}^{(+)}(r) r^2 \, dr$$

$$(17)$$

Eqs. (16) and (17) are essentially identical except for the radial integrals. Thus a comparison of the phases of one- and two-step amplitudes need involve only the radial integrals. Considering that for collective states the nuclear form factors V_λ are peaked at the nuclear surface, and that the absorption by optical potentials further emphasizes the region of the nuclear surface, the Green's function Eq. (13) will be approximately equal to its form just beyond the nuclear potential, Eq. (14), and furthermore r and r' in Eq. (14) will be nearly equal in the important region of integration. Thus the phase of Eq. (16) relative to Eq. (17) will be the phase of $V_0^{4,3}$ g $V_\lambda^{3,1}$ minus the phase of $V^{4,1}$. Since in the Lane model $V^{4,1}$ is minus β_1 times the derivative of $V_0^{4,3}$, their phases should be about the same. The phase of g is negative imaginary, and that of $V_\lambda^{3,1}$ in the collective model is minus β_0 times the derivative of the optical potential $\hat{V} + i\hat{W}$, so the overall phase factor of Eq. (16) relative to Eq. (17) should be

$$-i(V + iW)/\sqrt{V^2 + W^2} \quad . \qquad (18)$$

Taking a typical strength ratio of $W/V = \frac{3}{4}$, this phase is about $104°$, almost incoherent but slightly destructive, as mentioned above.

Note that this result assumes that β_0 and β_1 have the same sign. If the signs are opposite, the relative phase of one- and two-step amplitudes is reversed, and the two remain primarily incoherent, but what coherence there is will now be constructive, adding to instead of subtracting from the incoherent part of the one-step mechanism. If we simply reverse the signs of the phase factor, the phase turns out to be about $-76°$.

It has been seen that the 3-step amplitudes of Fig. 2a are expected to be destructive to the dominant 1-step analog transition. To understand this effect one can also use perturbation theory for example for the sequence $0^+ \to 2^+ \to 2^+$ analog $\to 0^+$ analog, which has an amplitude

$$A^{(3)} = \iiint X_f^{(-)}(r'') V^{2,4}(r'') g_4^+(r'',r') V^{4,3}(r')$$

$$\times g_3^+(r',r) V^{3,1}(r) X_i^{(+)}(r) d^3r \, d^3r' \, d^3r''$$

$$= (4\pi)^{\frac{1}{2}} a_3 a_4 \sum_{\ell_1 \ell_3} \ell_3 ||Y_\lambda|| \ell_1{}^2$$

$$\times P_{\ell_1}(\hat{k}_i \cdot \hat{k}_f) \iiint R_{\ell_1}^{(-)*}(r'') V_\lambda^{2,4}(r'') g_{\ell_3}(r'',r')$$

$$\times V_0^{4,3}(r') g_{\ell_3}(r',r) V_\lambda^{3,1}(r) R_{\ell_1}^{(+)}(r) r^2 r'^2 r''^2 dr \, dr' \, dr'' \quad (19)$$

The first order amplitude by comparison is

$$A^{(1)} = \int X_f^{(-)*}(r) V_0^{2,1}(r) Y_{00}(r) X_i^{(+)}(r) d^3r$$

$$= (4\pi)^{\frac{1}{2}} \sum_{\ell_1} P_{\ell_1}(\hat{k}_i \cdot \hat{k}_f) \int R_{\ell_1}^{(-)*}(r) V_0^{2,1}(r) R_{\ell_1}^{(+)}(r) r^2 dr \quad (20)$$

These expressions Eqs. (19) and (20) are again very similar, differing only in the form of the radial integral and the extra sum over $\lambda + 1$ partial waves ℓ_3 in the inelastic channel for each partial wave ℓ_1 in the target channel. Since $V_0^{4,3}$ of Eq. (19) and $V_0^{2,1}$ of Eq. (20) are equal, the first and third order amplitudes differ in phase only by the factor $g_\ell g_\ell V_\lambda^{2,4} V_\lambda^{3,1}$, which is roughly $(-i)^2$ times the phase of the inelastic interaction functions. This relative phase is simply -1 if the inelastic interaction is real. For $\hat{W} = \frac{1}{4}\hat{V}$ the phase would be about $152°$, still highly destructive.

The other two routes shown in Fig. 2a for the 3-step mechanism have slightly different third-order amplitudes.[6] However they behave very similarly to that of Eq. (19), and in particular the phases are rather similar, although there is some small degree of

destructive interference among the three. The difference in phase
with the dominant one-step mechanism of the three as a function of
scattering angle lie almost entirely within a small band about 180^0
which is at least 80% destructive.

Extending the ideas to excited analogs, it seems that the 4-step
process shown in Fig. 2b ought also to interfere destructively in
the $2^+ \rightarrow 2^+$ analog stage of the dominant two-step mechanism by comparison with the one of the three routes of Fig. 2a. This expectation has been verified[8] by coupled-channel calculations; however,
it is inconsistent now not to include the two-phonon vibrational
states, which give routes such as the one shown in Fig. 2c. Alto-
gether there are 8 routes involving destructive 4-step amplitudes
to the 2^+ analog when one includes a single two-phonon state. These
have a similar effect to that of the 3-step interference on the 0^+
analog, to decrease the cross section with increasing strength of
the $0^+ \rightarrow 2^+$ inelastic transition.

All these results on destructive interference and multistep
processes for low energy scattering point out that one is dealing
with an effective interaction. If all nuclei were equally collective it might never have been noticed how important the three-step
interference can be. The question of how free one is of multistep
processes in the (p,n) reaction is at higher energy will be addressed in the next section.

ENERGY DEPENDENCE OF MULTISTEP EFFECTS

In view of the recent experiments with protons at higher energies, it is of interest to try to understand how the effects of
multistep amplitudes vary with increasing energy. The impulse
approximation, which assumes a single collision with each target
nucleon, is better at higher energy. This is partially due to the
increased validity of the approximation of a free nucleon-nucleon
collision taking place within the nucleus, but also due to a decreasing importance of multiple-scattering effects.

In Table II we present results for total cross sections of
calculations for protons on ^{26}Mg made at 20, 40, 80, and 135 MeV.
At the lower two energies the optical parameters used were those
of Patterson, et al[9] and at the upper energies V_0 and W_0 were taken
as those[10] for ^{28}Si. The Lane potential $V_1 + iW_1$ at the upper energies is somewhat uncertain, so table II cannot be considered a
prediction of analog strengths. V_1 and W_1 were calculated using
the volume integrals of the G-matrix parameters of Bertsch, et al[11]
for the real direct contribution, the pseudopotential method[12] for
the real exchange contribution, and the impulse-approximation parameters of Love, et al[13] for the 135 MeV imaginary potential W_1. The
zero-range pseudopotentials thus obtained were then folded with

Table II. Energy dependence of the ^{26}Mg(p,p')2^+, ^{26}Mg(p,n)0^+ analog, and ^{26}Mg(p,n)2^+ analog cross sections for weak coupling (DWBA), full coupling of these three states with the ground state, and the "2 step", obtained leaving out the direct $0^+ \to 2^+$ charge exchange coupling.

State		2^+ inelastic	0^+ analog	2^+ analog
Energy	case			
20 MeV	Weak	36.8	7.73	1.42
	2-step	33.6	5.53	2.05
	Full	35.4	5.48	4.48
40 MeV	Weak	28.0	4.43	.678
	2-step	26.6	3.87	.524
	Full	27.1	3.73	1.60
80 MeV	Weak	12.83	1.63	.205
	2-step	12.67	1.58	.0554
	Full	12.78	1.61	.302
135 MeV	Weak	4.77	1.03	.146
	2-step	4.76	1.03	.0114
	Full	4.80	1.06	.190

densities having optical-potential[10] geometry. The value of $\hat{W}_1$ at 80 MeV was obtained by using the value at 135 MeV and assuming proportionality with the $\hat{W}_0/\hat{V}_0$ ratio as a function of energy. The imaginary potentials obtained in this way are especially uncertain because the impulse approximation doesn't take into account the Pauli-principle inhibition of the reactions. These potentials are presented here for comparison purposes only. They are not recommended for the reasons given above.

$$V_0 = -(42.57 - .177E) \quad , \tag{21a}$$

$$V_1 = 76.25 - .356E \quad , \tag{21b}$$

$$W_0 = -(3.05 + .029E) \quad , \tag{21c}$$

$$W_1 = 13.13 + .051E \quad , \tag{21d}$$

the energy dependence being simply an interpolation between 80 and

135 MeV. The neutron and proton potentials can then be obtained from Eq. (1) as

$$V_{\{\begin{smallmatrix}n\\p\end{smallmatrix}\}} = V_0 \pm \frac{N-Z}{4A}\, V_1 \tag{22}$$

There are several interesting points about table II. First, the 2^+ analog 1-step cross section is fairly large. Further, it interferes constructively with the 2-step mechanism. These features are due to the relatively large, negative value of $\beta_1 = -2.2\beta_0$ for ^{26}Mg. This ratio was obtained[14] from analysis of E2 transitions in isobars of the mass-26 nuclei. A rough measure of the interference is given by the "phase angle" ϕ defined by

$$\cos \phi = \frac{\sigma_{full} - \sigma_1 - \sigma_2}{2\sqrt{\sigma_1 \sigma_2}}$$

It takes on the value 73°, 71°, 79°, 66° for the four energies 20, 40, 80, and 135 MeV. These values are in reasonable agreement with the estimate of $\phi = 76^\circ$ obtained from Eq. (16) with an effective ratio $W/V = \frac{1}{4}$. Second, it is evident from these results that all the cross sections decrease with increasing energy, and this is due largely to decrease in the exchange amplitudes.[12] Not surprisingly the cross section for the two-step mechanism falls off faster than the one-step process, so at higher energies the one-step mechanism dominates all reactions. In particular, the amplitude for direct excitation of the 2^+ analog is significant at all energies and dominant at the higher energies. However even at 135 MeV the effects of interference of 1- and 2- step mechanisms is rather large and has to be taken into account for an accurate measurement of β_1. On the other hand, the 3-step destructive interference of the 0^+ analog is a 3% effect at 80 MeV and .5% at 135 MeV. The increase of the full-coupling result at 135 MeV over the weak coupling is due to the primarily incoherent contribution of the 2-step sequence $0^+ \to 2^+ \to 0^+$ analog, which is negligible compared to the 3-step amplitudes at lower energies.

Is there a simple explanation of the energy dependence of the multistep relative to the 1-step mechanism? Intuitively the 2-step amplitude Eq. (16) can be looked at as the product of two 1-step amplitudes. Strictly speaking this is not true because the projectile in the intermediate state is represented by the optical Green's function Eq. (8) and (9) which is not quite the product of two regular radial waves. If one ignores this fact, there is still another important way in which Eq. (16) differs from the product of two 1-step amplitudes: for each partial wave the intermediate angular momentum is restricted to a few small values by the angular momentum transfer λ. For the purpose of simplifying the argument, consider a case $J_i = J_f$ with a $J = 0$ intermediate state, so $\lambda = 0$ and $\ell_1 = \ell_4$ in Eq. (16). Integrating over directions of $\hat{k}_f$ to get the total cross section and imagining that g_ℓ separates into a product of regular functions makes it possible to write

$$\sigma^{2\text{-step}} \approx c \sum_\ell \sigma_\ell^{CE} \sigma_\ell^{in} \quad , \tag{23}$$

where c is some constant taking care of the dimensionality of the two sides of the equations. For inelastic scattering of strongly absorbed particles[15] a small range of angular-momentum waves $\Delta\ell \approx k\,\Delta R$ participates in the scattering process, where ΔR is the interaction region of the nuclear surface. Lower ℓ waves are lost to absorption, and higher ℓ waves miss the nucleus. For both charge exchange and inelastic direct scattering processes the ℓth partial wave within this range is

$$\sigma_\ell \approx \frac{1}{\Delta\ell} \sigma \quad . \tag{24}$$

Thus the two step cross section Eq. (24) is

$$\sigma^{2\text{-step}} = \frac{c}{\Delta\ell^2} \sum_\ell \sigma^{CE}\sigma^{in} \approx \frac{c}{\Delta\ell} \sigma^{CE}\sigma^{in} = \frac{c}{k\,\Delta R} \sigma^{CE}\sigma^{in} \quad . \tag{25}$$

From this crude argument one expects the 2-step cross section for excitation of the 2^+ analog to fall off as the product of the inelastic cross section and the $0^+ \rightarrow 0^+$ analog cross section divided by the square root of the energy. Table III shows the quantity $\sigma^{2\text{-step}}\sqrt{E}/\sigma^{CE}\sigma^{in}$, which Eq. (25) shows should be constant as a function of energy. It appears from these numbers that Eq. (25) roughly accounts for the energy dependence of the two-step cross section.

From table II, it has been seen that the $0^+(p,n)2^+$ analog cross section at the lower energies is highly sensitive to the one-step mechanism and therefore that for this nucleus a measurement at 40 MeV, for example, along with a coupled-channel analysis should constitute a measurement of β_1. While this possibility is rather promising, the analysis is complicated by interference from the two strong 4^+ states in the spectrum, which, through the 4-step interference mechanism of Fig. 2c interfere destructively with the 2-step contribution. These states were not included in the calculations of table III, which was done as an example, because of uncertainties in the structure and transition strengths of the 4^+ states.

In table II only total cross sections have been presented. In practice, of course, one would compare differential cross sections

Table III. Energy dependence of the 2-step mechanism for excited analog excitation for ^{26}Mg(p,n). σ^{2-step} is the 2-step cross section for 2^+ analog excitation, and σ^{in} and σ^{CE} are direct inelastic scattering and charge exchange cross sections.

E	$\sigma^{2-step}\sqrt{E}/(\sigma^{in}\sigma^{CE})$
20	.0322
40	.0266
80	.0233
135	.0256

to try to determine transition strengths. It would be possible in principle to determine the separate 1- and 2-step transition strengths from differences in angular distribution. Although this remains a possibility, the results obtained until now with our coupled-channels analyses indicate that the individual contributions to the amplitude are not really different enough in character to allow one to distinguish them by use of angular distributions or even polarization.

DISCUSSION AND SUMMARY

Our work on multistep processes has concentrated on charge-exchange transitions to analog states. Multistep processes of various kinds will also be important for other charge-exchange transitions, particularly in states for which the direct transition is inhibited in some way. This includes not only inelastic and charge-exchange couplings, but also coupling to pickup and stripping channels.[16] It is, however, in the analog transitions that clear multistep routes are to be found with strengths large enough to be competitive with the direct 1-step mechanism. This is due to the combined effect of very strong isoscalar inelastic amplitude and to the large $J^+ \to J^+$ analog transitions.

Coupling to pickup and stripping channels has been[16] much more difficult to assess than for inelastic channels. These and other rearrangement channels are not entirely orthogonal to inelastic ones, nor can they be adequately replaced by a small number of inelastic channels. Energy dependence of pickup-stripping amplitudes is expected to be similar to that of inelastic multistep amplitudes. Clarification of the role of these mechanisms is clearly very important for the application of charge exchange to studies of nuclear structure or effective 2-nucleon interactions.

Some remarks about charge exchange with other projectiles, the pion and ^{3}He particle, are appropriate. Multi-step processes in (^{3}He,t) are relatively weaker[5] than in (p,n), which seems to be due simply to the fact that both inelastic ^{3}He scattering and (^{3}He,t)0^+ analog charge exchange are weaker than the corresponding nucleon reactions. The interference between the 1-step and 2-step amplitudes is here rather different because the phenomenological interaction is largely imaginary. In normal cases where β_1/β_0 is positive, the phase would be that of $-i(V + iW)$ which for a large negative imaginary W is nearly completely destructive. For a case like ^{26}Mg where β_1/β_0 is negative, the interference would be constructive and the experimental data[5] would be compatible with a value $\beta_1 \gtrsim \beta_0$. Unfortunately the possible contamination by the pickup-stripping mechanism[16] leaves the conclusions of such analysis rather uncertain. Good measurements with pions, which also have mostly imaginary interactions with nucleons, might be interesting from the point of view of 1- and 2-step interference.

In summary, we have found that at low energies (several tens of MeV) multistep inelastic transitions play a very important role in analog charge-exchange transitions. The excitation of 2^+ analog states proceeds primarily by 2-step mechanisms, and the 1- and 2-step amplitudes are nearly incoherent. The 0^- analog state is itself reduced by the combined effect of three 3-step mechanisms, which interfere destructively with the dominant 1-step mechanism. The phases of these amplitudes is surprisingly well explained by multistep perturbation theory and rough arguments using approximate properties of the optical Green's function. At higher energies the multistep amplitudes fall off more rapidly than the 1-step, and this behavior is explained on the basis of very crude development based again on multistep perturbation theory. At 135 MeV, the 0^- analog transition is essentially free of multistep inelastic effects, and the 2^+ analog is mostly excited by the 1-step mechanism, although even at this energy the 2-step mechanism needs to be taken into account.

We wish to acknowledge the help of Drs. F. Petrovich and G. Love in supplying various essential pieces of information used in preparation of this paper.

REFERENCES

1. J.D. Anderson and C. Wong, Phys. Rev. Lett., 7:250 (1961); J.D. Anderson, C. Wong, and J.W. McClure, Phys. Rev., 126:2170 (1962).
2. G.R. Satchler, R.M. Drisko, and R.H. Bassel, Phys. Rev., 136:B637 (1964).

3. J.D. Anderson, C. Wong, J.W. McClure, and B.D. Walker, Phys. Rev., 136:B118 (1964).
4. J.S. Blair, Inelastic Excitation of Collective Modes, in Direct Reactions and Nuclear Reaction Mechanisms, E. Clementel and C. Villi, eds., Gordon and Breach, New York, (1962).
5. V.A. Madsen, M.J. Stomp, V.R. Brown, J.D. Anderson, Luisa Hansen, Calvin Wong, and J.J. Wesolowski, Phys. Rev. Lett., 28:629 (1972).
6. V.A. Madsen, V.R. Brown, S.M. Grimes, C.H. Poppe, J.D. Anderson, J.C. Davis, and C. Wong, Phys. Rev., C13:548 (1976).
7. N. Vin Mau and A. Bouyssy, Nucl. Phys., A257:189 (1976).
8. C. Wong, V.R. Brown, V.A. Madsen, and S.M. Grimes, to be published.
9. D.M. Patterson, R.R. Doering, and Aaron Galonsky, Nucl. Phys., A263:261 (1976).
10. P. Schwandt, A. Nadasen, P.P. Singh, M.D. Kaitchuck, W.W. Jacobs, J. Meek, A.D. Bacher, and P.T. Debevec, Indiana University Cyclotron Facility Annual Report, 79 (1977).
11. G. Bertsch, J. Borysowicz, H. McManus, and W.G. Love, Nucl. Phys., A284:399 (1977).
12. F. Petrovich, H. McManus, V.A. Madsen, and J. Atkinson, Phys. Rev. Lett., 22:895 (1969).
13. W.G. Love, Alan Scott, and F. Todd Baker, Phys. Lett., 73B:277 (1978).
14. Aron M. Bernstein, V.R. Brown, and V.A. Madsen, Phys. Rev. Lett., 42:425 (1979).
15. E. Rost, Phys. Rev., 128:2708 (1962).
16. L.A. Charlton and P.D. Kunz, Phys. Lett., 72B:7 (1977).

THE SEQUENTIAL TRANSFER MECHANISMS IN (p,n) REACTIONS

P. D. Kunz

Nuclear Physics Laboratory
University of Colorado
Boulder, CO   80309

INTRODUCTION

The first order distorted wave Born approximation (DWBA) has provided a good description for many nuclear reactions. In the study of charge exchange reactions such as (^{3}He,t) and (p,n) one would initially expect to understand these processes in terms of an effective two-body charge exchange potential which acts between the incident projectile and the target nucleons. For the low spin transitions, especially the analogue state transitions, one could approximately explain[1,2] the data in terms of effective nucleon-nucleon force strengths. These successes were tempered by the failure of the theory to account for the excitation strength[3-5] of the higher spin states such as the 6^+ and 8^+ states. (The strength of the effective interaction was found to be about an order of magnitude larger for the 6^+ state than for the 0^+ analogue state.) These results were for cases where simple shell model configurations such as $(\pi f_{7/2})(\nu f_{7/2})^{-1}$ and $(\pi g_{9/2})(\nu g_{9/2})^{-1}$ could be used. Schaeffer[6] has shown that exchange effects cannot remove the high spin renormalization difficulty although the corrections were significant and in the right direction.

The inclusion of second order effects with the intermediate state being a rearrangement different from the initial or final states was proposed by Schaeffer and Bertsch[7] and Toyama.[8] The reaction was assumed to proceed via a (^{3}He,α)(α,t) or (^{3}He,d)(d,t) mechanism. Calculations by Toyama using the second order DWBA with strengths for the particle transfers taken from experiment accounted quantitatively for the shape and size discrepancies of the differential cross sections to all the natural parity states. Since these original suggestions there have been many efforts[9-13] to apply the

451

particle transfer mechanism to other reactions such as (p,d)(d,t) and (p,d)(d,n). Here the application of the new effect has been less than satisfactory. While the excitation of unnatural parity states such as the first-order-forbidden ^{208}Pb(p,t)^{206}Pb(3$^+$) case is explained, the presence of a strong direct term in the allowed states gives rise to several difficulties. In the (p,t) reaction, a persistent phasing problem between the direct and second order term depends upon the choice of the intermediate deuteron optical potential as pointed out by deTakascy.[12] This may possibly be due to non-orthogonality corrections[14] which arise from the manner in which the calculations are performed. For the (p,n) reactions the strength of the direct part of the reaction is taken to be adjustable[10] so that when a second order term of fixed strength is admixed, the magnitude of the data can be fitted. The resulting strength of the two-body interaction is twice the strength of that obtained from a two-body G matrix calculation.

Finally, in the reactions mentioned so far, the calculation of the transfer interaction for each step of the second order process is performed using zero-range approximations. Extensive full finite range calculations[15-20] have shown that off-energy-shell effects give large corrections to the zero-range results which must be properly taken into account. Further, the previous work has neglected the continuum states of the intermediate projectiles. While this may not be a bad approximation for (^{3}He,α) and (d,t) reactions where the projectile ground state accounts for 80-90% of the sum-rule estimate, it is clearly a less valid procedure to use only the ground state for the (p,d)(d,n) reaction where the deuteron ground state accounts for 25% or less of the sum rule.

This paper will attempt to summarize the present understanding of the (p,n) charge exchange reaction as a sequential transfer process for second order effects.

THEORY

The derivation of the equations of motion which give sequential transfer terms will differ from the transition matrix expansion method of Toyama[8] and the coupled reaction channels (CRC) methods.[9,21] Although our equations are similar to the CRC method, our method differs in that we do not incur the non-orthogonality terms unless we make a post-prior interchange of the interactions as discussed in ref. 14. However, the equations are what would be derived using the channel coupling array method of Hahn, Kouri and Levin[22] with a suitable choice of the channel coupling array matrix.

The description of the sequential transfer reaction as a series of coupled differential equations starts by defining the Hamiltonian in the different fragmentations

$$H = H_\alpha + V_\alpha = H_\beta + V_\beta = H_\gamma + V_\gamma = \cdots. \tag{1}$$

The H_i, $i = \alpha, \beta, \gamma$ define the two separated clusters containing perhaps an optical potential depending only on the relative separation between the two cluster and the V_i are the residual interactions between the clusters. The solution to the total Hamiltonian H with the boundary condition of incoming waves in the fragmentation α can be written as a sum of two parts

$$\Psi_\alpha = \psi_\alpha + \Delta\psi_\alpha \tag{2}$$

where $\psi_\alpha = \chi_{\alpha o}\phi_{\alpha o}$ and ϕ_α describes the internal motion of the clusters α in the initial state and $\chi_{\alpha o}$ is a function of their relative motion with incoming and outgoing scattered waves boundary condition. If we define ψ_α by

$$H_\alpha\psi_\alpha = E_\alpha\psi_\alpha, \tag{3}$$

then the equation of motion for $\Delta\psi_\alpha$ is

$$(E-H)\Delta\psi_\alpha = V_\alpha\psi_\alpha. \tag{4}$$

We now expand $\Delta\psi_\alpha$ in a complete set of states $\varphi_{\gamma n}$ of the intermediate fragmentation

$$\Delta\psi_\alpha = \sum_n \Delta\chi_{\gamma n}\varphi_{\gamma n} \tag{5}$$

where the $\Delta\chi_{\gamma n}$ are functions of the relative cluster motion. The equation of motion of the center-of-mass of the intermediate state with the outgoing wave boundary conditions is obtained by solving

$$(E_{\gamma n} - T_\gamma - U_\gamma)\,\Delta\chi_{\gamma n} = \langle\varphi_{\gamma n}|V_\alpha|\psi_\alpha\rangle + \langle\varphi_{\gamma n}|V_\gamma|\Delta\psi_\alpha\rangle. \tag{6}$$

The amplitude for the transition to the final state in fragmentation β is given by the usual procedure of expanding the total solution Ψ_α in the complete set of states of the fragmentation β

$$\Psi_\alpha = \sum \chi_{\beta n}\varphi_{\beta n}, \tag{7}$$

which gives

$$(E_{\beta n} - T_\beta - U_\beta)\chi_{\beta n} = \langle\varphi_{\beta n}|V_\beta|\psi_\alpha\rangle + \sum_{n'}\langle\varphi_{\beta n}|V_\beta|\varphi_{\gamma n'}\rangle\Delta\chi_{\gamma n'}. \tag{8}$$

The first term on the right is the usual first order term while the second term contains the higher order parts. If $\Delta\psi_\alpha$ in (6) is neglected than (8) gives the first and second order DWBA terms where the second term describes sequential transfer amplitudes.

If in the (p,n) reaction the deuteron fragmentation is taken for the intermediate state, the interactions V_α and V_β in (6) and (8) are of the form $V_{np}+V_p-U_p$ and $V_{np}+V_n-U_n$ respectively, where V_p and V_n are the microscopic interactions of the proton and neutron with the core and U_p and U_n are their respective optical potentials. With the neglect of the V_p-U_p and V_n-U_n terms in the interaction we are left with the short range potential V_{np} which lends itself to the use of the zero range approximation.

The (p,n) reaction has been calculated[10] with the sequential transfer term for the deuteron ground state. This calculation uses the zero range approximation and an adjustable strength V_τ for the isovector part of V_{np}. The strength of V_τ for a Yukawa potential of 1 fm range is about 33 MeV which is about double the strength one might expect from nucleon-nucleon forces and the value of V_τ which is needed to fit the (p,n) data from the direct term alone. The larger value of V_τ required for the sequential transfer is due to the opposite phase of the sequential transfer amplitude compared to the direct term amplitude.

Since these calculations are performed with the zero range approximation one needs to assess the effects of finite range on the sequential transfer process. In the usual single-step transfer reaction the corrections to lowest order in the range parameter R can be expressed by the operator in the transition amplitude of the form

$$\theta = 1 + \frac{2}{\hbar^2} \frac{m_p m_n}{m_d} R^2 \left[-\frac{\hbar^2}{2\mu_d} \nabla_d^2 + \frac{\hbar^2}{2\mu_p} \nabla_p^2 + \frac{\hbar^2}{2\mu_n} \nabla_n^2 \right], \qquad (9)$$

where one can make use of the Schrödinger equation for the distorted waves and bound state

$$-\frac{\hbar^2}{2\mu} \nabla^2 \chi = (E-U)\chi, \qquad (10)$$

to replace the Laplacian operator. This results in a modulation function

$$F(r) = 1 + \frac{2}{\hbar^2} \frac{m_p m_n}{m_d} R^2 \left[|SE_d| + U_p + U_n - U_d \right], \qquad (11)$$

where the U_i are the optical potentials for the proton, neutron and deuteron. For the case where the potentials are matched and the separation energy for the projectile is small, eq. (11) gives a small correction (typically a few percent) for typical (d,p) or (d,n) reactions. This is primarily because of energy conservation between the initial and final states. However, for a two-step process the intermediate state can be off the energy shell and the energy cancellation no longer holds so large corrections can occur.

In order to evaluate the effect of finite range corrections in the two-step processes we have used the intermediate state Green function

$$\frac{\hbar^2}{2\mu_D} \nabla^2 G\ (\underset{\sim}{r}_D - \underset{\sim}{r}'_D)\ =\ (E_D - U_D)\ G_D(\underset{\sim}{r}_D - \underset{\sim}{r}'_D)\ - \delta(\underset{\sim}{r}_D - \underset{\sim}{r}'_D)$$ (12)

instead of (10) for the deuteron distorted wave. The first term on the right of (12) gives the usual small first order correction term. The second term is the energy non-conserving part where the δ function for the deuteron removes the dependence of this portion of the correction term on the intermediate state coordinates thereby giving an equivalent one-step transition amplitude. The second correction term in (12) is opposite in sign to the two-step amplitude for a pickup-stripping process and one thus obtains a smaller sequential amplitude when calculated in finite range than in zero range. Calculations[16] show that the sequential amplitude is reduced by finite range effects to about half the zero range result. When the single step amplitude is now added to the sequential transfer amplitude, the value of V_τ needed to fit the data is reduced to about 26 MeV.

All the calculations mentioned above have been restricted to an intermediate deuteron in its ground state. To estimate the contributions from the continuum states we use a sum rule. We begin by noticing that the second order amplitude via the deuteron ground state is proportional to D_0^2 the particle transfer strength constant in zero range. An s wave continuum state φ_k of momentum k contributes an amplitude proportional to $|d_k|^2$ where

$$d_k = \int V_{np}\ \varphi_k d\tau .$$ (13)

From the closure property of the functions φ_k a sum rule for the triplet s state strength can be formed,

$$\begin{aligned} S_1 &= D_0^2 + \int d\underset{\sim}{k} |d_k|^2 \\ &= \int |V_{np}|^2 d\tau . \end{aligned}$$ (14)

If we make the additional assumption that the sequential transfer amplitude is independent of the Q-value of the intermediate state, then one finds that the second order term is enhanced over the ground state contribution by the factor S_1/D_0^2. For simple potential forms we find

$$S_1 \approx 4.5\ D_0^2 \qquad \text{Hulth\'en potential}$$ (15)

$$S_1 \approx 7\ D_0^2 \qquad \text{Yamaguchi separable potential.}$$

The resulting sum rule, S_o, for the singlet deuteron continuum states is about the same in magnitude as for the triplet deuteron but the spin factors in calculating the amplitude effectively reduce the contribution to about $-1/3$ of that for the triplet state. The total sum rule for the second order amplitude, including both the singlet and triplet detueron states, is then about $2/3$ the triplet strength. Thus the addition of the continuum states of the deuteron will give an effectively larger sequential transfer amplitude than has been calculated for the ground state alone.

An interesting observation may be made at this point. If the nucleon-nucleon forces were spin and isospin independent so that the singlet and triplet states were identical, the two states would not cancel their second order contributions and a resulting amplitude for the (p,n) reaction would still be obtained. This is analogous to the (p,n) reaction still taking place in first order via the knock-out amplitude. This fact suggests that these higher order effects which are independent of the isospin nature of the nuclear forces may play an important role in the (p,n) reaction.

Another effect that has not been taken into account and may reduce considerably the sequential transfer amplitude is the Pauli principle between the nucleons in the deuteron and the nucleons in the target. While this may be a small correction for the ground state of the deuteron for large projectile energies, the correction may reduce significantly the contribution of the deuteron continuum states where the center-of-mass motion is near or below zero energy. Thus one would expect serious corrections to the sum rule for low energy (p,n) reactions and a smaller enhancement of the sequential contribution from the deuteron continuum states than the sum rule indicates.

At this point we wish to consider the interpretation of sequential transfer in terms of inelastic excitation processes in the initial channel. We begin by considering a set of coupled equations for the initial state in (3) in which N excited states of the initial nucleus are coupled to the ground state. The coupled equations are written as

$$(E_n - H_\alpha)\chi_{\alpha n} = \sum_{n=0}^{N} <\varphi_{\alpha n}|V_\alpha|\varphi_{\alpha n'}> \chi_{\alpha n'} \tag{16}$$

with $\psi_\alpha = \sum_{n=0}^{N} \chi_{\alpha n}\varphi_{\alpha n}.$

This solution when inserted into (4) to obtain the equation of motion for $\Delta\psi_\alpha$ gives

$$(E-H)\Delta\psi_\alpha = \sum_{n=0}^{N} (V_\alpha\varphi_{\alpha n} - \sum_{n=0}^{N}\varphi_{\alpha n'}<\varphi_{\alpha n'}|V_\alpha|\varphi_\alpha>) \chi_{\alpha n}. \tag{17}$$

As is apparent, when the number N is increased to encompass the complete set of initial states $\varphi_{\alpha n}$, the closure relation $\Sigma_{n'} |\varphi_{\alpha n'}><\varphi_{\alpha n'}| = \delta(\underline{r}-\underline{r}')$ causes the right hand side of (17) to vanish and the solution $\Delta\psi_{\alpha}$ then vanishes for the boundary condition of outgoing wave only. Thus the inclusion of the sequential transfer terms is another method of describing the second order processes in the initial channel. In figure 1 the qualitative situation for the ground state deuteron contribution is shown. Since the deuteron is loosely bound it can be considered as a free proton and neutron at half the deuteron kinetic energy with a small energy spread due to the deuteron internal motion. In the case of an incoming proton at 23 MeV energy, the deuteron typically will have 14 MeV energy; hence the proton and neutron will each have 7 MeV energy corresponding to a neutron excitation in the initial nucleus of about 18 MeV. In this picture one can make a qualitative correspondence between the sequential transfer process and second order contributions through excited states of the target nucleus.

Although no calculations have been made for finite range and Pauli principle effects for the deuteron continuum states, one expects contributions from the full second order terms to be larger than for the deuteron ground state and V_r for first order terms will be equal to or greater than the 26 MeV found from the finite range calculation of ref. 16. There is a puzzle here. The missing ingredient may be the inadequacy of the first order term. Customarily an effective potential is used to account for the singlet-triplet potential difference where the triplet potential is more attractive than the single potential. However, the phenomenological potentials[23] which are much more complicated than effective central potentials contain a short range repulsive behavior and a tensor component. Nuclear matter calculations may serve as a guide by using the simple ideas of the Moskowski and Scott[24] separation method. To a first approximation a considerable fraction of the attractive part of the nucleon-nucleon potential is cancelled by the short ranged repulsive region leaving a weak attractive part beyond the separation distance as a first order interaction. In addition, for triplet-even states the tensor potential reduces the need for a strong central attraction. In the case of a strong tensor force the long-range part of the potential is actually weaker than the singlet-even, long range part. This is in contrast to the usual conclusion one would draw from purely central forces. While these arguments are qualitative and may not completely apply to the case where an incoming particle is above the filled levels of the nucleus they do point out that one must be careful in defining the expansion of the amplitude into first and second order and higher orders and what one means by the potential in the first order term.

In summary, the understanding of the higher order effects in the (p,n) reaction requires much more sophistication in the

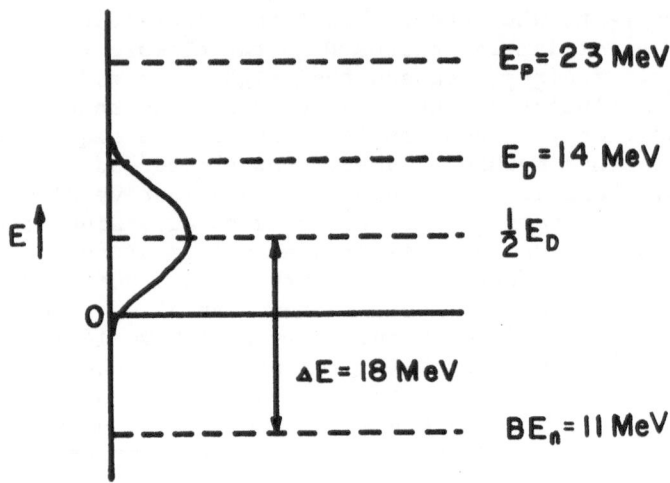

Fig. 1.   Diagram showing the position of the excited states of the
neutron at $\frac{1}{2}E_D$ which are represented by the sequential
transfer processes for the deuteron ground state.

calculations before one can hope to quantitatively fit the data.
One must be able to calculate the continuum state contributions
in finite range and with Pauli principle effects included.   A
promising method of analyzing the (p,n) reaction may be the nuclear
matter approach of Brieva and Rook[25] for proton elastic scattering
which has been extended to the (p,n) reaction by von Geramb and
Rook.[26] Another is the microscopic form factor approach of Baur,
Madsen and Osterfeld which uses the coupling of the excited states
of a finite well nucleus.   These methods also require a large amount
of computation.   Further work is required to obtain a simple
effective interaction which one can use in a calculation such as the
first-order distorted wave Born approximation.

REFERENCES

1.   J. J. Wesolowski, E. H. Schwarcz, P. G. Roos, and C. A.
     Ludemann, Phys. Rev. 169, 878 (1968).
2.   P. D. Kunz, E. Rost, R. R. Johnson, G. D. Jones, and S. I.
     Hayakawa, Phys. Rev. 185, 1528 (1969).
3.   S. I. Hayakawa, W. L. Fadner, J. J. Kraushaar, and E. Rost,
     Nucl. Phys. A139, 465 (1969).
4.   A. Richter, J. R. Comfort, N. Anantaraman, and J. P. Schiffer,
     Phys. Rev. C 5, 821 (1972).

5. P. Kossanyi-Demay, P. Roussel, H. Faraggi, and R. Schaeffer, Nucl. Phys. A148, 181 (1970).
6. R. Schaeffer, Nucl. Phys. 158, 321 (1970).
7. R. Schaeffer and G. Bertsch, Phys. Lett. 38B, 159 (1972).
8. M. Toyama, Phys. Lett. 38B, 147 (1972); Nucl. Phys. A211, 254 (1973).
9. W. R. Coker, T. Udagawa, and H. H. Wolter, Phys. Rev. C 7, 1154 (1973).
10. L. D. Rickertsen and P. D. Kunz, Phys. Lett. 47B, 11 (1973).
11. W. R. Coker, T. Udagawa, and J. R. Comfort, Phys. Rev. C 10, 1130 (1976).
12. N. B. de Takacsy, Nucl. Phys. A231, 243 (1974).
13. F. Iachello and P. P. Singh, Phys. Lett. 48B, 81 (1974).
14. P. D. Kunz and E. Rost, Phys. Lett. 47B, 136 (1973).
15. L. A. Charlton, Phys. Rev. Lett. 35, 1495 (1975); Phys. Rev. C 14, 506 (1976).
16. P. D. Kunz and L. A. Charlton, Phys. Lett. 61B, 1 (1976).
17. L. A. Charlton and P. D. Kunz, Phys. Lett. 72B, 7 (1977).
18. M. Igarishi, Phys. Lett. 78B, 379 (1978).
19. J. D. Burch, M. J. Schneider, and J. J. Kraushaar, Nucl. Phys. A299, 117 (1978).
20. W. R. Zimmerman, J. J. Kraushaar, and F. E. Cecil, Nucl. Phys. A309, 34 (1978).
21. T. Udagawa, H. H. Wolter, and W. R. Coker, Phys. Rev. Lett. 31, 1507 (1973).
22. Y. Hahn, D. J. Kouri, and F. S. Levin, Phys. Rev. C 10, 1615 (1974).
23. R. V. Reid, Ann. Phys. 50, 411 (1968).
24. B. L. Scott and S. A. Moskowski, Ann. Phys. 14, 107 (1961); Ann. Phys. 11, 65 (1960).
25. F. A. Brieva and J. R. Rook, Nucl. Phys. A291, 299 (1976); Nucl. Phys. A291, 317 (1976).
26. H. von Geramb and J. R. Rook,
27. G. Baur, V. A. Madsen, and F. Osterfeld, Phys. Rev. C 17, 819 (1978).

DISCUSSION

Lind: Do multistep contributions in charge exchange reactions become less important with increasing A?

Wong: Perhaps I can answer Lind's question. We see the same multistep interference effects in the samarium isotopes as we see in the Mo isotopes. The importance of multistep contributions in (p,n) reactions depends on the degree of collectivity of the target nucleus and not on A.

ENERGY DEPENDENCE OF V_τ IN THE (p,n) REACTION 10-30 MeV

C. H. Poppe

Lawrence Livermore Laboratory
Livermore, CA   94550

The (p,n) reaction on spin-zero nuclei leading to the target isobaric analog state is the most promising way to extract the $V_{ST} = V_{01} = V_\tau$ part of the effective nucleon-nucleon interaction.[1,2] In order to do this a microscopic calculation is required. However many authors[3-9] have analyzed this reaction in terms of the macroscopic Lane model[10] in order to determine V_1, the isovector part of the optical model potential. Madsen[11] has shown that in the absence of exchange forces, V_τ and V_1 are related by the expression

$$V_1(r) = \frac{U_1(r)\vec{T}\cdot\vec{t}}{A} = V_\tau \sum_j \langle N_j - Z_j \rangle \, R_j^2(r)/2\pi \sqrt{N-Z} \qquad (1)$$

where R_j is the radial wave function for the bound particle and $\langle N_j \rangle$ and $\langle Z_j \rangle$ are the expectation values of neutron and proton numbers in the j-shell. In principle, then, one may determine V_τ from such a macroscopic calculation.

In recent years, several laboratories have shown that for a large range of mass number there is a strong energy dependence for the (p,n) analog reaction from several MeV above threshold to approximately 30 MeV.[4-9,12-16] The origin of this behaviour with energy is not understood. Several analyses[5,7,13,14] in this energy range have extracted an isovector potential which reflects this variation by an anomalous energy dependence of the potential, which according to (1) leads to a corresponding behavior for V_τ. Other studies have attempted to explain this energy dependence in terms of the energy dependence of the optical model potentials, in particular that of the neutron channel,[16] but we have found this difficult to do over the broad range of masses studied at the

Lawrence Livermore Laboratory ($40 \lesssim A \lesssim 209$). The effects of
strongly excited collective states have been included using coupled-
channels calculations[17],[18] and although this has been successful in
accounting for deviation from the expected N–Z dependence of the
simple Lane model, it has not been able to account for the energy
dependence. Recently, this author[19],[20] has extended the coupled-
channels formalism to include a giant resonance in the compound system, but although it is possible to fit the energy dependence of the
(p,n) analog cross section the resulting A–dependence of the resonance energy is not very systematic[20] and no understanding as to the
nature of the supposed resonance has yet been achieved.

With the inclusion of exchange in DWBA calculations, relation
(1) no longer obtains, and we were led to investigate what a proper
exchange calculation would yield with regard to the energy dependence of the (p,n) analog reaction and V_T in this energy range. For
our first attempt ^{92}Mo was chosen as the target, because of its
closed $1g_{9/2}$ neutron shell and the weakness of excited analog states
for the (p,n) reaction on this target.[7],[17] The code DWBA70[21] was
used for the reaction calculation, and the effective interaction
used was the even central part of the MSU potential[22] derived from
fitting the G–matrix elements of the Reid potential. For the central terms this potential uses a sum of three Yukawas of different
strengths and ranges, but the results here will be discussed in
terms of the strength V_T^{EFF} of a single Yukawa of 1 fm range which
yields the same volume integral in order that a comparison to previous work[1],[2] may be made. (For this MSU potential V_T^{EFF} = 17.8 MeV).

Information about the target and residual nuclear wave functions
is supplied to the code by spectroscopic amplitudes

$$Z^J_{TM_T} (j_2 j_1) = \hat{J}^{-1} \hat{J}_1 \left\langle T_2 M_{T_2} J_2 \middle|\middle| A^J_{TM_T} (j_2 j_1) \middle|\middle| T_1 M_{T_1} J_1 \right\rangle \qquad (2)$$

for transferred angular momentum J and transferred isospin T with
projection M_T. The shell model orbits of the particle in the initial and final state are given by j_1 and j_2, respectively, and $J_{2,1}$
$T_{2,1}$ $M_{T_{2,1}}$ are the quantum numbers of the residual and final nucleus. The operator $A^{JM_J}_{TM_T}$ is given by

$$A^{JM_J}_{TM_T} (j_2 j_1) = \sum_{\substack{m_1 m_2 \\ \alpha_1 \alpha_2}} (-)^{j_1 - m_1 + 1/2 - \alpha_1} \left\langle j_2 j_1 m_2 - m_1 \middle| J M_J \right\rangle$$

$$\qquad (3)$$

$$\times \left\langle 1/2 \; 1/2 \; \alpha_2 - \alpha_1 \middle| T M_T \right\rangle a^\dagger_{j_2 m_2 \alpha_2} a_{j_1 m_1 \alpha_1}$$

For T = 1 transitions with $(J_1, T_1, j_1) = (J_2, T_2, j_2) = (J_0, T_0, j)$,

Madsen has derived a reduced spectroscopic amplitude[23]

$$S_T^J(j_2 j_1) = S_1^0(jj) = \left[\frac{(T_0+1)(2T_0+1)}{T_0} \right]^{1/2} \frac{\hat{J}_0}{\sqrt{6} \, \hat{j}} \langle N_j - Z_j \rangle \qquad (4)$$

which may be related to (2) by an appropriate change in normalization and application of the Wigner-Eckart theorem. One obtains

$$Z_{TM_T}^J(j_2 j_1) = \hat{J}_1 \, \hat{T} \, \hat{T}_2^{-1} \, \langle T_1 \, T \, M_{T_1} M_T \mid T_2 M_{T_2} \rangle \, S_T^J(j_2 j_1) \qquad (5)$$

which for the (p,n) reaction to the isobaric analog state yields

$$Z_{1-1}^0(jj) = \frac{\langle N_j - Z_j \rangle}{\sqrt{2T_0} \, \hat{j}} \qquad (6)$$

Because of the relative insensitivity of the (p,n) analog reaction to details of the nuclear wave functions, we have chosen a simple description for ^{92}Mo(p,n) in which a proton is created and a neutron destroyed in the $1g9/2$ orbit.

The optical model potentials used for the distorted waves were the best-fit global potential of Becchetti and Greenlees[24] for protons and the recent neutron potential of Lagrange[25] derived from neutron scattering and reactions on the Mo isotopes. This latter potential does an excellent job in reproducing the energy dependence of the neutron total cross section, which was not the case for the global potentials used in some of our previous macroscopic studies.[16]

The result of a complete exchange calculation for a bombarding energy of 26 MeV is shown by the solid curve of Fig. 1. Although no parameters were adjusted, the fit is excellent both in shape and absolute magnitude and the value of the angle-integrated cross section agrees with that for the experimental data[7] to within the quoted error. With the parameters of this calculation held fixed and varying only the strengths of the optical model potentials according to the authors' prescriptions,[24,25] the calculation was repeated at a series of lower energies. Figure 2 shows the results by comparing the angle-integrated analog cross sections with the data. The energy dependence of the measured cross section is seen not to be reproduced by the MSU potential with constant strengths (V_T^{EFF} = 17.8 MeV). In addition, as the energy was decreased from 26 MeV, the fit to the angular distribution at forward angles deteriorated, although the shape for angles greater than 40° was maintained very well. Specifically, the calculated forward-angle cross section developed a pronounced dip, while the data remain forward-peaked.

In order to reproduce the energy dependence of the integrated

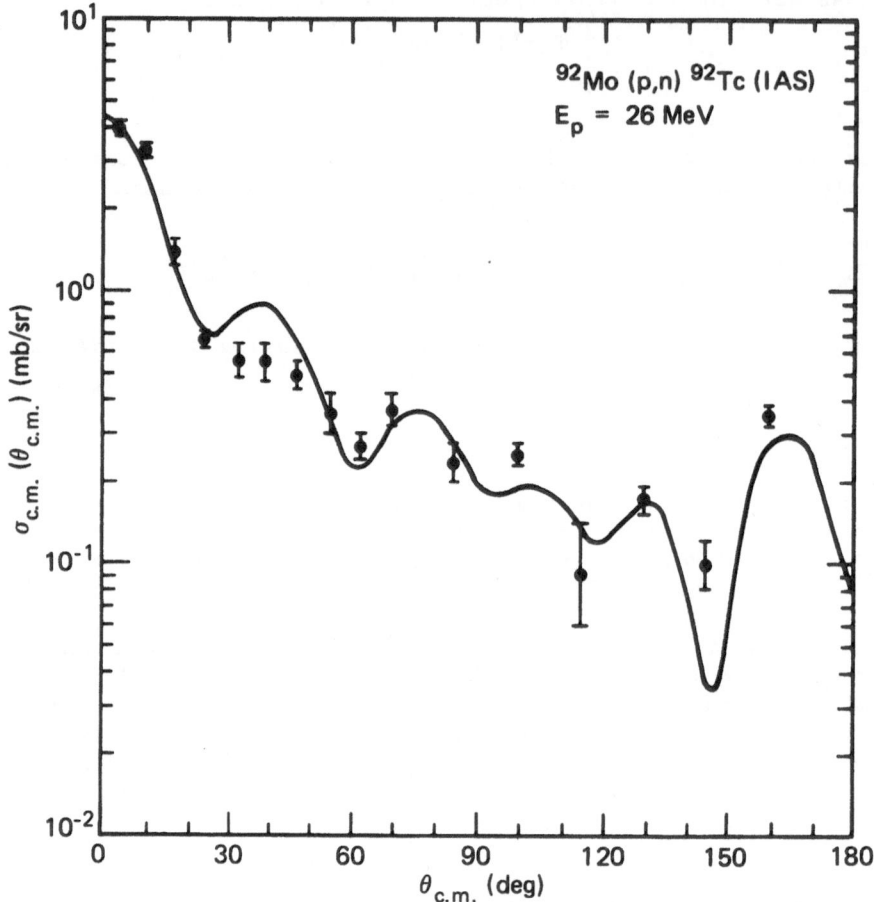

Figure 1.   26-MeV angular distribution for ^{92}Mo(p,n) to the
            isobaric analog state. The solid curve is a distorted
            wave microscopic calculation with exchange using the
            MSU effective nucleon-nucleon potential as explained
            in the text.

cross section, the strengths of the V_τ parts of the three Yukawas
of the even central part of the MSU potential were increased by
the same fractional amount until agreement with the experimental
angle-integrated cross sections was achieved. The results are
shown in Fig. 3 for the equivalent 1 fm Yukawa, V_τ^{EFF}. Error bars reflect the uncertainty of the data and the dashed line indicates the
constant V_τ^{EFF} for the MSU potential. One observes that the energy
dependence of the cross section requires an energy-dependent V_τ,
however the maximum variation in strength is only about 20% over

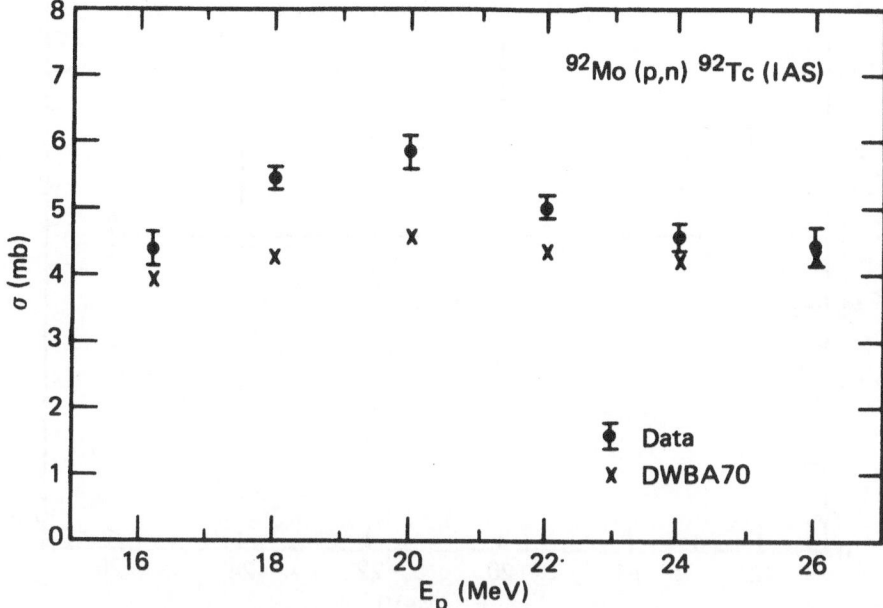

Figure 2. Angle-integrated ^{92}Mo(p,n) analog cross sections
compared with distorted-wave calculations.

the energy range studied, considerably less than that required for
the isovector potential in our previous Lane-model analysis.[7]

     The energy dependence required of V_τ to fit the data is hard
to explain, and the behavior of the forward angle cross section
leads one to believe that, perhaps, some other mechanism is occur-
ring in this energy range which is not included in the model used
for this calculation. In a microscopic analysis of the ^{90}Zr(p,n)
analog reaction Rikus et al.[26] included resonance terms which suggested the existence of giant quadrupole and octupole resonance
strength in the 25 to 30 MeV excitation energy region. Unfortunately, the data were too sparse for any clear statement to be made and
the resulting coupling strengths had large standard deviations.
Nevertheless, with the inclusion of such resonance effects it was
possible to remove the need for any residual energy dependence of
the effective two-nucleon interaction. A strong isovector E2 resonance near 27 MeV excitation has been reported by Fukuda et al.,[27]
which corresponds to an incident energy of about 22 MeV for ^{92}Mo+p,
near the region of the anomalous behavior for V_τ^{EFF} illustrated in
Fig. 3. Consequently, the data and analysis presented here would

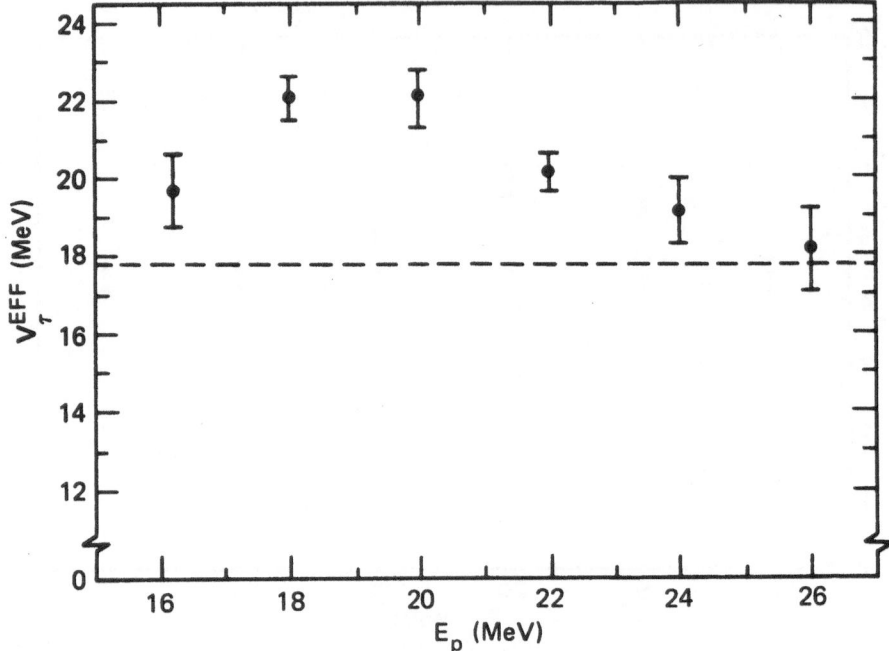

Figure 3.   Values of the effective V_τ for a 1 fm range extracted
            from fitting the integrated cross sections of Fig. 2.
            The dashed line is the value inferred for the MSU
            potential.

tend to support the approach of Rikus et al.[26] and to stress the
importance of completely understanding the reaction mechanisms before
it is possible to extract reliably V_τ from (p,n) analog reactions in
the 10-30 MeV range.

REFERENCES

1.   S. M. Austin, The Two-Body Force in Nuclei, edited by S. M.
     Austin and G. M. Crawley (Plenum Press, New York-London, 1972)
     p. 285.
2.   V. A. Madsen, Nuclear Spectroscopy and Reactions, Part D (Aca-
     demic Press, New York-London, 1975), p. 251.
3.   J. D. Carlson, C. D. Zafiratos and D. A. Lind, Nucl. Phys. A249,
     29, (1975).
4.   S. D. Schery, D. A. Lind, H. W. Fielding, and C. D. Zafiratos,
     Nucl. Phys. A234, 109 (1974).
5.   R. K. Jolly, T. M. Amos, A. Galonsky, R. Hinrichs, and R. St.Onge,
     Phys. Rev. C 7, 1903 (1973).

6. C. Wong, J. D. Anderson, J. W. McClure, B. A. Pohl, and J. J. Wesolowski, Phys. Rev. C 5, 158 (1972).
7. S. M. Grimes, C. H. Poppe, J. D. Anderson, J. C. Davis, W. H. Dunlop, and C. Wong, Phys. Rev. C 11, 158 (1975).
8. C. Wong, J. D. Anderson, J. C. Davis, and S. M. Grimes, Phys. Rev. C 7, 1895 (1973).
9. D. M. Patterson, R. R. Doering, and A. Galonsky, Nucl. Phys. A263, 261 (1976).
10. A. M. Lane, Phys. Rev. Lett. 8, 171 (1962); Nucl. Phys. 35, 676 (1962).
11. V. A. Madsen, Nuclear Isospin, edited by J. D. Anderson, S. D. Bloom, J. Cerny, and W. W. True (Academic Press, New York-London, 1969), p. 149.
12. P. S. Miller and G. T. Garvey, Nucl. Phys. A163, 65 (1971).
13. G. W. Hoffmann, W. H. Dunlop, G. J. Igo, J. G. Kulleck, J. W. Sunier, and C. A. Whitten, Jr., Nucl. Phys. A187, 577 (1972).
14. G. W. Hoffmann, W. H. Dunlop, G. J. Igo, J. G. Kulleck, C. A. Whitten, Jr., and W. R. Coker, Phys. Lett. 40B, 453 (1972).
15. C. H. Poppe, S. M. Grimes, J. D. Anderson, J. C. Davis, W. H. Dunlop, and C. Wong, Phys. Rev. Lett. 33, 856 (1974).
16. D. H. Fitzgerald, G. W. Greenlees, J. S. Lilley, C. H. Poppe, S. M. Grimes, and C. Wong, Phys. Rev. C 16, 2181 (1977).
17. V. A. Madsen, V. R. Brown, S. M. Grimes, C. H. Poppe, J. D. Anderson, J. C. Davis, and C. Wong, Phys. Rev. C 13, 548 (1976).
18. D. H. Fitzgerald, J. S. Lilley, C. H. Poppe, and S. M. Grimes, Phys. Rev. C 18, 1207 (1978).
19. C. H. Poppe, Bull. Am. Phys. Soc. 21, 560 (1976).
20. C. H. Poppe, paper presented at "The (p,n) Reaction and Its Connection to the Nucleon-Nuclear Optical Model," Workshop held at Lawrence Livermore Laboratory, Oct. 1978, unpublished.
21. R. Schaeffer and J. Raynal, unpublished.
22. G. Bertsch, J. Borysowicz, H. McManus, and W. G. Love, Nucl. Phys. A284, 399 (1977).
23. C. Wong, J. D. Anderson, J. McClure, B. Pohl, V. A. Madsen, and F. Schmittroth, Phys. Rev. 160, 769 (1967); V. A. Madsen, private communication.
24. F. D. Becchetti, Jr. and G. W. Greenlees, Phys. Rev. 182, 1190 (1969).
25. Ch. Lagrange, paper presented at the "Specialist Meeting on Neutron Data of Structural Materials for Fast Reactors," Geel, Belgium, Dec. 1977, unpublished.
26. L. Rikus, R. Smith, I. Morrison, and K. Amos, Nucl. Phys. A286, 494 (1977).
27. S. Fukuda and Y. Torizuka, Phys. Rev. Lett. 29, 1109 (1972).

Work performed under the auspices of the U.S. Department of Energy by Lawrence Livermore Laboratory under Contract No. W-7405-ENG-48.

POLARIZATION OBSERVABLES IN (p,n) REACTIONS

Richard L. Walter and Roger C. Byrd

Department of Physics, Duke University,
Durham, NC 27706
and Triangle Universities Nuclear Laboratory,
Duke Station, NC 27706

INTRODUCTION

In (p,n) reactions, three parameters have been measured that involve polarized incident proton beams, polarized outgoing neutron beams, or both. The present paper will review the data that have been obtained and the conclusions that have been based on that data. The discussion will be limited to experiments below 30 MeV and therefore will ignore some of the earliest work,[1] that is, the measurements performed at the Rutherford Laboratory with 30 to 200 MeV polarized proton beams. With the advent of polarized proton beams at the Indiana cyclotron, one would expect that this intermediate energy range soon will be active again.

The three parameters to be discussed here are: 1) the polarization transfer coefficient $K_y^y(0°)$, which connects the transverse polarization in the outgoing neutron beam to the transverse polarization in the incident proton beam; 2) the (transverse) polarization $P^y(\theta)$ of the outgoing neutron beam, which is produced by incident unpolarized protons; and 3) the asymmetry parameter $A_y(\theta)$, which is a measure of the ratio of the cross section at the reaction angle on the left side of the reaction plane to that on the right side for a reaction induced with an incident polarized beam. The primary goal of the $K_y^y(0°)$ measurements was to provide data on the two-body spin-dependent terms and on (p,n) reaction mechanisms. Most of the P^y measurements were performed to provide tests for nuclear reaction models; a few others were intended to give spectroscopic information. Experiments to determine A_y have been used primarily in attempts to test the Lane model. Most of the A_y data has been obtained in the last five years and its value is just emerging.

Except for one experiment, all the P^y and the K_y^y measurements involve light nuclei as targets. This unfortunate result has occurred because the inherent energy resolution in neutron polarization measurements is poor, and only in light nuclei can one clearly separate the neutron groups leading to the various states in the residual nuclei.

POLARIZATION-ANALYZING POWER THEOREM

From basic symmetry arguments, for elastic scattering of nucleons the polarization parameter $P^y(\theta)$ is identical to the asymmetry parameter $A_y(\theta)$. The quasi-elastic scattering nature of (p,n) reactions to analogue states suggests that P^y might be equal to A_y in such reactions also. This question has been addressed by Arnold,[2] Conzett,[3] and in this volume by Philpott and Halderson.[4] To our knowledge, Arnold was the first to point out that for a spin-zero target, the basic symmetry arguments require that $P^y(\theta) \equiv A_y(\theta)$ for (p,n) reactions also. Looking further to explain observed differences between P^y and A_y, he developed the concept of asymmetric spin-flip in quasi-elastic scattering and showed that $P_y \equiv A_y$ in such reactions unless there is a difference between the interaction for a spin-up proton to come out as a spin-down neutron and the interaction for a spin-down proton to come out as a spin-up neutron. He suggested that observing differences between P_y and A^y would be a sensitive technique to detect regions of isospin mixing. Philpott and Halderson[4] investigated P^y and A_y for the $^3H(p,n)^3He$ and $^{15}N(p,n)^{15}O$ reactions using a microscopic model as will be mentioned below. In the review presented in the following sections, several comparisons between P^y and A_y will be exhibited; however this will be done only after each type of experiment is described individually.

MEASUREMENTS OF THE POLARIZATION PARAMETER

Most of the polarized beam experiments for (p,n) reactions are of the $(p,\vec{n})$ type, where the incident beam is unpolarized and the outgoing neutron polarization is determined by scattering the neutrons from some convenient analyzer like ^4He. Much of the existing data was obtained prior to 1970 and is tabulated in a review paper[5] delivered at the Madison Polarization Conference. All of the data is for targets with A<16 and for E_p<12 MeV. Since 1970, the experiments have improved in statistical accuracy and precision. A large portion of this later data has been summarized in the proceedings of the 1976 Lowell Neutron Conference.[6]
To indicate the quality of P^y data and some of the difficulties in early P^y measurements, results for $^3H(p,n)^3He$ are shown in Fig. 1 and 2 which are taken from Donoghue et al.[7] The newest

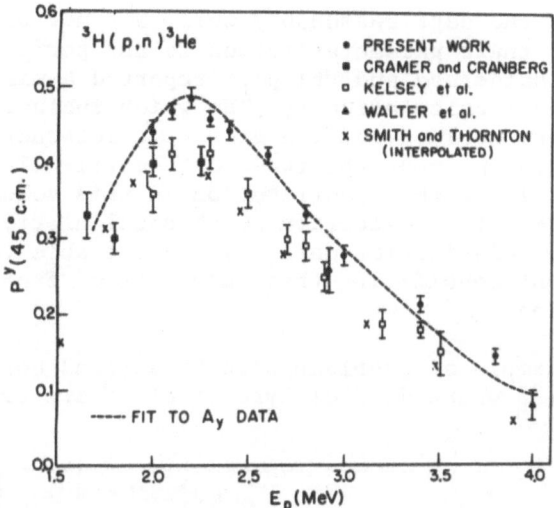

Fig. 1. Results for P^y(45° c.m.) for the ^{3}H(p,n)^{3}He reaction reported by Donoghue et al.[7]

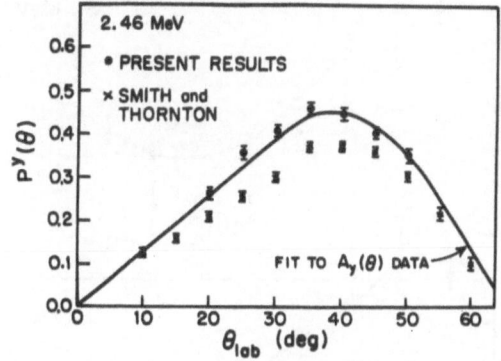

Fig. 2. Results for $P^y(\theta)$ for the ^{3}H(p,n)^{3}He reaction for E_p=2.46 MeV reported by Donoghue et al.[7]

results, which are plotted as solid circles ("Present Results"), are 20% higher than most of the previous work reported by several laboratories. The authors[7] claim that the earlier work had an un-polarized background that was not taken into account properly in the data analysis. This example clearly illustrates that P^y is a difficult parameter to measure and that there have been problems in data reduction and in making corrections for the ever-present neutron background. The new P^y results for ^{3}H(p,n)^{3}He are now part of a massive data base for the 4-nucleon system which is being employed by Hale and Dodder[8] in their attempts to define the level structure of ^{4}He with an R-matrix approach. Presently, they

are able to fit the data reasonably well, but not exactly, indicating to us that their parameterization is not perfect yet. In a
recent letter[9] Halderson and Philpott reported a microscopic continuum shell model calculation for ^{4}He which included noncentral
components in the nucleon–nucleon effective interaction. This
interaction is taken to be the two-body g-matrix elements of
Bertsch et al. In another contribution in this volume Halderson
and Philpott[4] report an extension of this calculation to describe
the ^{3}He(p,n)^{3}He polarization data. They were able to obtain reasonable agreement considering that there are no free parameters in
their calculation.

Another example of problems with background corrections is exhibited in Fig. 3 where data of Byrd et al.[10] are compared to

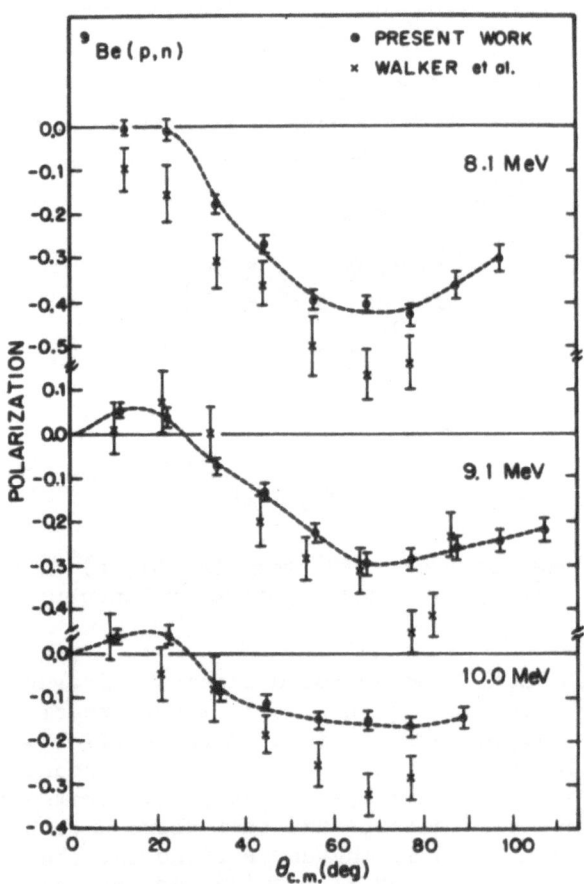

Fig. 3.   Results for $P^y(\theta)$
from ref. 10 and 11.

earlier Livermore[11] results for the ^{9}Be(p,n)^{9}B reaction. Here the
difficulty apparently was an inability to resolve the neutrons
leading to the ground state of ^{9}B from neutrons produced in 3- and
4- body breakup reactions. The data analysis of Byrd et al. incorporated careful scrutiny on an interactive computer system.
These data have been used in the detailed multi-channel Lane model
analysis described by Byrd et al.[12] in another contribution in this
volume.

Careful measurements of P^y have also been made by Byrd et al.
for ^{15}N(p,n)^{15}O from 3 to 10 MeV. Considerable resonance structure
exists here. Some of the data is shown on the right-hand side of
Fig. 4 and as crosses in Fig. 5. (The curves in Fig. 4 and the A_y
data in the figures are discussed below.)

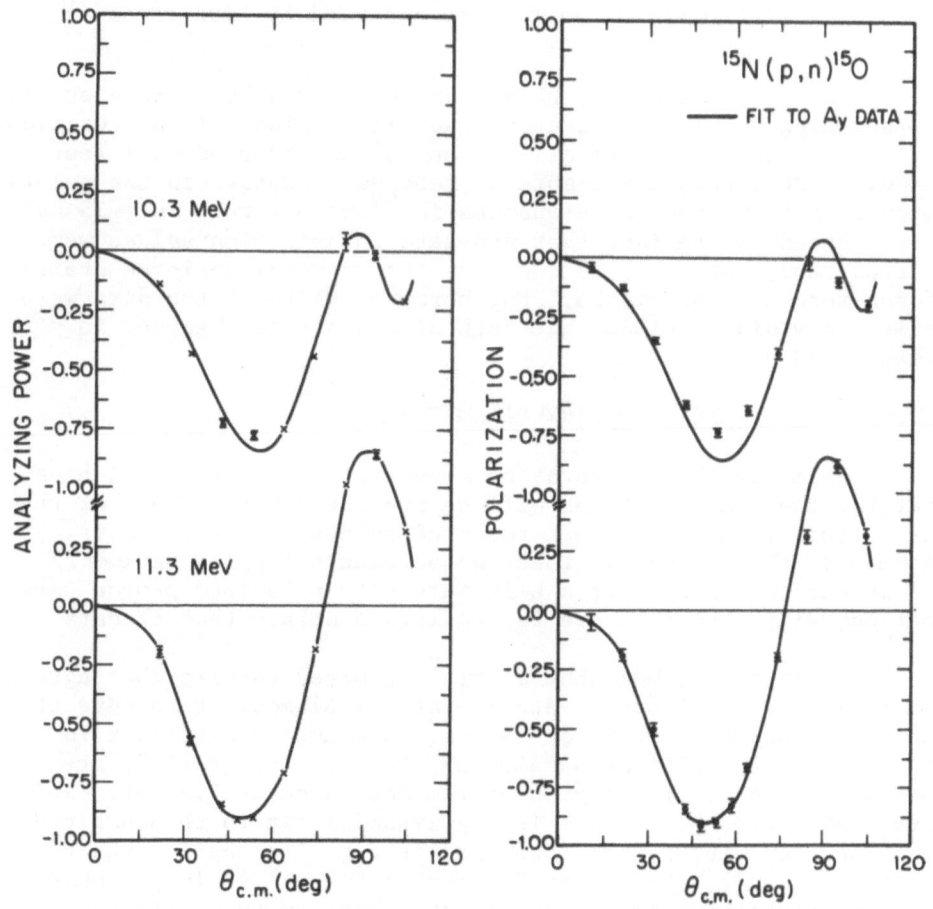

Fig. 4. Results for $P^y(\theta)$ and $A_y(\theta)$
reported by Byrd et al.[10]

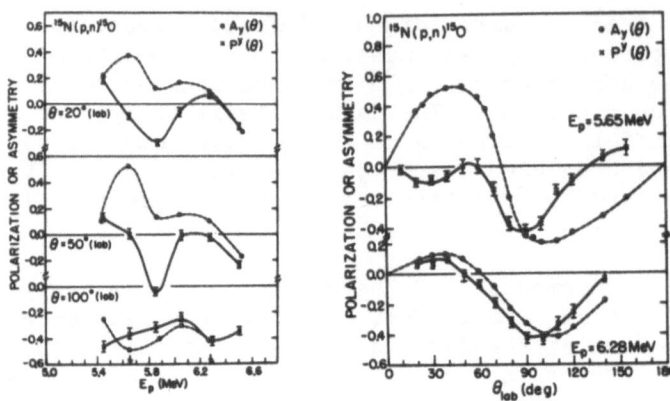

Fig. 5. Results for $P^y(\theta)$ and $A_y(\theta)$ for the
$^{15}N(p,n)^{15}O$ reaction as reported by Byrd et al.[10]

In the only low energy P^y measurement for A>16, Sexton et al. obtained high resolution data for isospin-forbidden (p,n) reactions on ^{51}V and ^{55}Mn. Although all of the values observed were less than 0.15, they were still able to resolve structure in the polarization across two analog resonances in ^{52}Cr that were only 15 keV apart. Based on the fact that non-zero polarization values were observed, they were able to conclude that several analogue states had non-zero neutron widths. Furthermore, although the data were sparse, they did conclude that both of the states observed in ^{52}Cr probably had $J^\pi=4^+$.

MEASUREMENTS OF THE ASYMMETRY PARAMETER

Paralleling the presentation above, results for the A_y parameter for the same reactions will be presented next. Most of the data in this section has been reported in the last five years. Presumably, the reason for these measurements lagging behind P^y ones is the requirement of a high intensity polarized proton beam; experimentally, the data are far easier to obtain than P^y data.

For $^3H(p,n)^3He$, Donoghue et al.[7] reported results that agreed with earlier work of Haight et al.[7] at Los Alamos. Both sets of data are accurate to ±0.01 or better, an accuracy difficult to achieve reliably in P^y measurements. The data for $A_y(45°)$ are shown in Fig. 6 for the region of the resonance at 2.2 MeV. The curve drawn through the data is a polynomial fit to the combined data. The same curve was reproduced in Fig. 1 alongside the P^y results. Clearly, the newest P^y data of Fig. 1 are in agreement with the curve; that is, $P^y=A_y$ in this reaction to within the experimental uncertainties. This result is further born out by comparing the $A_y(\theta)$ data of Donoghue et al. to the new $P^y(\theta)$ results, as is shown in Fig. 2 where the curve represents a fit to

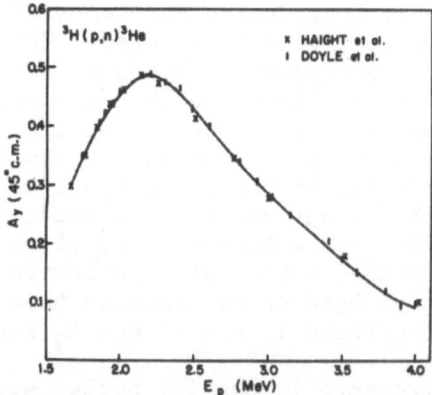

Fig. 6.   Results for A_y for ^{3}H(p,n)^{3}He as reported by Donoghue
          et al.[7]   The curve is a polynomial fit to the data.

the $A_y(\theta)$ data at 2.48 MeV.   The earlier differences between P^y
and A_y stimulated interest in finding their cause, and the suc-
ceeding calculations led Arnold[2] and Conzett[3] to propose their
postulates about charge-symmetry breaking terms in the case of
$P^y \neq A_y$.   At the present, both groups, Hale and Dodder[8] and Philpott
and Halderson[4], predict that $P^y(\theta) \approx A_y(\theta)$ for ^{3}H(p,n)^{3}He, although
both still predict a slight difference in magnitude for the two
functions.

    Additional similarities are known to exist for these parameters
in the reactions ^{7}Li(p,n)^{7}Be and ^{9}Be(p,n)^{9}B.   In the former some
slight differences may exist, but in the latter $P^y = A_y$ to within
experimental errors.[10]   This agreement for ^{9}Be has permitted Byrd
et al.[12] to pursue Lane model calculations without introducing the
complexity of target spin into the problem since a description of
differences in P^y and A_y requires an "asymmetric spin-flip", which
in turn necessitates a non-zero target spin.   (The Lane model conclusions are the topic of another paper[12] in this volume.)

    Byrd et al. also studied A_y for the ^{15}N(p,n)^{15}O reaction be-
tween 3 and 15 MeV.   The distributions for 10.3 and 11.3 MeV are
shown on the left hand side of Fig. 4.   The solid curves on the
left and the right are identical. They are polynomial fits to the
A_y data.   Here again, $P^y = A_y$ within the uncertainties.   However,
as is exhibited in Fig. 5, around 5.7 MeV a clear violation of the
rule $P^y = A_y$ exists.   (Both the A_y and the P^y measurements here were
repeated to verify this feature.)   This reaction provides the only
clear evidence of such differences in (p,n) reactions.   Just recently, Philpott and Halderson[4] studied this reaction in a microscopic continuum shell-model approach to determine if the differences

could be explained without introducing any "exotic" (charge-symmetry breaking) components of the nuclear force. They were able to obtain "good qualitative agreement" with the differences in P^y and A_y with a standard two-nucleon force which contained central, spin-orbit and tensor components. It is noteworthy that in 1956 Wilkinson[14] predicted that this energy region in the compound nucleus of ^{16}O would show the effects of isospin mixing, whereas at the lower energies, where the resonances are isolated, and at the higher energies where there is a high density of overlapping states, the effects of isospin mixing would not be observed in (p,n) reactions. Interestingly, Byrd et al. seem to have discovered a sensitive probe of such effects in the P^y and A_y measurements.

The first A_y measurements in heavier nuclei were reported in 1972 by Moss et al.[15] The experiments were performed at 25 MeV and involved five targets ranging from ^{11}B to ^{120}Sn. Their purpose was to obtain information on the isospin-dependent spin-orbit term of the Lane potential. It was expected, and proven by calculation, that the $A_y(\theta)$ function would be more sensitive to this term than cross section measurements. In fact, at the time of the experiments there was no firm evidence for the need of such a term. Using Becchetti and Greenlees proton optical potential parameters, Moss et al. calculated $A_y(\theta)$ in DWBA for Ni and Sn and concluded that the symmetry spin-orbit strength is opposite in sign compared to the standard optical model term, but that the data was too inprecise to make more quantitative assignments.

One feature of the Moss et al. calculation that was not reproduced in a later paper from the same laboratory was the very high sensitivity of $A_y(\theta)$ to changes in the strength of the spin-orbit term. That is, in similar calculations performed a few years later, Gosset et al.[16] showed results that were about an order of magnitude less sensitive to spin-orbit variations, although they did not comment on this aspect. (We have been able to reproduce the later calculations with our code at TUNL.) Gosset et al. were able to extend the experiments initiated by Moss et al. to more targets and larger angles. Half of their $A_y(\theta)$ data is reproduced in Fig. 7. The cross sections exhibited are from the Colorado laboratory.[16] These superb $A_y(\theta)$ measurements should provide the opportunity to determine the spin-orbit term, both in the macroscopic Lane model and in microscopic models. The curves in Fig. 7 are Lane model calculations of Gosset et al. with parameters from the analysis of Patterson et al.[16] of (p,n) and (p,p) data. The solid curves are DWBA calculations with a 4.3 MeV isospin spin-orbit term, while the dashed curves are the results with this term set to zero. The results are quite encouraging, considering the constraints on the parameters; the general trends in the $A_y(\theta)$ and $\sigma_{pn}(\theta)$ data seem to be fairly well represented by the calculations. Gosset et al. conclude from these fits that

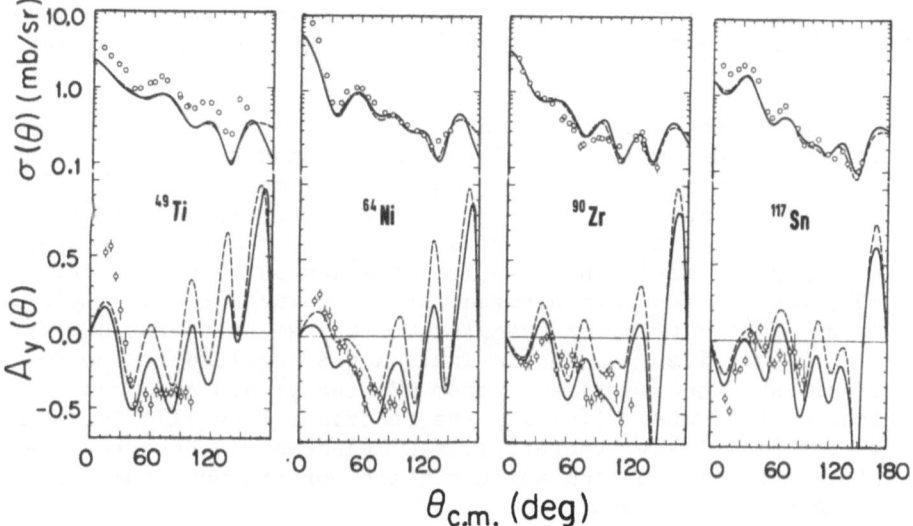

Fig. 7. Lane model fits to σ and A_y for several (p,n) reactions.[16]

the data favor a mean value of 4.3 MeV for the depth of the spin-orbit potential. It is unfortunate that this analysis has not been carried farther, since the conclusions about the success of the Lane model in describing both cross-section and polarization data simultaneously will not be known until someone explores parameter space and searches for better detailed fits. Byrd et al.[12] attempted such a multichannel search, but until now, only for the light nucleus ^{9}Be. In this work the fits were nearly as good with no spin-orbit term as they were with the optimum value of about 1 MeV.

Gosset et al.[16] also tried to describe their data with microscopic calculations which used g-matrix elements based on a Reid potential. The fits were poorer than those shown in Fig. 7. However, they did conclude that $A_y(\theta)$ is not sensitive in this model to the magnitude of the symmetry spin-orbit term. Here again, it would be interesting to see the calculations carried to a more complete conclusion.

POLARIZATION TRANSFER COEFFICIENTS

The parameters that relate the polarization state of an outgoing beam to the polarization state of an incident beam are called polarization transfer coefficients. Wolfenstein introduced the D, R, and A parameters, but these were not suitable to easily characterize the variety of conditions that one can create in a scattering situation. Currently the parameters have the notation K_i^j where i represents one of three orthonormal axes (x,y,z) defined by the incident momentum and the normal to the reaction plane. The j is similarly defined for the outgoing frame of reference, (x',y',z').

For (p,n) reactions, the incident proton beam can be longitudinally or transversely polarized, and one can study projections of the outgoing neutron polarization vector along any of the x'-, y'- or z'-axes. The only widely measured coefficient is $K_y^{y'}(\theta)$, a measure of the change in polarization of the transverse component of the beam polarization between incoming and outgoing states. For $\theta=0^\circ$, the relationship between the two polarizations simplifies to $p_n^y = K_y^{y'}(0^\circ)p_p^y$. For example, if the two polarizations are equal, $K_y^{y'}(0^\circ)=1$, i.e., no depolarization has resulted in the interaction. When one moves away from 0°, the p_n^y depends upon the polarization parameter $P^y(\theta)$ for the reaction, as well as upon $K_y^{y'}(\theta)$. For simplicity, therefore, few measurements have been made at angles other than 0° and neither these nor the non-transverse K_i^j will be discussed here.

To determine polarization transfer coefficients, one must conduct the equivalent of a triple-scattering experiment. The bulk of the data for (p,n) has appeared in the last few years, reflecting the development of intense sources of polarized proton beams. Measurements of $K_y^{y'}(0^\circ)$ have been reported by Lisowski et al.[17] for $E_p<15$ MeV for ^{9}Be, ^{11}B, and ^{13}C, and for ^{11}B by Graves et al.[18] at 21 and 26 MeV. Considerable structure exists over the entire range for ^{13}C and near 5 MeV for ^{9}Be. The experimental results are shown in Fig. 8 for ^{11}B, where a dashed curve has been drawn through the data to guide the eye. The data appears to monotonically approach a value of +0.7 above 12 MeV. In other words, the neutrons emitted at 0° reach 70% of the polarization of the incident protons.

One purpose of the above measurements is to investigate the nuclear spin-spin interaction and other depolarization mechanisms. Before the above data existed, Madsen et al.[19] studied the terms in the two-body interaction which have an important role in determining the magnitude of $K_y^{y'}(0^\circ)$ in charge-exchange transitions. The calculations used a simple effective $\vec{t}\cdot\vec{T}$ interaction with central spin-spin and tensor forces, and were done for ^{15}N, ^{11}B and ^{3}H targets. Some of their results for ^{11}B for the region between 16 and 28 MeV are illustrated by the three curves in Fig. 8: Curve I is from a simple model for L=0 orbital angular momentum transfer only; curve II is with the tensor part in the effective interaction set to zero; and curve III includes both the central spin-spin and tensor terms in the interaction. Clearly there is a strong dependence on the form of the interaction. Curiously, the ^{11}B data have an energy dependence that is opposite to the calculations, and approach the 0.7 value of the simple model.

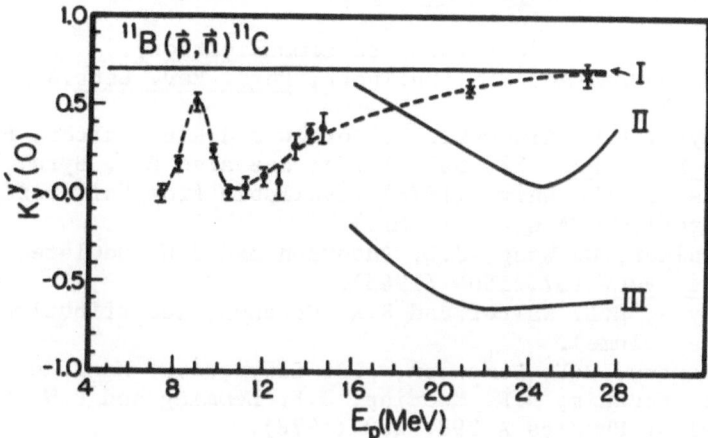

Fig. 8.   Polarization transfer coefficient $K_y^{y'}(0^o)$. Data are from Lisowski et al.[17] and Graves et al.[18] The curves are described in the text.

    In conclusion, a considerable amount of (p,n) polarization data now exists, but some of it is perhaps either too low in energy or for targets that are too light to be of much value. Although in most cases the current model calculations do not satisfactorily describe details of the data, we believe that much will be learned about spin-dependent nuclear interactions as progress is made in fitting the observed polarization quantities.

References

1. L.P. Robertson, R.C. Hanna, K. Ramarataram, D.W. Devins, T.A. Hodges, Z.J. Mowry, S.J. Hoey and D.J. Plummer, Nucl. Phys. A134, 545 (1969).
2. L.G. Arnold, Bull. Am. Phys. Soc. 22, 588 (1977); L.G. Arnold, R.G. Seyler, T.R. Donoghue, L.Brown and U. Rohrer, Phys. Rev. Lett. 32, 310 (1974).
3. H.E. Conzett, Phys. Lett. 51B, 445 (1974).
4. R.J. Philpott and D. Halderson, contribution to this volume and references therein.
5. R.L. Walter, in "Polarization Phenomena in Nuclear Reactions," H.H. Barschall and W. Haeberli, ed., Univ. of Wisconsin Press, Madison, WI (1971), p. 317.
6. R.L. Walter, in "Proc. of the Int. Conf. on the Interactions of Neutrons with Nuclei," Eric Sheldon, ed., N.T.I.S. CONF-760715-P2 (1976), p. 1061.

7. T.R. Donoghue, Sr. M.A. Doyle, H.W. Clark, L.J. Dries, J.L. Regner, W. Tornow, R.C. Byrd, P.W. Lisowski and R.L. Walter, Phys. Rev. Lett. 37, 981 (1976) and references therein.

8. G. Hale and D. Dodder (private communication).

9. Dean Halderson and R.J. Philpott, Phys. Rev. Lett.42, 36 (1978).

10. R.C. Byrd, P.W. Lisowski, W. Tornow and R.L. Walter, Bull. Am. Phys. Soc. 22, 587 (1977); see also R.C. Byrd, Ph.D. thesis, Duke Univ. (1978) (available from University Microfilms, Ann Arbor, MI.).

11. B.D. Walker, C. Wong, J.D. Anderson and J.W. McClure, Phys. Rev. 137, 1504 (1965).

12. R.C. Byrd, R.L. Walter and S.R. Cotanch, (contribution to this volume).

13. E.H. Sexton, R.E. Benenson, A.J. Elwyn, J.E. Monahan, F.T. Kuchnir, F.P. Mooring, J.F. Lemming and R.W. Finlay, Nuclear Physics A 298, 269 (1978).

14. D.H. Wilkinson, Phil. Mag. 1, 279 (1956).

15. J.M. Moss. C. Brassard, R. Vyse and J. Gosset, Phys. Rev. C6, 1698 (1972).

16. J. Gosset, B. Mayer and J.L. Escudié, Phys. Rev. C14, 878 (1975) and references therein.

17. P.W. Lisowski, R.C. Byrd, R.L. Walter and T.B. Clegg, Nucl. Phys. A259, 188 (1976).

18. R.G. Graves, J.C. Hiebert, J.M. Moss, E.P. Chamberlin, R. York and L.C. Northcliffe, in "Proc. of the Fourth Int. Symp. on Pol. Phen." W. Grüebler and V. König, ed., Birkhäuser Verlag, Basel Switzerland (1976), p. 642.

19. V.A. Madsen, J.D. Anderson and V.R. Brown, Phys. Rev. C9, 1253 (1974).

SELF-CONSISTENT APPLICATION OF THE LANE MODEL

Roger C. Byrd and Richard L. Walter
Dept. of Physics, Duke Univ., Durham, N.C. 27706
and Triangle Universities Nuclear Laboratory,
Duke Station, N.C. 27706

Steve R. Cotanch
Dept. of Physics, N.C. State Univ., Raleigh, N.C. 27607
and Triangle Universities Nuclear Laboratory

The Lane model has been an appealing and successful technique for coupling the (p,p), (n,n), and (p,n) channels in a single calculation. However, detailed description has been more difficult with the additional constraints of fitting either all three channels or the (p,n) analyzing power. In our attempts to examine the ability of the Lane model to describe a complete set of data using a new search code, we find that consistent application of the model leads us to choose an approach which we believe to be more fundamental than those usually employed in similar calculations. In this paper we will review the more significant aspects and applications of the Lane model and indicate the developments which led us to our present position. We include an application of the model to the ^{9}Be(N,N) system, for which a fairly complete set of data has been obtained at our laboratory.

As described by Lane,[1] the optical potential when written as $U = U_o + 4(t \cdot T)U_1/A$ demands isospin and multi-channel consistency on two levels. For an individual target nucleus C and its analog state A, the potentials U_{pC}, U_{nC}, and U_{nA} for elastic scattering C(p,p)C, C(n,n)C and A(n,n)A are given by the diagonal $(t_3 T_3)$ terms, while the quasi-elastic scattering C(p,n)A is generated by the off-diagonal $(t_+ T_-)$ term. These constraints require no further assumption than isospin conservation in evaluating the $(t \cdot T)$ operator. Furthermore, the assumption that the Lane derivation extends over a range of individual nuclei then leads to global systematics in both single-channel and multi-channel studies of (p,p), (n,n), or (p,n) scattering. Such global analyses must trade off detailed individual description to obtain general overall trends. The relationships between the individual analyses, such as (p,n) calculations[2] using DWBA, and global studies, such as the Perey (p,p) analysis,[3] are reviewed in the excellent 1969 article by Satchler.[4]

Individual multi-channel consistency is based on the nuclear isospin-conserving potentials obtained[4,5] from the basic $U=U_0+4(t \cdot T)U_1/A$ Lane potential which can be written as:

C(p,p)C: $U_{pC}=U_0-2TU_1/A$

C(n,n)C: $U_{nC}=U_0+2TU_1/A$

A(n,n)A: $U_{nA}=U_0+2(T-1)U_1/A$

C(p,n)A: $U_{pn}=2\sqrt{2T}\ U_1/A.$

Note the distinction between the neutron potentials U_{nC} for scattering from the target and U_{nA} for scattering from the analog state. U_{nA} is physically coupled to U_{pC} and U_{pn} by the (p,n) reaction, while U_{nC} is connected only by the isospin framework. On the other hand, since U_{nA} cannot be directly determined, the observable potentials are instead related to U_{pC}, U_{pn}, and U_{nC}. The equations imply that specification of any two potentials, by convention U_{pC} and U_{nC} or U_0 and U_1, will determine all others, so a two-channel analysis will in principle constrain the three-channel problem. However, the isospin-conserving potential U_{pC} is not directly observable because of the symmetry-breaking coulomb interaction in the p+C channel. The observable is instead the Coulomb-corrected proton potential, $U_p=U_{pC}+U_c$. Full constraint then requires a three-channel analysis to determine all three potentials, for example, U_0, U_1 and U_c. A (p.n) analysis might yield U_1, (n,n) scattering would then give U_0, and (p,p) results could determine U_c. In practice, optical potential ambiguities suggest that all channels should be analyzed together, thus giving a consistent, isospin-conserving, multi-channel analysis of an individual nucleus.

By sacrificing detailed descriptions for global systematics, the individual approach could be averaged over different nuclei to give a global multi-channel optical model. However, such analyses evolved instead as separate single-channel global studies which had to be adapted into the multi-channel Lane equations. To illustrate the resulting conflicts, we compare the (p,p) and (n,n) potentials obtained by Becchetti and Greenlees[6] (BG) to the effective nuclear-plus-coulomb Lane forms:

Becchetti-Greenlees

$$U_p^{BG}(E_p) = U_0^p + \Delta U_0(E_p) - \frac{2T}{A} U_1^p + \Delta U_c$$

$$U_n^{BG}(E_n) = U_0^n + \Delta U_0(E_n) + \frac{2T}{A} U_1^n$$

$$U_p = U_{pC} + \Delta U_c = U_0 - 2TU_1/A + \Delta U_c$$

Lane

$$U_n = U_{nC} \qquad = U_0 + 2TU_1/A.$$

Two difficulties are immediately encountered in attempts to identify corresponding terms in the above equations. First, the Coulomb correction ΔU_c and the proton symmetry potential U_1 vary similarly throughout the periodic table and might not be properly separated in global models. Second, the different functional forms, e.g., U_1^p and U_1^n, make it difficult to maintain both isospin and multi-channel consistency in a three-channel analysis.

The conventional wisdom[7] on separation of Coulomb effects suggests two approaches based on the reduced kinetic energy of the proton in the Coulomb field of the nucleus. The "potential-correction" method increases the proton potential by adding a ΔU_c term, which is nominally of strength $0.4Z/A^{1/3}$MeV in global analyses. The "energy-shift" approach reduces the proton bombarding energy E_p by $E_{pC} = E_p - \Delta E_c$, thus giving $E_p = E_n + E_c$ for equal energies $E_{pC} = E_{nC} \equiv E_n$. Data and optical potentials for $C(n,n)C$ and $C(p,p)C$ scattering are therefore given at offset energies. Note that the effective Coulomb energy shift ΔE_c between proton and neutron nucleon-target states is distinct from the well-defined Coulomb displacement energy $\Delta_c \equiv -Q_{pn}$ between target and analog states. The potential correction ΔU_c and energy shift ΔE_c are related to the optical potential energy dependence $\Delta U_0(E)$ by $\Delta U_c \approx (\Delta U_0(E)/\Delta E)\Delta E_c$.

The second difficulty with the BG potentials U_p^{BG} and U_n^{BG} results from their asymmetric form even after separation of Coulomb effects, as seen in comparing the resulting BG potentials with the corresponding Lane forms:

Becchetti-Greenlees	Lane
$U_p^{BG} = U_0^p + \Delta U_0 - 2TU_1^p/A$	$U_{pC} = U_0 - 2TU_1/A$
$U_n^{BG} = U_0^n + \Delta U_0 + 2TU_1^n/A$	$U_{nC} = U_0 + 2TU_1/A.$

It is desirable to preserve the global potentials U_p^{BG} and U_n^{BG} while obtaining the other potentials needed for multi-channel and analyses. For example, the potentials U_{pC}, U_{nA}, and U_{pn} are required in DWBA calculations, although some early studies[8] may have instead used U_{pC}, U_{nC}, and U_{pn}. To identify the required U_1 from the U_p^{BG} and U_p^{BG} potentials, one approach[9] averages the potential parameters for U_1^p and U_1^n, e.g., $\bar{r} = (r^p + r^n)/2$, thus obtaining an apparently symmetric potential set of U_p, U_n, and $\bar{U}_1$. However, as was pointed out in 1973 by Cotanch and Robson,[10] these three independent potentials are not consistent with the Lane equations and hence do not conserve isospin. Use of the averaged potentials $\bar{U}_0$ and $\bar{U}_1$ alone would do so, but at the expense of consistency with the original U_p^{BG} and U_n^{BG} potentials.

One solution to the dilemma of conserving isospin while preserving the global potentials was suggested by Hoffman[11] in 1973. The U_{pC} and U_{nC} Lane potentials are directly identified with the global potentials U_p^{BG} and U_n^{BG}. These potentials are thus taken as independent variables, and U_0, U_1, and all other potentials are predicted from the inverted Lane equations:

$$U_0 = (U_{nC} + U_{pC})/2 \qquad\qquad U_1 = (U_{nC} - U_{pC})A/4T.$$

This solution avoids the ambiguities of separating the various terms in the U_p and U_n global potentials and suggests a basic technique for transformation of any potential. The different geometries of U_{pC} and U_{nC} result in decidedly ambiguous parameters for the derived potentials, especially when evaluated at different energies as in energy-shift Coulomb correction. Hoffman accepts this result as a necessary consequence of multi-channel consistency and transforms the radial potentials themselves point-by-point. All potentials other than the original ones then have non-Woods-Saxon shapes.

The BG analysis does not generally provide satisfactory (p,n) descriptions since no (p,n) results were included to constrain the symmetry potential. To combine good global elastic predictions with (p,n) searching capability, modified global DWBA analyses were presented by Carlson et al.[12] in 1975 and Patterson et al.[13] in 1976. The former approach fixed U_p at the BG values, varied U_1 to describe (p,n) cross sections $\sigma_{pn}(\theta)$, derived a self-consistent U_n from U_p and the new U_1, and then recalculated $\sigma_{pn}(\theta)$ and $\sigma_{nn}(\theta)$. This approach is isospin-consistent but forces all the departure from the BG parameters into the neutron channel. Patterson et al. instead systematically adjusted both U_p and U_n away from their BG values while consistently varying U_1 to fit $\sigma_{pn}(\theta)$. Both analyses yielded reliable global U_1 potentials and generated U_{nC} potentials which satisfactorily predicted (n,n) results. By modifying the BG potentials, fitting $\sigma_{pn}(\theta)$, and predicting $\sigma_{nn}(\theta)$ for a range of nuclei and energies, the Patterson et al. approach provides strong support for a consistent global multi-channel Lane potential.

Since these global successes, difficulties have been encountered in Lane model attempts to describe detailed results. In the first global study of (p,n) analyzing powers $A_{pn}(\theta)$ by Gosset et al.[14] in 1976, somewhat unsatisfactory descriptions were obtained despite the use of the successful parameters of Patterson et al., the addition of a spin-orbit symmetry interaction $V_{s.o.}^1$ $(\vec{\ell}\cdot\vec{s})(\vec{t}\cdot\vec{T})$, and additional searching on $A_{pn}(\theta)$ and $\sigma_{pn}(\theta)$. Although (p,n) polarization observables are likely to be very sensitive to the characteristics of individual nuclei, their poor global description is nevertheless a serious deficiency of the Lane model.

Further problems were encountered by Schery et al.[15] in a 1974 extension of the approach of Carlson et al. to heavier nuclei. In an attempt to improve unsatisfacory (n,n) predictions, these authors relaxed global constraints and individual searches were made on the (p,p) and (p,n) channels. Surprisingly, improved fits to the included data only worsened the (n,n) predictions. These results were confirmed in a 1976 study by Lovas,[16] who found that simultaneous two-channel searching on either (p,p) and (n,n) or on (p,p) and (p,n) inevitably worsened the third channel prediction. The conclusion drawn was that the Lane model is valid only at the global level and is incorrect when applied to an individual target.

We have drawn heavily on the experience on these previous analyses in designing a program of Lane model studies at our laboratory. Our basic assumption, which was encouraged by the success of global cross-section studies, is that multi-channel isospin consistency is inherently valid for individual nuclei. It does not seem reasonable that the isospin constraints can be valid globally but generally invalid in the more basic individual studies on which global results are built. As explained in reference to the individual multi-channel problem, the ambiguities of optical potentials may make success in two-channel predictions of the third reaction somewhat fortituous. The previous analyses have not simultaneously searched to describe all three channels, nor have they included complete polarization constraints. Such tests are essential before validity in individual cases can be dismissed. Our immediate goal is thus to consistently describe a complete set of data in all three channels with a quality of agreement comparable to that of single-channel analyses.

The code TWAVE[10] has been expanded to emphasize these philosophies. Multi-channel completeness is achieved using a flexible search routine capable of handling all potential parameters and all related cross-section and polarization observables. Isospin conservation is guaranteed by coupled-channels calculations in which any two independent potentials can be specified and the remainder are internally generated. Coulomb correction, form-factor, and spin-orbit symmetry assumptions are addressed with appropriate options.

We have introduced a useful concept to handle the form factor ambiguity demanded by multi-channel isospin consistency. In using the Hoffman prescription to generate radial form factors point-by-point, only two potentials are independent and have characteristic Woods-Saxon shapes. We refer to these chosen potentials as a representation, e.g., specifying the (U_{pC}, U_{nC}) representation requires that U_{nA}, U_{pn}, U_0, and U_1, as generated by the Lane equations, will inevitably have more complicated shapes. In choosing between representations two considerations dominate, one pragmatic and one theoretical. Since (p,n) results are generally more sensitive than elastic scattering ones, direct inclusion of U_1 potentials simplifies data fitting. A global potential is often convenient for describing a particular channel, usually that for proton scattering. Practical considerations have therefore usually designated (U_p, U_1) as the most convenient representation. Theoretical support for a particular representation is obtained from microscopic calculations[17] of optical model behavior, which have generally yielded Woods-Saxon shapes for U_{pC} and U_{nC}. The potential U_0, essentially the average of U_{nC} and U_{pC}, is then also roughly Woods-Saxon; U_1, derived from their difference, changes from surface-peaked in light nuclei to a surface-plus-volume shape in heavy nuclei. The crucial point is that in any representation the resulting U_1 should have roughly the above behavior, since U_0, U_{nC}, U_{nA}, and U_{pC} will always have approximately Woods-Saxon shapes.

We briefly illustrate our approach for the ^{9}Be nucleus, chosen for its complete, consistent data set in the 11 to 15 MeV region. Smooth behavior of observables with energy and previous theoretical analyses[18] support optical model validity in all channels. Working in the (U_p, U_1) representation with starting proton potentials from the literature, after considerable searching a smooth parameter set was obtained with the reasonable values shown in Fig. 1. The simultaneous predictions of the complete data set are shown in Fig. 2; highlights are appropriate energy dependence in $\sigma_{pn}(\theta)$, correct magnitudes for $\sigma_{nn}(\theta)$, and satisfying predictions of both $A_{pp}(\theta)$ and $A_{pn}(\theta)$. Although ^{9}Be is certainly not the best nucleus for this type of study, the complete data set allows all five types of data to be calculated with one potential.

Although no conclusions can be drawn concerning Coulomb corrections for this low-Z target, alternate potential assumptions for V_1 form factors, spin-orbit symmetry terms, and energy dependence have been considered. The solution used in Fig. 2 employs a volume V_1 shape, yet surface-peaked V_1 potentials obtained comparable results for all except the $A_{pn}(\theta)$ distributions. While at most only a small symmetry spin-orbit potential ($\sim$1 MeV) is necessary when using a volume V_1 shape, the surface-peaked solution requires a $V_1^{s.o.}$ strength of 3.5 MeV for even minimal agreement with the $A_{pn}(\theta)$ data. Finally, the energy dependence in the proton channel is not avoidable, since constant-geometry solutions for this target deteriorate over energies of a few MeV.

We conclude with some comments addressed specifically to previous difficulties with detailed polarization and multi-channel descriptions and more generally to the larger problem of Lane model consistency. We have extended the constraints of previous analyses by the inclusion of polarization results and a simultaneous search on all three channels. In point of fact, parameters obtained from a two-channel search on both cross-section and polarization data

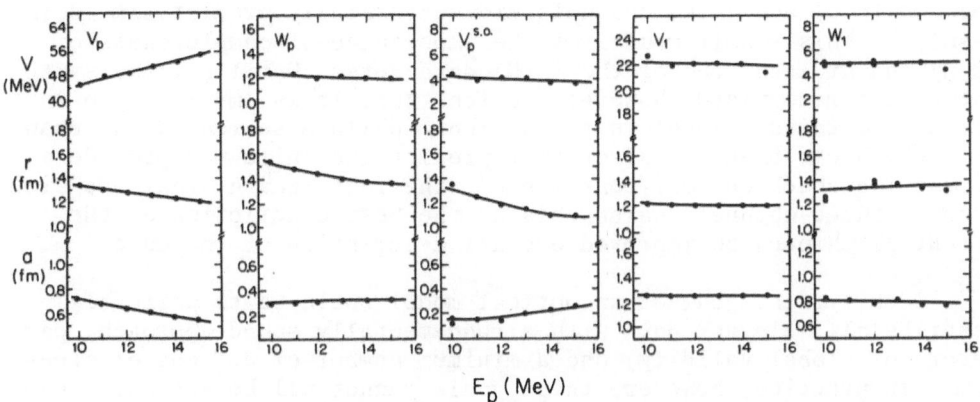

Fig. 1.   Energy dependence of Lane potential parameters.

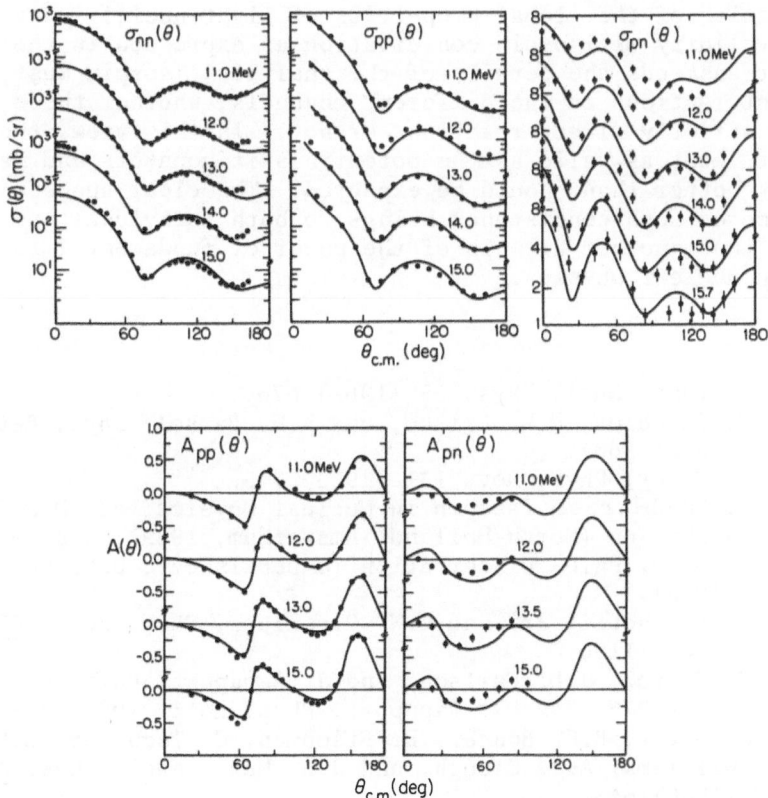

Fig. 2.   Simultaneous Lane model description of ^{9}Be + nucleon data.

for the (p,p) and (p,n) channels did successfully predict the ^{9}Be(n,n) results. This result reaffirms the importance of completeness of the (p,n) and at least one of the (N,N) data sets. Until all the systematics are recognized, however, we feel that it is important to include the third channel in the search. While a search on two channels which can then satisfactorily predict the third may provide a reassuring check on self-consistency, the fact remains that the simultaneous three-channel search obtains the best description of the potential through an improved overall description of the data.

Ideally, the goal of an optical model would be to perfectly describe all relevant data with a fundamentally sound approach, parameters of global validity, and a minimum number of degrees of freedom. In practice, however, these goals cannot all be satisfied and priorities must be set. We have insisted first on maximum consistency with fundamental isospin symmetry and then weighed the other criteria more or less equally. We recognize that use of optical potentials for the ^{9}Be + nucleon system is perhaps more a means of conveniently describing an individual effective interaction than a true reflection of the global properties of light nuclei. Nevertheless, the validity of isospin conservation as expressed in the Lane equations transcends the details of the analysis; isospin must connect the interactions in the different channels, whether these processes are strictly direct reactions or not. In this view, the Lane model is at least as valid as the potentials it connects and may indeed work better than should be expected. The clear success that the reliance on such consistency brings to both individual and global studies is thus another example of the power of fundamental isospin symmetry in nuclear physics.

References

1.   A.M. Lane, Nucl. Phys. 35 (1962) 676.
2.   G.R. Satchler, R.M. Drisko, and R.H. Bassel, Phys. Rev. 136 (1964) B637.
3.   F.G. Perey, Phys. Rev. 131 (1963) 745.
4.   G.R. Satchler, in Isospin in Nuclear Physics, ed. D.H. Wilkinson (North-Holland, Amsterdam, 1969) Chap. 9.
5.   R.C. Byrd, Ph.D. Dissertation (unpublished), Duke University (1978).
6.   F.D. Becchetti, Jr. and G.W. Greenlees, Phys. Rev. 182 (1969) 1190.
7.   J.C. Ferrer, J.D. Carlson, and J. Rapaport, Phys. Lett. 62B (1976) 399; J. Rapaport, Phys. Lett. 70B (1977) 141.
8.   C.J. Batty, B.E. Bonner, E. Friedman, C. Tschaler, L.E. Williams, A.S. Clough, and J.B. Hunt, Nucl. Phys. A116 (1968) 643.
9.   C. Wong, J.D. Anderson, J.W. McClure, B.A. Pohl, and J.J. Wesolowski, Phys. Rev. C5 (1972) 158.
10. S. Cotanch and D. Robson, Phys. Rev. C7 (1973) 1714.

11. G.W. Hoffman, Phys. Rev. C8 (1973) 761.
12. J.D. Carlson, C.D. Zafiratos, and D.A. Lind, Nucl. Phys. A249 (1975) 29.
13. D.M. Patterson, R.R. Doering, and A. Galonsky, Nucl. Phys. A263 (1976) 261.
14. J. Gosset, B. Mayer, and J.L. Escudié, Phys. Rev. C14 (1975) 878.
15. S.D. Schery, D.A. Lind, H.W. Fielding, and C.D. Zafiratos, Nucl. Phys. A234 (1974) 109.
16. R.G. Lovas, Nucl. Phys. A262 (1976) 356.
17. C.B. Dover and Nguyen van Giai, Nucl. Phys. A190 (1972) 373.
18. M.F. Werby, S. Edwards, and W.J. Thompson, Nucl. Phys. A169 (1971) 81; H.J. Votava, T.B. Clegg, E.J. Ludwig and W.J. Thompson, Nucl. Phys. A204 (1973) 529; H.H. Hogue, Ph.D. Dissertation (unpublished), Duke University (1977).

DISCUSSION

Baer: Have you been able to describe K_y^y results?

Byrd: In this model K_y^y takes on the trivial value of 1/2 since there is no target spin in the optical mode. This is similar to our inability to calculate P=A.

Madsen: What is the effect of coupling on the (pp) and (p,n) channels?

Byrd: The DWBA approach reproduces these results very well. The only appreciably sensitive region is for back-angle Apn(θ).

Bloom: What about averaging over resonances?

Byrd: We have simply drawn a straight line through the parameters obtained at discrete energies. The results have been quite good, especially for the (p,n) analyzing powers, which seem to be less sensitive to resonances than the cross sections.

Wong: Do you have similar sucess in fitting the ^{15}N(p,n) and ^{13}C(p,n) ground state data?

Byrd: The ^{9}Be results are non-resonant -- this is not true of ^{13}C and ^{15}N. We have attempted to describe the ^{15}N results and obtain at least a description of the optical model background beneath the resonances. In simultaneous searches on (p,p) cross sections and polarizations, as well as our (p,n) $\sigma(\theta)$ and A(θ) results, we obtain the overall characteristics of the data. The best we have been able to do thus far is to average parameters to get the average background under the resonances. We are looking at ways to handle the resonances in a consistent manner.

THEORETICAL STUDIES OF THE POLARIZATION-ANALYZING POWER

DIFFERENCE IN ^{3}H(p,n)^{3}He AND ^{15}N(p,n)^{15}O*

R. J. Philpott and D. Halderson

Department of Physics
Florida State University
Tallahassee, Florida 32306

INTRODUCTION

Differences between the polarization (P) and the analyzing power (A) in (p,n) reactions connecting corresponding states in mirror nuclei have been of particular interest since it was pointed out by Haight et al.[1] that such differences can arise only as a result of charge dependent effects. Comparison[1] of early measurements of P and of A in the ^{3}H(p,n)^{3}He reaction at first suggested the existence of a small but significant (~17%) difference between P and A in the energy region E_p ~ 1.7 to 4 MeV. Subsequent remeasurement of both P and A, however, has shown[2,3] that there is actually no difference to within experimental uncertainty. In sharp contrast to this result, sizable P-A differences have recently been observed[3,4] in the ^{15}N(p,n)^{15}O reaction, extending over the energy region E_p ~ 4.5 to 9.5 MeV. The observed P-A difference is a rapidly fluctuating function of the incident energy, and appears to be intimately associated with some of the resonance structures which are prevalent in this energy region.

While there have been some theoretical attempts to determine the mechanisms underlying the observed effects, progress has largely been limited to an elaboration of the consequences of the known conservation laws. Thus, Conzett[5] has formalized the contention that P and A must be equal in the absence of an isospin symmetry breaking component in the interactions responsible for the reaction. Arnold[6] has shown that a non-vanishing P-A difference requires, in addition, the presence of transverse spin-flip

* Research supported in part by the National Science Foundation.

transitions which yield a spin-flip asymmetry. A possible impli-
cation of this two-fold requirement is that the existence of large
P-A differences might reflect the presence of some exotic component in the nuclear interaction which is simultaneously spin dependent and non-scalar in isospin space.

   In order to clarify the situation, we have performed continuum shell model calculations for the two reactions mentioned
above. The one-particle one-hole (1p-1h) approximation is employed[7] and the center of mass is treated correctly.[8] We note
that this is perhaps the simplest model which is capable of em-
bracing the various essential features of the problem. The model
is microscopic, so that the results relate back to the nucleon-
nucleon interaction itself. The model has a multichannel capability which is required in order to encompass both the (p,n)
reaction and the asymmetric spin-flip requirement (non-zero target
spin). In addition, the model is able to handle non-central interactions and isospin mixing.

   It is the purpose of this paper to present the calculated
results and to comment further on the mechanisms which lead to
appreciable P-A differences. We also seek to explain the marked
qualitative difference which is found in the observed effects in
the ^{3}H(p,n)^{3}He and ^{15}N(p,n)^{15}O reactions. An adequate understanding of this qualitative difference is needed in order to help
one to predict whether P-A differences are likely to be observable
in (p,n) reactions between other mirror nuclei. It is perhaps
worthwhile to state at the outset that all calculations reported
here are based on standard representations of the nucleon-nucleon
interaction: we find no need for exotic components in the nuclear
force.

CALCULATIONS

   As mentioned above, the calculations employ the recoil corrected continuum shell model in the 1p-1h continuum approximation.
All interactions are calculated directly from an assumed form of
the nucleon-nucleon interaction. The two-body interaction is
represented by standard relative g-matrix elements, a potential
prescription for which has been given by Bertsch et al.[9] It includes central, spin-orbit and tensor components which conserve
isospin and a standard Coulomb term. This potential prescription
has been employed with considerable success in a variety of reaction studies involving heavier nuclei. Extensive CSM calculations[7] for the A=4 system have also yielded excellent agreement
with a variety of data.

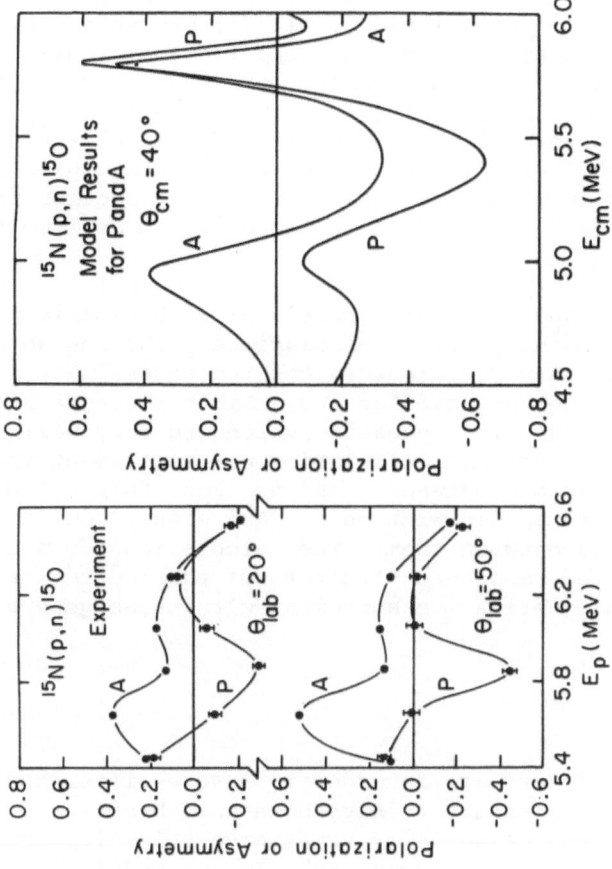

Fig. 2. Calculated polarization (P) and analyzing power (A) for the $^{15}N(p,n)^{15}O$ reaction compared with experiment.[4]

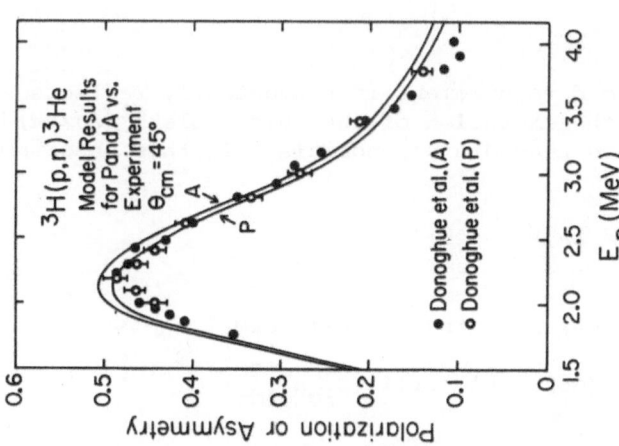

Fig. 1. Calculated polarization (P) and analyzing power (A) for the $^{3}H(p,n)^{3}He$ reaction compared with experiment.[2]

Two minor modifications were made to the two-body potentials used for the $^{15}N(p,n)^{15}O$ calculations, viz (i) the spin-orbit odd strength was increased by 15% and the triplet-even potential was decreased by about 10%. The adjustments were made in order to improve the threshold energy of the inelastic channels $(p_{1/2}^{-1} \rightarrow p_{3/2}^{-1})$ and to improve the general location of the 1p-1h levels in ^{16}O relative to the elastic threshold energy. These modifications are well within the known uncertainties of the interaction potentials and in any case do not affect the conclusions derived from the present study.

Figures 1 and 2 show typical comparisons with experiment for P and A obtained from the present calculations. The character of these observables is very different in the two cases. For $^{3}H(p,n)^{3}He$, P and A are very similar and exhibit a single broad peak close to threshold. This peak is reproduced very well by the present calculations. Neither calculation nor experiment indicate any significant difference between P and A. For $^{15}N(p,n)^{15}O$, P and A both vary quite rapidly with energy and significant differences are observed between them. The calculations do not reproduce the detailed resonance structure of this energy region, but they are in good qualitative agreement with these general trends.

DISCUSSION

As was mentioned earlier, in order to have a difference between P and A, it is necessary to have both some breaking of charge symmetry and the presence of transverse spin-flip transitions which yield a spin-flip asymmetry. In our model, the charge symmetry is broken only by the Coulomb interaction.

P-A Splitting Strengths

The above two-fold requirement is conveniently reflected in the coefficients of the expansion of the observable $k^2(d\sigma/d\Omega)(P-A)$ in terms of associated Legendre polynomials. If this observable is written

$$k^2(d\sigma/d\Omega)(P-A) = \sum_L A_L P_L^1(\cos\theta)$$

then the coefficients A_L have the general form[10]

$$A_L = \sum_{\alpha\beta J\alpha'\beta'J'} C(\alpha\beta J;\alpha'\beta'J';L) \operatorname{Im}(S_{\beta\alpha}^J A_{\beta'\alpha'}^{J'*})$$

where αJ and $\alpha'J'$ and, similarly, βJ and $\beta'J'$ are proton or neutron channel labels, $C(\alpha\beta J;\alpha'\beta'J';L)$ is a geometrical coefficient and

$$S^J_{\beta\alpha} = \tfrac{1}{2}(T^J_{\beta\alpha} + T^J_{\alpha\beta}) \qquad A^J_{\beta\alpha} = \tfrac{1}{2}(T^J_{\beta\alpha} - T^J_{\alpha\beta})$$

are linear combinations of T-matrix elements for the (p,n) reaction. This expansion shows that any P-A differences must arise as a direct consequence of the non-vanishing of the $A^J_{\beta\alpha}$'s. We shall refer to these quantities as "P-A splitting strengths". If either of the requirements for non-vanishing P-A is not met, then the splitting strengths vanish identically.

This representation of the coefficients A_L shows that P-A differences (splittings) are generated by a relatively small number of splitting strengths. For $^{15}N(p,n)^{15}O$, for example, with $\ell \leq 3$ in all channels, the possible strengths are

$$A^{1-}_{s_{1/2}d_{3/2}} \, , \quad A^{2-}_{d_{3/2}d_{5/2}} \, , \quad A^{1+}_{p_{1/2}p_{3/2}} \, , \quad A^{2+}_{p_{3/2}f_{5/2}} \, , \quad A^{3+}_{f_{5/2}f_{7/2}} \, .$$

Of these strengths, only the first two contribute significantly to our calculated results.

Association with Resonant Structures

The 1^- and 2^- splitting strengths obtained in our $^{15}N(p,n)^{15}O$ calculations are shown in Fig. 3 together with the dominant contributions to the calculated (p,n) reaction cross section. It is clear that there is a correspondence between the resonant structures seen in these two quantities: but the relationship between them is rather subtle. Differences in the relative peak positions and magnitudes arise not only because the T-matrix elements contribute differently to the two quantities but also because there are several overlapping resonances in each J^π. The calculated 3^- resonance near 5.8 MeV provides an example of a resonance which strongly affects both P and A but which itself has no inherent P-A splitting strength.

Role of Non-Central Forces

Let the orbital, spin, and total angular momenta of the target and residual nuclei be described by $L_A S_A J_A$ and $L_B S_B J_B$. Then it is possible to transform the T-matrix of the (p,n) reaction from our initial ℓj channel coupling scheme to an L-S coupling scheme and further to represent the T-matrix elements in the L-S scheme as linear superpositions of contributions of rank K in the two

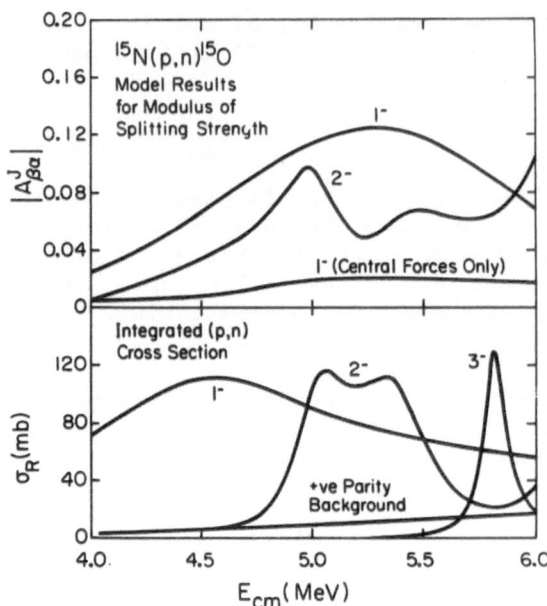

Fig. 3. Calculated P-A splitting strength moduli and various con-
         tributions to the integrated (p,n) cross section for the
         reaction $^{15}N(p,n)^{15}O$.

separate (L and S) spaces. The structure of the L-S coupled T-
matrix element can be visualized as follows.

$$\left\langle \begin{array}{c} \ell_\beta \quad S_B \\ L_B \quad L_\beta \quad S_\beta \quad \frac{1}{2} \\ J \end{array} \middle| \middle| T \middle| \middle| \begin{array}{c} \ell_\alpha \quad S_A \\ L_A \quad L_\alpha \quad S_\alpha \quad \frac{1}{2} \\ J \end{array} \right\rangle = \sum_K [K] W(JS_\beta L_\alpha K; L_\beta S_\alpha) T^K_{\overline{\beta}\overline{\alpha}}$$

with

$$\overline{\alpha} \equiv L_A \ell_\alpha L_\alpha S_A \tfrac{1}{2} S_\alpha, \qquad \overline{\beta} \equiv L_B \ell_\beta L_\beta S_B \tfrac{1}{2} S_\beta.$$

According to the Wigner-Eckart theorem, the T^K's are all zero
for $K \neq 0$ unless the nucleon-nucleon interaction contains an explicit
non-scalar spin dependence, i.e. a spin-orbit or a tensor component. In the absence of non-central forces, the triangle rules in
the Racah coefficient on the right, with K=0, require $L_\alpha = L_\beta$ and
$S_\alpha = S_\beta$. Now, in order to have non-vanishing splitting strengths
in the L-S coupling scheme, it is necessary to be able to find
sets of quantum numbers such that $\overline{\alpha} \neq \overline{\beta}$. Since $L_A S_A J_A$ and $L_B S_B J_B$
are equal by virtue of the choice of reaction, for K=0, the only

possibility for $\overline{\alpha} \neq \overline{\beta}$ is $\ell_\alpha \neq \ell_\beta$. For the $^{15}N(p,n)^{15}O$ reaction, possible non-vanishing strengths with $\ell \leq 3$ are the 1^- and 2^+ strengths, of which only the 1^- strength is significant in our calculations. For $^3H(p,n)^3He$, there are no non-vanishing strengths for K=0 because the triangle $L_A \ell_\alpha L_\alpha$ with $L_A = 0$ yields $\ell_\alpha = L_\alpha$. Similarly, ℓ_β and L_β must also be equal. Hence $L_\alpha = L_\beta$ forces $\ell_\alpha = \ell_\beta$.

Thus, the difference in the target orbital angular momenta, $L_A = 0$ for ^3H and ^3He and $L_A = 1$ for ^{15}N and ^{15}O, could provide a reason for the qualitative difference observed between the two reactions: the splitting in $^{15}N(p,n)^{15}O$ could arise simply as a consequence of the dominant central forces. In fact, however, our calculations suggest that this is not the important difference. There are two reasons. Firstly, our original calculations with central, spin-orbit and tensor forces indicate that the 2^- splitting in $^{15}N(p,n)^{15}O$ is just as important as the 1^- splitting. The 2^- splitting does not exist in the absence of the non-central components. Secondly, other calculations in which central forces alone were employed yielded only very weak P-A splittings, see Fig. 3 above. Therefore, we conclude that, as well as the Coulomb interaction, a non-central term in the nucleon-nucleon interaction is a necessary ingredient to obtain a qualitatively reasonable result.

Reasons for the Qualitative Difference in the P-A Splittings

The interactions employed in the $^3H(p,n)^3He$ and $^{15}N(p,n)^{15}O$ calculations are essentially the same, yet the P-A splittings are distinctly different for the two reactions. There are several contributing reasons for this qualitative difference. Firstly, the different positions in the nuclear mass table imply that different quantum numbers are involved. Thus, the ^{16}O system is characterized by s and d-wave resonances whereas the ^4He system is characterized by p-wave resonances. The penetrations of the d-waves are smaller than those of the p and s waves. The existence of p orbits in the mass 15 target states may help to enhance the influence of the non-central forces even though they do not yield large P-A splittings with central forces alone. Secondly, the increased complexity of the A=16 system allows closed channels to exist which have no counterpart in the A=4 system. The consequent fragmentation of the states which can decay into the open channels yields resonances whose widths can be much narrower than they would be if the closed channels were not available. Thirdly, because there are more protons in ^{16}O than in ^4He, the Coulomb interaction has a stronger influence in the A=16 system. This difference is manifest in the Q values of the reactions (-0.76 and -3.54 MeV, respectively) where it affects penetrations and

"external" isospin mixing[11] and possibly also in the extent of
"internal" isospin mixing in the states of the compound nucleus.

In Tables 1 and 2, we list some of the properties of the
R-matrix[12,13] levels of appropriate spin and parity which lie close
to the respective energy regions of Figs. 1 and 2. The 1^- splitting strength in ^{4}He is dominated by the three overlapping levels
of Table 1. (The 1^- spurious state is identically a null state[7]
in these calculations.) These three levels have isospin T as shown
and isospin mixing P(T') of a few percent. Although the isospin
mixing in ^{4}He is comparable to the mixings in ^{16}O shown in Table 2
(with exception of 2^- levels #6 and 7) the corresponding P-A splitting strength is two orders of magnitude smaller. However, if one
or two of the ^{4}He levels is arbitrarily deleted from the calculation, a significant splitting strength is obtained. This result
can be understood as follows. The three 1^- levels in ^{4}He together
exhaust the entire 1^- strength of $1\,\hbar\omega$ states and so produce no
splitting on average. In order to obtain significant splitting,
it is necessary for the levels to be not only partially mixed
with respect to isospin, but also sufficiently well separated in
energy. A similar increase in the 1^- splitting strength can be
obtained by arbitrarily reducing the **reduced** widths of the contributing resonance levels.

In contrast to those in ^{4}He, the ^{16}O levels in Table 2 do not
exhaust the $1\,\hbar\omega$ strength corresponding to their respective spins
and parities and their widths are considerably smaller. Two of
the 2^- levels have extremely large isospin admixtures. Inasmuch
as the amplitudes of these states are concentrated in either the
proton or the neutron channels but not both, the strong isospin
mixing does not in itself guarantee that these states dominate the
2^- splitting strength.

The 1^- splitting strength in ^{16}O is quite sensitive to differences in the channel wavefunctions associated with the Q-value
of the reaction. If the proton threshold is artificially raised
by 3 MeV to bring the Q value closer to that of the ^{3}H$(p,n)^3$He
reaction, the 1^- peak of the splitting strength is raised in
energy and reduced in magnitude by 70%. Another large reduction
occurs if the inelastic channels are dropped. Similarly, the 2^-
splitting strength is considerably reduced in magnitude when these
two modifications are made. Despite these reductions in strength,
the ^{16}O calculation still yields a measurable P-A splitting even
when the threshold and channel structure have been made as similar
as possible to those in ^{4}He. The remaining splitting strength is
associated with $d_{3/2}$ particle resonances whose widths are still
small enough to provide the necessary departures from a statistical
average of $1\,\hbar\omega$ splitting strength.

Table 1. Some Calculated Resonance Levels in ^{4}He. Level energies are cm energies relative to the proton threshold.

Spin-Parity	Level #	E_μ (MeV)	E_R^a (MeV)	Γ_R^a (MeV)	T	P(T') %
1^-	1	3.52	3.57	3.82	1	3.3
	2	4.06	4.23	4.66	1	0.9
	3	4.58	4.95	4.88	0	4.1

^{a}Resonance parameters extracted with single level formulae.[14] Interpret with caution.

Table 2. Some Calculated Resonance Levels in ^{16}O. Level energies are cm energies relative to the proton threshold.

Spin-Parity	Level #	E_μ (MeV)	E_R^a (MeV)	Γ_R^a (MeV)	T	P(T') %	Structure	
1^-	3	3.12	4.87	1.76	0	1.4	mixed	$s_{1/2}$
	4	4.77	6.64	0.75	1	1.3	inelastic	$s_{1/2}$
	5	6.33	5.77	2.04	1	2.6	elastic	$d_{3/2}$
2^-	4	3.28	5.16	0.94	1	7.0	inelastic	$s_{1/2}$
	5	4.08	5.30	1.17	0	5.4	inelastic	$s_{1/2}$
	6	5.81	5.84	1.06	1	45.6	proton	$d_{3/2}$
	7	6.19	5.41	0.64	0	49.7	neutron	$d_{3/2}$

^{a}Resonance parameters extracted with single level formulae.[14] Interpret with caution.

CLOSING COMMENTS

In the foregoing, we have employed a simple continuum model in order to clarify the relationship between the occurrence of P-A differences in (p,n) reactions connecting mirror nuclei and the underlying nuclear dynamics. Several interesting facets of this relationship have emerged. In particular, the requirement that there be both an isospin symmetry breaking component in the inter-action[5] and a transverse spin-flip mechanism yielding a spin-flip asymmetry[6] can be met by standard Coulomb and non-central (spin-orbit and tensor) forces. Differences in the manifestation of the

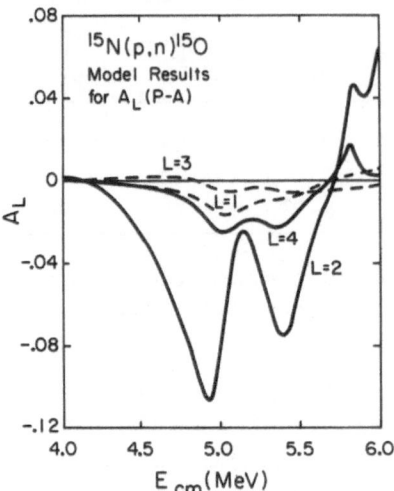

Fig. 4. Calculated coefficients A_L obtained from the Legendre
expansion of $k^2(d\sigma/d\Omega)$ (P-A) for the reaction $^{15}N(p,n)^{15}O$.

P-A splitting effect in different nuclei can be traced to specific
differences in the structure of these nuclei.

    Another aspect of the phenomenon which is immediately suggested
by the above analysis in terms of a small number of splitting
strengths is its possible use as a spectroscopic tool. In Fig. 4,
we show the coefficients A_L obtained from the $^{15}N(p,n)^{15}O$ calculation. The dominance of a single parity in the model calculations
(seen also in Fig. 3) is reflected in the relatively large size of
the A_L's with even L. Experimentally, the resonance structure in
the energy region where P-A differences are observed contains contributions from both parities. Interferences between resonances
of opposite parity can be expected to lead to increased structure
in the A_L's with odd L. More speculatively, if a narrow resonance
which itself has no splitting strength occurs in an energy region
dominated by slowing varying splitting strengths, the A_L's are
linear in the resonating amplitudes. Under suitable circumstances,
therefore, one might expect to obtain a characteristic signature
which would enable the spin and parity of the resonance to be
determined.

REFERENCES

1. R. C. Haight, J. J. Jarmer, J. E. Simmons, J. C. Martin, and T. R. Donoghue, Phys. Rev. Lett. 28:1587 (1972).
2. T. R. Donoghue, Sr. M. A. Doyle, H. W. Clark, L. J. Dries, J. L. Regner, W. Tornow, R. C. Byrd, P. W. Lisowski, and R. L. Walter, Phys. Rev. Lett. 37:981 (1976).
3. R. C. Byrd, Ph.D. Thesis (Duke University, 1978), unpublished.
4. R. L. Walter and P. W. Lisowski, Proc. Internat. Conf. on the Interaction of Neutrons with Nuclei, E. Sheldon, ed., (USERDA, Washington, 1976), p. 1061.
5. H. E. Conzett, Phys. Lett. 51B:445 (1974).
6. L. G. Arnold, Bull. Am. Phys. Soc. 22:588 (1977) and unpublished preprint.
7. D. Halderson and R. J. Philpott, Nucl. Phys. (to be published).
8. R. J. Philpott, Nucl. Phys. A289:109 (1977).
9. G. Bertsch, J. Borysowicz, H. McManus and W. G. Love, Nucl. Phys. A284:399 (1977).
10. L. G. Arnold, R. G. Seyler, T. R. Donoghue, L. Brown, and U. Rohrer, Phys. Rev. Lett. 32:310 (1974).
11. D. Robson, Phys. Rev. 137:B535 (1965).
12. A. M. Lane and R. G. Thomas, Rev. Mod. Phys. 30:257 (1958).
13. R. J. Philpott, Nucl. Phys. A243:260 (1975).
14. R. J. Philpott, Phys. Rev. C12:1540 (1975).

EFFECTIVE INTERACTIONS AT LOW AND MEDIUM ENERGIES[†]

S. A. Moszkowski

Department of Physics
UCLA
Los Angeles, CA 90024

ABSTRACT

The effective interactions in nuclei are strongly density dependent and, in fact, this feature is vital for understanding the theoretical basis of the nuclear shell model. Especially because of this density dependence, the effective interaction between nucleons on top of the Fermi sea in nuclear matter is quite weak, even though this same interaction leads to a large total nuclear binding energy. Thus, at low energies the effective interaction (other than that included in the average single-particle potential) occurs only between nucleons at the nuclear surface. It is further argued that this interaction has essentially zero range, i.e., it is a surface delta interaction. Thus we obtain nuclear deformations as well as the usual pairing effects. We conclude with a very simple model for estimating the effect of the nuclear medium at higher energies. We find that the effective interaction should rapidly approach the free particle scattering matrix with increasing energy.

1. DENSITY DEPENDENT INTERACTION

The nucleon-nucleon interaction (in free space) is quite well known by now. Using many body theory it can be shown that the _effective_ interaction in nuclei which should be used in a relatively small model space ($E_x \lesssim 100$ MeV) is strongly density dependent. This occurs since a part of the effective interaction behaves in some way like a three body interaction. A rough estimate yields a contribu-

---

[†]This work was partially supported by the NSF.

tion of around 15 MeV per nucleon for the three body term for the
energy of nuclear matter at normal density. This is quite large,
indeed of the same magnitude (but opposite sign) as the volume
energy of nuclear matter.

What about the physical origin of this effective three body
interaction? In principle, some of it could come from multimeson
exchange terms, but meson theoretical calculations suggest these
are relatively unimportant, producing only $\sim$ 1 or 2 MeV per nucleon
at normal density.

Thus the effective three body interaction probably arises as a
higher order effect of the two body interaction. With a realistic
nucleon-nucleon potential, such as the Reid Soft Core, we find that
roughly half of the 15 MeV comes from the effect of the tensor
forces, a fourth from the short range repulsion, and the rest from
excitation of virtual Δ isobar states, but the details of this are
probably not important for the theory of the independent particle
model. What matters is that, to a good approximation, the three
body part of the interaction can be represented as a density dependent two body delta interaction $\rho\delta(r)$ where ρ is the local density.
This is true to the extent that the correlations involved in virtual
excitations to high energy states are relatively short ranged. (This
holds well for the short range repulsion but less so for tensor
forces.)

The equivalence between

$$\delta(r_{123}) \qquad \text{and} \qquad \rho\delta(r_{12})$$

holds only if exchange terms in the energy are neglected. Thus we
may adjust the relative strengths of the two interactions so as to
reproduce the energy for N = Z nuclei. But the equivalence breaks
down for N $\neq$ Z. For example: if Z = 0 (neutron gas) the three
body δ-interaction has no effect at all, since we cannot have three
neutrons at the same point (only two spin directions). Yet the
$\rho\delta(r_{12})$ interaction can give a finite interaction energy, even for
a neutron gas. We will not discuss this point further in this talk,
since we will only consider N = Z nuclei.

There is another more basic difference between two body and
three body interactions, however, which is of crucial importance
for the validity of independent particle motion.

To illustrate this difference, consider a very simplified model
involving only two body and three body delta interactions.

$$V = -\alpha\delta(r_{12}) + \gamma\delta(r_{123}) \quad .$$

The effective two body interaction (neglecting exchange effects) is

$$V^{(2)} = (-\alpha + \gamma\rho)\ \delta(r_{12}).$$

The potential $V^{(2)}$ may be a good first approximation to the spin- and isospin-independent part of a density-dependent effective inter- action for shell-model calculations. Note that the ratio of the three body to the two body term is

$$R^{(2)} = -\gamma\rho/\alpha\ .$$

On the other hand, the potential energy per nucleon is given by

$$\frac{PE}{A} = -\frac{1}{2}\ \alpha\rho + \frac{1}{6}\ \gamma\rho^2\ .$$

The factors 1/2 (and 1/6) arise so as to avoid over counting pairs (and triplets). This time, the ratio of three body to two body terms is

$$R^{(PE)} = -\frac{1}{3}\ \frac{\gamma\rho}{\alpha} = \frac{1}{3}\ R^{(2)}\ .$$

Thus an effective two body interaction which reproduces the <u>potential energy</u> is:

$$V^{(PE)} = (-\alpha + \frac{1}{3}\ \gamma\rho)\ \delta(r)\ .$$

$V^{(PE)}$ is just the conventional Brueckner-Bethe matrix. The difference between $V^{(2)}$ and $V^{(PE)}$ is very important in nuclear structure. We know that $V^{(PE)}$ is strongly attractive (nuclei are bound!). Yet $V^{(2)}$ can vanish because of the greater importance of the three body term. As we will see, this is basically what is responsible for the validity of the independent particle model in nuclei.

The above treatment could also be given without ever referring to three body interactions. Thus if we start out with $V^{(PE)}$ and treat the density dependence correctly, using Landau theory, we ar- rive at the same expression for $V^{(2)}$ as that used here. I.e., the density dependent term is three times as large in $V^{(2)}$ as in $V^{(PE)}$.

2. INDEPENDENT PARTICLE MOTION IN NUCLEAR MATTER

a) <u>Average interaction energy of valence nucleons.</u>

Let us now extend the δ-function model used in the last sec- tion slightly so as to allow the two body interaction to have a finite range. This form is known as the Skyrme interaction. But we specialize to the case where it acts only in even states. Thus we find that the effective two body interaction can be written as

$$V^{(2)} = (-\alpha + \beta k^2 + \gamma\rho)\ \delta(r)$$

(where β is proportional to the mean square radius of the interaction), while the potential energy is determined by summing the interaction

$$V^{(PE)} = [- \alpha + \beta k^2 + (\gamma \rho / 3)] \ \delta (r)$$

over all nucleon pairs.

This potential energy (per nucleon) is given by

$$\frac{PE}{A} = - \frac{3}{8} \alpha \rho + \frac{3}{8} \times \frac{3}{10} \beta k_F^2 \rho + \frac{3}{8} \frac{\gamma \rho^2}{3} \quad .$$

The factor 3/8 is a product of the 1/2 (correcting for counting pairs twice), and 3/4 (to correct for exchange in N = Z nuclear matter). $3/10 \ k_F^2$ is the average value of k^2 for nucleons in nuclear matter.

The average two body interaction energy (neglecting exchange) between valence nucleons $\langle \vec{k}_i \vec{k}_i ' | V^{(2)} | \vec{k}_i \vec{k}_i ' \rangle$ (with $k_i = k_i ' = k_F$), denoted by F_o, is

$$F_o = - \alpha + \beta \frac{1}{2} k_F^2 + \gamma \rho$$

$1/2 \ k_F^2$ is the average k^2 for valence nucleons averaged over all angles.

Using the saturation condition for nuclear matter

$$\left. \frac{dE}{d\rho} \right|_{\rho_o} = 0$$

we can show that

$$F_o = \frac{1}{\rho_o} \left(- \frac{4}{5} T_F + 2W^{(3)} \right)$$

where T_F = Fermi kinetic energy

$W^{(3)}$ = 3 body energy = $1/8 \ \gamma \rho^2$.

Now since

$$T_F = 36 \text{ MeV}$$

and

$$W^{(3)} \simeq 15 \text{ MeV}$$

we see that the two contributions to F practically cancel

$$F_o \sim 0 \quad .$$

This vanishing of the average interaction energy suggests, though it does not prove, the validity of independent particle motion in nuclear matter.

The value of F_0 can be determined rather indirectly from properties of nuclear states by using a theory developed by Migdal. It is found that, indeed, F_0 averaged over spins (as was done here--in fact, we never mentioned spin) is actually quite small.

While $F_0 \sim 0$ in nuclear matter, it is expected to be attractive at the nuclear surface where ρ and k_F are lower. Thus for the Skyrme interaction we can write

$$F(\rho,\theta) = \sum_L F_L(\rho) \, P_L(\cos\theta)$$

where

$$F_o(\rho) = -\alpha + \frac{1}{2}\beta k_F^2(\rho) + \gamma\rho$$

$$F_1(\rho) = -\frac{1}{2}\beta k_F^2(\rho)$$

other

$$F_L(\rho) = 0 \quad .$$

However, the effective interaction strengths are further increased by higher order effects. The most important of these is core polarization, in which there is a particle-hole excitation in the intermediate state. If only small momentum transfers (i.e., particle hole pairs of momentum $q \to 0$) are considered, we find that the F_L are replaced by "renormalized" constants F_L^R given by

$$F_L^R = \frac{F_L}{1 + [CF_L/(2L+1)]}$$

where C is a positive constant, which is related to the Fermi momentum. Clearly $F_L^R < F_L$. Thus the renormalization weakens the repulsive terms and increases the magnitude of the attractive ones. It follows from the occurrence of the factor $1/(2L+1)$ that the renormalization is strongest for the $L = 0$ case. Thus if F is generally attractive to begin with (i.e., at the nuclear surface), then F^R will be even _more_ attractive and also more nearly isotropic than F. Such isotropy with respect to momentum angle is equivalent to zero range

in coordinate space. This is a rough argument for the validity of
the surface delta interaction.

   b)   Pairing

      Pairing is a very important aspect of the nuclear shell model.
It means that in the ground state, each pair of identical nucleons
will tend to couple their angular momenta to a resultant $J_0 = 0$.
This pairing energy Δ related to the odd-even mass differences, is
roughly given by

$$\Delta_{expt} \sim 12\ A^{-1/2}\ \text{MeV}$$

i.e., ~ 1 MeV for medium heavy nuclei. The pairing energy in
nuclear matter can only be roughly deduced from the observed pairing
energies, but it is probably less than 0.5 MeV.

      Δ can be roughly related to the two body interaction. A simple
minded theory gives for Δ in nuclear matter

$$\Delta_{NM} = 2\ T_F\ \exp[-\pi/(2\delta_s(k_F))]$$

$\delta_s(k_F)$ is the S wave phase shift in free space at $k = k_F$. Δ is a
non-analytic function of the strength of the interaction. Using
known $^1S_0$ phase shifts we have

$$\delta_s(k_F) = 15°$$

(quite small because of the action of the short range repulsion) and
$\Delta_{NM} = 0.17$ MeV, quite small. At the nuclear surface, ρ and thus k_F,
is smaller. Thus $\delta_s(k_F)$ is larger and so is Δ. In fact, in this
way, one can roughly account for the variation of Δ with A.

      Let us now try to relate the pairing effect to the effective
interaction discussed here. Instead of $\delta_s(k_F)$, which refers to free
nucleon-nucleon scattering, we should be using the pairing matrix
element G in nuclear matter. Now G = average matrix element for
scattering of a pair of nucleons in conjugate states m, m* into
m', m'*. In nuclear matter, we have plane waves. Thus we have
scattering of a pair k, -k into k', -k', where

$$k = k' = k_F\ .$$

Thus

$$G = \langle \vec{k}', -\vec{k}' | V^{(2)} | \vec{k}, -\vec{k} \rangle$$

averaged over all angles. G can involve large momentum transfer (unlike F_0), but the center of mass momentum of the nucleons must vanish. For the interaction discussed above, this is

$$G = -\alpha + \beta k_F^2 + \gamma\rho \, ,$$

$$= F_0 + \frac{1}{2} \beta k_F^2 \, .$$

As we have seen, $F_0 \sim 0$. Thus, since β is positive, due to the finite range of the interaction, we find $G > 0$, which means that the pairing in nuclear matter is repulsive; i.e., no pairing. Of course, at the nuclear surface, the density is lower, and G may be attractive. Thus the pairing effect occurs mainly at the nuclear surface.

As we have seen, the pairing interaction involves nucleon pairs with zero center of mass momentum. The scattering of a pair $\vec{k}$, $-\vec{k}$ into $\vec{k}'$, $-\vec{k}'$, with both k and k' = k_F, is unique in the sense that both the Pauli principle and energy conservation are satisfied.

Although in quantum mechanics it is permissible to violate energy conservation for a short time, we expect that energy conserving scatterings should be the ones of the greatest importance for low energy nuclear structure. These are precisely a) the forward scatterings (average interaction energy) described by F, in which both momentum and energy remain unchanged, and b) the pairing interaction G discussed here.

3. SURFACE DELTA INTERACTION

We have shown that the average interaction F_0 between valence nucleons as well as the pairing interaction G are both small, but probably somewhat repulsive, in nuclear matter. Thus they will not have much effect there. (If anything, nucleons will simply stay away from the region of the repulsion.) But at the nuclear surface where ρ and k_F are lower, both are attractive. Furthermore, we can argue that at least the pairing interaction G is, in fact, isotropic, with respect to the angle of the momentum, i.e., it acts like a delta function in the angle (in coordinate space). Thus we can assume that the effective interaction has zero range and that vanishes in the interior. This is the physical argument for the surface delta interaction.

Now, in a finite nucleus, the valence nucleons all have approximately the same energy and the same radial dependence $\exp[-cE^{1/2}r]$ at large distances, i.e., at the surface. Detailed calculations show that the amplitudes at the surface are also not very different for different valence states. Of course, different valence wavefunctions are quite different in the interior, and even have different numbers of nodes there, but this is not relevant if the effective interaction acts only at the surface. Thus we can set all radial integrals equal. This assumption leads to the surface delta interaction,

$$V_{SDI} = 4\pi \, F_0 \, \delta(\Omega)$$

where Ω is the angle between the nucleons. The surface delta interaction is quite easy to treat, all matrix elements are simply Clebsch-Gordon coefficients. Further, with such an interaction, it has proven possible to fit a large variety of nuclear structure data.

The required strength F_0 turns out to be about $-25A^{-1}$ MeV, if a model space of about one oscillator shell on each side of the Fermi surface is used. This guarantees that the wavefunction has full <u>angular</u> but no <u>radial</u> degrees of freedom. Approximately this value of F_0 can also be derived theoretically by using an interaction of the Skyrme form discussed in an earlier section.

4.   IF THE INTERACTION WERE NOT DENSITY DEPENDENT

Suppose now that the effective interactions were not dependent on density. This could happen, for example, if the interactions are attractive in even states but equally repulsive in odd states. ($V_{odd} = - V_{even}$.) In this case we can get nuclear saturation even without any repulsive or tensor forces. Indeed, in the early days of nuclear physics the nuclear forces were thought to be more or less of this type. In this case we find, using a Skyrme interaction,

$$F_0 = G = -\alpha + \beta k_F^2 \ .$$

In this case, F_0 and G are equal. (The previous expression for F_0 holds only for even state interactions.) But since $W^{(3)} = 0$ we find, using the relations given before

$$F_0 = -\frac{4}{5} \, \frac{T_F}{\rho_0}$$

which leads to

$$\delta_s(k_F) = \frac{3}{10} \, \frac{\pi}{2}$$

and the following pairing energy of nuclear matter

$$\Delta_{NM} = 2T_F \, e^{-10/3} \sim 3 \text{ MeV}$$

much larger than the experimental value.

We obtain an even larger result if renormalization is taken into account and also for finite nuclei. In any case, Δ is larger than the average spacing between single particle levels. Thus we would find that for even-even nuclei the ground state $J_0 = 0$, but for odd A nuclei $J_0 \neq J$ (single particle), i.e., the single particle shell model would lose its validity. Another way of saying this is that the Fermi surface would be unstable. The pairing correlations would trigger all kinds of instabilities, such as breaking into alpha particles (where all interactions are even state, i.e., attractive). Fortunately, thanks to the density dependence, all this does not happen and so the shell model lives!

5. EFFECTIVE INTERACTION AT MEDIUM ENERGIES

The density dependence discussed decreases with increasing energy. At about 100 MeV, it is expected to be only about half as important as at low energies. At about 1 GeV, the density dependent effects essentially disappear, and we approach the impulse approximation

$$V^{(2)} \to t_o$$

where t_0 is the scattering matrix in free space. Still, the effect of the nuclear medium cannot be ignored entirely. First of all, in a nuclear reaction we cannot scatter a nucleon into a state occupied by another nucleon, i.e., a state with momentum < Fermi momentum.

The Pauli correction decreases rapidly with energy. Thus for a nucleon of incident kinetic energy $T > 2T_F$; the Pauli effect reduces the phase space available for elastic scattering by a fraction $f(T)$ where

$$f(T) = \frac{7}{5} \frac{T_F}{T} \quad .$$

Now the density dependence discussed in previous sections is due **mainly not to the Pauli principle, but to the fact that the single** particle potential U felt by a nucleon outside the Fermi sea is much less attractive than one felt by a nucleon inside. This is known as the dispersion effect. Let us define

$$\Delta U = U_{i'} + U_{j'} - U_i - U_j$$

when nucleons in states ij scattering into virtual states i'j'.

The dispersion correction can, in fact, be roughly represented by the density dependent δ-function interaction $\gamma\delta(r)$ discussed earlier. In order to make a rough estimate of the energy dependence

of the dispersion effect, we take a very crude but simple model of
the nucleon-nucleon interaction, namely, we assume that V is attractive and separable, i.e., V acts only in S states and the off-
diagonal elements of V are related to diagonal elements V(T) (T =
kinetic energy) by:

$$V(T',T) = -\big(V(T') \ V(T)\big)^{1/2}$$

Then the Schroedinger equation can be solved exactly, and we can obtain an explicit expression for the effective interaction in free
space which is just the scattering matrix t_0 (as used in the impulse
approximation).

We can rewrite this expression in the form

$$\frac{V(T)}{t_o(T)} = 1 - \phi(T)$$

where

$$\phi(T) = \sum_{T'} \frac{V(T')}{(T'-T)}$$

$\phi(T) > 0$ for an attractive interaction. In this model, the effect
of dispersion is very simple. It is simply to add a contribution
ΔU to the energy denominator. Thus we find

$$\frac{V(T)}{t(T,\Delta U)} = 1 - \phi(T-\Delta U) \qquad .$$

But presumably $\phi(T)$ is known from solving the free particle problem. Thus to first order in ΔU we obtain

$$\frac{t(T,\Delta U)}{t_o(T)} = \frac{1 - \phi(T)}{1 - \phi(T-\Delta U)} = 1 - \frac{\Delta U}{\big(1 - \phi(T)\big)^2} \frac{d\phi(T)}{dT} \qquad .$$

We would expect $\phi(T)$ to decrease rapidly with T. Thus, if, for
example,

$$\phi(T) = T_o/T , \qquad \text{where } T_o \text{ is } \simeq 100 \text{ MeV} .$$

Then for $T \gg T_o$,

$$\frac{t_o(T)}{V(T)} = 1 + \frac{T_o}{T} \quad \cdots$$

$$\frac{t(T,\Delta U)}{t_o(T)} \sim 1 + \frac{T_o \Delta U}{T^2} \qquad .$$

As T increases, V approaches t_0; but $t(T,\Delta U)$ approaches t_0 even faster. This would be true even if ΔU were independent of energy (actually ΔU is expected to decrease somewhat with increasing T), which further speeds up the convergence toward the impulse approximation.

6.   SUMMARY

I have tried to connect some features of effective interactions in nuclei.

1.) The important role of the density dependence, which leads to the smallness of both average interaction (F_0) and pairing interaction (G) between valence nucleons in nuclear matter. These are the most important matrix elements of the effective interaction, i.e., that part of the interaction not included in the average potential.

2.) However, the effective interaction is important at the nuclear surface, where the density is lower. Higher order corrections have the effect of shortening its range. This is the physical origin of the surface delta interaction.

3.) As the energy increases, many body effects and thus the density dependence become less important. At high energies, the effective interaction to be used in nuclear reaction calculations rapidly approaches the free nucleon-nucleon scattering matrix.

ACKNOWLEDGMENT

I am very grateful to Dr. K. F. Liu for many stimulating discussions and helpful comments concerning this work.

HADRONIC AND ELECTROMAGNETIC NUCLEAR PROBES

G. E. Walker[*]

Physics Department
Indiana University
Bloomington, IN 47405

I. INTRODUCTION

Both at this conference and historically there have been three major (overlapping) motivations given for studying the (p,n) reaction. These include

A. Using the (p,n) reaction on nuclei as a way of studying certain spin-isospin components of the nucleon-nucleon interaction. Of course before this approach can be quantitatively useful considerable uncertainties in the appropriate nuclear structure and associated many-body reaction mechanism must be clarified.

B. The (p,n) reaction, because of its utility in isospin identification could be used to study nuclear strucutre. Although there are problems associated with energy resolution inherent in (p,n) experiments this should not present fundamental difficulties for studies involving:

1. the location, width and variation with Z and A of isovector giant resonances,

2. the structure of analogue, Gamow-Teller, and high spin states. Such types of states are either well isolated in energy or are predicted to be dominant in a given region of the (q,ω) surface in the linear response of a nuclear target. Such states are particularly attractive to study using several nuclear probes because each probe has its own nuclear spin-isospin flip enhancement selection rules (both overlapping and unique). In addition we shall see that the phase relations between particle-hole and hole-particle transition matrix elements, both of which arise in the RPA and more

[*]Work supported in part by the National Science Foundation.

complicated nuclear structure models, are different for different
projectiles. The nucleon can play a special role in this regard
because of the importance of antisymmetrization and the practicality
of polarized beams.

       C. Use of the (p,n) reaction to eliminate the $\Delta T_z = 0$ back-
ground in reaction mechanism studies (especially for reactions in
the continuum). In general for hadronic-nucleus reactions one is
faced with the problem of reducing a multichannel set of equations
for a strongly interacting system using various approximations.
Speaking loosely, one might wish to know in what energy region for a
given probe one can with confidence use the DWIA (or a few coupled
channels) and a proposed probe-nuclear constituent t matrix to
analyze differential cross section and polarization data with the
aim of learning about nuclear structure. To obtain this goal one
must make comparative reaction mechanism studies using several
probes where the strength and variation with energy of the elementary interaction plays a different role in each case.

       For example if we consider figure 1, it is clear there is a
significant variation in the energy dependence of the mean free path
of various projectiles in nuclear matter. In addition to the
obvious conclusion that the different probes will therefore "see"
different parts of the nucleus, it is also true that, as a result,
the energy dependence of, for example, two step processes should be
quite different also. Our reaction theories should be able to
correctly predict the experimental energy dependence of such complications before we can hope to learn much about nuclear structure.

       In the next section we shall discuss meson-nucleus inelastic
scattering and charge exchange and give some examples of how meson
results could be a useful tool for studying nuclear structure even
though there are substantial reaction mechanism uncertainties. In
section III we discuss the states predicted to be strongly excited
for various projectiles and how these results can be exploited. We
discuss the phase relations between p-h and h-p diagrams for transition operators associated with electromagnetic and hadronic probes
in section IV. We outline the manner in which (p,p′) and (p,n)
reactions are especially interesting for studying nuclear structure
because of deviations for these reactions from the phase relations
usually obtained. Finally in section V we make some suggestions for
theoretical and experimental studies using (p,n) reactions. Some of
these investigations are already being pursued.

II. MESON-NUCLEUS INELASTIC SCATTERING AND CHARGE EXCHANGE

       Although there are still considerable uncertainties with
respect to the reaction mechanism for pion-nucleus reactions (as
examples, the appropriateness of the usual multiple scattering

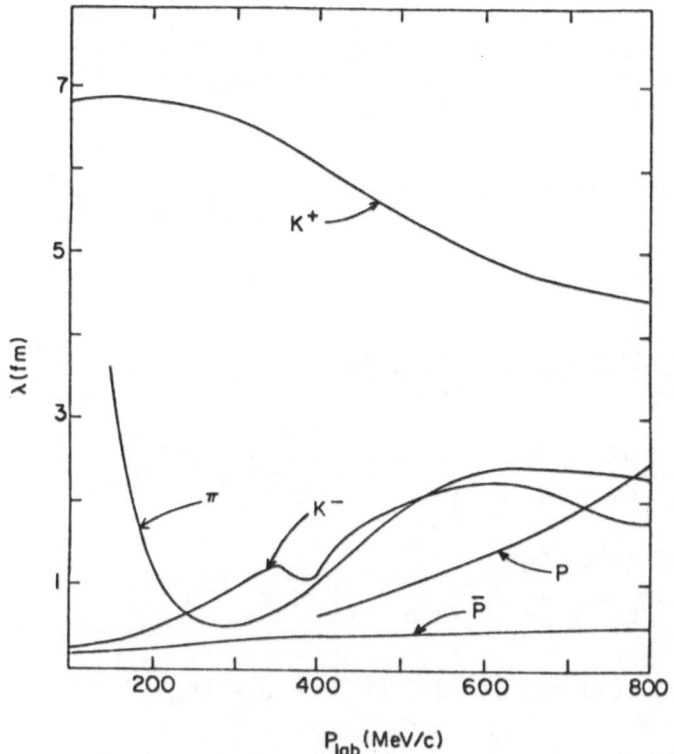

Fig. 1 - Probe mean free path vs a function of lab momentum

assuming $\lambda = \dfrac{1}{\rho \, \sigma_{tot}}$.

expansion and the importance of crossing and relativistic effects needs further study) there exists enough preliminary information to indicate that the application of the DWIA for making qualitative predictions is a worthwhile exercise.

A procedure commonly adopted to obtain the differential cross section for meson-nucleus inelastic scattering and charge exchange from an initial state J_0 to a final nuclear excited state J is discussed in detail in references 1-3. The resulting expression is given by

$$\frac{d\sigma_{JJ_0}}{d\Omega} = \frac{E(p_f)p_f}{E(p_i)p_i} \left| F_{JJ_0} \left(p_f, p_i \right) \right|^2 \tag{1}$$

where $E(p_{i,f}) = (M^2 + p_{i,f}^2)^{\frac{1}{2}}$ and p_i and p_f are the initial and final meson momenta, respectively. Using the DWIA, assuming the transition operator can be written as a sum of a single meson-nucleon amplitudes f_{Ei}, normalized so that $(d\sigma/d\Omega)_{\text{two-body}} = \left| f_{Ei} \right|^2$, treating the nuclear wavefunctions in the TDA approximation, assuming a separable form for the elementary transition operator, and making liberal use of standard techniques for angular momentum recoupling allows one to write $|F_{JJ_0}(p_f, p_i)|^2$ as follows;

$$|F_{JJ_0}(p_f, p_i)|^2 = (2\pi)^{-4} [1/2](8/9) \sum_{Jz} \Big| \sum_{\ell_p \ell_h j_p j_h} \alpha_{j_p j_h}^{\ell_p \ell_h} M_{\ell_p \ell_h j_p j_h}^{JJ_z} \Big|^2 \quad (2)$$

where

$$M_{\ell_p \ell_h j_p j_h}^{JJ_z} = \sum_i a_{n\ell_a} \, a_{n'\ell_b} \, \alpha\beta\gamma(N+\hat{N}) \quad (3a)$$

$$\alpha = \begin{pmatrix} \ell & \ell_3 & \ell_b \\ o & o & o \end{pmatrix}\begin{pmatrix} \ell & \ell_4 & \ell_a \\ o & o & o \end{pmatrix} \hat{\ell}_3 \hat{\ell}_4 (\hat{\ell}_a \hat{\ell}_b \hat{\ell}_p \hat{\ell}_h \hat{J})^{\frac{1}{2}} \, i^{\ell_3 - \ell_4}(-1)^\ell \quad (3b)$$

$$\beta = (-1)^{m_b}\begin{pmatrix} J & \ell_a & \ell_b \\ -J_z & -m_a & m_b \end{pmatrix} Y_{\ell_b m_b}^+(r_{p_f}) \, Y_{\ell_a m_a}(r_{p_i}) \quad (3c)$$

$$\gamma = \int_o^\infty r^2 dr \, R_{\ell_p}(r) \, R_{\ell_h}(r) \, j_{\ell_3}(k_{n'}r) \, j_{\ell_4}(k_n r) \quad (3d)$$

$$N = (-1)^{j_h + \frac{1}{2}}\begin{pmatrix} J & \ell_3 & \ell_4 \\ o & o & o \end{pmatrix}\begin{pmatrix} \ell_p & \ell_h & J \\ o & o & o \end{pmatrix}\begin{Bmatrix} \ell_p & j_p & \frac{1}{2} \\ j_h & \ell_h & J \end{Bmatrix}$$

$$\times \begin{pmatrix} A_{\Delta T=0}^{\Delta S=0} + B_{\Delta T=0}^{\Delta S=0} \end{pmatrix} \quad (3e)$$

$$\hat{N} = \sqrt{6} \sum_{\bar{J}} \begin{pmatrix} \bar{J} & \ell_3 & \ell_4 \\ o & o & o \end{pmatrix}\begin{pmatrix} \ell_p & \ell_h & \bar{J} \\ o & o & o \end{pmatrix}\begin{Bmatrix} \ell_p & \frac{1}{2} & j_p \\ \ell_h & \frac{1}{2} & j_h \\ J & 1 & J \end{Bmatrix}\begin{Bmatrix} \ell & \ell_4 & \ell_a \\ \ell & \ell_3 & \ell_b \\ 1 & \bar{J} & J \end{Bmatrix}$$

$$\times \hat{\bar{J}}\hat{\ell}^{\frac{1}{2}}(-1)^{\ell_p}\begin{pmatrix} C_{\Delta T=0}^{\Delta S=1} + D_{\Delta T=1}^{\Delta S=1} \end{pmatrix} \quad (3f)$$

where $i = \ell, \ell_3, \ell_4, \ell_a, \ell_b, n, n', m_a, m_b$ and $\hat{\ell} \equiv 2\ell + 1$. The $(8/9)[1/2]$ factor is appropriate for (pions)[kaons]. In eq. (3a-f), the factor β contains the angular dependence and γ contains the radial overlap integrals. The single nucleon bound state wavefunctions, $R(r)$, are normalized by $\int_o^\infty r^2 R^2(r) \, dr = 1$. The factors $a_{n\ell_a}, a_{n'\ell_b}$ arise because the configuration space meson distorted

waves, $X_E^{\ell}(p_i r_k)$, have been expanded in terms of spherical Bessel functions $j_0(k_n r_k)$ where the k_n are chosen so that $j_{\ell}(k_N r_k)\big| = 0$.
$$r_k = R$$

After taking the Fourier Transform, $X_E^{\ell}(p_i, k)$ can be written

$$X_E^{\ell}(p_i, k) = \sum_n a_{n\ell}(p_i) \frac{\pi}{2k^2} \delta(k_n - k) \tag{4}$$

The $\alpha_{J_p J_h}^{\ell_p \ell_h}$ are the admixture amplitudes of the pure p-h states obtained for the configuration mixed p-h state.

For pions the transition amplitudes A, B, C, D are given by

$$A_{\Delta T=0}^{\Delta S=0} = \ell\left(f_{\ell-}^{1/2} + 2f_{\ell-}^{3/2}\right) + (\ell+1)\left(f_{\ell+}^{1/2} + 2f_{\ell+}^{3/2}\right) \tag{5a}$$

$$B_{\Delta T=1}^{\Delta S=0} = \ell\left(f_{\ell-}^{1/2} - f_{\ell-}^{3/2}\right) + (\ell+1)\left(f_{\ell+}^{1/2} - f_{\ell+}^{3/2}\right) \tag{5b}$$

$$C_{\Delta T=0}^{\Delta S=1} = \sqrt{\ell(\ell+1)}\left(f_{\ell+}^{1/2} - f_{\ell-}^{1/2} + 2f_{\ell+}^{3/2} - 2f_{\ell-}^{3/2}\right) \tag{5c}$$

$$D_{\Delta T=1}^{\Delta S=1} = \sqrt{\ell(\ell+1)}\left(f_{\ell+}^{1/2} - f_{\ell-}^{1/2} + f_{\ell-}^{3/2} - f_{\ell+}^{3/2}\right) \tag{5d}$$

For kaons the amplitudes A, B, C, D are given by

$$A_{\Delta T=0}^{\Delta S=0} \equiv \ell\left(f_{\ell-}^{0} + 3f_{\ell-}^{1}\right) + (\ell+1)\left(f_{\ell+}^{0} + 3f_{\ell+}^{1}\right) \tag{5a$'$}$$

$$B_{\Delta T=1}^{\Delta S=0} \equiv \ell\left(f_{\ell-}^{0} - f_{\ell-}^{1}\right) + (\ell+1)\left(f_{\ell+}^{0} - f_{\ell+}^{1}\right) \tag{5b$'$}$$

$$C_{\Delta T=0}^{\Delta S=1} \equiv \sqrt{\ell(\ell+1)}\left(f_{\ell+}^{0} - f_{\ell-}^{0} + 3f_{\ell+}^{1} - 3f_{\ell-}^{1}\right) \tag{5c$'$}$$

$$D_{\Delta T=1}^{\Delta S=1} \equiv \sqrt{\ell(\ell+1)}\left(f_{\ell+}^{0} - f_{\ell-}^{0} + f_{\ell-}^{1} - f_{\ell+}^{1}\right) \tag{5d$'$}$$

The quantities $f_{\ell\pm}^{I}$ are the appropriate off-shell meson-nucleon amplitudes for isospin I and $J = \ell \pm 1/2$. The fully off-shell t-matrix required for calculating inelastic scattering is related to the on-shell t-matrix (assuming a separable form) by

$$T_{\alpha}(k_1, k_2; E(k)) = \frac{g_{\alpha}(k_1)\, g_{\alpha}(k_2)}{g_{\alpha}^2(k)}\, T_{\alpha}(k, k; E(k)) . \tag{6}$$

In order to use eqs. (2), (3a-f) for meson-charge exchange one simply sets A = C = O, since ΔT = 1 only, and for kaons (pions) multiplies the remaining cross section by a factor of two (one).

The ΔS and ΔT labels on A-D (eqs. 5a-d) indicate the final spin and isospin (spin-flip, iso-spin flip character) of the final nuclear excited state reached via that part of the transition opera- tor if the ground state has S = T = O. The terms A-D are found to provide a simple and accurate means for predicting the spin and isospin structure of the dominant states excited by meson probes. For example, the J = 3/2, t = 3/2, p wave (ℓ = 1) pion-nucleon amplitude is the dominant component in the pion nucleon interaction in the region below 250 MeV, thus, for the purposes of obtaining qualitative estimates of the most strongly excited states we ignore the other amplitudes. This means we retain only the $f_{\ell+}^{3/2}$ term in eqs. 5a-d. We immediately obtain the result for the amplitudes

$$A = 4f_{\ell+}^{3/2}, \quad B = -2f_{\ell+}^{3/2}, \quad C = 2\sqrt{2}f_{\ell+}^{3/2}, \quad D = \sqrt{2}f_{\ell+}^{3/2}.$$

If we consider meson charge exchange then as discussed above A = C = O. By examining B and D for the π and $K^{\pm}$ we can predict what kinds of states should dominate the meson-nuclear charge ex- change spectrum. For pions $|B|^2 = 2|D|^2$ for p wave amplitudes (for s wave amplitudes, B, the non-spin-flip amplitude is the only term that survives) thus one predicts that normal parity states should dominate the charge exchange response spectrum. This prediction is in general reinforced by more detailed calculations carried out by Gupta and Walker.[2]   Although in many important ways the K^+ and K^- have different properties from each other (because of the strangeness quantum number) and from the pion, it turns out that over a wide range of kaon incident kinetic energies the non-spin flip term, B, is dominate over D for $K^{\pm}$. Thus all these mesons are predicted to preferentially excite normal parity states in the charge-exchange reaction. More detailed predictions using the DWIA are fairly consistent with this conclusion as is shown in figs. 2, 3, 4 taken from reference 3.   Note in figs. 2 and 3 only the solid lines which represent T = 1 states should be included for charge exchange.

While meager experimental data currently exists for meson- nucleus interactions apparently some pion results have already been useful in helping clarify a situation puzzling from the standpoint of inelastic electron scattering results.

The pion can be used to elucidate isospin mixing in selected excited states of self conjugate nuclei. Because the J = 3/2, t = 3/2 partial wave is dominant in the pion-nucleon interactions, the π^+-proton (π^--neutron) interaction is much stronger than the

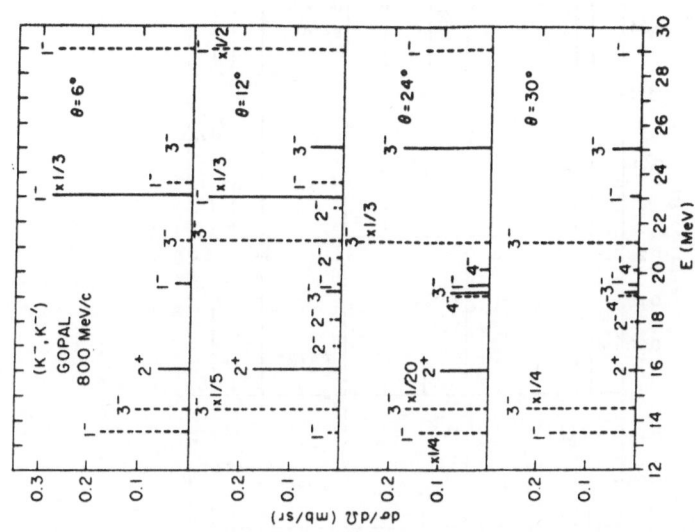
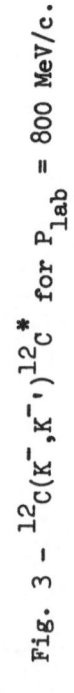

Fig. 3 – $^{12}C(K^-,K^{-\prime})^{12}C^*$ for P_{lab} = 800 MeV/c.

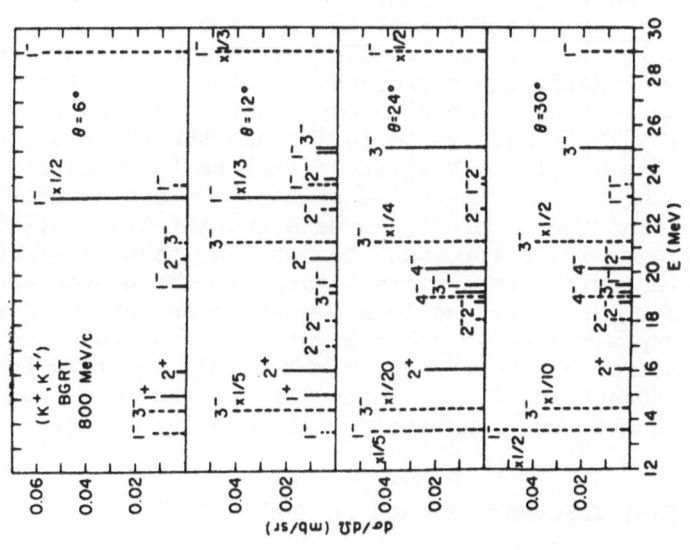

Fig. 2 – $^{12}C(K^+,K^{+\prime})^{12}C^*$ at P_{lab} = 800 MeV/c.

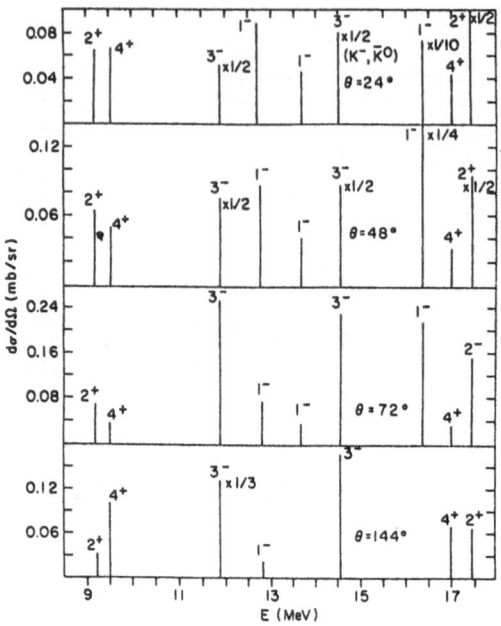

Fig. 4 - Differential lab cross sections at selected angles for ^{30}Si$(K^-, \bar{K}^0)^{30}$Aℓ at P_{lab} = 300 MeV/c.

π^+-neutron (π^--proton) interaction. This result can be exploited to study isospin mixing in self-conjugate nuclei by subtracting the π^+ results from the π^- results for a given set of states. A specific example is shown in fig. 5. If isospin were a good quantum number there would be no significant difference for the $\pi^\pm$ results for a given state. The fact that the π^+ cross section is significantly larger for the 4^- (19.25 MeV) state in ^{12}C and the π^- cross section is dominant for the 4^- (19.65 MeV) state implies there is significant isospin mixing between these two states. We are currently in the process of using these results to make quantitative estimates of the isospin mixing. It is important to consider several effects of coulomb mixing including the different tails of the proton and neutron single particle wavefunctions. This technique shows promise for studying isospin mixing in other light self conjugate nuclei. In inelastic electron scattering one might simply identify both 4^- states as T = 1 because of the dominance of the isovector magnetic moment at large angles and reasonably high momentum transfer.

III. COMPLEMENTARY INFORMATION FROM SEVERAL PROBES

It is useful to compare the results predicted for the nuclear response to several probes in a region where one step processes are

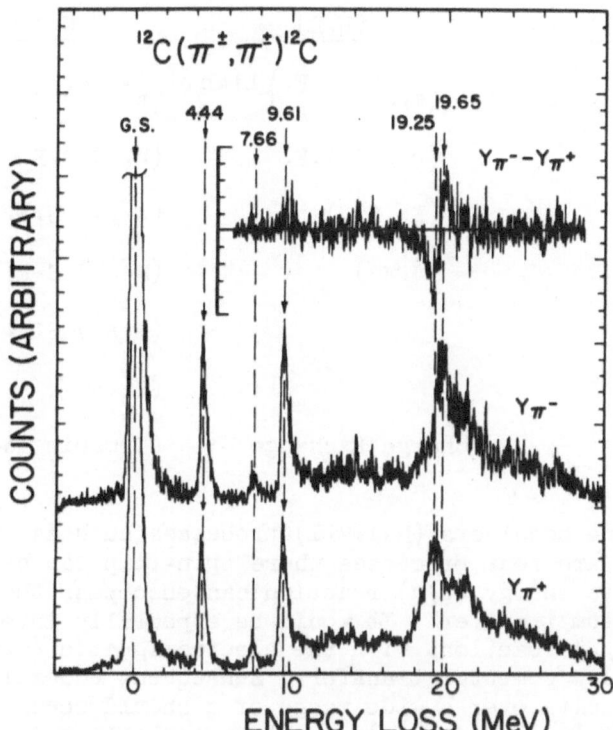

Fig. 5 - Difference of π^+ and π^- cross sections in ^{12}C conjectured to be due to isospin mixing in the two 4^- states. See reference 5.

assumed to be dominate (i.e. the nuclear response is linear and/or the DWIA is valid). The general results are given below for various cases. The presentation must be brief here (for a more detailed discussion see ref. 4) but the point is there are regions of over-lap and uniqueness for each projectile (not surprising) depending on the region of momentum transfer under consideration. The fact that the hadronic probe results can be compared with electron results for some cases (low-lying T = 0 normal parity states and high spin T = 1 states) can be used to test the reaction mechanism. After these comparisons have been completely made then one may proceed, if the hadronic reaction mechanism has been shown to be understood, to quantitatively study nuclear structure with hadronic probes. Those processes below, where spin flip is quite important in a given domain have S.F. written beside them. The inelastic processes where there are important $\Delta T = 1$ contributions are denoted by I.F. Of course these spin-iso-spin flip rules are important both for studying individual states and the spin and isospin character of giant resonances.

(N-1, Z+1)	(N, Z)	N+1, Z-1

(π^+,π^0) S.F. high q only $(e,e')\begin{Bmatrix}S.F.\\I.F.\end{Bmatrix}$ high q (π^-,π^0) S.F. high q only

(p,n) S.F. (p,p') I.F. (n,p) S.F.

(γ,π^-) S.F. $(K^\pm,K^{\pm\prime})$ (π^-,γ) S.F.

(π^+,γ) S.F. (π,π') (μ^-,ν) S.F.

(K^+,K^0) (γ,π^+) S.F.

$(K^-,\overline{K}^0)$

Charge Exchange ⇒ (Isospin Identifier)

Note if one considers (N-1, Z+1) processes such as (p,n) that although there are four processes where spin-flip can be important, the intermediate energy (p,n) reaction can encompass the widest range of momentum transfer. It would be especially interesting to compare the (p,n) reactions with the electromagnetic (γ,π) reaction at relatively low momentum transfer. Subsequent comparison with the (π^+,π^0) results over a wide range of q should occur. Of course comparison of (π^+,π^0) and (p,n) reactions to analogue and Gamow-Teller states for a wide range of nuclei is now being done (because the experimental data is becoming available). The detailed theoretical experimental comparison currently being looked at by an Indiana-Los Alamos collaboration involves the ^{13}C charge exchange results.

IV. PHASE RELATIONS BETWEEN p-h AND h-p TRANSITIONS

If it is appropriate to treat the probe-nucleus interaction using the DWIA, then the single nuclear particle transition density contains the required (or desired) nuclear structure information. Ignoring for the moment finite range exchange effects, this means that one can reduce the many-body target co-ordinate integrals to a sum of single nucleon integrals of the form

$$\sum_{J_p J_h} \alpha_{J_p J_h} \langle J_p \mid \mathcal{O} \mid J_h \rangle \tag{7}$$

where the $\alpha_{J_p J_h}$ are configuration-mixing parameters obtained from nuclear structure calculations, J_p and J_h are single nucleon bound stat orbitals, and $\mathcal{O}$ is a (perhaps complicated!) single nucleon operator.

If one considers a closed shell ground state, adopts the DWIA, and employs the Tamm-Dancoff approximation (TDA) for the configuration-mixed particle-hole nuclear final states then J_p (J_h) labels can be associated with the quantum numbers of the excited particle (hole) orbital. Often the TDA approximation is inadequate for treating the nuclear states and one must adopt the random phase approximation (RPA) or carry out large basis shell model calculations. In such situations, if the nuclear response can still be treated as linear, equation (7) becomes only slightly more complicated and can be written as

$$\sum_{J_p J_h} \left(\alpha_{J_p J_h} \langle J_p | \mathcal{O} | J_h \rangle + \alpha'_{J_p J_h} \langle J_h | \mathcal{O} | J_p \rangle \right) \tag{8}$$

where α and α' are determined from nuclear structure calculations as before. Note that the single nucleon matrix elements appearing in equation (8) have the order of J_p and J_h reversed. Past experience, based mainly on inelastic electron scattering results and RPA calculations, has indicated that for particular states significant enhancement or reduction factors can result if one compares the TDA reaction predictions with those using wavefunctions obtained from slightly more involved nuclear structure calculations.[6,7] Of course one determining feature in obtaining the transition rate renormalization is the size and phase relation of the nuclear structure constants α and α'. Note that these will be the same independent of the type of projectile. Another determining feature in obtaining the renormalization is the relation between the two types of matrix elements appearing in equation (8). For example most often one obtains[7]

$$\left(J_p | \mathcal{O} | J_h \right) = \text{phase} \times \left(J_h | \mathcal{O} | J_p \right) \tag{9}$$

where the phase depends on J_p, J_h, the multipolarity and nature of the operator $\mathcal{O}$. A more detailed discussion will be submitted for publication[8] but here we briefly summarize some implications for protons. If we specialize to non-normal parity final states reached via a spin-flip operator the phase relation is

$$\left(J_p | \mathcal{O} | J_h \right) = (-1)^{J_p + J_h + 1} \left(J_h | \mathcal{O} | J_p \right) \tag{10}$$

for the electron transverse magnetic, muon, pion, kaon spin-flip and proton direct spin flip terms. Thus a similar renormalization may be expected for all these projectile operators if the DWIA is valid. The finite range proton exchange term can significantly alter this phase relation so that this projectile can be used to give another check on proposed nuclear wavefunctions which have been adjusted to give the correct experimental transition rate for all the other projectiles.

The situation for normal parity transitions is somewhat more complicated because both spin-flip and non-spin flip operators can contribute. In this case the special role to be played by the proton in renormalization effects is not only due to antisymmetrization but also because the spin flip and non-spin flip operators can interfere with each other (for example, in <u>polarization</u> experiments which can be performed at intermediate energies at IUCF).

One final note on this phase relation section may be useful for those interested in inelastic electron scattering. Since the phase relation for the Longitudinal Coulomb matrix element is given by $(-1)^{J_p+J_h+1}$ and the phase relation for the Transverse Electric matrix element is $(-1)^{J_p+J_h}$, if there has been a significant enhancement of the Longitudinal term one should not be surprised if the Transverse Electric term is significantly <u>reduced</u>. These two terms both contribute to normal parity transitions but do not interfere for an unpolarized target. However their relative contributions to the differential cross section can be determined by using a Rosenbluth plot.

V.   CONCLUSIONS AND SUGGESTIONS FOR (p, n)

From a theoretical perspective a great deal of work remains to determine the importance of multistep or coupled channels effects in the energy region from 60-300 MeV. For example what are the theoretical error bars necessary in angular distribution and <u>polarization</u> predictions for a given q region because of their neglect. This important study may require a microscopic coupled channels code allowing for realistic non-central forces.

It is important to use experimental results using other probes (especially involving electromagnetic interactions) so that cross comparisons can be made to attempt to <u>experimentally</u> deduce the validity of the DWIA for the (p, n) reaction. In carrying out these comparisons we need to modify some existing reaction codes to use more realistic trial nuclear wavefunctions.

The expected polarization data will be more useful if there are accompanying theoretical studies of the sensitivity of $(\vec{p}, n)$ predictions to the spin-orbit term in the optical potential and to the various complex spin-dependent terms in the assumed transition operator. In carrying out these theoretical studies one should exploit the ability to vary q, E^i, and θ much as is done in (e, e') using a Rosenbluth plot. Finally, in addition to the theoretical and experimental studies anticipated for

1)   low q-analogue and spin analogue states, and

2)   high q-stretched (high spin states)

it would be useful to study quasi-elastic $(\vec{p}, n)$ and $(\vec{p}, p')$ reactions. By mapping out the (q, ω) plane for the quasi-elastic reaction for significant variations in ω, and averaging over, perhaps 5 MeV bins one should be able to study the appropriateness of the DWIA and the assumed t matrix while minimizing the importance of uncertainties in nuclear wavefunctions.

VI.   REFERENCES

1. T. S. H. Lee and F. Tabakin, Nucl. Phys. A226, 253 (1974).
2. M. K. Gupta and G. E. Walker, Nucl. Phys. A256, 444 (1976).
3. C. B. Dover and G. E. Walker, accepted for publication in Phys. Rev. C.
4. G. E. Walker, "Nuclear Studies Involving Intermediate Energy Projectiles," NATO Advanced Institute in Nuclear Theory, Banff, Alberta, Canada, August 21-September 2, 1978, Proc. to be published.
5. Data and figure private communication by W. Braithwaite, W. B. Cottingame, and C. L. Morris (EPICS data).
6. V. Gillet and M. Melkanoff, Phys. Rev. 133, B1190 (1964).
7. G. E. Walker, Phys. Rev. C5, 1540 (1972).
8. A. Picklesimer and G. E. Walker, to be submitted for publication.

B. D. Anderson
Department of Physics
Kent State University
Kent, Ohio 44242

Sam M. Austin
Department of Physics
Michigan State University
East Lansing, Michigan 48824

Helmut W. Baer
Los Alamos Scientific Laboratory
MP-4, MS 846
Los Alamos, New Mexico 87545

Bruce R. Barrett
Department of Physics, Bldg. 81
University of Arizona
Tucson, Arizona 85721

Monique Bernas
University of Paris
Orsay, France 91406

U. E. P. Berg
Institut fur Kernphysik
Leihgesterner Weg 217
D6300 Giessen
WEST GERMANY

Stewart D. Bloom
Lawrence Livermore Laboratory
Livermore, California 94550

F. Paul Brady
Department of Physics
University of California
Davis, California 95616

Virginia Brown
Lawrence Livermore Laboratory
P.O. Box 808
Livermore, California 94550

Roger C. Byrd
Triangle Universities Nuclear
  Laboratory
Department of Physics
Duke University
Durham, North Carolina 27706

Bunny C. Clark
Ohio State University
Department of Physics
174 W. 18th Street
Columbus, Ohio 43210

Douglas Cline
Nuclear Structure Research
  Laboratory
University of Rochester
Rochester, New York 14627

Joseph R. Comfort
Department of Physics
University of Pittsburgh
Pittsburgh, Pennsyvlania 15260

Stephen R. Cotanch
Department of Physics
North Carolina State University
Raleigh, North Carolina 27612

Bill J. Dalton
Ames Laboratory
Iowa State University
Ames, Iowa 50011

Paul T. Debevec
Department of Physics
University of Illinois
Urbana, Illinois 61801

Raymond P. DeVito
Cyclotron Laboratory
Michigan State University
East Lansing, Michigan 48824

Roger W. Finlay
Physics Department
Ohio University
Athens, Ohio 45701

C. C. Foster
Indiana University Cyclotron
Facility
Milo B. Sampson Lane
Bloomington, Indiana 47405

Aaron I. Galonsky
Cyclotron Laboratory
Michigan State University
East Lansing, Michigan 48823

Charles D. Goodman
Oak Ridge National Laboratory
P.O. Box X, Bldg. 6000
Oak Ridge, Tennessee 37830

Charles A. Goulding
LAMPF Users Group, MS 831
Los Alamos Scientific Laboratory
P.O. Box 1663
Los Alamos, New Mexico 87545

John C. Hiebert
Cyclotron Institute
Texas A and M University
College Station, Texas 77845

Steven D. Howe
Los Alamos Scientific Laboratory
MS-805, GRP P-9
Los Alamos, New Mexico 87544

George A. Keyworth
Los Alamos Scientific Laboratory
Physics Division, Mail Stop 434
P.O. Box 1663
Los Alamos, New Mexico 87545

N. S. P. King
Los Alamos Scientific Laboratory
P.O. Box 1663
Los Alamos, New Mexico 87545

Peter D. Kunz
Department of Physics
University of Colorado
Boulder, Colorado 80302

David A. Lind
Department of Physics and
   Astrophysics
University of Colorado
Boulder, Colorado 80309

Paul W. Lisowski
Los Alamos Scientific Laboratory
MS 442 Group P-3
Los Alamos, New Mexico 87544

W. G. Love
Department of Physics
University of Georgia
Athens, Georgia 30602

R. Madey
Department of Physics
Kent State University
Kent, Ohio 44240

Victor A. Madsen
Department of Physics
Oregon State University
Corvallis, Oregon 97331

J. B. McGrory
Oak Ridge National Laboratory
P.O. Box X, Bldg. 6003
Oak Ridge, Tennessee 37830

Hugh McManus
Department of Physics
Michigan State University
East Lansing, Michigan 48824

Michael J. Moravcsik
Department of Physics
University of Oregon
Eugene, Oregon 97403

Steven A. Moszkowski
Department of Physics
University of California
  Los Angeles
Los Angeles, California 90024

Fred L. Petrovich
Department of Physics
Florida State University
Tallahassee, Florida 32306

John Philpott
Physics Department
Florida State University
Tallahassee, Florida 32306

Alan Picklesimer
Department of Physics
Indiana University
Bloomington, Indiana 47405

Carl H. Poppe
Lawrence Livermore Laboratory
P.O. Box 808, L-405
Livermore, California 94550

J. Rapaport
Department of Physics
Ohio University
Athens, Ohio 45701

Philip G. Roos
Department of Physics
University of Maryland
College Park, Maryland 20742

Takane Saito
Osaka University
Research Center for
  Nuclear Physics
Suita, Osaka, 565 JAPAN

G. R. Satchler
Oak Ridge National Laboratory
P.O. Box X, Bldg. 6003
Oak Ridge, Tennessee 37830

S. D. Schery
Department of Marine Science
Moody College
Texas A and M University
P.O. Box 1675
Galveston, Texas 77553

Peter Signell
Department of Physics
Michigan State University
East Lansing, Michigan 48824

William Sterrenburg
Cyclotron Laboratory
Michigan State University
East Lansing, Michigan 48824

Terry Taddeucci
Department of Physics
McCormack Road
University of Virginia
Charlottesville, Virginia 22901

John Ullmann
Crocker Nuclear Laboratory
University of California
Davis, California 95616

George E. Walker
Physics Department
Indiana University
Bloomington, Indiana 47405

Richard Walter
Triangle Universities Nuclear
  Laboratory
Department of Physics
Duke University
Durham, North Carolina 27706

B. H. Wildenthal
National Science Foundation
1800 G. Street, N.W.
Washington, D.C. 20550

Calvin Wong
Lawrence Livermore Laboratory
Livermore, California 94550

John G. Woodworth
Lawrence Livermore Laboratory
Livermore, California 94550

Stanley Yen
Department of Physics
University of Toronto
Toronto, Ontario
Canada M5S 1A7

C. D. Zafiratos
Department of Physics
University of Colorado
Boulder, Colorado 80309

INDEX